Modern Physics

An Introductory Text

2nd Edition

Modern Physics

An Introductory Text

2nd Edition

Jeremy I. Pfeffer
Shlomo Nir

Hebrew University of Jerusalem, Israel

Imperial College Press

ICP

Published by

Imperial College Press
57 Shelton Street
Covent Garden
London WC2H 9HE

Distributed by

World Scientific Publishing Co. Pte. Ltd.
5 Toh Tuck Link, Singapore 596224
USA office: 27 Warren Street, Suite 401-402, Hackensack, NJ 07601
UK office: 57 Shelton Street, Covent Garden, London WC2H 9HE

British Library Cataloguing-in-Publication Data
A catalogue record for this book is available from the British Library.

MODERN PHYSICS
An Introductory Text
(2nd Edition)

ISBN 978-1-84816-878-7
ISBN 978-1-84816-879-4 (pbk)

Typeset by Stallion Press
Email: enquiries@stallionpress.com

Printed in Singapore.

Preface to the Second Edition

This second edition of *Modern Physics – An Introductory Text* preserves the blend of readability, scientific rigour and clarity that characterised its predecessor and made it such a highly recommended text book for non-physics science majors. As in the first edition, the authors set out to present 20th century physics in a form accessible and useful to undergraduate and graduate students of the life, agricultural, earth and environmental sciences. The text has also proved valuable as a first reader and source for students majoring in the physical sciences and engineering.

Several revisions and additions have been incorporated into this second edition, prompted, on the one hand, by the feedback received from teachers and students who have used the text and, on the other, by the advances that have taken place in technology and research.

Two new chapters have been added on the following subjects:

- The Reality of Atoms: Einstein's Elucidation of Brownian Motion and Perrin's Experiments.
- Quantum Electrodynamics: The Electron's Anomalous Magnetic Moment and the Fine Structure Constant.

A number of new topics have also been incorporated into the text at appropriate places. These include:

- The Solar Neutrino Problem; Neutrino Oscillations.
- Particle Accelerators; The Large Hadron Collider.
- The Standard Model; Weak Charge.
- The Fukushima Catastrophe.
- Solid State Storage; Flash Memory Devices.

As in the first edition, all mathematics other than that required for descriptive or illustrative purposes has been omitted from the main body of the text and incorporated into the 47 worked examples and 11 appendices.

The authors acknowledge their indebtedness to the many students of the course 'An Introduction to Modern Physics' given at the Faculty of Agriculture, Food and Environment of the Hebrew University of Jerusalem, whose participation and searching questioning over the years has contributed so much to the honing of this text and to our sense of fulfilment. Our thanks also to the staff of World Scientific in Singapore and of Imperial College Press in London, in particular the book's editor, Tasha D'Cruz, without whom this second edition would never have come to fruition.

Jeremy I. Pfeffer
Shlomo Nir
Rehovot, 2012

Preface to the First Edition

This book originated from the need for a suitable student text for the course 'An Introduction to Modern Physics' given at the Faculty of Agriculture, Food and Environment of the Hebrew University of Jerusalem. The primary goal of this course is to produce graduates who, whatever their field of specialisation, are 'modern physics literate'. It is open to all students who have successfully completed their first-year physics and mathematics studies.

The course sets out to recount, in terms amenable to non-physics majors, the development of the three seminal ideas – Special and General Relativity, Quantum Theory and the Nuclear Atom – out of which modern physics grew in the first half of the 20th century. These topics constitute the principle subject-matter of this book.

However, in addition to a final examination on these subjects, the students participating in the course are required to submit a term-paper on any topic of their choice, providing it falls within the scope of modern physics or involves one or other of the industrial, technical or research applications derived from it. Accordingly, the scope of the book was widened beyond the narrow limits of the course material and chapters or sections devoted to such topics were added. Among those given a more detailed treatment are:

* Magnetism as a Relativistic Effect.
* The Interaction of Radiation with Matter – Spectroscopy.
* Fluorescence in Biological Systems and Membrane Research.
* Nuclear Structure and Elementary Particles – the Standard Model.
* The Design of Nuclear Weapons: fission and fusion weapons.
* Nuclear Reactors: the events at Chernobyl are described in detail.
* The Design and Use of Lasers.
* The Mössbauer Effect.
* Nuclear Magnetic Resonance and Magnetic Resonance Imaging.
* The Conduction of Electricity in Solids and Semiconductor Devices.

In each case, the underlying theory is presented together with a general description of the practical aspects of the application. For the sake of completeness we have also added:

* Quantum Electrodynamics – QED.
* Invariance, Symmetry and Conservation Laws.

In general, the presentation of the material emphasises the physical aspects of the phenomena. Problem solving is not a major or primary objective. Thus, the text presumes upon the spirit of the physical method of J.J. Thomson*:

> '*The physical method has all the advantages in vividness which arise from the use of concrete quantities instead of abstract symbols... we shall be acting in accordance with Bacon's dictum that the best results are obtained when a research begins with Physics and ends with Mathematics...*
>
> *The use of a physical theory will help to correct the tendency – which I think all will admit is by no means uncommon – to look on analytical processes as the modern equivalent of the Philosopher's Machine in the Grand Academy of Lagado, and to regard as the normal process of investigation in this subject the manipulation of a large number of symbols in the hope that every now and then some valuable result may happen to drop out.*'†

Notwithstanding, this book is not popular science; it does not avoid the 'hard bits'. Studying physics requires a mental effort and there is no reason to hide this fact. Mathematics is incorporated into the main body of the text, but only to the extent required for descriptive or illustrative purposes. Most of the mathematical proofs have been separated from the main body of the text so as not to interrupt its flow; they appear in the worked examples and appendices and can be skipped at a first reading. Questions, exercises and problems for student assignments will be found at the end of each section; answers to these are to be found at the end of the book.

The techniques by which trigonometric functions, phasors (rotating vectors) and complex numbers are used in the mathematical description of wave motion are summarised in a supplementary section. A comprehensive index is also included.

In addition to its suitability as a student text for courses similar to that for which it was originally written, we would recommend the book as a first reader and source-text for students majoring in the physical sciences and engineering.

The authors acknowledge their deep indebtedness to the many excellent standard reference works to which they had recourse during the writing of this

---

*Sir Joseph J. Thomson 1856–1940, English physicist who was awarded the Nobel prize in 1906 for his research on the conduction of electricity through gases at low pressures.

†J.J. Thomson (1893), *Notes on Recent Researches in Electricity and Magnetism*, Oxford: Clarendon Press, p. vi.

book. They also wish to thank Dr Zvi Kalberman who read and commented on parts of the text, and Daniel Pfeffer, who prepared the computer simulations and provided the technical advice without which this camera-ready manuscript could never have been completed.

Jeremy I. Pfeffer
Shlomo Nir
Rehovot, 2000

Table of Contents

Part One The Birth of a New Physics

Classical physics has its origins in the 17th century, with Galileo's[1] experiments on falling bodies, Kepler's[2] calculations of the planetary orbits and Newton's[3] postulates and mathematical laws of motion. During the following two centuries, the Newtonian system of mechanics, together with its theory of gravitation, dominated scientific thinking; its achievements were unprecedented. Perhaps its most impressive feat was predicting the existence of the planet Neptune. Calculations based on Newtonian mechanics indicated that the slight perturbations that had been observed in the orbit of the planet Uranus could be accounted for by the presence of an additional planet in the solar system, one previously unknown. In 1846, this new planet, Neptune, was found at the exact position in space indicated by the calculations.[4]

By the middle of the 19th century, the kinetic theory of matter and the science of thermodynamics had between them solved the ancient mystery of the nature of heat and had provided an understanding for the 'arrow of time' inherent in the workings of nature. Classical physics reached its zenith in the second half of the 19th century with the publication of Maxwell's[5] theory of electromagnetism and the discovery of the electromagnetic waves whose existence it had predicted. This theory summarised and unified everything that was known at the time about electrical and magnetic phenomena and provided the first comprehensive conceptual basis for the science of optics.

However, towards the end of the 19th century, it became clear that there were important physical phenomena for which classical physics had no satisfactory explanation. The electron, X-rays and radioactivity, all of which were discovered within a few years of each other in the last decade of the century, were beyond the competence of classical physics to explain. Moreover, there were even instances where the hypotheses and laws of classical physics were found

[1] Galileo Galilei 1564–1642, the Italian physicist and astronomer who first asserted that 'the book of nature is written in the language of mathematics'.
[2] Johannes Kepler 1571–1630, German astronomer.
[3] Sir Isaac Newton 1642–1727, English mathematician, astronomer and physicist.
[4] In a similar technique currently being used in the search for planetary systems other than the solar system, the 'wobbling' observed in the motion of certain stars and the slight periodic variations in their intensity is attributed to large orbiting bodies (planets).
[5] James Clerk Maxwell 1831–1879, Scottish mathematician and physicist.

to be totally incompatible with the results of experiments in more conventional fields, such as studies of the speed of light relative to different observers and the emission and absorption of heat radiation. It became apparent that despite its great achievements, some of the most fundamental principles underlying classical physics were incorrect. Clearly, fresh ideas and theories were needed.

At the beginning of the 20th century, three new hypotheses were put forward that changed the face of physics. They were:

1. The theory of relativity.
2. Quantum theory.
3. The nuclear model of the atom.

This triad laid the foundations of modern physics. They, and the effects derived from them, are the subject of this book.

The new theories of modern physics not only resolved the problems left unanswered by classical physics but extended the reach of the physical sciences into previously unknown fields. In general, the familiar and well-established laws of classical physics remain valid for dealing with phenomena that occur on a 'normal' scale, in the human sense of the term. However, when dealing with phenomena occurring on the cosmic scale on the one hand, and on the atomic scale on the other, only the more comprehensive laws of modern physics can be employed.

Part One of our study of 20th century physics opens with the discovery of the electron, the first elementary particle to be identified; no other single discovery has ever led to the understanding of so many physical phenomena previously thought to be unrelated. This is followed by a short review of the Maxwellian theory of electromagnetism and of the empirical and theoretical advances that led to the general acceptance of the reality of atoms. These three developments provided the background for the scientific revolution ushered in by Einstein's *theory of special relativity* and Planck's *quantum hypothesis* during the first decade of the new century. Part One concludes with a brief account of the theory of general relativity.

Chapter 1.1

The Electron

When first investigated in the 18th and 19th centuries, electrical and magnetic phenomena were generally construed in terms of *æthereal fluids*, as were those associated with heat and light. These fluids were thought to comprise minute mutually repelling particles. Thus, heat was either thought to be vibrations in the fluid *caloric* or an accumulation of this fluid in the interstices of materials. Light was either a flux of particles emitted at high speed from luminous bodies or the vibrations of a ubiquitous fluid æther. Electric fluids – *electricity* – flowed readily through metals and other conductors but did not penetrate insulators such as paper and glass. Opinion was divided as to whether there was just one electric fluid or two – a *positive* fluid and a *negative* fluid.

The possibility that electricity might not be a continuous fluid was first raised in the middle of the 19th century following Faraday's[6] quantitative researches on electrolysis. These showed the existence of a systematic relationship between the amount of electricity passed through an electrolytic cell and the quantity of material that undergoes chemical reaction (electrolysis) in the cell. Thus, the passage of a certain amount of electricity – 96,500 coulombs in modern terms – always liberates a gram-equivalent of substance from the electrolyte, whether it is the metal released at the negative cathode or the non-metal at the positive anode. Putting aside his misgivings about atomism, Faraday recognised that this suggested electricity might be *atomic* in nature and that a natural indivisible unit of electricity exists. In 1891 Stoney[7] suggested that this natural unit of electricity be called an *electron*.

On this view, every ion carries an integer multiple of this natural unit. For example, a silver ion, Ag^+, carries a single natural unit of positive charge; a typical copper ion, Cu^{++}, carries two such units. Given that a gram-equivalent of monovalent ions comprises a mole, the magnitude of the natural unit of

[6] Michael Faraday 1791–1867, English chemist and physicist.
[7] George Stoney 1826–1911, Irish physicist.

electricity, e, can be calculated by dividing the 96,500 coulombs by Avogadro's number, $6.02 \cdot 10^{23}$:

$$e = \frac{96{,}500}{6.02 \cdot 10^{23}} = 1.6 \cdot 10^{-19}\text{C}. \tag{1.1}$$

The term *electron* is now used to designate the elementary particle that carries the natural unit of negative charge and which was first identified towards the end of the 19th century in experiments on the conduction of electricity through gases at very low pressures. At atmospheric pressure, gases do not usually conduct electricity. However, at reduced pressures of 0.5 mmHg to 10 mmHg and with applied potentials of several thousand volts, they can be made to pass a current. These greatly reduced gas pressures were first achieved at the end of the 19th century following the advances made at that time in vacuum pump technology. The gases were contained in narrow glass tubes, called *discharge tubes*, into which suitable electrodes had been inserted. The passage of the current is accompanied by the appearance of striking colours in the tubes (Fig. 1.1).

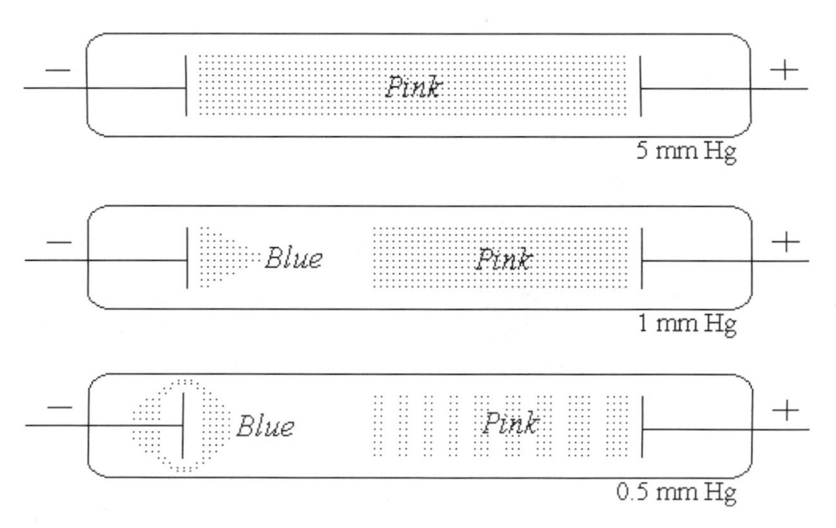

Fig. 1.1. The conduction of electricity through air at low pressures in a discharge tube. The colours result from the excitation and ionisation of the atoms of the gas in the discharge tube. The bands and the coloured and dark regions arise from the variations in the electric field strength throughout the tube.

At still lower pressures, ~0.02 mmHg, the colours disappear but the glass tube itself begins to glow with a green hue. An object placed in front of the cathode (the negative electrode) casts a shadow on the opposite wall of the discharge tube (Fig. 1.2). Certain minerals, when placed in front of the cathode, fluoresce with brilliant colours. It appears that something is being emitted from the cathode; this emanation was given the name *cathode rays.*

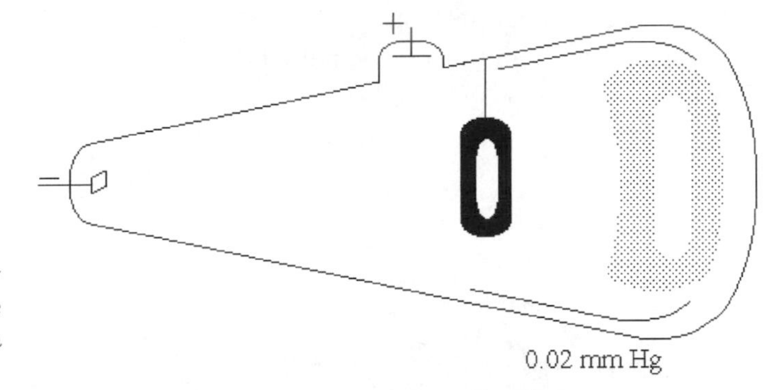

Fig. 1.2. The shadow cast by an object in the path of the rays in a cathode ray tube.

In further experiments it was found that the cathode rays were deflected by a magnetic field as would a stream of negative charge (Fig. 1.3a). Furthermore, a small paddle wheel positioned between the electrodes rotated under their impact; switching the polarity of the electrodes reversed the direction of the rotation (Fig. 1.3b). These two phenomena suggested the cathode rays might be negatively charged particles. Nevertheless, many physicists at the time still considered them to be of an æthereal rather than a material nature.

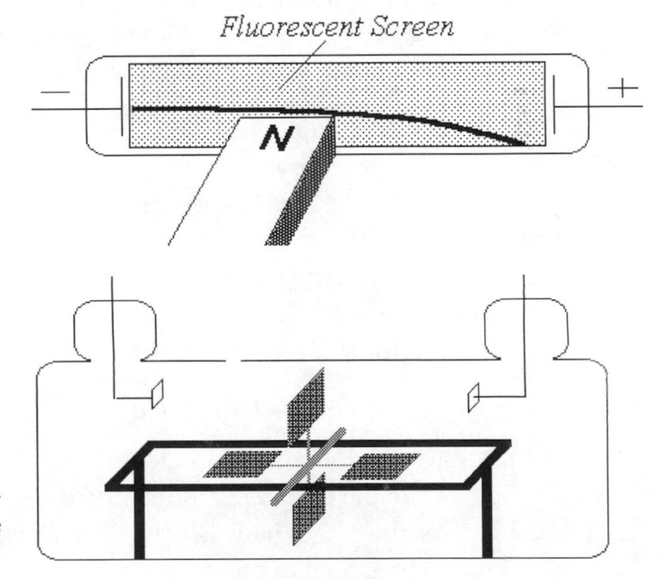

Fig. 1.3a. The deflection of cathode rays in a magnetic field. The direction of the deflection shows that it is negative charge that flows with the cathode rays.

Fig. 1.3b. The paddle wheel rotates under the impact of the cathode rays.

Convinced that the cathode rays were in fact charged particles of matter, J.J. Thomson set out in 1897 to measure their velocity, v, and the ratio, q/m, between their charge, q, and their mass, m. In one of the experiments he conducted, a narrow collimated beam of cathode rays was aimed along the length of a very low pressure glass discharge tube. After emerging from the hole in the anode, the beam passed through a thin slit, between a pair of vertical coils and, finally, between the horizontal parallel plates of a condenser. The green spot that appeared on the glass at the far end of the tube indicated where the beam impinged upon it (Fig. 1.4).

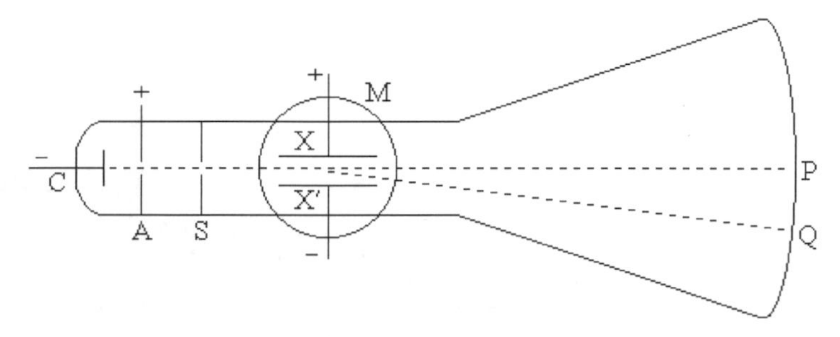

Fig. 1.4. The type of very low pressure discharge tube used by Thomson for the determination of the ratio between the charge and the mass of the particles (electrons) in cathode rays. A collimated beam was produced by the arrangement of the cathode, C, the pierced anode, A, and the slit, S. The circle M represents the pair of coils, one on each side of the tube. Passing a current through these coils produces a uniform horizontal magnetic field in the gap between them. X and X′ are the parallel plates of the condenser. Connecting the plates to a source of a potential produces a uniform vertical electric field between them. An undeflected particle beam strikes the point P.

The cathode ray beam could be deflected vertically by two different means: (1) magnetically; (2) electrically:

1. Magnetically – by passing a current through the coils. This induces a uniform horizontal magnetic field, B, in the space between them which, in turn, exerts a force, F_B, on the cathode ray particles such that

$$F_B = Bqv, \tag{1.2}$$

where v is the velocity of the particles. This force acts at right angles to the direction of the particle's motion and so constitutes a centripetal force. Thus, as they pass between the pair of coils, the cathode ray particles move in a circular path – along the arc of a circle – such that

$$F_B = Bqv = \frac{mv^2}{r}, \tag{1.3}$$

where r is the radius of the arc.

2. Electrically – by connecting the condenser plates to a potential source; this produces a vertical electric field, E, in the space between them that exerts a vertical force, F_E, on the cathode ray particles such that

$$F_E = Eq. \tag{1.4}$$

Initially, Thomson applied just the magnetic field. This had the effect of moving the green spot down from P to Q. He then activated the electric field and adjusted the potential between the condenser plates until the field was just sufficient to return the green spot from Q to P. At this point, the two forces, F_B and F_E, cancelled each other out, such that:

$$\begin{aligned} F_E &= F_B \\ qE &= Bqv = \frac{mv^2}{r}. \end{aligned} \tag{1.5}$$

From which, the particles' charge to mass ratio is given by

$$\frac{q}{m} = \frac{E}{rB^2}. \tag{1.6}$$

Thomson calculated the value of the magnetic field, B, from the dimensions of the coils and the current flowing through them; the arc radius, r, from the dimensions of the apparatus and the distance PQ; and the electric field, E, from the potential applied between the condenser plates and the distance between them.

In modern SI units, the value Thomson found for the charge to mass ratio was $\sim 2 \cdot 10^{11}$C/kg. Thomson repeated the experiment using different metal cathodes and with different gases in the low pressure tube, but in each case he found approximately the same value for the ratio.

The value found by Thomson for the charge to mass ratio of the cathode ray particles was three orders of magnitude greater than the largest previously known value of this ratio, namely, that found for the aqueous hydrogen ion, $H^+_{(aq)}$. This could be attributed either 'to the smallness of m or the largeness of q, or a combination of these two'. Thomson opted for the 'smallness of m' and assumed that the magnitude of the charge, q, carried by the cathode ray particles was equal to the smallest charge known to be carried by any ion, i.e. $q = -1.6 \cdot 10^{-19}$C. On this assumption, he calculated the mass of the cathode ray particles and obtained the result, astonishing at the time, that their mass was only 1/1836 that of the hydrogen atom, itself the smallest of all atoms.[8]

[8] This is the modern value.

In an attempt to verify Thomson's assumption, direct measurements of the magnitude of the charges carried by gaseous ions were made. Although these experiments suffered initially from many sources of error, their results appeared to confirm Thomson's supposition. The issue was finally and unequivocally settled in 1906 when, in a series of accurate and careful experiments, Millikan[9] measured the magnitude of the electrical charges carried by both positively and negatively charged oil droplets. In none of the hundreds of measurements he made, did he detect a droplet that carried a charge whose magnitude was less than $1.6 \cdot 10^{-19}$C, nor one that carried a charge whose magnitude was a fractional multiple of this amount. In every case, the magnitude of the charges was an integer multiple of $e = 1.6 \cdot 10^{-19}$C. The clear inference was that electrical charge – positive or negative – always appears as an integer multiple of the natural unit e.

The cathode ray particles Thomson had discovered are the elementary particles we now call *electrons*; the electron was the first elementary particle to be discovered. Electrons carry the natural unit, $-e$, of negative charge. They are the smallest of the three constituent particles – *protons, neutrons and electrons* – of ordinary matter. Though electrons have no discernible internal structure or dimensions, they nevertheless possess an intrinsic angular momentum (spin) and an associated magnetic moment.

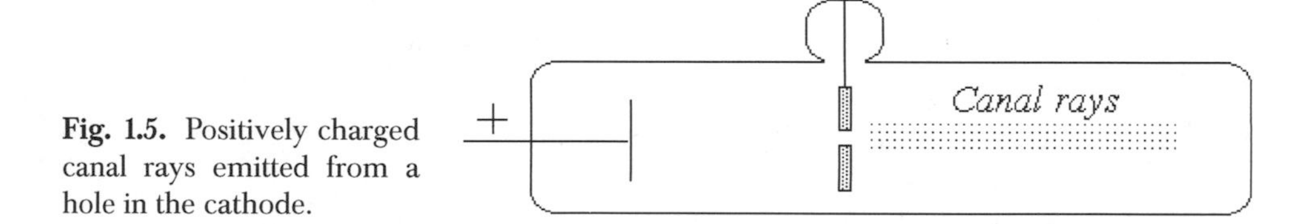

Fig. 1.5. Positively charged canal rays emitted from a hole in the cathode.

Cathode rays were not the only emanations observed in discharge tubes. If the cathode in the discharge tube was pierced, rays were seen to emerge from the hole in the direction away from the anode (the positive electrode), i.e. in the opposite direction to the cathode rays (Fig. 1.5). These emanations were called *canal rays*; they were found to be positively charged.

Having identified the cathode rays, Thomson proceeded to investigate the nature of the canal rays. These too proved to be particles of matter. By measuring the ratio between their charge and their mass he identified them as positive ions of the gas present in the discharge tube. He construed that these positive ions were produced in the space between the cathode and the anode from

[9]Robert Millikan 1868–1953, American physicist who was awarded the Nobel prize in 1923 for his experimental work.

the bombardment of the gaseous atoms by the cathode rays (electrons). Since they carried a positive charge, they were attracted to and accelerated towards the negative cathode, passing through the hole that had been pierced in it and emerging from the other side. In one notable experiment, in which the gas in the discharge tube was a pure sample of the noble gas neon, Thomson discovered two different species of positive ions in the canal rays whose masses differed by about 10%. The ions could only be neon ions, the purity of the sample ensured this. This discovery of different species of neon ions first demonstrated the existence of atoms that share the same chemical identity but have different masses; what we now refer to as *isotopes.*

In 1932, a particle was discovered that is identical to the electron in all its properties except that it is positively and not negatively charged; it was designated the *anti-electron* or *positron.* When a positron and electron collide, the pair of particles self-destruct, releasing a burst of high-frequency electromagnetic radiation – γ-rays.

Chapter 1.2

Electromagnetic Waves

The notion that an æther pervades all of space, including even that occupied by material bodies, was popularised in the 17th century by the philosopher Descartes.[10] Traditionally, space had been regarded as passive and unchangeable; an inert background, like a cinema screen, upon which any action can be projected but without which no action can be perceived. Descartes argued that in order to account for many physical phenomena, a dynamic active medium – an *æther* – that transmitted the forces exerted at a distance by one object on another, had to be postulated.

It was also generally accepted that the propagation of light involved an æther. However, two conflicting hypotheses held sway at various times concerning the nature of light and the role of this æther in its propagation. One theory asserted that light is a particle or corpuscular phenomenon: a flux of æthereal particles emitted at high speed from luminous bodies. The other, that light is a wave phenomenon: the vibrations of a ubiquitous *luminiferous æther*.

According to the corpuscular theory – the one advocated by Newton – light propagates at a greater speed through a dense medium than through a less dense one. On this view, the speed of light in water should be greater than in air. The wave theory proposed by his 17th century contemporary Huygens[11] posited just the opposite: that light should propagate faster in air than in water. Fairly good estimates of the speed of light – about 200,000 km/s – had been obtained at the time by Römer[12] from astronomical observations. However, the technical means required for a direct terrestrial comparison of the speed of light in air and water and for carrying out an *experimentum crucis* that would decide between the two theories would not be available for another 200 years. Until then, the dispute between the two hypotheses was conducted as much on the basis of the eminence of their respective supporters as on their relative capacities to explain the phenomena.

The strength of Newton's corpuscular theory lay in the simple explanation it provided for the rectilinear propagation of light, i.e. for the formation of

---

[10]René Descartes 1596–1650, French philosopher and mathematician.

[11]Christaan Huygens 1629–1695, Dutch physicist and astronomer.

[12]Olaus Römer 1644–1710, Danish astronomer.

well-defined shadows. It postulated that the material corpuscles that constitute a beam of light are emitted from luminous bodies in straight lines, as required by Newton's first law of motion, and hence the shadows; that they are reflected elastically from mirrors, which accounts for the familiar laws of reflection; and that they vary in size according to the colour of the light. In rejecting this approach, Huygens pointed to the extreme speed at which light propagates and to the fact that when light rays intersect they cross without hindrance and without leaving any lasting impression on one another. Material particles could not possibly travel at such a speed; nor could their paths cross without the corpuscles colliding and deflecting one another. Hence, he argued, it must be by some other mechanism that light propagates. In his opinion, the propagation of light was to be understood by analogy to the propagation of sound.

Throughout the 18th century, Newton's corpuscular theory of light was the one embraced by most scientists. However, by the middle of the 19th century, opinion had completely changed and the issue appeared to have been settled once and for all in favour of the wave hypothesis. The turning point was the landmark experiment conducted by Thomas Young[13] in 1801. Young observed that when light from a single source is projected onto a white screen through two adjacent pin-holes or narrow slits, a sequence of bright and dark bands – *fringes* – appears on the screen (Fig. 1.6).

In order to account for the appearance of the fringes, Young proposed the simple hypothesis that if two waves are superposed, their joint effect is the combination of the two. In so doing, he provided the key concept – *interference* or *superposition* – needed for the establishment of the wave theory of light (Fig. 1.7). He wrote:

> '*Suppose a number of equal water waves move upon the surface of a stagnant lake at a constant velocity and enter a narrow channel leading out of the lake. Suppose another similar series of waves arrive at the same channel with the same velocity. Neither series of waves will destroy the other, but their effects will be combined: if they enter the channel such that the elevations of the one coincide with those of the other, they must together produce a series of greater joint elevations [constructive interference]; but if the elevations of one series correspond to the depressions of the other, they must exactly fill up these depressions and the surface of the water must remain smooth [destructive interference]. Now, I maintain that similar effects take place whenever two portions of light are thus mixed.*'[14]

[13]Thomas Young 1773–1829, English polymath.

[14]A. Wood and F. Oldham (1954) *Thomas Young, Natural Philosopher, 1773–1829*, Cambridge: Cambridge University Press, p. 431.

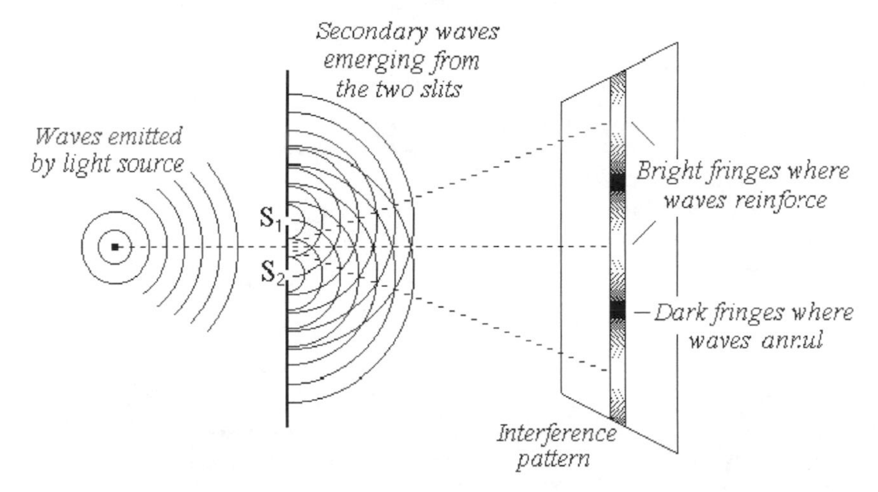

Fig. 1.6. Young's double-slit experiment. Light from the source passes through the narrow slits, S_1 and S_2, and emerges as two coherent trains of secondary waves. Young explained that bright fringes appear on the screen where the secondary waves reinforce each other (the wavetrains are in phase with each other) and dark fringes where they annul one another (the wavetrains are out of phase).

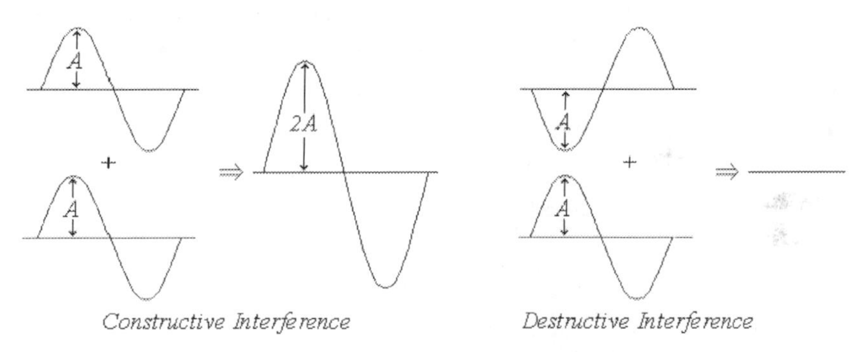

Fig. 1.7. The destructive and constructive interference of two waves. If light waves are in phase when or where they join, they reinforce each other and produce light of greater intensity (constructive interference); if they are out of phase, they annul one another and produce darkness (destructive interference).

Following Young's experiment and the subsequent mathematical elucidations of the interference and diffraction of light based on his principle of superposition, it was generally accepted that 'light is in the æther the same thing as sound in the air', i.e. light is a mechanical wave in the æther, just as sound is a mechanical wave in the air. However, further efforts to clarify the actual physical nature of light waves were confounded by the phenomenon of light polarisation.

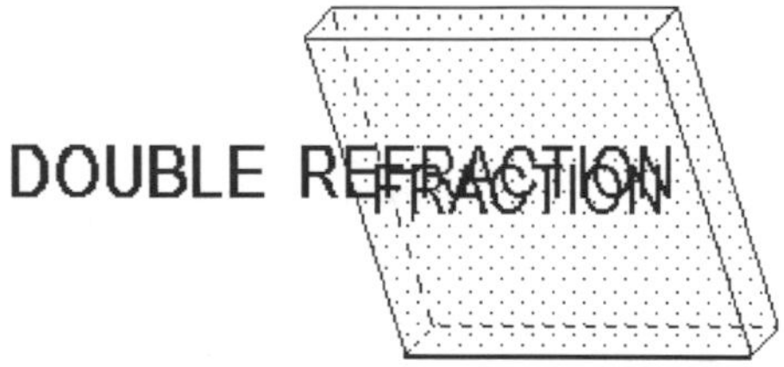

Fig. 1.8a. Double refraction. When looking at a printed page through a calcite crystal, two images are seen.

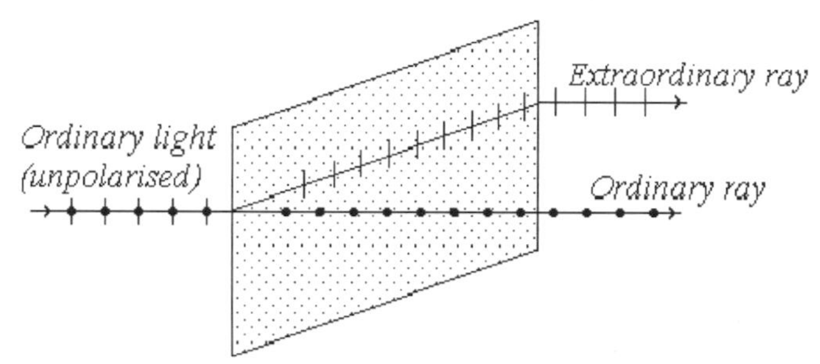

Fig. 1.8b. An incident ray of ordinary (unpolarised) light is split into two rays on passing through a calcite crystal. The ordinary and extraordinary rays are polarised at right angles.

It had been known for some time that when a beam of ordinary light passes through a calcite crystal it splits into two beams; this is the phenomenon known as *double refraction* or *birefringence.* If the two beams are then passed through a second calcite crystal they do not behave like ordinary light. Furthermore, the two beams are in turn extinguished when viewed through a rotated tourmaline crystal (or polaroid). We say that the two light-beams are *polarised at right angles* (Fig. 1.8). In 1808, Malus[15] discovered that such polarised light can also be obtained by the reflection of ordinary light from a glass surface.

When differently polarised beams of light are combined, no interference pattern is produced. In order to account for this and for the other phenomena associated with light polarisation, Young and Fresnel[16] determined that light waves must be transverse æther waves, i.e. the æther particles vibrate at right angles to the direction of light propagation. Sound waves in air are longitudinal waves, i.e. the air particles vibrate parallel to the direction of the sound propagation (Fig. 1.9). Polarisation is not exhibited by longitudinal waves. Thus, light polarisation has no sound analogue.

Transverse mechanical waves cannot propagate through fluids such as water or air but only through solids.[17] Consequently, if the light waves that propagate through space are indeed transverse æther waves, the æther must be described in terms of an elastic solid. Transverse waves propagate through elastic solids

[15] Etienne Louis Malus 1775–1812, French physicist.
[16] Augustin Jean Fresnel 1788–1827, French physicist.
[17] Transverse waves can propagate on the surface of liquids but not through them.

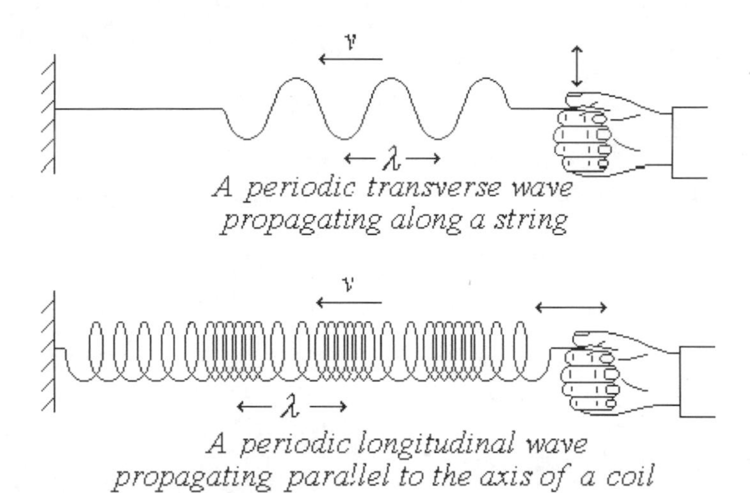

Fig. 1.9. Transverse and longitudinal periodic waves. Moving the hand up and down produces vertical transverse waves along the string; moving it sideways produces horizontal transverse waves. Indeed, transverse waves can be generated in any plane. Such transverse waves are ***plane polarised.*** However, if the coil is pulled to and fro, the longitudinal waves produced can only propagate in one way, parallel to the axis of the coil. In both cases, the frequency of the waves, f, is set by that at which the hand oscillates and their wavelength, λ, by the ratio between the velocity at which they propagate, v, and their frequency: $\lambda = v/f$.

with velocity

$$v = \sqrt{\frac{S}{\rho}}, \tag{1.7}$$

where S is the shear modulus of the solid and ρ is its density. The speed of light was known to be of the order 10^8 m/s. For the velocity of the æther waves to be of this magnitude, the æther had to be more rigid than steel (its shear modulus S had to be very large) whilst at the same time its density had to be less than that of air (ρ very small)! Although the existence of a material with such contradictory properties was quite inconceivable, the concept of a luminiferous æther was still not abandoned.

The æther concept was further complicated when Faraday, having conceived the notion of electric and magnetic fields, successfully employed the idea of lines of force in explaining electrical and magnetic interactions and in formulating the laws of elecromagnetic induction. Faraday's ideas were not universally accepted. There were those who argued that the forces acting at a distance between charges do so directly, being the result of an instantaneous unmediated interaction, the same way as it was thought the force of gravity acted. This notion can be represented schematically as:

$$\text{Charge} \leftrightarrow \text{Charge}$$

However, in his revolutionary elucidation of electrical and magnetic interactions, Faraday placed the emphasis on the space between the interacting bodies. He asserted that the space surrounding a magnet is 'a part of the true and complete magnetic system'. In Faraday's view, space plays an active role in the propagation and transmission of the interactions. As Maxwell wrote:

> *'Faraday, in his mind's eye, saw lines of force traversing all space where the mathematicians saw centres of force attracting at a distance; Faraday saw a medium where they saw nothing but distance; Faraday sought the seat of the phenomena in real actions going on in the medium.'*[18]

This idea can be represented schematically as:

$$\text{Charge} \leftrightarrow \text{Space} \leftrightarrow \text{Charge}$$

According to this approach, electrical and magnetic forces act through the medium of a *field* that pervades the space between the charges or the magnets. In order to sustain this field and its lines of force, it was assumed that something must be present in the space. *Faute de mieux*, Faraday presumed this to be the æther.

In support of his view, Faraday cited the fact that though polarised light is normally unaffected by its passage through ordinary glass, when a beam of polarised light passes through a glass block in a magnetic field, the form of its polarisation changes. Presuming light to be an æther wave, the conclusion Faraday drew from this was that the lines of force of electric and magnetic fields could be regarded as elastic deformations in the æther. Thus, both electrical and magnetic properties were now added to the exotic mechanical properties already required of the æther.

Convinced by his reading of Faraday's *Experimental Researches in Electricity* that 'his method of conceiving the phenomena was also a mathematical one, though not exhibited in the conventional form of mathematical symbols', Maxwell took upon himself the task of providing a firm mathematical basis to Faraday's intuitive ideas. In an immense enterprise, he developed a mathematical system that described and accounted for all the then known electrical and magnetic phenomena in terms of electric and magnetic fields. In doing so, Maxwell formulated two fundamental postulates:

1. that a moving or changing magnetic field sets up an electric field;

[18]J.C. Maxwell (1954) *A Treatise on Electricity and Magnetism*, Vol. 1, New York: Dover Publications, p. ix.

2. that a moving or changing electric field sets up a magnetic field.

The mathematical functions and equations Maxwell developed from these two symmetrical assumptions predicted the existence of transverse electromagnetic waves, i.e. disturbances in the electric and magnetic fields that would propagate through empty space with a velocity, c, given by the formula:

$$c = \sqrt{\frac{1}{\mu_0 \varepsilon_0}}, \tag{1.8}$$

where μ_0 is the permeability of empty (free) space and ε_0 is its permittivity. The numerical values of these fundamental constants were already known from electromagnetic experiments. Hence, calculating the velocity of propagation of the electromagnetic waves just required their substitution in Eq. (1.8). Maxwell wrote:

> '*If it should be found that the velocity of propagation of electromagnetic disturbances is the same as the velocity of light... we shall have strong reasons for believing that light is an electromagnetic phenomenon.*'[19]

Substituting the numerical values for the permeability and permittivity of empty space, *viz.* $\mu_0 = 4\pi \cdot 10^{-7}\,\mathrm{m \cdot kg \cdot C^{-2}}$ and $\varepsilon_o = 8.85 \cdot 10^{-12}\mathrm{C^2 \cdot N^{-1} \cdot m^{-2}}$, in Eq. (1.8) does indeed give a wave velocity equal to the measured speed of light *in vacuo*, $c = 3 \cdot 10^8\,\mathrm{m/s}$.

$$\sqrt{\frac{1}{\mu_0 \varepsilon_0}} = \sqrt{\frac{1}{(4\pi \cdot 10^{-7}) \cdot (8.85 \cdot 10^{-12})}} = 3 \cdot 10^8\,\mathrm{m/s}.$$

Maxwell inferred from this identity that light itself is an electromagnetic wave.

Maxwell summarised his ideas in *A Treatise on Electricity and Magnetism*, published in 1873. Fifteen years later, Hertz[20] succeeded in transmitting and receiving the periodic electromagnetic waves that, according to Maxwell's theory, should be produced by oscillating electrical circuits. His experiments confirmed all of Maxwell's predictions.

Maxwell's equations summarise all the laws of electromagnetism just as Newton's laws do for classical mechanics. It should be noted that these equations do not involve any actual assumptions about the existence of the æther. However, in his total rejection of unmediated action at a distance, Maxwell continued to believe in its existence. He regarded it as being the medium that

[19] J.C. Maxwell (Reprinted 1954) *A Treatise on Electricity and Magnetism, Vol. 2*, New York: Dover Publications, p. 431.

[20] Heinrich Hertz 1857–1894, German physicist.

carried the potential and kinetic energy associated with electromagnetic waves and even visualised electromagnetism as a kind of æther mechanics, imagining space to be filled with 'æthereal' interlocking gear wheels that transmitted the electromagnetic forces.

1.2.1 The Production and Properties of Electromagnetic Waves

In general, all accelerated electric charges radiate electromagnetic energy in the form of electromagnetic waves. If the charge constitutes an electric oscillator, the energy is radiated as periodic electromagnetic waves whose frequency is equal to that of the oscillating charge. Similarly, electric oscillators possessing a natural frequency that matches that of the electromagnetic radiation incident upon them, i.e. when there is *resonance* between the oscillators and the incident radiation, will absorb energy from the radiation.

The electric and magnetic disturbances that constitute an electromagnetic wave propagate in phase with one another but at right angles to each other (Fig. 1.10).

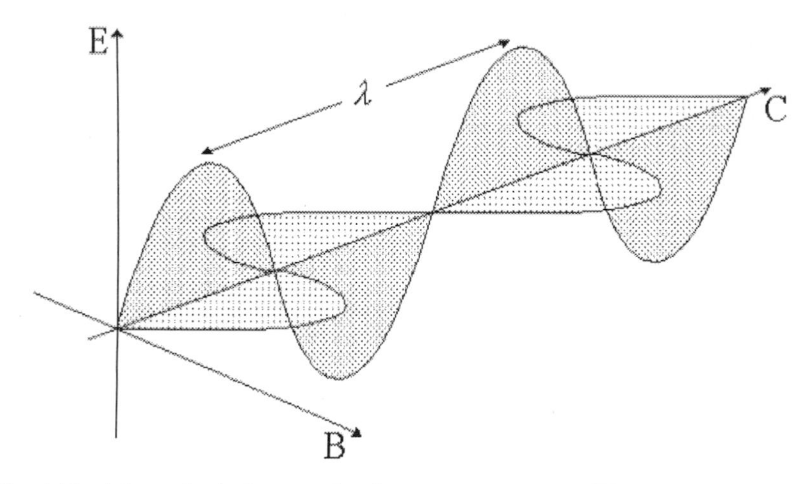

Fig. 1.10. A 'snapshot' of a periodic electromagnetic wave. The wave is composed of a periodic electric field and a periodic magnetic field at right angles to one another but in phase with one another. E indicates the electric field, B the magnetic field and C the direction of propagation of the wave.

The characteristic property of a periodic electromagnetic wave is its frequency; this is the parameter that fixes its *colour.* A continuous spectrum of periodic electromagnetic waves has been identified with frequencies ranging from about 10^4 Hz to 10^{22} Hz (Fig. 1.11). The waves comprising the electromagnetic spectrum all propagate through empty space at the speed of light, $c = 3 \cdot 10^8$ m/s.

The lower frequencies of the electromagnetic spectrum, such as radio waves, are generated by oscillating electrical circuits; these frequencies are in turn detected (absorbed) by the use of aerials in which the incident radiation sets up corresponding electrical oscillations. Microwaves, such as those used for radar detection and rapid cooking (microwave ovens), are generated by the velocity modulation (acceleration) of electron beams. The emission and absorption of infra-red radiation (heat radiation), visible light and ultra-violet radiation usually involves the transfer of electrons between the energy levels associated with the outer structure of atoms, molecules or ions. X-rays are generated by the deceleration of high-energy electrons or by electron transfers between the energy levels associated with the inner structure of atoms. The high-frequency γ-rays are emitted during the decay of radioactive nuclei.

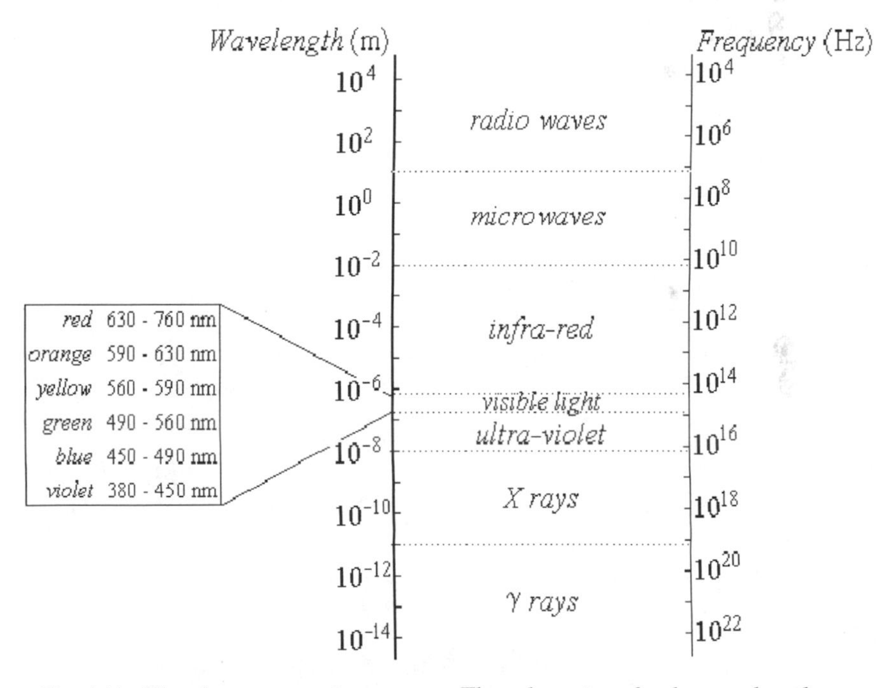

Fig. 1.11. The electromagnetic spectrum. The values given for the wavelengths are those exhibited by the radiation when it propagates *in vacuo.*

1.2.2 The Limits of Electromagnetic Theory

Despite its outstanding success in elucidating electrical, magnetic and optical phenomena, it was subsequently found that there were certain physical phenomena for which Maxwell's theory could not account. In particular, it failed to explain many of the phenomena involving interactions between electromagnetic radiation and matter.

The discovery of the electron as the 19th century drew to its close was also totally unexpected. When considering what happens in a wire carrying an electric current Faraday had written:

> '*By a current I mean anything progressive, whether it be a fluid of electricity or two fluids moving in opposite directions or merely vibrations, or speaking more generally, progressive force.*'[21]

This nebulous view of an electric current was at the root of Maxwell's postulate that an electric current (a *displacement current*) flows through the space between the plates of a condenser as it is charged and gives rise to magnetic effects equivalent to those produced by an ordinary current flowing through a wire. The shock of the discovery of the electron to the followers of Faraday and Maxwell is well illustrated by the following contemporary remark attributed to the physicist Arthur Schuster:

> '*The separate existence of a detached atom of electricity never occurred to me as possible. And if it had, and I had openly expressed such heterodox opinions, I should hardly have been considered a serious physicist. For the limits to allowable heterodoxy in science are soon reached.*'[22]

There remains to this day a troubling and still unresolved aspect of Maxwell's equations, *viz.* their indifference to the distinction between past and future. The electromagnetic waves whose existence the equations posit can travel backwards or forwards in time; the equations do not tell us whether waves arrive before or after they were emitted. Because we never seem to observe radio waves travelling backwards in time, the backward-in-time solutions of the equations are usually dismissed as being 'unphysical'. Nature exhibits an intrinsic awareness of past-future, the so-called *arrow of time*, that has yet to be satisfactorily incorporated into scientific theories.

---

[21] M. Faraday (Reprinted 1914) *Experimental Researches in Electricity*, London: J. M. Dent & Sons, p. 6.
[22] Quoted in Stuart M. Feffer (1989) 'Arthur Schuster, J.J. Thomson and the Discovery of the Electron', *Historical Studies of the Physical and Biological Sciences*, **20**, 33–61.

Chapter 1.3

The Reality of Atoms – Brownian Motion

The physical sciences advanced along two parallel conceptual tracks during the 19th century. The first was predicated on the *atomic hypothesis*, which sought to explain physical phenomena in mechanical terms; even light was conceived of as a wave-motion in a quasi-mechanical æther. The second was based on the new *science of thermodynamics*, which viewed both matter and energy as continuous substances; it dealt with entities such as pressures, volumes, temperatures and energies but without seeking to elucidate their fundamental nature.

As the century grew to a close, it was clear that both the atomic hypothesis and the science of thermodynamics had proved their usefulness in the advance of scientific knowledge. However, questions had begun to be asked as to their compatibility and which of them gave the better description of physical reality.

The *atomists*, who included most of the chemists for whom atoms and molecules had become the very building blocks of matter, could point to the success of their approach in giving an insight into *how* things happen. It had produced the kinetic theory of gases which, in turn, had provided a quantitative elucidation of the transport phenomena in gases – viscosity, heat-conduction and diffusion – and which, in turn, had also furnished a measure of the magnitude of the imputed molecules.

The *anti-atomists* or *energeticists* as they called themselves, could point to the success of thermodynamics in explaining *why* certain things do happen and others do not; why chemical reactions proceed spontaneously in one direction but not in the other. Moreover, the fundamentals of the *heat engines* that had powered the Industrial Revolution could only be explained by the science of thermodynamics; 19th century atomism had nothing to say on the subject. And above all, thermodynamics concerned itself with 'the economical description of observed facts' and not with unobservable entities. Boyle's law, $PV = \text{constant}$, was such an 'economical description', and to explain it in terms of unseen particles was objectionable. Atomism was no more than a useful working hypothesis (Example 1.1).

The dispute came into the open in 1876, when Loschmidt[23] put forward a theorem to the effect that, given Newton's laws, an explanation of the second law of thermodynamics on the basis of kinetic theory was impossible. His argument was that the collisions between gas molecules are all essentially reversible and so they can lead nowhere; each collision can be undone by its reverse. Should such a system reach an equilibrium, there was nothing to prevent the reverse collisions from returning it in the direction from which it had come. In 1890, Poincaré[24] further showed, with certain assumptions, that any initial state of a three-dimensional system will eventually recur.

However, it is an observed fact that all naturally occurring processes are irreversible; a mixture of gases does not spontaneously unmix into its component gases. In its own terminology, thermodynamics expresses this truth by the maxim: 'Any change that actually occurs spontaneously in Nature is accompanied by a net increase in *entropy*.' Which led Clausius[25] to his famous summary of the first two laws of thermodynamics: 'The energy of the universe is a constant; the entropy of the universe tends always towards a maximum.'

Clearly, if atomism was to prevail, it had to find a way to accommodate the concept of entropy.

1.3.1 Statistical Mechanics

The first atomist to take up this challenge was Boltzmann.[26] Abandoning Newtonian mechanics, he turned to probability theory and treated a gas as an assembly (*macro-state*) of identical but distinguishable particles (*micro-states*). Mathematically, the system's total energy, E, could be distributed in a myriad of different ways between its multitude of micro-states. However, Boltzmann postulated that at equilibrium, the actual distribution would be that which was the most probable of all. On this view, what is first needed to specify the state of a gas is the number of different ways that a particular macro-state can be assembled from its component micro-states.

In principle, the total number of different ways that n micro-states can be permutated (rearranged in a different order) is $n!$. However, suppose that numbers of the micro-states possess the same energies, such that n_1 micro-states have energy E_1, n_2 have energy E_2 or, in general, n_K each have energy E_K ($\sum n_K = n$ and $\sum n_K E_K = E$). Permuting the micro-states within a particular energy value

[23]Josef Loschmidt 1821–1895, Austrian chemist and physicist.
[24]Jules Henri Poincaré 1854–1912, French mathematician.
[25]Rufus Clausius 1822–1888, German physicist and mathematician.
[26]Ludwig Boltzmann 1844–1906, Austrian physicist.

does not produce a new distribution. Accordingly, the number of different ways, W, that the n micro-states can actually be assembled is:

$$W = \frac{n!}{n_1!n_2!\dots n_K!}. \tag{1.9}$$

Starting from this expression and using only standard mathematical procedures, Boltzmann determined that for the most probable distribution, i.e. that found at equilibrium, the number of micro-states, n_K, possessing a particular energy, E_K, was given by the function

$$n_K = \alpha e^{-\beta E_K}, \tag{1.10}$$

where α and β are two constants, both characteristics of the particular system. This is the Maxwell–Boltzmann distribution of which the familiar exponential *barometric formula*, which gives the variation of atmospheric pressure with height, is a special case (Example 1.2).

A dynamic system with any distribution, W, other than the most probable will tend towards the most probable, whereupon it will have achieved equilibrium. Likewise, a system's entropy, S, increases spontaneously and reaches a maximum at equilibrium. All of which suggests a correspondence between these two properties. However, whereas entropy is additive, $S_{12} = S_1 + S_2$, the probability of a combined event is obtained by multiplying the probabilities of the individual events: $W_{12} = W_1 \times W_2$. Accordingly, the mathematical relationship between probability, W, and entropy, S, must be logarithmic. Boltzmann showed it to be

$$S = k \ln W, \tag{1.11}$$

where the constant $k = 1.38 \cdot 10^{-23}\text{J}\cdot\text{K}^{-1}$, is called *Boltzmann's constant.*

Boltzmann's probabilistic thesis provided a molecular interpretation for what had previously been a purely thermodynamic function. The maxim that any spontaneous change is accompanied by a net increase in entropy could now be understood as a probability law. Mixing gases produces a more disordered system, one that has more randomness; and the more random an assembly of molecules, the more ways they can be arranged. Thus, entropy could be viewed as a measure of the degree of disorder or randomness in a system.

But all this still did not prove that the molecules actually existed.

Example 1.1: The Kinetic Theory of Gases

Consider a molecule of mass m moving with velocity v inside a hollow sphere of radius r (Fig. 1.12). Since the collisions between the molecule and the interior wall of the sphere are assumed to be elastic, the angle of incidence equals the angle of rebound and the molecule's velocity changes in direction but not in magnitude. To put it another way, during the collisions with the interior wall of the sphere, the molecule is accelerated centripetally (radially) but not tangentially.

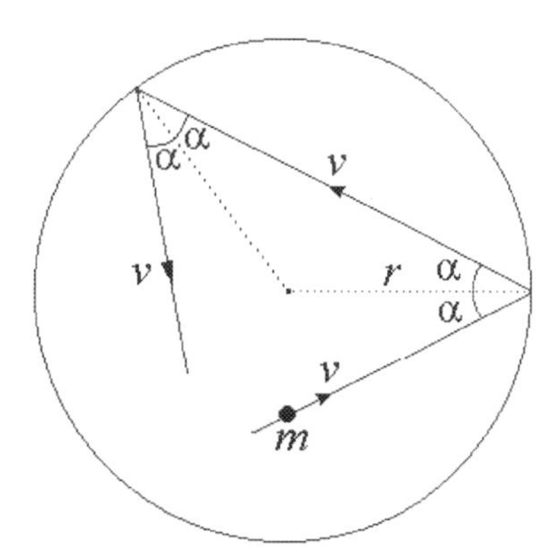

Fig. 1.12. The elastic collision of a gas molecule of mass m and velocity v with the inside wall of a spherical container of radius r.

The molecule strikes and rebounds from the inside of the sphere repeatedly and ceaselessly. Hence, it can be regarded as moving essentially at a constant velocity v in a circular path of radius r around the inside walls of the sphere. The centripetal force required for this circular motion is provided by the normal force exerted by the walls on the molecule. The reaction to this force is the source of the pressure the gas molecule exerts on the interior of the sphere. The centripetal force exerted on the molecule is

$$F_C = \frac{mv^2}{r}. \tag{1.12}$$

The surface area of the interior walls of the sphere is $A = 4\pi r^2$ and so the pressure, $P = F_C/A$, exerted by the gas molecule is given by

$$P = \frac{mv^2}{4\pi r^3}. \tag{1.13}$$

The volume of the sphere is $V = 4\pi r^3/3$, which from substitution gives the familiar expression

$$PV = \frac{mv^2}{3}. \tag{1.14}$$

In terms of the translational kinetic energy of the gas molecule, $K = mv^2/2$:

$$PV = \frac{2}{3}K. \tag{1.15}$$

From which, for a sphere of volume V containing n gas molecules (distinguishable but not necessarily identical) with kinetic energies $K_1, K_2, K_3, \ldots, K_n$, the following expression is obtained:

$$P_1V + P_2V + \cdots + P_nV = \frac{2}{3}(K_1 + K_2 + \cdots + K_n)$$
$$PV = \frac{2}{3}n\overline{K}, \tag{1.16}$$

where $P = P_1 + P_2 + P_3 + \cdots + P_n$ is now the total pressure exerted by the gas molecules and $\overline{K} = \frac{1}{n}\sum_i K_i$ is the average translational kinetic energy of the molecules. Comparing the expression $PV = \frac{2}{3}n\overline{K}$ with the *ideal gas equation*, $PV = nkT$, gives the familiar fundamental identity

$$\overline{K} = \frac{3}{2}kT, \tag{1.17}$$

where k is Boltzmann's constant and T the absolute temperature. This equation can also be written by reference to the usual orthogonal coordinate system as:

$$\begin{aligned}\overline{K} &= \overline{K}_x + \overline{K}_y + \overline{K}_z \\ &= \left(\frac{mv_x^2}{2}\right)_{Average} + \left(\frac{mv_y^2}{2}\right)_{Average} + \left(\frac{mv_z^2}{2}\right)_{Average} \\ &= \frac{3}{2}kT.\end{aligned} \tag{1.18}$$

But since $(v_x^2)_{Average} = (v_y^2)_{Average} = (v_z^2)_{Average}$, we have $\overline{K}_x = \overline{K}_y = \overline{K}_z = kT/2$. This equal sharing of the system's average kinetic energy at equilibrium between its orthogonal *degrees of freedom* is attributed to the energy exchanges that accompany the ceaseless collisions between the gas molecules. It is a special case of the ***principle of the equipartition of energy***.

Example 1.2: The Barometric (or Laplace) Formula

It is common knowledge that the density of the air, ρ, falls with increasing altitude. By making the simplifying (if incorrect) assumption that the air temperature does not change with altitude, a formula for the variation of the atmospheric pressure with altitude, known as the *barometric* (or *Laplace*) *formula*, can be derived.

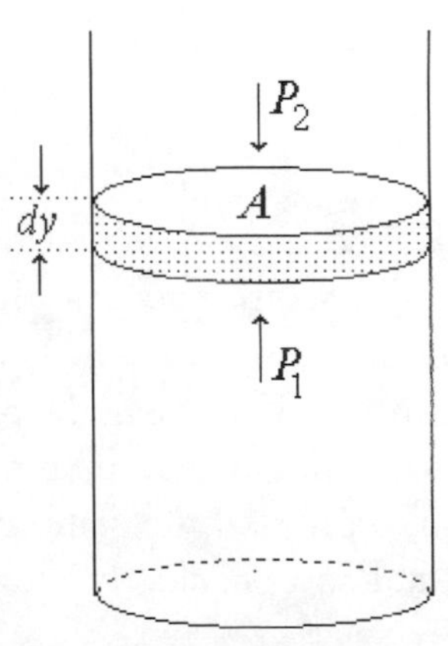

Fig. 1.13. A column of atmospheric air of cross-section A. The pressure at the base of the layer must be greater than that at its top, $P_1 > P_2$, as it has to support its weight.

Taking the positive direction of y upward (Fig. 1.13), the two forces acting on the layer are its weight, $-(\rho Ady)g$, and that which arises from the pressure difference, $-dP = P_2 - P_1$, across its faces, namely, $-AdP$. At equilibrium this gives,

$$-AdP + (-(\rho gA)dy) = 0. \tag{1.19}$$

From which

$$dP = -\rho gdy. \tag{1.20}$$

The density of an ideal gas of volume V and comprising n molecules each of mass m is:

$$\rho = \frac{nm}{V} = \frac{nm}{nkT} \cdot P = \frac{m}{kT} \cdot P. \tag{1.21}$$

Combining Eqs (1.20) and (1.22) gives:

$$-\frac{dP}{P} = \frac{mg}{kT} \cdot dy. \tag{1.22}$$

Integrating between the limits $P = P_0$ at $y = 0$ and $P = P$ at $y = y$:

$$\ln \frac{P}{P_0} = -\frac{mg}{kT} \cdot y$$

$$P = P_0 e^{-mgy/kT}. \tag{1.23}$$

But mgy is simply a particle's gravitational potential energy, E_y, at the height y:

$$P = P_0 e^{-E_y/kT}. \tag{1.24}$$

This is a special case of the Maxwell–Boltzmann distribution which states that for any assembly of distinguishable but not necessarily identical particles, the ratio between the number of particles, n_0, in a given energy state and the number, n, in a state whose energy is E higher than that of the given state is given by:

$$n = n_0 e^{-E/kT}. \tag{1.25}$$

1.3.2 Brownian Motion

Robert Brown, after whom the phenomenon of Brownian motion is named, was a Scottish botanist. In 1827, while studying an aqueous suspension of pollen under a microscope, he noticed that the grains executed an incessant jigging motion. The size of the pollen grains was of the order of a micrometer ($1\,\mu\text{m} = 10^{-6}\,\text{m}$). He examined pollen from other types of flowers and found that they all exhibited the same motion. His first hypothesis was that the motion was vital and peculiar to the male plant cells. Let Brown himself describe how this view came to be rejected:

> '*It occurred to me to appeal to this peculiarity as a test in certain Cryptogamous plants, namely Mosses... in which the existence of sexual organs had not been universally admitted... I found that on bruising the floral leaves of Mosses, I readily obtained similar particles, not in equal quantity indeed, but equally in motion. My supposed test of the male organ was therefore necessarily abandoned.*'[27]

He went on to examine suspensions of finely powdered mineralised vegetable remains, glass, soot and stone:

> '*Rocks of all ages, including those in which organic remains have never been found, yielded the molecules [the fine incessantly jiggling particles] in abundance. Their existence was ascertained in each of the constituent minerals of granite, a fragment of the Sphinx being one of the specimens observed.*'[28]

Evidently, the incessant jiggling motion was a property of all fine particle suspensions – *colloids* – irrespective of their composition.

Various suggestions were put forward by his contemporaries as to the cause of the motion: convection currents in the strongly illuminated suspensions, evaporation effects, capillary action and the release of volatile fractions from the suspensions or minute air bubbles, were just a few. However, he refuted all these explanations when he demonstrated that the particles in a microscopic drop of aqueous suspension immersed in oil also exhibited the unabated motion:

> '*This experiment consists in reducing the drop of water containing the particles to microscopic minuteness, and prolonging its existence by immersing it in a transparent fluid of inferior [lower] specific gravity, with which it is not miscible, and in which evaporation is extremely slow... Of these, the most minute necessarily contain but few particles, and some may be occasionally observed with one particle only. In this manner minute drops, which if exposed to the air would be dissipated in less than a minute, may be retained for more than an hour. But in all the drops thus formed and protected, the motion of the particles takes place with undiminished activity, while the principal causes assigned for that motion, namely, evaporation, and their mutual attraction and repulsion, are either materially reduced or absolutely null.*'[29]

After a period of 30 years, during which little was published on the subject, interest in the phenomenon was renewed and it was suggested that the zig-zag

---

[27] R. Brown (1828), A brief account of microscopical observations made on the particles contained in the pollen of plants, *Phil. Mag.* **4**, p. 165.
[28] R. Brown (1828), A brief account of microscopical observations made on the particles contained in the pollen of plants, *Phil. Mag.* **4**, p. 167.
[29] J.J. Bennett (ed.) (1866) *The Miscellaneous Botanical Works of Robert Brown, Esq., D.C.L., F.R.S. Vol. 1*, pp. 433–461.

motion of the colloidal particles was caused by their continuous bombardment by the molecules of the suspension liquid (the solvent). This hypothesis was supported by certain of the observed features of the motion: it was more active the smaller the particles and the higher the temperature. It also depended on the viscosity of the solvent; no motion was observed in glycerine. Notwithstanding, a number of serious objections were raised against this hypothesis.

1. It was argued that because of their random nature, the net result of a relatively large particle's bombardment by a myriad of much lighter solvent molecules, coming at it from all directions, would be zero. Each collision would barely move the particle and even if, at a particular instant, more molecules struck it on one side than on the other, their collective overall effect would be zero. On average, the collisions would cancel each other out, leaving the particle where it was.
2. Attempts were made to measure the velocity of the particles, though, as we shall see, the traditional concept of velocity is an inappropriate parameter in Brownian motion. The principle of the equipartition of energy implied that, at equilibrium, the colloidal particles and the solvent molecules should have the same average translational kinetic energy, $\sim kT$. However, it was found that the velocities of the particles were an order of magnitude smaller than would be expected if they were in thermal equilibrium with the solvent molecules.
3. When a particle gains kinetic energy, there must be a corresponding loss of heat energy by the contiguous solvent, which thus undergoes a localised cooling. However, the second law of thermodynamics prohibits the conversion of heat energy into mechanical energy unless there is an accompanying transfer of heat from a hot to a cold reservoir.

If Brownian motion did indeed result from the bombardment of the suspension (colloidal) particles by the randomly moving solvent molecules, its elucidation would require a totally new approach.

1.3.3 Sedimentation Equilibrium

No single person contributed more to 20th century physics than Albert Einstein.[30] In 1905, when just 26 years of age, in addition to completing his doctoral dissertation, he produced four original scientific papers that set a new agenda for physics. Long used to describe the year 1666 during which Isaac Newton laid the foundations for much of classical physics – his theories of

[30] Albert Einstein 1879–1955, German–Swiss–American physicist who was awarded the Nobel prize in 1921 for discovering the law of the photoelectric effect.

calculus, colours, motion and gravitation – the term *annus mirabilis* has also been ascribed to the year 1905, in which Einstein laid the foundations for the break with Newtonian concepts that would revolutionise 20th century science.

The problem of the compatibility of atomism and the science of thermodynamics had interested Einstein from early on. His earliest published papers all touch on this issue. However, after reading Boltzmann's *Gastheorie* early in his student days, he had become a convinced follower of his principles.

In his doctoral dissertation, Einstein set out to show how an estimate of Avogadro's number and of the size of the solute molecules in a solution, could be obtained from measurements of the pure solvent's viscosity and that of the solution. In developing his argument, Einstein asserted that 'with respect to its mobility within the solvent and its effect on the latter's viscosity, such a molecule behaves approximately like a solid body suspended in a solvent.'

This was the starting point for his 1905 paper on 'The Motion of Small Particles Suspended in Liquids at Rest Required by the Molecular-Kinetic Theory of Heat' – the term Brownian motion does not appear in the paper's title[31] – in which he asserted that:

> *'From the standpoint of the molecular-kinetic theory of heat... a dissolved molecule differs from a suspended body only in size, and it is difficult to see why suspended bodies [colloidal particles] should not produce the same osmotic pressure as an equal number of dissolved molecules. We have to assume that the suspended bodies perform an irregular, albeit very slow, motion in the liquid due to the latter's molecular motion; if prevented by a wall from leaving a given volume V, they will exert a pressure on the wall just like molecules in solution.'*

Although, from the classical theory of thermodynamics, no pressure should be exerted on the wall by the suspended colloidal bodies, Einstein argued that as a consequence of the solvent's thermal molecular motions, all the particles in a suspension, irrespective of their size, should exert an osmotic pressure just like the solute molecules in a solution. Before Einstein, no one appears to have realised that this provided a touchstone for the kinetic theory. Van t'Hoff[32] had already shown that in dilute solutions, the osmotic pressure, P, exerted by z moles of solute molecules in a solution of volume V is given by the *ideal gas law*, $PV = zRT$. Einstein contended that the same should apply to colloidal

[31]Einstein would later claim that he had no knowledge of Brownian motion when he first examined the subject of particles in fluid suspensions. All he added to the original paper when his attention was drawn to it was: 'It is possible that the motions to be discussed here are identical with so-called Brownian molecular motion; however, the data available to me on the latter are so imprecise that I could not form a judgment on the question.'

[32]Jacobus Henricus Van t'Hoff 1852–1911, Dutch chemist who, in 1901, was the first person to be awarded the Nobel prize for chemistry. The award was made in recognition of his discovery of the laws of chemical dynamics and osmotic pressure in solutions.

particles: 'Equal numbers of solute molecules and suspended particles behave identically as regards osmotic pressure.'

In conventional treatments of osmotic pressure the 'wall' is usually a semipermeable membrane through which solvent molecules can pass but not solute particles. In his analysis of diffusion equilibria, Einstein replaced this 'wall' by a force which, at equilibrium, balances the osmotic pressure exerted by the solute particles. In the case of a suspension of colloidal particles in a gravitational field, this force is gravity (Fig. 1.14).

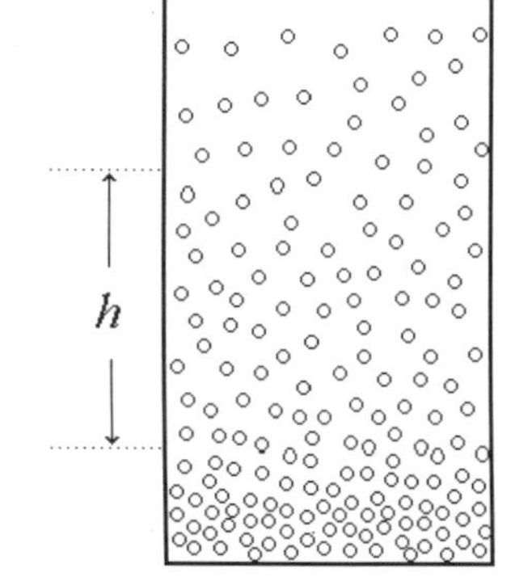

Fig. 1.14. Sedimentation equilibrium. The downward gravitational attraction on the suspended particles is balanced by the upward osmotic pressure, acting from the region of higher particle concentration towards the region of lower concentration. Dynamically, the rate at which particles fall is equal to the rate at which they diffuse upwards due to the effect of the thermal motion of the solvent molecules upon them.

In 1907, Perrin[33] used Einstein's radical concepts in a landmark experiment to determine the value of Avogadro's number, N. By careful fractional centrifuging, Perrin first obtained samples of spherical gamboge[34] particles of known density and uniform diameter, between $0.5\,\mu$m and $1\,\mu$m, with which he prepared aqueous suspensions.

Assuming that these colloidal particles behave like gas molecules in a gravitational field, the ratio between the number of particles at a given level in the suspension, n_1, and the number, n_2, at a height, h, above it should be given by the Maxwell–Boltzmann distribution:

$$\begin{aligned} n_2 &= n_1 e^{-(E_2 - E_1)/kT} \\ &= n_1 e^{-mgh/kT}. \end{aligned} \tag{1.26}$$

Substituting $m = \frac{4}{3}\pi r^3(\rho_{particle} - \rho_{fluid})$, where r is the radius of the particles, $\rho_{particle}$ the density of gamboge and ρ_{fluid} that of the suspending liquid gives

$$\begin{aligned} \ln \frac{n_1}{n_2} &= \frac{\frac{4}{3}\pi r^3(\rho_{particle} - \rho_{fluid})}{T} \cdot \frac{gh}{k} \\ &= \frac{\frac{4}{3}\pi r^3(\rho_{particle} - \rho_{fluid})}{T} \cdot \frac{gh}{R} \cdot N, \end{aligned} \tag{1.27}$$

[33]Jean Baptiste Perrin 1870–1942, French physicist who was awarded the Nobel prize for physics in 1926 for his work on the discontinuous structure of matter and his discovery of sedimentation equilibrium.
[34]Gamboge is a gum-like material obtained from the desiccation of natural latex.

where R is the molar gas constant and N is Avogadro's number $(k = \frac{R}{N})$.

Using a microscope, Perrin counted the number of particles at two different levels, separated by a distance h, in a column of gamboge suspension and thus, knowing the numerical values of all the other variables in Eq. (1.27), he could calculate the value of Avogadro's number.[35] The result he obtained by this method was $6.8 \cdot 10^{23}$.

1.3.4 The Reality of Atoms

Even the most convinced energeticists would finally be convinced of the reality of atoms following a second series of experiments, predicated on Einstein's further ideas on the motion of colloidal particles, in which Perrin tracked the actual movement of the particles in a suspension.

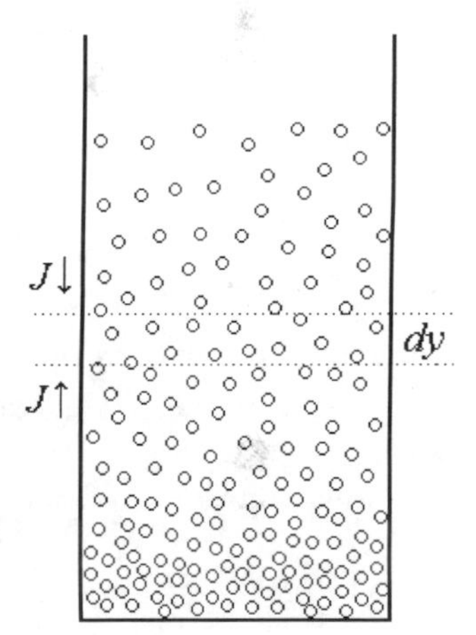

Fig. 1.15. There are $dn = \kappa dy$ particles between each unit area of the upper and lower faces of the horizontal lamina. The effective weight of these particles, $mg\kappa dy$, is balanced by the difference, dP, between the osmotic pressure on the lower face and the upper face.

Consider the dynamic equilibrium across a horizontal lamina of thickness dy in a suspension of particles containing κ particles per unit volume (Fig. 1.15). The downward flow of particles as a result of gravity across a unit area of the layer, $J\downarrow$, will be matched by an equal upward flow, $J\uparrow$, driven by osmotic pressure:

$$J\uparrow = J\downarrow = D\frac{d\kappa}{dy}, \tag{1.28}$$

where D is a constant of the system called the *diffusion coefficient* and $\frac{d\kappa}{dy}$ is the *concentration gradient.* This is an instance of *Fick's first law of diffusion.*[36]

[35] The term 'Avogadro's number' for the number now called 'Avogadro's constant' or a *mole* was originally proposed by Perrin.

[36] Fick's first law postulates that the flux goes from regions of high concentration to regions of low concentration, with a magnitude that is proportional to the concentration gradient.

When a spherical particle of radius r falls under gravity through a viscous medium, it achieves a terminal velocity, v_0, given by Stoke's law:

$$v_0 = \frac{mg}{6\pi\eta r}, \tag{1.29}$$

where mg is its effective weight and η the coefficient of viscosity of the medium. Einstein assumed that this relationship could be applied to the motion of suspension particles. Accordingly, the downward flow of particles through each unit area of the layer in unit time is

$$J\downarrow = \kappa v_0 = \kappa \cdot \frac{mg}{6\pi\eta r}. \tag{1.30}$$

The osmotic pressure, P, attributable to the particles in the suspension is given by the general relationship

$$P = \kappa \cdot \frac{RT}{N} \tag{1.31}$$

from which simple differentiation gives

$$\frac{dP}{d\kappa} = \frac{RT}{N}. \tag{1.32}$$

At equilibrium, the effective weight, $mg\kappa\, dy$, of the particles contained between each unit area of the upper and lower faces in the horizontal lamina is balanced by the difference, dP, between the osmotic pressure on its lower face and upper face.

$$dP = mg\kappa\, dy = \frac{RT}{N} \cdot d\kappa, \tag{1.33}$$

from which

$$\frac{d\kappa}{dy} = \kappa \cdot \frac{mgN}{RT}. \tag{1.34}$$

Substitution and elimination in Eqs (1.28), (1.30) and (1.34) gives

$$D = \frac{RT}{N} \cdot \frac{1}{6\pi\eta r}. \tag{1.35}$$

Having obtained this expression for the diffusion coefficient of the particles from considerations of osmotic pressure and the internal viscosity of the fluid, Einstein now turned to a closer examination of the actual disordered particle motions.

Classical mechanics employs the concept of continuous trajectories in order to predict where a body will get to, or where it was situated, at a certain time. Clearly, the Brownian particles' trajectories are not smooth but jagged, and so their motion cannot be correctly described by continuous functions; this is what was meant by the observation that 'the traditional concept of velocity is an inappropriate parameter in Brownian motion'. Recognising that this must be so, Einstein wrote:

> '*Evidently, we must assume that each individual particle executes a motion that is independent of the motions of all the other particles; the motions of the same particle in different time intervals must also be considered as mutually independent processes, so long as we think these time intervals as chosen not to be too small.*'[37]

What was 'evident' to the 26-year-old Einstein had evidently escaped the notice of others. His radical hypothesis was that Brownian motion is a random step-wise occurrence, in which each particle's motion is unrelated to that of the others and each step taken by a particle is unrelated to the previous one it took.[38]

At first sight this might appear to solve nothing. Clearly, by virtue of the random nature of the steps, the expected average finishing position of the particles must be back at the starting point. However, that might well take almost an infinity of steps and time. The interesting question is how far from the starting point a particle can be expected to be, on average and without regard to direction, after a particular number of steps, n, or a certain finite time. To determine this, in the case of a random step-wise motion restricted to a particular axis where the steps can only be in the positive or negative direction, we first find the squares, $x_1^2, x_2^2, x_3^2 \ldots$ of the particle's distance from the starting point after each step and average them; this gives $\overline{x^2} = \frac{1}{n}\sum_{i=1}^{n} x_i^2$, which is called the *mean square distance.*[39] We then take its square root, $\sqrt{\overline{x^2}}$, to obtain the expected distance from the starting point after the n steps; this is called the *root mean square* or *rms distance.*

Einstein determined that the displacement, λ_x, in the direction of the X-axis that a particle experiences on the average after a time t, its *root mean square displacement*, is given by (Example 1.3):

$$\lambda_x = \sqrt{2Dt}. \tag{1.36}$$

Combining this result with Eq. (1.35) gives:

$$N = \frac{t}{\lambda_x^2} \cdot \frac{RT}{3\pi\eta r}. \tag{1.37}$$

[37] R. Fürth (ed.) (1956) *Albert Einstein: Investigations on the Theory of the Brownian Movement*, New York: Dover.

[38] In modern parlance we would say that Einstein's insight was that Brownian motion is a *stochastic* (random) phenomenon, akin to the 'random walk'.

[39] *mean* [of the] *square* [of the] *distance.*

Perrin employed this relationship in a second series of experiments he conducted to determine Avogado's number, N. Using a microscope with a *camera lucida* set-up, which enabled him to observe both the particle and its projection on a sheet of paper at the same time, Perrin recorded the positions of particles of various sizes, in a number of different fluid suspensions, every 30 seconds (Fig. 1.16).

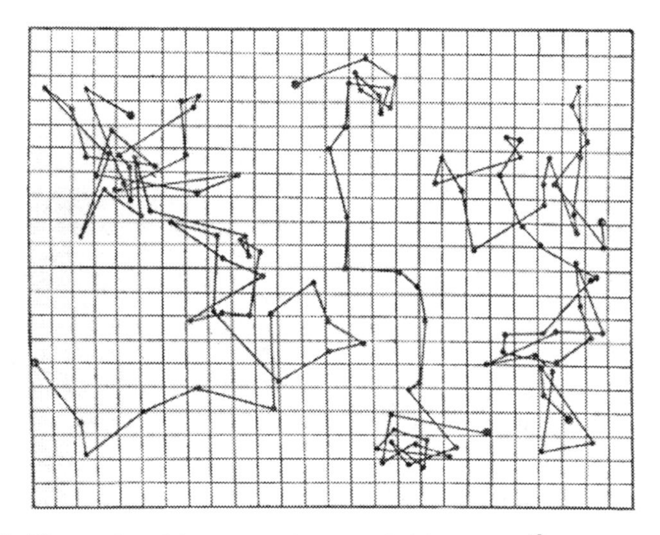

Fig. 1.16. The tracks of three particles recorded by Perrin.[40] The dots show the particle positions at 30 second intervals. The scale is 1 division $= 3.125 \times 10^{-6}$ m. The radius of the particles was $0.52\,\mu$m.

From the tracks he recorded, the value of λ_x^2, the mean square displacement, could be obtained by measurement and simple calculation. Perrin found an average value for N of $6.9 \cdot 10^{23}$, in good agreement with his previous experiment. Perrin concluded, 'This remarkable agreement proves the rigorous accuracy of Einstein's formula and, in a striking manner, confirms the molecular theory.'

Atoms are for real!

Example 1.3: The Mean Square Displacement of a Brownian Particle

Restricting himself to the X-coordinates and starting from the definition that $f = f(x, t)$ is the probability density that a Brownian particle is at the point x at time t, Einstein showed how the random step-wise motion he had postulated led to a differential

[40] J. Perrin (1909) *Annales de Chimie et de Physique*, 8th series, September.

equation involving the diffusion coefficient,[41]

$$\frac{\partial f}{\partial t} = D\frac{\partial^2 f}{\partial x^2} \tag{1.38}$$

whose solution for a suspension of n particles diffusing outwards from a point is

$$f(x,t) = \frac{n}{\sqrt{4\pi Dt}} \cdot e^{-\left(x^2/4Dt\right)}. \tag{1.39}$$

The distribution of the particle displacements after a time t is thus seen to have a similar form to the Gaussian or normal distribution of random errors

$$f(x) = \frac{1}{\sqrt{2\pi\sigma^2}} e^{-\left((x-\mu)^2/2\sigma^2\right)}.$$

Accordingly, making the correspondence with the standard deviation, σ, the root mean square displacement of a particle after time t is given by:

$$\lambda_x = \sqrt{\overline{x^2}} = \sqrt{2Dt}. \tag{1.40}$$

---

[41]This is the familiar diffusion equation known as *Fick's second law* which predicts how diffusion changes the concentration with time.

Chapter 1.4
The Special Theory of Relativity

The æther enigma was first resolved in 1905 with the publication of Einstein's special theory of relativity. This revolutionary hypothesis redefined some of the most basic assumptions of the physical sciences. To fully appreciate the nature of these new definitions and of the circumstances that made them necessary, we must return to the very fundamentals of science and reconsider the rationale that underlies our individual and collective awareness of the physical world around us.

Does a physical world, independent of humanity, really exist? Do mountains, trees, land, sea and sky all actually exist whether or not anybody is looking at them? After all, our awareness of the external world occurs within us, in our minds; perhaps it is all an illusion.

> '*When we do our utmost to conceive the existence of external bodies, we are all the while contemplating our own ideas. But the mind, taking no notice of itself, is deluded into thinking it can and does conceive bodies existing with aught of or without the mind.*'[42]

Furthermore, since no experiment can prove the physical existence of an object that is indiscernible to human observers, what meaning can *physical existence* have in the absence of such observers? On the other hand, does everything we observe or perceive really exist? Can we always distinguish between real and virtual reality? Where is the border between reality and illusion?

The prototypes of the resolutions Western Civilisation has contrived for this mystery are the two hypotheses associated with the classical Greek philosophers, Plato and Aristotle, depicted in Raphael's Vatican fresco *The School of Athens*. The one linked with Plato's name asserts that all true knowledge is to be found through the mind's contemplation of abstract ideas.

[42]Bishop George Berkeley (1708) *A Treatise Concerning the Principles of Human Knowledge*, paragraph 23. Though he probably intended it only to apply to physical entities, Berkeley's solipsism denies the existence of anything outside the mind.

'Plato admitted the existence of an external world but came to the conclusion that the world perceived by the senses is motley, protean, ever-changing and unreliable. The true world is the world of ideas, which is unchangeable and incorruptible. Yet this world of ideas is not accessible to the senses but only to the mind. Observations are useless.'[43]

In Plato's view, physical evidence proves nothing; it cannot overturn pure reasoning. The reality behind appearances, what is inherently true of them, is the pure reason of mathematics. Man interprets and remoulds the imperfect world he experiences in terms of his vision of perfection. The *idea* of lines, circles and squares is more real than physical nature, in which there are few pure examples of these perfect forms.

The alternative view, that associated with Aristotle, holds that it is material things that are the primary substance and source of reality.

'Physics, and science generally, must study the physical world and obtain truths from it. Thus, genuine knowledge is obtained from sense experience by intuition and abstraction. These abstractions have no existence independent of the mind. To arrive at truths, Aristotle used what he called universals – general qualities that are abstracted from real things.'[44]

Using intuition and the human faculty for abstraction, Aristotle elevated the observed qualities of objects into mental concepts. For example, since every material thing he saw around him undergoes corruption, decay or mortality, Aristotle concluded that *ceaseless change* must be a universal attribute of the sublunar world.

After dominating Western thinking for almost two millennia, the inadequacy of these ancient hypotheses became apparent in the Renaissance. In their classical conceptions, neither the *Ideas* of Plato nor the *Universals* of Aristotle could cope with the new knowledge and the profusion of discoveries, inventions and innovations that, from the 15th century on, swept through Western Europe. The new arts and technologies were simply beyond the scope and grasp of these hallowed doctrines.

Perceiving no limit or end to the flood of new ideas, the 17th century philosopher Descartes was forced to acknowledge that all the beliefs he presently held might perhaps be false. He found just one fact of which he could still be sure – *cogito ergo sum* – I think, therefore I am.

This truth made him aware of his own limitations. But, by the same token, it also allowed him to conceive the existence of an Omnipotent

[43]Morris Kline (1985) *Mathematics and the Search for Knowledge*, Oxford: Oxford University Press, p. 4.
[44]*Ibid.*

Perfection. Descartes argued that this Perfection would not be a deceiver; He would not make us believe there was a material universe if it did not really exist.[45] Descartes averred that the material universe really does exist. Moreover, and most important of all, he asserted that this universe works like a great machine, according to laws accessible to human thought and understanding.

Assured of its consonance and reality, the Europeans now set out in earnest to learn all they could about the physical world. Their findings appeared to confirm Descartes' beliefs: the physical world was evidently both fathomable and governed by natural laws. Thus, tradition and scholarship were replaced by reason and experiment as the arbiters of truth.

As Newton wrote in the opening passage of his *Opticks*, the first science text to be written in a language that could be understood by non-specialists:

> '*My design in this book is not to explain the properties of light by hypotheses, but to propose and prove them by reason and experiments.*'[46]

If the phenomenon of light, the *fiat* of the very first day of Creation, could be explained by reason and experiment, was there anything in the world that could not? The agenda of modern times had been set.

Measurement became the yardstick of reality. A concept was considered physically real if something corresponding to it in the physical world could be measured. Heat, though unseen, is physically real because we can measure temperature with a thermometer. Similarly, colour acquired a physical reality when Newton declared that 'Lights which differ in Colour, differ also in Degrees of Refrangibility – their disposition to be refracted or turned out of their Way in passing out of one transparent Body or Medium into another'. In short shrift, every concept that did not satisfy the measurement criterion was purged from the corpus of science. Only those for which a method or instrument of measurement could be devised were allowed to remain.

1.4.1 The Principle of Covariance

The foundation-stone upon which the new sciences were established was the belief that the physical world around us is real and that its existence is independent of any percipient entity. On this view, physical changes are events, not the actions of an Agency; the physical world is driven by detectable intrinsic forces, inherent in its nature, not by unverifiable external causes;

[45] On this view, God's existence is as important for science as for religion.

[46] (Reprinted 1914), New York: Dover Publications, p. 1.

and all true knowledge is near at hand, reproducible and just waiting to be learnt.[47]

A corollary of this new vision of reality was that all observers of the physical world are, in principle, equally authoritative and legitimate. All have the same access to knowledge. No observer, whoever or wherever he is, has a better view of physical reality than any other; all observers are equivalent. For example, each player in a football team constantly monitors the movements of the other players and of the ball. But is any one player's description of these motions superior to those of the others? Obviously not. Similarly, the description of planetary motions from the point of view of a stationary sun given by the Copernican heliocentric system, is not inherently superior to that of the Ptolemaic geocentric system which views things from a stationary Earth. Nor is the opposite true: the Ptolemaic description is no better than the Copernican. In fact, any motion described by the one can be replicated and accounted for by the other. Given sufficiently complex mathematical devices, the celestial motions can be accurately and precisely described from anywhere.

As a consequence of their underlying equivalence, all observers should discover that the physical world is governed by the same laws; the laws of nature must take the same form for all observers.[48] Physical laws which take the same form for all observers, irrespective of their position or motion, are said to be *covariant.* A quantity whose measured value is unaffected by the position or motion of the observer is said to be *invariant.* Thus, the laws of nature must be covariant. Furthermore, if all observers are subject to the same laws of nature, then these laws cannot be the basis for distinguishing between them; observers who are subject to the same laws are indistinguishable by reference to those laws.[49]

---

[47]Before uncritically accepting this maxim, the reader might like to ponder the fact that in modern information theory, which is concerned with that area where physical science and psychology overlap, the mathematical function that defines *information* is related to the thermodynamic function *entropy* in very fundamental ways. Although information is a measure related to the observer and entropy is related to the physical system, the distinction between them is rather artificial since thermodynamics always implies an observer who establishes or removes constraints on the system or otherwise manipulates it.

> '*Information deals with the usefulness of a set of symbols to an observer ... What appears to lie behind all this is the fact that our view of the physical universe is not a cold, lifeless abstraction but is structured by the human mind and the nature of life itself. While it has long been fashionable to inquire how biology depends on physics, we are being forced to inquire how physics depends on the underlying biology of the mind.*'
> Harold J. Morowitz (1970) *Entropy for Biologists*, Academic Press.

[48]This is the *principle of the uniformity of nature* or the cosmological *location principle* that asserts that no entity has a special location in the universe.

[49]Individualism, the sense of self, is essentially non-scientific.

Notwithstanding its apparent simplicity and self-evidence, the rigorous application of observer equivalence – *the principle of covariance* – to the formulation of the laws of physics has proved one of the greatest intellectual challenges of all times. As we shall see, it was in order to correct the failure of classical physics to meet this challenge that Einstein put forward his theories of relativity.

A limited form of covariance is found in the following corollary Newton added to his laws of motion:

> '*The motions of bodies included in a given space are the same among themselves [relative to one another], whether that space is at rest, or moves uniformly forwards in a right [straight] line ... A clear proof of this we have from the experiment of a ship; where all motions happen after the same manner, whether the ship is at rest, or is carried uniformly forwards in a right [straight] line.*'[50]

Relative to an observer on board the ship, the motion of an object is the same whether the ship is stationary or sailing at a uniform velocity in a straight line. In both cases, an object thrown vertically upwards falls straight back into the observer's hands; the motion of the object is subject to the same laws. Thus, by reference to the motions they observe in the immediate space around them, observers at rest cannot be distinguished from those in uniform motion in a straight line; the latter can never be sure whether it is they or their surroundings that is moving. Such indistinguishable observers are called *inertial observers*.[51]

Newton's laws of motion are true for inertial observers. However, in their canonical form, they are not valid for non-inertial observers such as passengers in an accelerating or breaking vehicle or persons riding on a rotating carousel. For example, according to *Newton's first law of motion*:

> '*Every body continues in its state of rest, or of uniform motion in a right [straight] line, unless it is compelled to change that state by forces impressed upon it*',

i.e. a body's motion remains unchanged unless an external force acts upon it. Observers on board a stationary ship or one that is sailing in a straight line at constant velocity do indeed find this to be true; persons and objects placed on deck stay where they are (so long as the ship remains on an even keel). However, a vehicle's passengers are thrown, respectively, backwards or forwards every time it accelerates or brakes, even though no actual external force pushes them. Similarly, relative to an observer on a rotating carousel, a free object moves

[50] *Principia.* Translation by Andrew Motte, revised by F. Cajori (1962), Berkeley: University of California Press, p. 20.

[51] The notion of indistinguishable inertial observers for whom neither absolute rest nor absolute uniform motion has any meaning is the underlying postulate of Newtonian mechanics.

in a curved path even though no actual external force acts upon it (Fig. 1.17). This is the familiar *Coriolis effect.* The atmospheric circulations – the cyclones and anti-cyclones – so important in interpreting terrestrial weather patterns, are the result of a Coriolis effect associated with the earth's rotation.

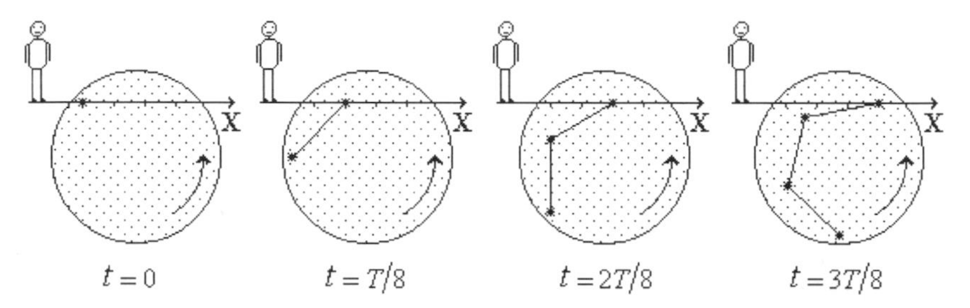

Fig. 1.17. Motion relative to an inertial and a non-inertial observer. The stationary inertial observer sees the free object moving away from him in a straight line along the X-axis across the face of the rotating disc. However, suppose the disc rotates with period T and that at regular intervals, the object squirts a drop of ink onto it. As the pattern of inkspots shows, the path taken by the object as seen from the disc (relative to the disc) is curved and not straight. Thus, a non-inertial observer riding on the rotating disc would not ascribe a straight path to the motion of the object but a curved one, even though no actual force acts upon it.

In order to account for these effects and to enable non-inertial observers to make sense of Newton's laws of motion, the existence of forces other than actual external forces must be postulated. It is these so-called *inertial* or *fictitious* forces that act on the passengers and are felt by them when their vehicle accelerates or brakes.[52] Their presence requires making modifications to the canonical form of the laws of motion that are particular to each non-inertial system; a modification valid in one non-inertial system will not necessarily work in another. Thus, each non-inertial system has its own formulation of the laws of motion. Evidently, Newton's laws of motion are truly covariant only for a special class of observers: inertial observers.

This usually presents no problem, as our mind's eye can always conjure up a hypothetical inertial observer, positioned at some convenient point in space, relative to whom the motion of the object in question can be analysed using Newton's laws of motion. Implicit in this approach is the assumption that all points in space are equivalent. Newton expressed this idea through his postulate of an *absolute space* which 'in its own nature, without regard to anything external, remains always familiar and immovable'. It is within this

[52] Actual forces are those exerted by one body on another; they are an expression of the interaction between the bodies. An electrical interaction gives rise to an electrical force; magnetic interactions to magnetic forces, and so on. The additional forces that appear in non-inertial systems are not the result of any interaction.

space that inertial observers can never be sure if it is they or the objects around them that are moving, and it was relative to this space that changes in motion occurred.

Newton identified an object's *mass* with the 'quantity of matter' that lies in it: its inertia (*vis inertiæ*) or tendency to persist in a state of rest or uniform motion in a straight line. This innate 'force of matter' is an absolute; an invariant quantity, that is manifested to the same degree whenever an external force is impressed upon the object. Thus, all inertial observers must exert the same force to effect a given change in an object's motion. It follows that, in the Newtonian conception, force too is an invariant quantity.

Unqualified use was made of the Newtonian conception of motion for some 200 years. However, when applied to electromagnetic theory, it was found to give rise to a profound paradox. It transpired that, under its application, the laws of electromagnetism take a different form for different inertial observers, depending on whether they are at rest or in motion. Thus, though stationary and moving inertial observers are indistinguishable by reference to Newton's laws, under the self-same conception of motion, they should be distinguishable by reference to laws of electromagnetism. The following two case studies illustrate the problem.

Case 1: Coulomb's law and magnetic forces

Consider the interaction between two identical positive charges, q at a distance r from one another, when the observer is

a. at rest relative to the charges;
b. in uniform rectilinear motion relative to the charges (Fig. 1.18).

When the observer is at rest relative to the charges, he finds that the magnitude of the repulsive force acting between them corresponds to that given by Coulomb's law, $F_E = kq^2/r^2$. However, when he moves in a straight line with velocity u relative to the charges or they move similarly in parallel relative to him, he finds that a weaker force, $F' < F_E$, acts between them. Coulomb's law is seemingly not valid when the observer and the charges are moving relative to one another. Thus, it should provide a basis for distinguishing between stationary and moving inertial observers.

Classical theory circumvents this difficulty by introducing a new phenomenon, *electromagnetism*. It postulates that by reason of the relative motion, a magnetic field B, is set up alongside the electric field. This, in turn, bears upon the charges such that an attractive force of magnitude $F_M = F_E - F' = Bqu$ acts between them.

However, this really solves nothing, for it is only by reason of the different forces, F_E and F', observed in the two instances – the one with the observer at rest relative to the charges, the other in uniform rectilinear motion relative to them – that the presence of the magnetic field is construed. There is no independent proof of its existence; no magnetic poles or magnetic charges have ever been isolated. The magnetic field appears to be no more than a contrivance – a mathematical fudge – that does not really solve the fundamental problem of the apparent non-equivalence of stationary and moving inertial observers under Coulomb's law.

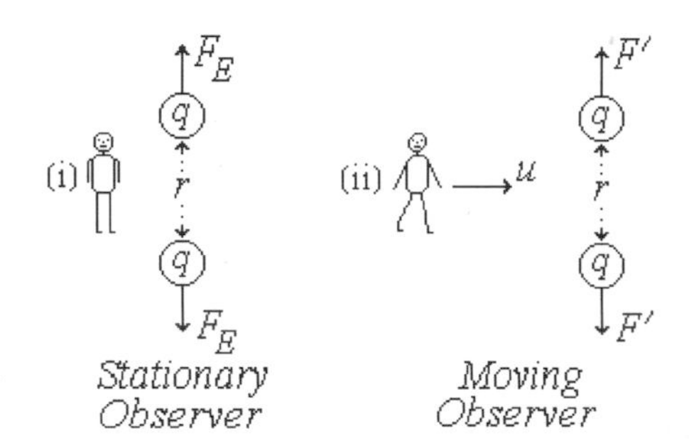

Fig. 1.18. The forces acting between two positive charges as viewed by two inertial observers. Observer (i), who is at rest relative to the charges, detects the Coulombic force, F_E. However, observer (ii), who is in uniform motion relative to the charges, detects a weaker force, F'.

In order to establish their equivalence, Coulomb's law must be shown to be valid for both inertial observers, without the need for any special postulates. The difference between the forces they measure must be explained in terms of a physical concept that has the same import for both of them, such as their relative motion, and not in terms of a phenomenon – the magnetic field – that only has meaning for one of them. This cannot be done under the Newtonian conception of motion.

Case 2: Electromagnetic induction

The interaction between a magnet and a conductor moving at a uniform velocity, u, relative to one another, can be viewed from two points of view (Fig. 1.19):

a. from that of an observer at rest relative to a metal conductor, who sees the magnet approaching the conductor at a uniform velocity, u;
b. from that of an observer at rest relative to the magnet, who sees the metal conductor approaching the magnet at a uniform velocity, u.

Under the Newtonian conception of motion, the explanations that Maxwell's theory of electrodynamics gives for the two cases are different. In the first case, it asserts that the motion of the magnet sets up an electric field of a certain energy in the space around it which, in turn, excites an electric current in the stationary conductor. In the second case there is no electric field.

Instead, an electromotive force, which of itself corresponds to no energy, is induced in the moving conductor as it moves through the stationary magnetic field; this electromotive force causes an electric current of the same magnitude as that realised in the first case to flow.

However, experiment belies these distinctions. In practice, electromagnetic induction experiments reveal no asymmetry between inertial observers; rest and uniform motion in a straight line are electromagnetically indistinguishable. Just like the sailor on Newton's ship, the inertial observers cannot tell whether they or their surroundings are moving.

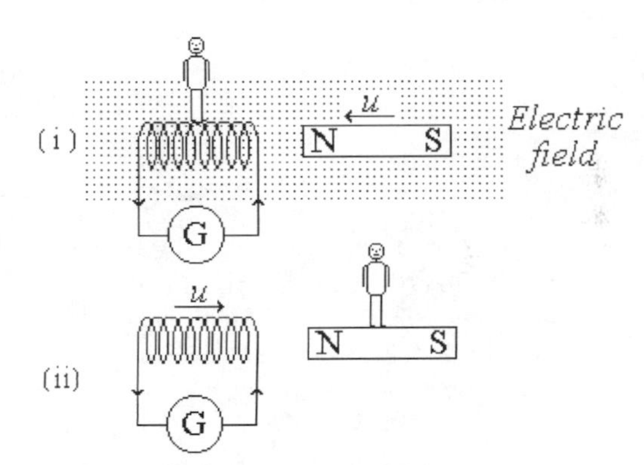

Fig. 1.19. Electromagnetic induction viewed from different inertial systems. The coil and magnet are closing at a relative velocity u. Under the classical approach observer (i), at rest relative to the coil, should detect an electric field arising from the motion of the magnet, whereas observer (ii), at rest relative to the magnet, detects no electric field.

Einstein recognised that the problem lay with the application of the Newtonian conception of motion to the interpretation of electromagnetic phenomena. Under the Newtonian rationale, magnetism cannot be shown to be just the effect of relative motion on electrical interactions, nor can electromagnetic induction be given a single explanation valid for all inertial observers. Despite its unprecedented achievements, Newtonian mechanics is, in fact, flawed. The theory of special relativity arose from the need to find a new conception of motion which would answer these needs. In order to accomplish this, Einstein had to overturn some of the most fundamental intuitive assumptions of classical physics, replacing them with new non-intuitive concepts that challenge our normal experiences. Thus, our study of the theory of relativity must start with a re-examination, in some depth, of the assumptions underlying the Newtonian conception of motion.

1.4.2 The Newtonian Conception of Motion

The science of mechanics is concerned with the motion of bodies or objects and the effect of forces upon them. It defines motion as *change in position*; a moving body is one whose position is changing. Classical (Newtonian) mechanics describes the motion of an object in terms of its trajectory through space. This is the continuous line obtained by plotting the object's instantaneous positions as it moves through space.

Positions in space are most usefully specified by means of a coordinate system in which every point in space is defined by its unique set of coordinates. The coordinate system constitutes the *reference frame* (or *frame of reference*) relative to which measurements of position are made (Fig. 1.20). Reference frames can be at rest, in uniform motion in a straight line or in an accelerated motion. Reference frames at rest or in uniform motion in a straight line are called *inertial reference frames.*

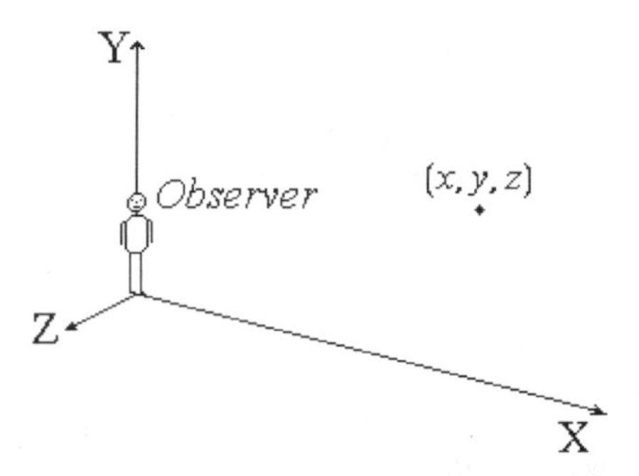

Fig. 1.20. An observer and the orthogonal coordinate system which is his chosen reference frame. Each point in space is defined by a unique set of coordinates (x, y, z).

There is no limit to the number of possible coordinate systems nor are there any restrictions on their choice. An observer may nominate any coordinate system he likes as the one relative to which the positions he observes are to be recorded. However, in most cases, the most convenient coordinate system for an observer to choose is the one at whose origin he himself is situated and whose motion corresponds to his own.

Suppose, that after choosing their respective reference frames, a number of inertial observers study a mechanical phenomenon that occurs in the space around them. Each records his observations and measurements of the phenomenon and each derives an abstract generalisation – *a physical law* – that rationally accounts for its occurrence. Observer equivalence requires that they all discover the same law; that they be indistinguishable by reference to it. This is expressed in classical mechanics by the *Newtonian* or *Galilean principle of relativity*:

The laws of classical mechanics have the same form in every inertial reference frame.

However, notwithstanding their fundamental equivalence, because of their different reference frames, each of the observers may well see things differently; they will not necessarily record identical observations and measurements even when studying the same phenomenon. How can their different sets of observations and measurements be shown to be essentially equivalent? How can the laws different observers formulate be shown to be identical? How, in fact, can the same law arise from different sets of data?

Consider two inertial observers, S and S′, whose chosen reference frames are in uniform motion relative to one another (Fig. 1.21). Observer S′ might be one of the passengers in a bus travelling at a constant velocity, u, along a flat straight road, the XX′-axis; observer S might be a person standing at the road-side.

Though both are inertial observers, the passenger sitting in the bus, S′, sees many things differently from the observer S standing at the road-side. For example, a point whose coordinates relative to S′ are (x', y', z') has a different set of coordinates (x, y, z) relative to the observer S. Moreover, whereas the passenger S′ sees all the other passengers sitting in the bus as being at rest, relative to observer S standing at the road-side, all the bus passengers including S′ are in motion, travelling with the same velocity, u, as the bus. How can these evidently different observations be shown to be equivalent and to result in the same physical law?

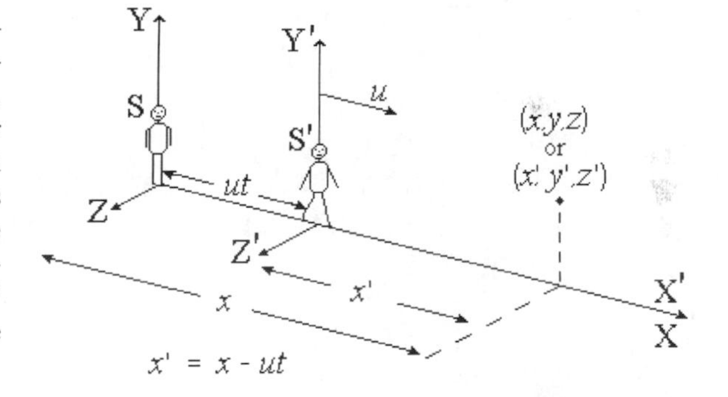

Fig. 1.21. Observer S′ is moving along the X-axis with velocity u relative to the observer S. An event that occurs in their common space is seen by both observers. Observer S records that the event took place at the point whose coordinates relative to him are (x, y, z); S′ records that it took place at the point (x', y', z').

In order to establish (or refute) the underlying equivalence of the sets of observations and measurements made from different reference frames and of the physical laws derived from them, a systematic independent procedure is required. In the language of the mathematicians, such a procedure is called a *transformation.*

If the application of the transformation to the observations and measurements made from one reference frame systematically converts them all into the corresponding observations and measurements made from a different reference frame, then the two sets of observations and measurements are essentially equivalent. If it doesn't, then they are not. The same applies to the laws formulated by the observers in each reference frame. They are covariant laws if the application of the transformation to the formulation made by one observer produces the formulation made by another; the laws are then said to be *form invariant.*

The laws of physics (and of the other 'exact' sciences) are typically formulated in terms of measurable physical quantities. In general, for physical laws to be covariant, the same transformation that reconciles measurements of the physical quantities made by one inertial observer with those made by another, must also convert the laws formulated by the one into those formulated by the other.

The transformation appropriate to the Newtonian conception of motion is called the *Galilean transformation* and, in the case where the relative motion of the reference frames is parallel to the X-axis, it can be derived as follows.

Suppose the two observers, S and S′, possess identical watches, which they synchronise and zero at the moment they pass one another. Some time later, the observer S notices the occurrence of an event and records that it took place at a time t according to his watch and at a point in space whose coordinates, relative to him, are (x, y, z).

The second observer, S′, who is travelling along the XX′-axis, notices the same event and records that it took place at a time t' according to his watch and at a point in space whose coordinates, relative to him, are (x', y', z'). What is the mathematical relationship between the space and time coordinates, x, y, z and t, recorded by observer S and those, x', y', z' and t', recorded by the observer S′? In particular, what is the relationship between t and t'?

The classical answer to this question was given by Newton in 1687 in the framework of the mechanical system described in his great work *Principia.* He wrote:

> '*Absolute, true, and mathematical time, of itself, and from its own nature, flows equably without relation to anything external... All motions may be accelerated and retarded, but the flowing of absolute time is not liable to any change ... the existence of things remains the same, whether the motions are swift or slow, or none at all.*'[53]

In Newton's world, time is innate and absolute: an inherent property, like the mass of an object, affected neither by motions nor events. In this world, 'time is the medium of narration' and *past, future, present* and *simultaneous* have clear and simple meanings that correspond to our intuitive understanding of them. The past comprises those events of which we may have knowledge but over which we have no influence. The future will be filled by events over which we may have some influence but of which we have no present knowledge. The

[53] *Principa.* Translated by Andrew Motte, revised by F. Cajori (1962), Berkeley: University of California Press, p. 6.

present is just a passing moment common to all observers, irrespective of their positions or motions; an infinitesimal that separates past events from future ones. And simultaneity is an inherent property of events themselves; events cannot be simultaneous relative to one observer but not so relative to another.

Thus, in the Newtonian conception, the passage of time is an invariant quantity. Hence, under the Newtonian conception, using their synchronised watches, the two observers, S and S′, must record that the event they both saw occurred at the same time, i.e.

$$t' = t. \tag{1.41}$$

Although the two observers, S and S′, assign identical time coordinates to the event, the same is not necessarily true of the space coordinates; these are *relative* measures whose values do depend on the reference frame from which they are taken. Because of their relative motion along the XX′-axis, x coordinates recorded by the observer S will be greater than the corresponding x' coordinates recorded by the observer S′ by an amount ut

$$x' = x - ut. \tag{1.42}$$

There being no relative motion parallel to the YY′ or ZZ′ directions, both observers record the same coordinate values in these directions

$$y' = y \tag{1.43}$$

$$z' = z. \tag{1.44}$$

These four equations, (1.41), (1.42), (1.43) and (1.44), constitute the *Galilean transformation.* They provide the means for the systematic conversion of the space and time coordinates determined by either of the inertial observers, S or S′, into the corresponding coordinates specified by the other. In this way, the underlying equivalence of their observations and measurements can be validated.

An important corollary of the Galilean transformation, is that although inertial observers like S and S′ do not assign the same coordinates to a position in space, they will always agree as to the distance between two points in space or as to an object's size and dimensions. Consider the rod AB in Fig. 1.17. Relative to observer S, the length of the rod is $x_B - x_A$, whereas relative to S′ it is $x'_B - x'_A$. However, from Eq. (1.42)

$$\begin{aligned} x'_B - x'_A &= (x_B - ut) - (x_A - ut) \\ &= x_B - x_A. \end{aligned} \tag{1.45}$$

Thus, in the Newtonian conception, the distance between two points (the length of the rod) is an invariant quantity whose value is unaffected by the motion of the inertial reference frame from which it is measured.

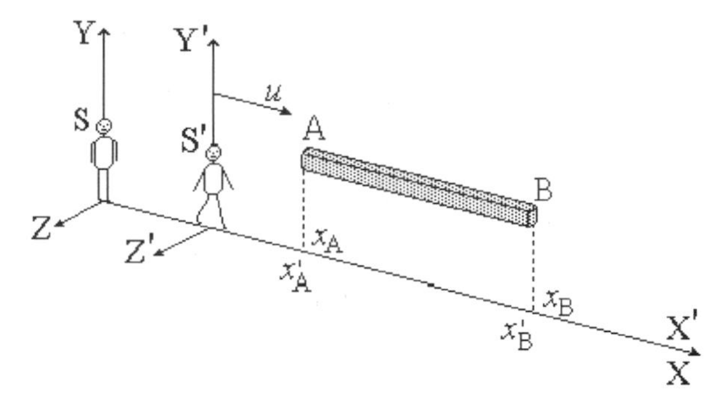

Fig. 1.22. Two points, A and B, marking the ends of a rod in the stationary reference frame S. The length of the rod according to the coordinate system of S is $x_B - x_A$, and according to the coordinate system S′ is $x'_B - x'_A$.

Now suppose that the two observers, S and S′, measure the velocity and acceleration of a moving object. What is the relationship, under the Newtonian conception, between the object's velocity, $\mathbf{v}$, as measured by the observer S, and its velocity, $\mathbf{v}'$, as measured by the observer S′? Likewise, what is the relationship between its acceleration, $\mathbf{a}$, relative to S and that, $\mathbf{a}'$, measured by the observer S′?[54]

The respective scalar components, v_x, v_y, v_z, and $v'_{x'}$, $v'_{y'}$, $v'_{z'}$, of the object's velocity that the two observers, S and S′, record are obtained by taking the derivatives of Eqs (1.42), (1.43) and (1.44) of the Galilean transformation with respect to time

$$\frac{dx'}{dt} = v'_{x'} = v_x - u \tag{1.46a}$$

$$\frac{dy'}{dy} = v'_{y'} = v_y \tag{1.46b}$$

$$\frac{dz'}{dt} = v'_{z'} = v_z. \tag{1.46c}$$

These three equations can be combined into the single vector equation,

$$\mathbf{v}' = \mathbf{v} - \mathbf{u} \tag{1.47}$$

known as the *velocity addition rule of the Galilean transformation*, which, like all vector relationships, has the same meaning in all coordinate systems.[55]

[54]Throughout this book, vector quantities are indicated by boldface.

[55]Relations between vectors remain unchanged no matter where the origin of the coordinate system is situated or how the coordinate axes are oriented. A vector relationship (physical law) that is true in one coordinate system (frame of referece) will be valid in all others.

A corollary of the velocity addition rule is that whenever two observers are moving relative to one another, i.e. $\mathbf{u} \neq 0$, the velocities, $\mathbf{v}$ and $\mathbf{v}'$, that they respectively assign to an object will be different. Thus, like its position, an object's velocity is a relative quantity, i.e. one whose value depends on the reference frame from which it is measured. To express an object's velocity or position without indicating the reference frame from which it was measured is to make a meaningless statement.

The relationship between the accelerations measured by the two inertial observers, S and S′, is obtained by taking the derivative of Eq. (1.47) relative to time. Since their relative velocity, $\mathbf{u}$, is constant, this gives:

$$\frac{d\mathbf{v}}{dt} = \mathbf{a}' = \mathbf{a}. \tag{1.48}$$

Thus, observers in different inertial reference frames will assign the same value to the acceleration of the object. Under the Galilean transformation, acceleration is an invariant quantity. Given that mass too is an invariant, the invariance of force follows from Newton's second law of motion, $\mathbf{F} = m\mathbf{a}$.

Returning to the passengers in the bus and the observer at the road-side, we can now reconcile their apparently conflicting observations:

* Relative to the bus passenger, S′, the velocity of all the other bus passengers equals zero, i.e. $\mathbf{v}' = 0$.
* Relative to the observer at the road-side, S, the passengers' velocity is $\mathbf{v} = \mathbf{u}$, where $\mathbf{u}$ is the velocity of the bus.
* However, substituting these two velocities, $\mathbf{v} = \mathbf{u}$ and $\mathbf{u} = \mathbf{u}$, in the transformation Eq. (1.47) gives for the velocity, $\mathbf{v}'$:

$$\begin{aligned} \mathbf{v}' &= \mathbf{v} - \mathbf{u} \\ &= \mathbf{u} - \mathbf{u} = 0, \end{aligned} \tag{1.49}$$

which is just what the passenger S′ determines. Thus, the conflicting observations are reconciled; they are shown to be essentially equivalent under the Galilean transformation.

To summarise the Newtonian conception of motion:

1. Measurements of position or velocity made from different inertial reference frames such as S and S′ will generally differ; position and velocity are relative measures. However, measurements of position and velocity made from one inertial frame can be reconciled with those made from another by means of the Galilean transformation.

2. Measurements of distance (length), time, mass and acceleration made from different inertial reference frames will be identical; distance, time, mass and acceleration are invariant quantities.
3. All inertial observers formulate the same laws of mechanics, which are covariant under the Galilean transformation. In order to convert a law discovered by the observer S′, into the form it must have for the observer S, the appropriate transformation equations for the relative quantities, x', y', z' and v', in terms of x, y, z and v are substituted in the formulation of the law made by S′.
4. In general, the Newtonian conception of motion corresponds to our intuitive conception of motion. It harbours no surprises.

Example 1.4: The Galilean Transformation

Figure 1.23 shows a plastic collision between two objects, A and B. The objects are moving parallel to the XX′-axis. At the collision, the objects stick together. The two Newtonian inertial observers, S and S′, investigate the conservation of linear momentum in the collision.

In classical mechanics, mass is an invariant quantity and so both observers assign the same values to the masses of the objects, m_A and m_B, respectively.

However, velocity is a relative quantity. The magnitude of the initial velocity of object A is v_A relative to observer S and v'_A relative to the observer S′. The corresponding velocities of object B are v_B and v'_B. Upon their collision the objects join together to form a new object of mass $(m_A + m_B)$ which moves off with velocity v relative to observer S and v' relative to observer S′.

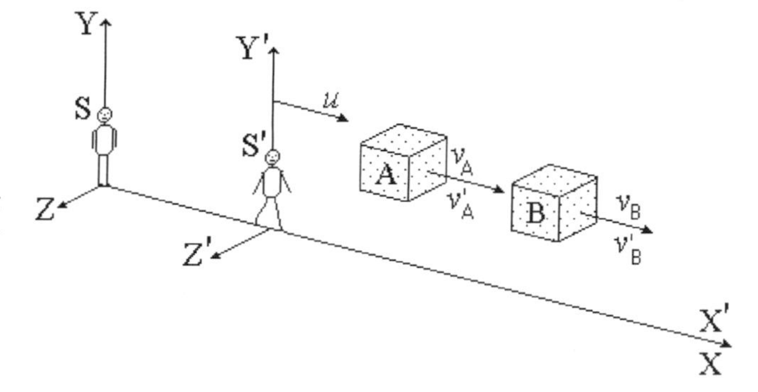

Fig. 1.23. A plastic collision. Observer S′ is moving along the XX′-axis at a velocity u relative to S. The objects A and B collide and join, producing an object of mass equal to the sum of theirs.

Relative to observer S, the linear momentum of object A before the collision is $p_A = m_A v_A$ and relative to observer S′ it is $p'_A = m_A v'_A$; the corresponding values for object B are $p_B = m_B v_B$ and $p'_B = m_B v'_B$. The linear momentum of the compound object produced in the collision is $p = (m_A + m_B)v$ relative to observer S and $p' = (m_A + m_B)v'$ relative to observer S′.

Each observer finds that linear momentum is indeed conserved in the collision and each formulates a mathematical law accordingly. The law formulated by S is:

$$m_A v_A + m_B v_B = (m_A + m_B)v.$$

The law formulated by S′ has the form:

$$m_A v'_A + m_B v'_B = (m_A + m_B)v'.$$

These two laws can be shown to be equivalent to one another under the classical Galilean transformation.

The classical velocity addition rule, Eq. (1.45), gives the following relationships between the objects' velocities:

$$v'_A = v_A - u$$
$$v'_B = v_B - u$$
$$v' = v - u.$$

Substituting these relationships in the law formulated by S′ gives:

$$m_A(v_A - u) + m_B(v_B - u) = (m_A + m_B)(v - u)$$
$$m_A v_A - m_A u + m_B v_B - m_B u = (m_A + m_B)v - m_A u - m_B u$$
$$m_A v_A + m_B v_B = (m_A + m_B)v$$

which is the law formulated by the observer S. Thus, the classical law of the conservation of linear momentum is covariant under the Galilean transformation.

Example 1.5: The Speed of Sound Relative to a Moving Observer

Calculate the magnitude of the velocity of sound (the *speed of sound*) relative to an observer moving through still air at a velocity of 30 m/s

1. away from a stationary sound source;
2. towards a stationary sound source;
3. such that the sound from the stationary source seems to him to be propagating along a path at right angles to his.

Take the speed of sound relative to still air to be $v = 340\,\text{m/s}$.

Calculation: Adopting the notation used above, we designate the still air as the S system and the moving observer as S′. The sound source is situated at the origin of the S coordinate system.
The observer S′ moves parallel to the XX′-axis at a velocity relative to the still air whose magnitude is $u = 30\,\text{m/s}$.

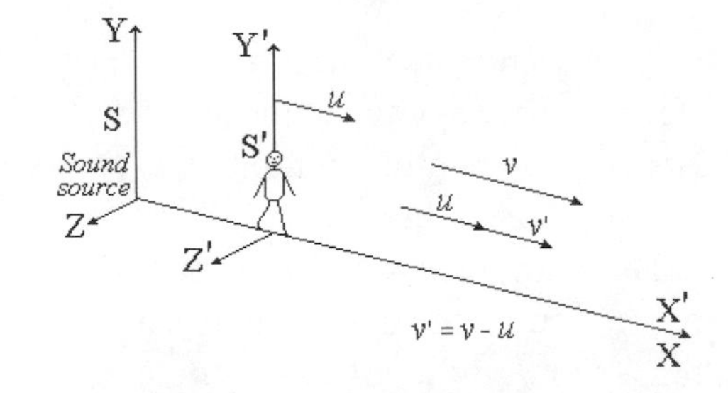

1. According to the classical rule for velocity addition, $\mathbf{v}' = \mathbf{v} - \mathbf{u}$, the magnitude of the velocity of sound (the speed of sound) relative to the observer S′ as he moves parallel to the XX′-axis in the positive direction is

$$\begin{aligned} v' &= v - (+u) \\ &= 340 - 30 \\ &= 310\,\text{m/s}. \end{aligned}$$

2. The speed of sound relative to the observer S′ as he moves parallel to the XX′-axis in the negative direction is

$$\begin{aligned} v' &= v - (-u) \\ &= 340 + 30 \\ &= 370\,\text{m/s}. \end{aligned}$$

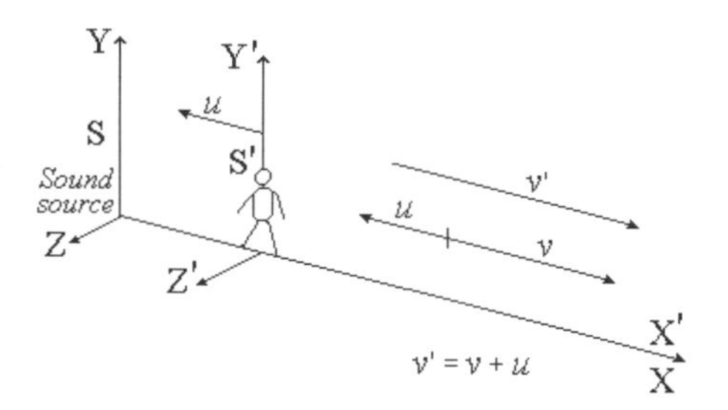

3. When the observer S′ moves such that the sound from the stationary source seems to him to be propagating along a path at right angles to his, the direction of the velocity $\mathbf{v}'$ is perpendicular to that of $\mathbf{u}$. In this case Pythagoras' theorem gives:

$$\begin{aligned} v' &= \sqrt{v^2 - u^2} \\ &= \sqrt{340^2 - 30^2} \\ &= 338.67\,\text{m/s}. \end{aligned}$$

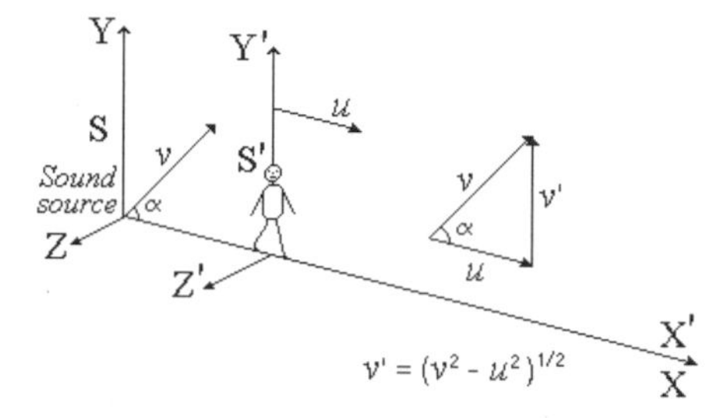

1.4.3 The Michelson–Morley Experiment

Although Maxwell's electromagnetic theory does not specify any particular or preferred reference frame for the velocity of light *in vacuo*, the existence of the æther was so deeply rooted in 19th century physics that it was generally presumed that it must be the reference frame for the velocity of light *in vacuo*. Just as the air is quite naturally taken to be the reference frame for the velocity of sound, it was simply taken for granted that the æther was the reference frame for the velocity of light (Chapters 2.1, 2.2 and 2.3 below).

It was likewise assumed that the propagation of light conformed to the Newtonian conception of motion. Thus, the velocity of light was presumed to be a relative quantity subject to the same velocity addition rule, *viz.* $\mathbf{v}' = \mathbf{v} - \mathbf{u}$, as the velocity of sound. On this assumption, the velocity of light relative to a reference frame other than the æther should be given by $\mathbf{c}' = \mathbf{c} - \mathbf{u}$, where $\mathbf{c}$ is its velocity relative to the æther and $\mathbf{u}$ is the velocity of the reference frame relative to the æther.

For example, an observer moving towards a stationary light source, or one towards whom the light source is moving, should measure a light speed greater than c. Similarly, an observer moving away from the source, or one from whom the source is receding, should measure a speed less than c. If their relative speed is u, the former observer should find the speed of light to be $c + u$ and the latter should find it to be $c - u$. Furthermore, should the light appear to the observer to be propagating along a path at right angles to his own, its speed relative to him should be $\sqrt{c^2 - u^2}$.[56]

The validity of these assumptions was investigated experimentally by Michelson[57] and Morley[58] in 1881. Their experiment set out to detect the effect of the motion of an observer through the æther on measurements he made of the speed of light *in vacuo*. The motion was provided by the earth itself, which was known from astronomical calculations to be orbiting the sun at a speed of approximately 30,000 m/s (Fig. 1.24). It was presumed that the earth passes through the all-pervading æther in its orbital motion around the sun. Accordingly, just as the motion of an observer through still air affects his measurements of the speed of sound, the earth's motion through the still æther was expected to affect terrestrial measurements of the speed of light.

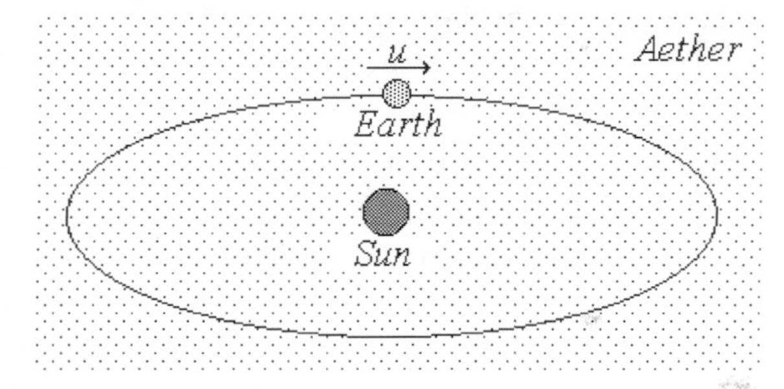

Fig. 1.24. The orbital motion of the earth around the sun through the supposed all pervading æther.

The orbital speed of the earth is four orders of magnitude smaller than the speed of light in vacuo, $c = 300{,}000{,}000\,\text{m/s}$, and so the maximum difference that Michelson and Morley could expect to find in their terrestrial measurements of the speed of light was about $0.0002c$. The detection of such a small difference in the speed of light required great accuracy and precision in the design and execution of the experiment. For this purpose, Michelson perfected a new and very accurate instrument – the *optical interferometer* (Fig. 1.25).

As its name implies, an optical interferometer is a measuring device that works on the basis of the interference patterns produced by visible light. In a typical instrument, monochromatic light is first split into two coherent beams. After travelling along different paths, the two beams are recombined

[56] See the calculations in Examples 1.5a, 1.5b and 1.5c.
[57] Albert Michelson 1852–1931, American physicist who was awarded the Nobel prize in 1907 for his work in optics.
[58] Edward Morley 1838–1923, American chemist.

producing an interference pattern. Where precisely the bright and dark fringes appear in the interference pattern depends on the phase difference between the recombined beams. This, in turn, depends on two factors: (1) the difference between their respective path lengths and (2) the difference between their travel times. Any change in the phase difference between the beams will show up as a shift in the position of the interference fringes.[59]

In the Michelson and Morley experiment, the whole apparatus was mounted on a heavy rotatable horizontal base so that observations could be made at different orientations relative to the direction of the earth's orbital motion. The paths of the two light beams, ABA and ACA, were set at right angles to one another and their lengths, l_1 and l_2, respectively, were fixed. They remained so throughout the experiment. Consequently, any shift in the position of the interference fringes as the orientation of the apparatus was changed could only be the result of variations in the difference, $\Delta t = t_{ABA} - t_{ACA}$, between the beams' travel times.

The travel time of each path depended in turn on the speed relative to the terrestrial observer, c', at which the light propagated along it. Thus, any shift in the position of the fringes would have revealed that the speed of light varied with the direction of its propagation relative to the earth's motion and, by inference, would have confirmed the æther's existence.

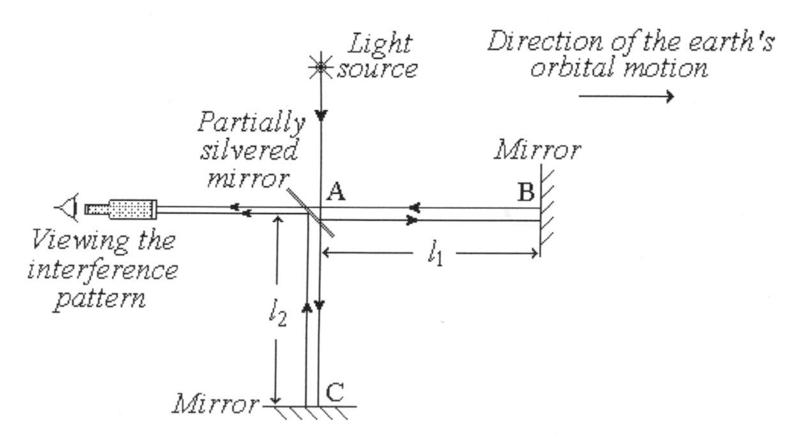

Fig. 1.25. The Michelson–Morley experiment. A light beam from a monochromatic source is split into two coherent beams by the partially silvered mirror. After reflection from the mirrors, the beams are recombined and the interference pattern produced is observed by means of a suitable telescope. The apparatus was mounted on a heavy rotatable horizontal base so that it could be set at different orientations relative to the direction of the earth's orbital motion.

[59]For a fuller explanation of the dependence of the interference pattern on the phase difference see Supplementary Topic A.

Suppose the apparatus is initially set up as shown in Fig. 1.24, i.e. with the ABA beam parallel to the direction of the earth's motion and the ACA beam at right angles to it. According to the Newtonian conception, as the light travels through the still æther from A to B, its speed relative to the terrestrial viewer is $c - u$; as it returns to A, after reflection from the mirror B, its speed relative to this same observer is $c + u$. As regards the other beam, ACA, the terrestrial viewer sees it as propagating along a path perpendicular to his own and so its speed relative to this observer is $\sqrt{c^2 - u^2}$. Hence, the difference, Δt, between the beams' travel times should be

$$\begin{aligned}\Delta t &= (t_{AB} + t_{BA}) - (t_{AC} + t_{CA}) \\ &= \left(\frac{l_1}{c-u} + \frac{l_1}{c+u}\right) - \left(\frac{l_2}{\sqrt{c^2-u^2}} + \frac{l_2}{\sqrt{c^2-u^2}}\right). \end{aligned} \tag{1.50}$$

Suppose the apparatus is now rotated through 90° so that the ACA beam turns parallel to the direction of the earth's motion and the ABA beam is now at right angles to it. The new value for the difference between the beams' travel times should be

$$\begin{aligned}\Delta t' &= (t_{AB} + t_{BA}) - (t_{AC} + t_{CA}) \\ &= \left(\frac{l_1}{\sqrt{c^2-u^2}} + \frac{l_1}{\sqrt{c^2-u^2}}\right) - \left(\frac{l_2}{c-u} + \frac{l_2}{c+u}\right). \end{aligned} \tag{1.51}$$

Clearly, $\Delta t \neq \Delta t'$ and so the rotation should result in a shift of the fringes in the interference pattern.

Michelson and Morley made repeated observations of the interference pattern with their apparatus set each time in a different orientation relative to the earth's orbital motion. However, in not a single case was a significant shift observed. The implications of this null result were awesome. Evidently, terrestrial measurements of the speed of light do *not* depend on the direction of the light's propagation relative to the earth's motion; the propagation of light is *not* analogous to the propagation of sound; there is no luminiferous æther; the earth does not move through the æther as it orbits the sun; and light waves are not æther waves.

In an attempt to 'save the æther' and to account for the null result of the Michelson and Morley experiment within the framework of the Newtonian conception of motion and the Galilean transformation, it was suggested that the æther is somehow dragged along with the earth in its motion through space, such that their relative velocity is zero. Experiments were carried out in an attempt to detect this *æther drag* but they were all inconclusive. Thus, the unavoidable conclusion was that the æther did not exist. This, however, posed a host of new questions:

1. If light waves are not æther waves, then just what sort of waves are they? How are we to understand the physical nature of light and of electromagnetic waves? Whereas sound waves require a material medium to carry their energy, electromagnetic waves appear to propagate without the need for any detectable medium. How can a wave exist and carry energy through absolutely empty space? What reality can we ascribe to a wave which propagates without the need for any physical medium at all?
2. Velocities are relative quantities; they are meaningless unless related to a reference frame. If the æther does not exist, what is the reference frame for the velocity of light *in vacuo*?
3. If, as the null result of the Michelson and Morley experiment indicates, the Galilean transformation does not correctly reconcile measurements of the velocity of light *in vacuo* from different inertial reference frames, how exactly is it flawed?
4. Is the flaw in the Galilean transformation specific to the velocity of light or is it a fundamental flaw that affects its application to all phenomena, both mechanical and electromagnetic?
5. If it is a fundamental flaw, why does the transformation still work so well in our everyday experience?
6. If the Galilean transformation is indeed fundamentally flawed, what is the correct transformation to use when reconciling observations made in different inertial reference systems?

All of these questions, except for that concerning the physical nature of light, were first answered by Einstein's *special theory of relativity*. The question of the actual physical nature of light would have to await the development of the quantum theory of light (Chapters 2.1, 2.2 and 2.3 below).

1.4.4 The Postulates of the Special Theory of Relativity

The special theory of relativity is derived from two postulates:

1. *the principle of special relativity;*
2. *the principle of the constancy of the speed of light.*

1. The principle of special relativity

The laws of physics have the same form in all inertial reference frames.

The Newtonian principle of relativity asserts that the laws of mechanics shall have the same form in all inertial reference frames. Einstein widened the scope of this principle to include all the laws of physics. He asserted that all the laws of physics, without exception, must be covariant. This extension of the Newtonian principle of relativity recognises the fact that no experiment, either mechanical or electromagnetic, has ever succeeded in determining whether an inertial reference frame is in absolute motion or at absolute rest. Experimentally, all inertial frames are indistinguishable. The laws of physics must acknowledge this experience.

2. The principle of the constancy of the speed of light

> **The speed of light *in vacuo*, *c*, has the same value in all directions in all inertial systems.**

Einstein's solution to the problem of the relativity of the speed of light was simple but shattering, for it challenged all our intuitive conceptions of space and time. Taking the results of the Michelson–Morley experiment at their face value, he simply declared that the speed of light *in vacuo*, c, has the same value relative to every inertial system, i.e. it is an invariant quantity.

In effect, Einstein raised the speed of light *in vacuo* to the status of a law of nature, true for all inertial observers. Consequently, instead of relating the speed of light *in vacuo* to the reference frame of the æther, it can be related to any inertial frame we choose; its value will always be c relative to the frame chosen. Thus, there is no need for the æther. Every inertial frame is its own reference frame for the velocity of the light *in vacuo*. Whether the inertial frame is moving towards the light source or away from it, measurements of the speed of light *in vacuo* made from the frame will all give the same value c.

Since the original Michelson and Morley experiment, the constancy of the speed of light *in vacuo* in inertial systems has been directly confirmed many times. In one particularly precise experiment carried out in 1964 at the CERN particle-physics laboratory, the speed of the high-frequency electromagnetic waves (γ-rays) emitted from the decay of unstable short-lived particles called *neutral pions* was measured. A high-velocity, $u = 0.99975c$, beam of neutral pions was produced in a particle accelerator. The researchers measured the speed relative to the laboratory of the electromagnetic waves emitted from these high-velocity particles as they decayed. The value they obtained was $c' = 2.9977 \cdot 10^8\,\text{m/s}$ as compared to the accepted value for the speed of light *in vacuo* of $c = 2.9979 \cdot 10^8\,\text{m/s}$. Plainly, the high velocity of their source, close itself to the speed of light, had no effect on the speed at which the electromagnetic waves they emitted propagated relative to the laboratory; had the source been stationary the same value would have been measured. The experiment was clear confirmation that $c' = c$ even if $u \neq 0$.

1.4.5 Simultaneity and the Relativity of Time

In developing his mechanical system of the world, Newton had postulated two absolutes: *absolute space* and *absolute time.* These were the pillars upon which the Newtonian conception of motion and its attendant Galilean transformation were established. However, as we have seen, this edifice is flawed; it does not pass the test of covariance. To find out why, we must re-examine the validity of Newton's original postulates. In particular, we must ask whether Newton's assertion of the absolute nature of time is compatible with the proven invariance of the speed of light? Our enquiry will be conducted through a series of *thought experiments.*

Thought Experiment 1

Imagine two spaceships, S and S′, somewhere deep in space. S is at rest and S′ is travelling in a straight line with a constant velocity u relative to S; the two spaceships are inertial reference systems (Fig. 1.26a).

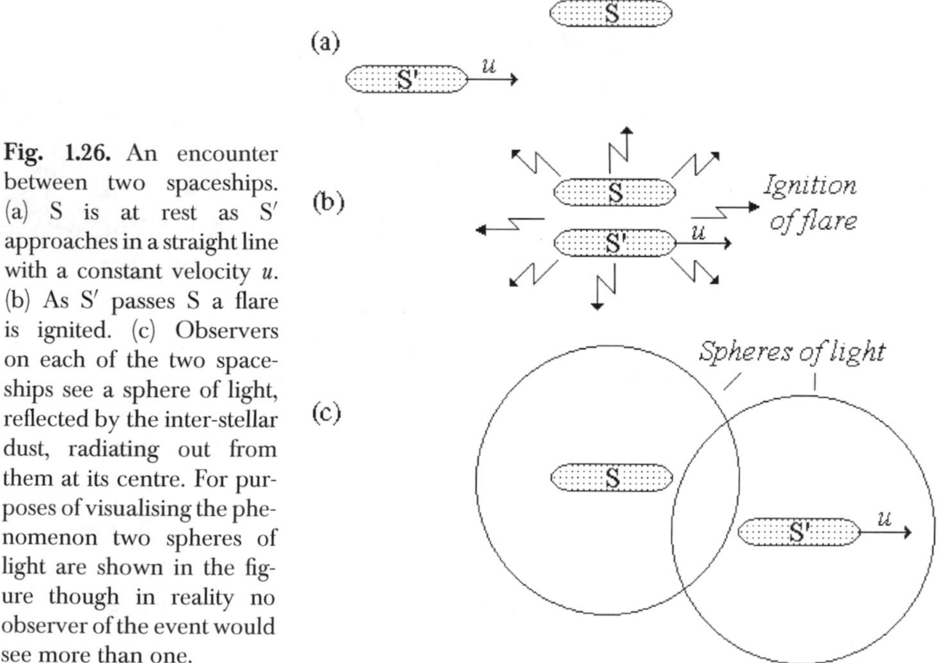

Fig. 1.26. An encounter between two spaceships. (a) S is at rest as S′ approaches in a straight line with a constant velocity u. (b) As S′ passes S a flare is ignited. (c) Observers on each of the two spaceships see a sphere of light, reflected by the inter-stellar dust, radiating out from them at its centre. For purposes of visualising the phenomenon two spheres of light are shown in the figure though in reality no observer of the event would see more than one.

Suppose, that at the moment spaceship S′ passes spaceship S, a flare is ignited from one of them. Its light spreads out into space uniformly in all directions (Fig. 1.26b). After the firing of the flare, the spaceships separate, S′ moving on away from S at the constant velocity u.[60]

As the light spreads out into space it is scattered by the fine inter-stellar dust. In accordance with the principle of the constancy of the speed of light, observers on each spaceship see a sphere of scattered light expanding away from them in all directions at speed c. The same light is seen by the observers on both spaceships, for only one flare was ignited. Moreover, each observer sees himself standing at the centre of the sphere of light. Despite the fact that the spaceships are moving apart at a relative velocity u, observers on both of them see identical phenomena (Fig. 1.26c). How is this possible? How can the light from the flare appear in exactly the same form to the two observers, S and S′, when they are at different points in space and are moving away from each other?

In order to elucidate this paradox, we will follow in Einstein's footsteps and start, as he did in his celebrated article on special relativity,[61] by reconsidering the very nature of time and space and the methods of measuring them. Einstein wrote:

> *'If an observer at point A has a clock, he can estimate the time of events occurring in the immediate neighbourhood of A, by looking for the position of the hands of the clock that is synchronous with the event. If an observer at point B possesses an identical clock to A's, he can similarly estimate the time of events occurring at B. But without further premises, it is not possible to compare, as far as time is concerned, the events at B with those at A. We have therefore an A time and a B time, but no time common to A and B.'*

Einstein asserted that in order to compare events at the two points we must first establish a time common to both of them, i.e. we must first synchronise the clocks at A and B. On this view, time is not a natural quantity that 'flows equably without relation to anything external' and that is present in nature whether measured or not. On the contrary, it is a 'manufactured' quantity that depends on how and where it is defined. The assertion that time and synchronisation are constructs and not inherent properties of the physical world is the underlying notion of the Einsteinian revolution.

---

[60] In reality, observers on the spaceships cannot discern which is in motion; they can only determine their relative velocity, u. But this does not affect our discussion.

[61] (1905) 'On the Electrodynamics of Moving Bodies', *Annalen der Physik*, **17**, 891, translated by M. Saha and S. Bose. Translations here and on page 62 appear in M.H. Shamos (ed.) (1959) *Great Experiments in Physics*, New York: Dover Publications.

If synchronism is not automatic and inherent, its existence must be experimentally confirmed in each case. Einstein argued that the synchronisation of clocks situated at the two different points, A and B, in the same inertial system could be established as follows:

> ‘*A common time can be defined [for any inertial system] if we establish by definition that the time which light requires to travel from A to B is equivalent to the time required for it to travel from B to A [Fig. 1.27]. For example, a ray of light proceeds from A at A time t_A towards B, arrives and is reflected from B at B time t_B and returns to A at A time t'_A. According to the definition, both clocks are synchronous, if*

$$t_B - t_A = t'_A - t_B\text{’} \qquad (1.52)$$

Implicit in this definition is the assumption that the speed of light from B to A equals that from A to B, as required by the principle of the constancy of the speed of light. Einstein continued:

> ‘*Thus with the help of certain physical experiences, we have established what we mean when we speak of clocks at rest at different points in space and synchronised with one another; and thereby we have arrived at a definition of synchronism and time.*’

Having established that the clocks are synchronous, they can now be used to determine whether events occurring at A and at B are simultaneous. They can also be moved to other places in the reference frame so as to check whether events occurring at them are simultaneous or not.

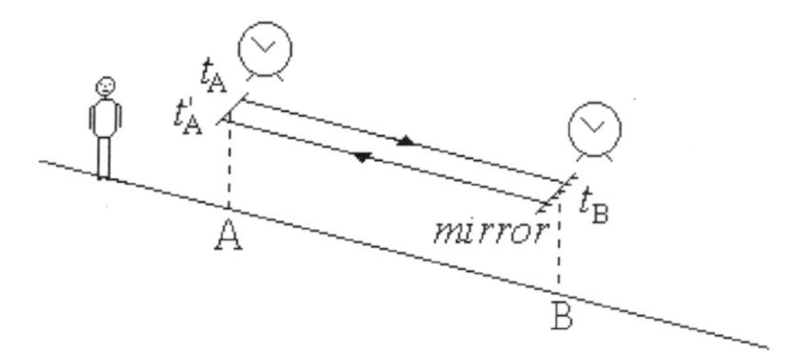

Fig. 1.27. Clocks at rest in the same inertial reference system, S. A ray of light proceeds from A at A time t_A towards B, arrives and is reflected from B at B time t_B and returns to A at A time t'_A. The clocks are synchronous if $t_B - t_A = t'_A - t_B$, i.e. if, according to them, the light takes the same time to travel from A to B as from B to A.

The next step is to examine whether clocks situated in different inertial reference frames can also be synchronised. For example, can a clock in the S inertial system be synchronised with one in the S′ inertial system such that a common time can be defined for the two reference frames?

Were it possible to signal instantaneously, i.e. to transmit a signal at an infinite speed, we could synchronise the clocks without difficulty, irrespective of the relative motion of S and S′. An observer in one of the systems would simply signal the time shown by his clock to an observer in the other system. The signal would be received instantaneously by the second observer who would set the time on his own clock accordingly. But no method of signalling faster than flashing a light has ever been discovered and the speed of light *in vacuo*, though extremely fast, is finite; it also has the same value in all inertial reference frames, irrespective of their motion. Thus, signals cannot be instantaneously transmitted.

Classical Newtonian mechanics ignores this simple fact. It acknowledges no upper limit on the speed at which objects or signals can move. Thus, it implicitly assumes instantaneous signalling is possible. Consequently, it can posit the synchronisation of all clocks and the existence of an absolute time common to all observers. This is the essential flaw in classical mechanics.

In order to illustrate the problem of finding a common time for two different inertial systems and of determining the simultaneity of events occurring in them, we will carry out two further thought experiments.

Thought Experiment 2

Suppose that flashes of light from two different light sources, each situated at the same distance, d, from an inertial observer, are seen by him at the same moment. Since the sources are equidistant from him and the speed of light in an inertial system has the same value c in all directions, the observer must conclude that the bursts of light were emitted by the sources simultaneously and at a time $t = d/c$ before he saw them. Whatever his or their motion, there is no way that the light from one source could have taken longer to reach the inertial observer than that from the other source.[62]

Thought Experiment 3

Figure 1.28 shows two spaceships that serve as inertial reference systems for two observers, Jack (S) and Jill (S′). Jill's ship is moving in a straight line with a constant velocity, u, relative to Jack's. The spaceships, which belong to rival galactic empires, pass close to one another.

Jack and Jill are the respective commanders of the ships and they each report the incident in their respective logs.

[62]This would not be the case with sound. If an inertial observer hears bursts of sound from equidistant sources, say from explosions, at the same moment, he cannot unequivocally conclude that the explosions occurred simultaneously. The time it takes for the sound of the explosions to reach him depends on the speed of the sound, which is not necessarily the same in all directions. It is affected by the motion of the observer towards or away from the sound sources and by winds and currents in the air.

Jack's log: As the enemy ship drew alongside, two of my cannons fired at point-blank range. I know from the dimensions of my vessel that the cannons are equidistant from the command bridge, one forward and one aft. Since I saw the flashes from the cannons' muzzles at the same moment, I conclude that the cannons fired simultaneously.

Jill's log: As we drew alongside the enemy ship, its cannons fired at point-blank range. My ship took two hits, one ahead of the command bridge and one astern. Measurements revealed that the hits were equidistant from the bridge.

Jill sees the flash from the cannon that struck the forward section of her ship before she sees the flash from the cannon-shot that hit the aft section. Since the light sources are equidistant from her and the speed of the light must be the same in both directions, she must conclude that the hits were not simultaneous. And so she writes:

Jill's log: I saw the flash from the cannon that hit my ship forward of the bridge before I saw the flash from the one that struck behind the bridge. Since the hits were equidistant from the bridge, I conclude that the enemy's forward cannon fired before his aft cannon.

Both commanders recorded accurate logs of the incident, each from their own view-point. But their reports contradict one another; they disagree as to the simultaneity of the firing of the cannons. According to Jack's inertial reference frame, S, they were fired at the same moment, whereas according to Jill's inertial frame, S′, they were not. Evidently, Jack and Jill do not share a common time; clocks that are synchronised in Jack's reference frame do not appear to be synchronised when viewed by Jill from her frame. In general, inertial observers, whose reference frames are in relative motion, will not agree about the simultaneity of events occurring at different points in space. The reference frame of each observer appears to the other to be affected by their relative velocity.

If the simultaneity of events occurring in different reference frames cannot be presumed, scientific theories that require or imply instantaneous actions and reactions must be discarded.[63] Contrary to Newton's third law:

---

[63]Certain interpretations of quantum theory might temper this assertion. Pairs of photons are emitted by atoms in opposite directions. Experiment has established that determining the polarisation state of one of the photons instantaneously fixes that of the other. In *realistic* terms, it is as though the pairs of photons interact instantaneously with one another at a distance; the phenomenon is called *locality*. However, since the actual outcome with each pair is, like tossing a coin, unpredictable, the effect cannot be used to transmit a meaningful signal instantaneously.

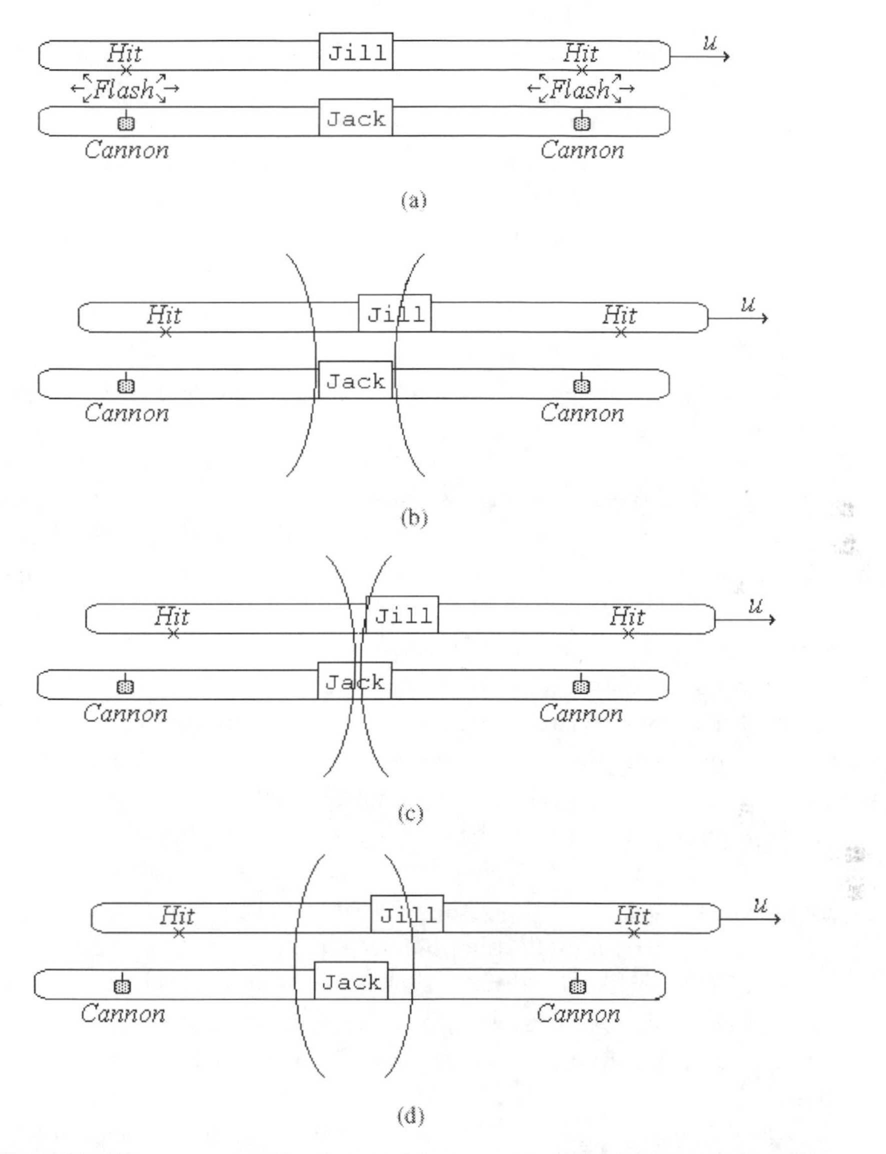

Fig. 1.28. The two spaceships from Jack's view-point. Jill is moving to the right with constant velocity, u, relative to Jack. (a) The cannons of Jack's ship fire as the two ships draw alongside. Wave-fronts from the muzzle flashes radiate outwards in all directions. (b) Jill sees the wave-front from the cannon that hit her ship forward of the bridge. (c) Jack sees the two flashes from his cannons simultaneously. (d) Jill sees the wave-front from the cannon that hit her ship behind the bridge.

The forces acting between two charges moving relative to one another are neither equal in magnitude nor opposite in direction.

In essence, this was Faraday's contention when he insisted that action at a distance must be mediated by a field. The field compensates for any imbalance in action or reaction that arises out of the fact that changes in one component of a system are not transmitted instantaneously to all the other components. For example, when considering the conservation of linear momentum in a dynamic system of interacting particles, such as a system of charged particles moving relative to one another, linear momentum must also be ascribed to the field in order that the system's total linear momentum be conserved at all times. The field acts as a flexible reserve of energy and momentum, available at all times to ensure their overall conservation whatever changes are occurring in the system. Thus, the relativity of time and simultaneity and the concept of the field can be regarded as corresponding concepts.

To return to the paradox of the observers on the two spaceships S and S′. The astonishment felt at the suggestion that observers at different points in space and who are moving away from each other both see the light from the flare in an identical form, arises from the imaginary reference frame high above the spaceships into which our mind's eye projects itself and from where it views the event. In this imaginary frame, time is absolute and does indeed 'flow equably without relation to anything external'; all observers, whatever they are doing and wherever they are, share the same common time. Viewed from such a frame, the only way the light from the flare can appear in an identical form to the observers on the two spaceships, S and S′, is if they are simultaneously at the same point in space. But in the physical world we inhabit, and with which science is concerned, such reference frames do not exist. In Einstein's *real world*, simultaneity is relative not absolute and observers cannot be distinguished by means of electromagnetic phenomena. In the physical reality described by science, it is not the constructs such as time that 'flow equably without relation to anything external' but the natural phenomena. It is the natural phenomena and not our mental constructs that are the invariants of the physical sciences.

The Newtonian conception of motion corresponds to the intuitive view of physical reality taken by the mind's eye. However, the symmetries of modern physics are different from those of the mind's eye. Whereas, in the world of physics observers are indistinguishable, in the world of the mind individuality does exist. Each mind has its own symmetries that remain unchanged whatever operations it performs or are performed upon it; the individual invariants by which it can be distinguished from the multitude of other minds.

1.4.6 The Lorentz Transformation

Some years before the publication of the special theory of relativity, Lorentz[64] had realised intuitively that in order for the laws of electromagnetism (Maxwell's equations) to be covariant and for all inertial observers, stationary or moving, to be electromagnetically equivalent and indistinguishable, the time measure, t', used in the moving system S′ must be different from that, t, employed in the stationary system S.

After many laborious calculations, Lorentz succeeded in developing a new transformation which incorporated this idea – the *Lorentz transformation* – the equations of which are[65]

$$x' = \gamma(x - ut) \qquad (1.53) \qquad\qquad x = \gamma(x' + ut') \qquad (1.54)$$

$$y' = y \qquad (1.55) \qquad\qquad y = y' \qquad (1.56)$$

$$z' = z \qquad (1.57) \qquad\qquad z = z' \qquad (1.58)$$

$$t' = \gamma\left(t - \frac{ux}{c^2}\right) \qquad (1.59) \qquad\qquad t = \gamma\left(t' + \frac{ux'}{c^2}\right) \qquad (1.60)$$

where the factor γ is given by the equation

$$\gamma = \sqrt{\frac{1}{1 - u^2/c^2}} \quad (u < c). \qquad (1.61)$$

In the second part of his 1905 paper, Einstein showed that the Lorentz transformation could be derived directly and quite simply from the two postulates of the special theory of relativity. This demonstrated the underlying compatibility of Maxwell's electromagnetism and special relativity; they both assume the relativity of time. It is Newtonian (classical) mechanics that is the odd man out.

Nevertheless, classical mechanics can still be used to describe everyday motions – in the human sense. The reason for this is that when the relative velocity, u, between the inertial observers is very much smaller than the speed of light *in vacuo*, c, as it is in everyday human experiences, the equations of the Lorentz transformation reduce very closely to those of the Galilean transformation. At these low velocities, the factor γ in the Lorentz transformation is practically equal to one (Fig. 1.29). In practice, the errors in the predictions

[64] Hendrik Lorentz 1853–1928, Dutch physicist who was awarded the Nobel prize in 1902 for his research on the effect of magnetism on radiation.

[65] The corresponding relativistic velocity addition rules can be found in Appendix 1.6.1.

of classical mechanics at these velocities are far too small ever to be detected. Only at relative velocities above $0.5c$, half the speed of light (150,000 km/s) do the flaws inherent of classical mechanics begin to be detectable.

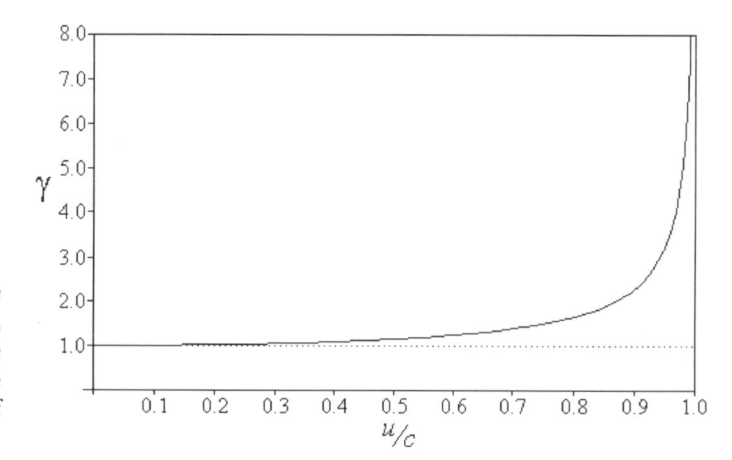

Fig. 1.29. Graph of the factor γ against the ratio, u/c, between the relative velocity of the reference frames u and the speed of light *in vacuo*, c.

Notwithstanding the utility of classical mechanics at low velocities, in order to retain the integrity and unity of physics, a new mechanics that is valid at all velocities and is compatible with special relativity is required. The laws of this new mechanics – *relativistic mechanics* – are covariant under the Lorentz transformation and are so formulated that the sciences of mechanics and electromagnetism concur.

Example 1.6: The Lorentz Transformation and Special Relativity

As the following discussion shows, the Lorentz transformation for two inertial reference frames, S and S′, moving along the XX′-axis with a relative velocity of u, can be shown, under certain reasonable premises, to be implicit in the postulates of special relativity.

It is a reasonable assumption that in any transformation the relationship between the coordinates x and x' will be of the form:

$$x' = k(x - ut), \tag{1.62}$$

where k is a constant of proportionality that is not a function of the space or time coordinates but which may be a function of the relative velocity, u. The choice of this form arises from a number of considerations:

1. The relationship between the coordinates x and x' is linear, such that a single event in frame S corresponds to a single event in frame S′.
2. The relationship is simple, and simple relations are preferable.
3. The relationship can reduce to the corresponding equation of the Galilean transformation, $x' = x - ut$.

The first postulate of special relativity, the principle of relativity, requires that the laws of physics have the same form in all inertial systems. Therefore, in order to obtain the appropriate relationship which expresses the coordinate x in terms of x' and t', just the sign of the velocity, u, in the above relationship has to be changed (because of the difference in the direction of their relative motion from the view-point of the other reference frame):

$$x = k(x' + ut'). \tag{1.63}$$

There is no relative motion parallel to Y or Z and so $y = y'$ and $z = z'$.

The second postulate of special relativity, the constancy of the speed of light, requires that a ray of light observed from each of the reference frames, S and S′, must propagate with the same velocity, c. Suppose, that at the moment the two observers in Fig. 1.21 meet and synchronise their clocks, a flare is ignited and the two observers measure the velocity of the propagation of the light from it along the XX′-axis. At a given subsequent moment, the wave-front from the flare will have reached point x in the coordinate system of observer S and point x' in that of observer S′. However, because of the constancy of the speed of light each observer must measure the same value for this velocity, i.e.

$$c = \frac{x}{t} = \frac{x'}{t'}. \tag{1.64}$$

Eliminating x, x', t and t' from Eqs (1.62), (1.63) and (1.64) gives

$$k = \sqrt{\frac{1}{1 - {u^2}/{c^2}}}. \tag{1.65}$$

Thus, the factor of proportionality, k, derived from the postulates of special relativity, is identical to the factor γ that appears in the equations of the Lorentz transformation. The proposed transformation equations, (1.62) and (1.63), are thus identical to the corresponding equations of the Lorentz transformation.

1.4.7 Relativistic Mechanics — Kinematics

The essential feature of the relativistic mechanics is its compatibility with the Lorentz transformation. In the realm of kinematics, we will investigate how this affects the measurement of three physical quantities: distances (spatial displacements) and time intervals; and a new quantity called the *space–time interval.*

1. The measurement of distances – *length contraction*

Under the classical Galilean transformation, the measured size of an object is unaffected by the relative motion of inertial observers; distances and displacements are invariants. However, as we shall see, under the Lorentz transformation, both of these measures are relative quantities.

Consider a rod, AB, that is at rest in the S reference frame (Fig. 1.29). The length of the rod, *viz.* the distance between the points A and B parallel to the XX′-axis, is measured by an observer in the reference frame S and by another in the moving frame S′ (Fig. 1.30). What is the relationship between the rod's length, L_0, measured by the observer in frame S and its length, L, measured by the moving observer from the S′ frame?

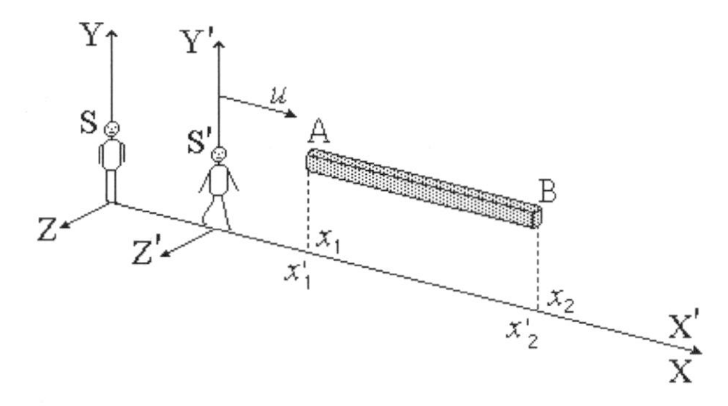

Fig. 1.30. Two points, A and B, marking the ends of a rod in the stationary reference frame S. The length of the rod according to the coordinate system S is $L_0 = x_2 - x_1$, and according to the coordinate system S′ is $L = x_2' - x_1'$.

According to the S coordinate system the length of the rod is:

$$L_0 = x_2 - x_1, \tag{1.66}$$

and according to the S′ coordinate system its length is:

$$L = x_2' - x_1'. \tag{1.67}$$

Substituting the appropriate values for x' and t' given by the Lorentz transformation – Eq. (1.54) – in place of the x-values in Eq. (1.66) gives:

$$\begin{aligned} L_0 &= \gamma(x_2' + ut') - \gamma(x_1' + ut') \\ &= \gamma(x_2' - x_1'). \end{aligned} \tag{1.68}$$

Combining Eqs (1.67) and (1.68) gives:

$$L_0 = \gamma L. \tag{1.69}$$

The factor γ is always equal to or greater than one. Consequently, in general, $L_0 \geq L$.

The rod has the same dimensions in the Y and Z directions for both observers, there being no relative motion in this direction.

The rod's length, L_0, measured from the reference frame at rest relative to it, is called its *proper length.* The length, L, measured from the moving reference frame in the direction of its motion, is always shorter than the proper length, L_0, measured from the stationary frame. Thus, stationary objects appear to be foreshortened when viewed from moving reference frames.

The phenomenon is called the *Lorentz–Fitzgerald contraction*[66] *contraction.* It should be noted that the contraction is a symmetrical phenomenon; if the rod was at rest in the reference frame S′, its proper length would be its length as measured in that frame and the observer S would measure the shorter length.[67]

Example 1.7: The Mystery of the Muons

Sub-atomic particles, called *muons*, are produced in the upper atmosphere by the impact of the high-energy cosmic rays that reach the earth from deep in space. The muons are unstable and decay, on average, just $2 \cdot 10^{-6}$ seconds after their formation. The speed of the muons relative to the earth is $2.994 \cdot 10^8$ m/s, i.e. 99.8% of the speed of light *in vacuo.* Accordingly, during an average life-time, a muon should travel only a short distance, $d = 2 \cdot 10^{-6} \cdot 2.994 \cdot 10^8 \approx 600$m. The muons are formed high in the atmosphere, some 6,000 metres above the earth's surface. Nevertheless, muons are detected in large quantities at sea level. How is this possible?

Calculation: Relative to the earth, the stationary reference frame in this case, the height above sea level of the atmospheric layer where the muons are formed is the proper height: $L_0 = 6000\,\text{m}$.

The muons' reference frame is moving downwards relative to the earth with velocity $u = 0.998c$ and so distances in this direction will appear to them to be foreshortened. For example, their initial height above the earth will appear to them to be $L = L_0/\gamma$ as required by the Lorentz transformation. The value of the factor γ is given by Eq. (1.61):

$$\gamma = \sqrt{\frac{1}{1 - {u^2}/{c^2}}} = \sqrt{\frac{1}{1 - {(0.998c)^2}/{c^2}}} = 15.82. \tag{1.70}$$

---

[66] George Fitzgerald 1851–1901, Irish physicist.

[67] Some years before Einstein proposed his special theory of relativity, Lorentz and Fitzgerald had suggested that the negative result of the Michelson and Morley experiment could be explained, within the framework of the classical conception of motion, by assuming that every object actually contracts in the direction of its motion through the æther. The magnitude of the contraction is such that it exactly cancels out the effect of the motion on the velocity of light. The moving (terrestrial) observer, S′, does not detect the contraction because his measuring rods contract to the same degree as any distances he measures. Thus, if he can fit the same number of rods between points A and B as between points A and C, then as far as he can tell the distance AB must equal AC whatever their orientations relative to his motion. However, relative to the stationary (æther) observer, S, if the line AB is in the direction of the moving observer's motion and the line AC perpendicular to it, then the measuring rods between the points A and B are actually foreshortened compared to those between A and C. Thus, in reality AB < AC.

This hypothesis gives the same mathematical formula for the contraction as special relativity. However, whereas the Lorentz–Fitzgerald hypothesis posits a *real* contraction attributable to the object's motion through the æther, special relativity speaks of a *measured* contraction that depends on the object's velocity relative to the observer. In special relativity, different observers measure different contractions.

Hence, the initial height of the muons relative to their reference frame is

$$L = \frac{6000}{15.82} = 379.2\,\text{m}. \tag{1.71}$$

Thus, the distance to the earth's surface, as measured from the reference frame of the moving muons, is less than the distance they can travel on average before decaying. As a result, muons are detected in large numbers at sea level.

---

2. The measurement of time intervals – *time dilation*

The second aspect of relativistic mechanics that we will examine is the effect of their relative motion on the time intervals that inertial observers measure.

Suppose that observers in the S and S′ reference frames possess identical clocks, which they use to measure the time that elapses between two events that occur in the S′ frame of reference at a point whose coordinates are (x', y', z') (Fig. 1.31).

In general, an observer at rest relative to a sequence of events can measure the time interval between them with a single stationary clock. This time interval, Δt_0, is called the ***proper time interval***. Relative to an observer in the S′ reference frame, the two events occur at the same point and so in order to measure

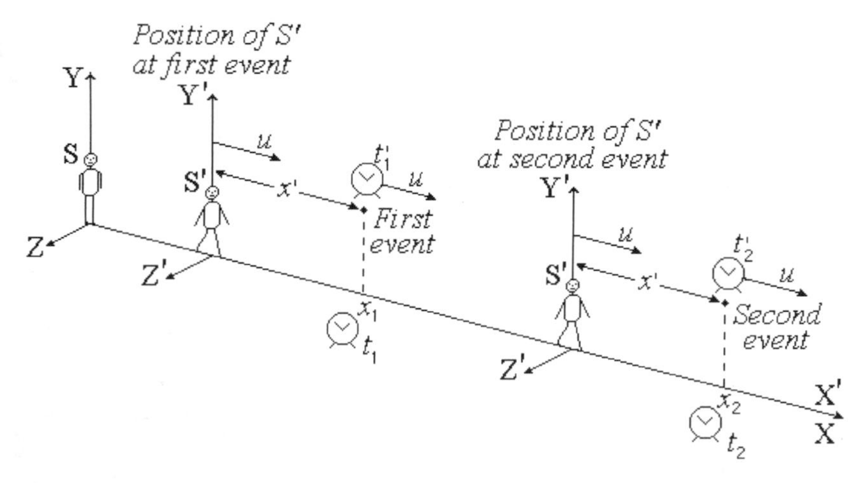

Fig. 1.31. Measuring the time interval between two events that occur in the S′ frame of reference. Both events occur at the same point whose coordinates are (x', y', z') in the S′ coordinate system but at different points, (x_1, y, z) and (x_2, y, z), in the S system. Observer S needs two synchronised clocks, one at x_1 and the other at x_2, in order to measure the time interval, $\Delta t = t_2 - t_1$, between the events. However S′, who is at rest relative to the events, and sees them both as occurring at the same point, can measure the time interval, $\Delta t_0 = t'_2 - t'_1$, between them with a single clock. S′ measures the proper time interval.

the time interval between them he requires just one clock, placed at the point (x', y', z') in his coordinate system. Thus, the time interval between the two events measured by the observer in the S′ reference frame is the proper time interval.

The situation of the observer in the other reference frame is different. Because the S′ reference frame is moving relative to him, the observer in the S reference frame finds that the events occur at two different points in his coordinate system, designated, respectively, by the coordinates (x_1, y, z) and (x_2, y, z). Consequently, the observer in the S system requires two synchronised clocks, one at each point, in order to measure the time interval, Δt, between the events.

According to the observer in the S coordinate system, the time that elapses between the events is:

$$\Delta t = t_2 - t_1, \tag{1.72}$$

and according to the observer in the S′ system it is:

$$\Delta t_0 = t_2' - t_1'. \tag{1.73}$$

Substituting Eq. (1.28) of the Lorentz transformation in Eq. (1.72) gives:

$$\begin{aligned} \Delta t &= \gamma\left(t_2' + \frac{ux'}{c^2}\right) - \gamma\left(t_1' + \frac{ux'}{c^2}\right) \\ &= \gamma(t_2' - t_1'). \end{aligned} \tag{1.74}$$

Hence,

$$\Delta t = \gamma \Delta t_0. \tag{1.75}$$

However, the factor γ is always equal to or greater than one; $\gamma \geq 1$. Consequently, in general, $\Delta t \geq \Delta t_0$.

Measured from a stationary reference frame such as S, the time, Δt, that elapses between events occurring at the same place in a moving reference frame, S′, is always longer than the proper time interval, Δt_0, measured by an observer in the S′ frame. The phenomenon is called *time dilation* and is exhibited by all physical processes including chemical reactions and biological growth.

The phenomenon of time dilation has been confirmed in numerous experiments. For example, in investigations of the average life-times of the unstable particles called *pions*, it is found that at rest, pions decay on average after 26 nanoseconds; this is their *proper average life-time.* However, their measured average life-time when moving at a speed of 0.99 c relative to the laboratory is, according to the laboratory's clocks, 7.1 times longer. Thus, whereas an observer riding one of the pions has to wait, on average, just 26 ns for it to decay, an observer in the laboratory has to wait 185 ns. To put this another way, relative to the stationary clocks in the laboratory, the moving pion's own clock is slow; it shows the passage of just 26 ns when, according to the laboratory clocks, 185 ns have already passed. This result is often expressed by the adage 'moving clocks run slowly'.

In general, the shortest time interval observed between events is the proper time interval measured from the reference frame in which the events are occurring, i.e. by a clock at rest relative to the events. The time intervals measured by clocks in reference frames moving relative to the events are always longer.

An important corollary of time dilation is that the *proper frequency*, f_0, of periodic events, i.e. their frequency as measured from the reference frame in which they occur, is always higher than the frequency measured from a frame moving relative to the events. For example, consider measurements of a patient's heart-beat made by two doctors, one who is at rest relative to the patient and one hurrying past. Whereas the stationary doctor measures the patient's proper heart-beat, the running doctor will measure a marginally lower heart-beat. If the process of ageing depends, among other things, on the rate of one's heart-beat, the patient is ageing faster relative to the stationary doctor than he is relative to the running doctor. As long as he keeps moving, the running doctor is not aware of this and it will not affect his diagnosis. Only when he returns and finds the patient to be older than he expected will he become aware of the difference between his measurement and that of his stationary colleague.

Though time is in fact a relative measure – a construct and not an absolute – not all our intuitive ideas about it are flawed. Although the laws of physics do not distinguish between forward and backward directions of time, time does not flow backwards for any observer.[68] Similarly, events occur in the same order for all observers, though each may measure a different time interval between them. By the same token, a distant observer, whatever his motion, cannot observe an event before an observer who is closer to it. And, finally, we cannot look into the future.

[68] Three arrows of time that distinguish the past from the future have been identified:

* The thermodynamic arrow – the time direction in which disorder increases.
* The psychological arrow – the time direction in which memory increases.
* The cosmological arrow – the time direction in which the universe expands.

Example 1.8: Stationary Clocks and Moving Clocks

At what speed, u, relative to a stationary observer, must a clock move in order that the rate at which it measures time be half that at which it is measured by a clock held by the observer?

Calculation: A time interval measured by the clock held by the stationary observer is a proper time interval, Δt_0; the observer is at rest relative to this clock. If the moving clock measures time at half the rate of a stationary one, then the time intervals it measures are twice as long as the proper time intervals; i.e. $\Delta t = 2\Delta t_0$. From Eq. (1.75), $\Delta t = \gamma\Delta t_0$, we have that in this instance $\gamma = 2$. Substituting this value for γ in Eq. (1.61) gives for the relative velocity $u = 0.866c$.

---

3. The fourth dimension

Eliminating the relative velocity, u, from the equations of the Lorentz transformation by simple algebra, gives the following relationship between the space and time coordinates of the two observers S and S′:

$$(ct)^2 - (x^2 + y^2 + z^2) = (ct')^2 - (x'^2 + y'^2 + z'^2). \tag{1.76}$$

The new physical quantity, I, defined by the equation[69]

$$I = (ct)^2 - (x^2 + y^2 + z^2) \tag{1.77}$$

is an invariant quantity called the space–time *interval*. Its significance in relativistic mechanics is analogous to that of distance or displacement in classical mechanics; it is the quantity that separates events in the Einsteinian conception of motion. On their own, neither space intervals (spatial distances) nor time intervals are invariant. However, the quantity I, which is compounded of both time and space, is invariant.

In determining space–time intervals, the time coordinate t has the same standing as the space coordinates x, y and z. Therefore, we say that space–time has four dimensions. Though in classical physics space and time coordinates can be considered separately, in relativistic physics they are treated together under the single heading of *space–time*.

---

[69]Some texts use the negative of the definition given here.

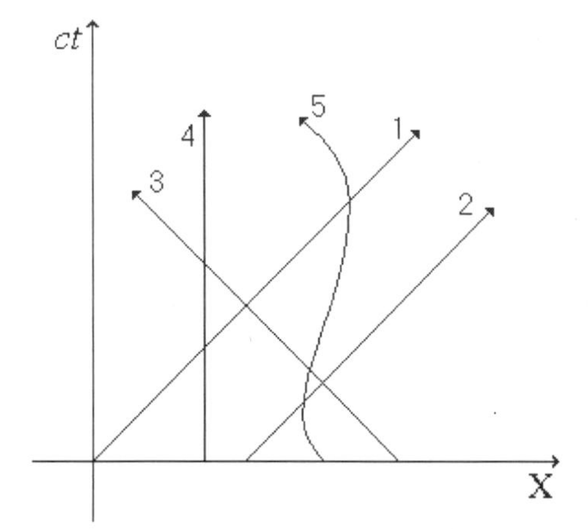

Fig. 1.32. Minkowski diagram for the observer S. The lines numbered 1, 2 and 3 are the world-lines of light signals propagated from different points in different directions. Line 4 represents a body at rest and line 5 one whose motion is changing with time, i.e. accelerated motion.

Events occurring in space–time, as well as motion through it, can be depicted in a graphical description, a *space–time diagram*, first suggested by Minkowski.[70] Figure 1.32 is the Minkowski space–time diagram of the stationary observer S; only motions parallel to the X-axis are shown. The horizontal axis in the diagram is the space axis and corresponds to the X-axis of the observer S. The vertical axis is the time axis; it is scaled in units of ct – units of length – so that the two axes should have the same units. A line or curve drawn on the diagram represents the motion of a particle or signal through space–time; such lines or curves are called *world-lines.*

The velocity, u, of a particle or a signal is inversely proportional to the slope, $\Delta(ct)/\Delta x$, of its world-line:

$$u = c \cdot \frac{\Delta x}{\Delta(ct)}. \tag{1.78}$$

In a Minkowski diagram, a light ray, i.e. a signal moving with velocity c, is represented by a line of slope 45°: the lines numbered 1, 2 and 3 in Fig. 1.32. A body at rest, $u = 0$, is represented by a straight vertical line: line number 4. A body whose motion undergoes change is represented by a curved or broken line: line number 5.

Minkowski diagrams can also be used to elucidate the nature of physical causality, i.e. what can and cannot be the cause or the result of a physical event. Clearly, one event cannot be the cause of another before information of its occurrence arrives at the point where the second event is to occur. The maximum speed at which information can be transmitted, c, is represented in a space–time diagram by the surface of a cone of slope 45°; the cone is called

[70]Hermann Minkowski 1864–1909, German mathematician.

the *light cone.* The light cone of an event is divided into the past light cone and the future light cone (Fig. 1.33).

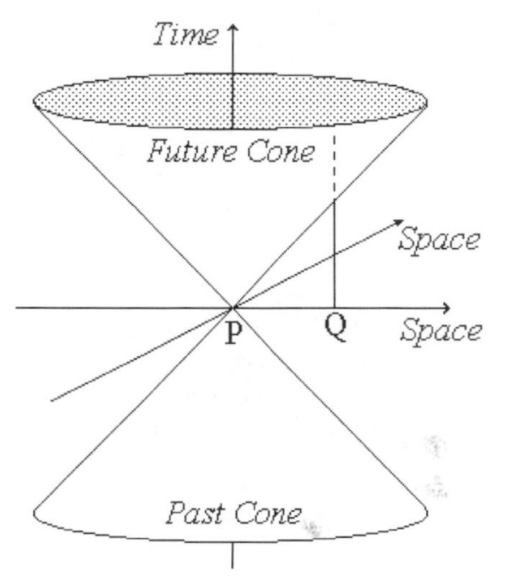

Fig. 1.33. The light cone of an event, P. The P event can be the cause of any event in its future cone or the result of any event in its past cone. It has no effect on, nor is it affected by, events outside its light cone. A stationary body at Q only learns of the P event when its vertical world-line enters P's future light cone.

The P event, represented in Fig. 1.33, can affect all the events inside its future light cone – its absolute future – and can be affected by all those in its past light cone – its absolute past. However, it cannot affect events outside its future light cone nor is it affected by those outside its past light cone. A stationary body at Q will not be affected by the P event until, after the passage of some time, its vertical world-line enters P's future light cone. Likewise, for an event occurring at Q to be the cause of the P event would require an instantaneous interaction, which is an impossibility.

Example 1.9: The Minkowski Diagrams of Different Observers — the Twins Paradox

We will start by explaining the construction of the Minkowski diagrams of the two inertial observers S and S′ and then use them to elucidate the twins paradox.

1. The Minkowski diagrams of different observers

The observer S′ moves with constant velocity, u, parallel to the X-axis. Hence, the slope of his world-line in the Minkowski diagram of the observer S is c/u. At every point on this line, $x' = 0$. Thus, it defines the orientation of the ct'-axis of the Minkowski diagram of the moving observer, S′ (Fig. 1.34).

Since the speed of light *in vacuo* has the same value for all inertial observers, a ray of light will be represented by the same world-line in the Minkowski diagrams of both S and S′. In the diagram of the observer S, the world-line of a light ray bisects the angle between the axes; it is reasonable to assume that the same will be true for the diagram of the observer S′. This defines the orientation of the X′-axis. Note, that the angle between the X′ and ct' axes is not a right angle.

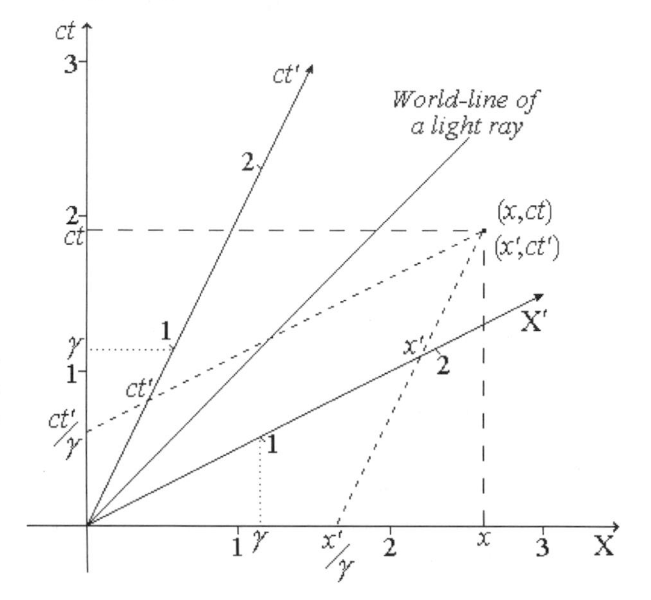

Fig. 1.34. Minkowski diagrams of the two observers, S and S′. In the orthogonal system of the observer S, the gradient of the X′-axis is u/c and that of the ct'-axis is c/u. A point with the space–time coordinates (x, ct) according to the observer S has the coordinates (x', ct') according to the observer S′.

Because of their relative motion, the scaling of the axes of the two observers differs. We employ the Lorentz transformation in order to scale the X′ and ct' axes. From Eq. (1.54), the x coordinate that corresponds to the point whose coordinates are $x' = 1$ and $t' = 0$ in the diagram of the observer S′ is

$$x = \gamma(x' + ct') = \gamma(1 + 0) = \gamma,$$

i.e. the point $x' = 1$ lies on the X′-axis immediately above the point $x = \gamma$ on the X-axis.

Similarly, from Eq. (1.60), the corresponding ct coordinate of the point having coordinates $x' = 0$ and $ct' = 1$ is

$$ct = c\gamma\left(t' + \frac{ux'}{c^2}\right) = c\gamma\left(\frac{1}{c} + 0\right) = \gamma,$$

i.e. the point $ct' = 1$ lies on the ct'-axis directly opposite the point $ct = \gamma$ on the ct-axis. In this way, the spacing of the integers on the axes of the S′ diagram is defined.

To find a point's coordinates (x', ct') relative to the moving observer S′, lines (the broken lines in Fig. 1.29) are drawn from the point parallel to the X′ and ct'-axes. The coordinates x' and ct' are, respectively, defined by the points where these lines cut the X′ and ct'-axes. The extension of these two lines crosses the X and ct-axes at the points x'/γ and ct'/γ, respectively.

2. The twins paradox

What has become known as the *twins paradox* was first propounded by Einstein in 1911:

> *'If we placed a living organism in a box [a spaceship] ... one could arrange that the organism, after any arbitrary lengthy flight, could be returned to its original spot in a scarcely altered condition, while corresponding organisms which had remained in their original positions had long since given way to new generations.'*[71]

If the travelling organism is a man and the one left behind on earth is his twin, the travelling twin finds, upon his return home, that his brother/sister has aged more than himself. It might appear at first that the phenomenon can be simply explained by time dilation; the travelling twin's clock – the rate at which his body ages – runs more slowly than that of the twin left on earth and so he has aged less than his stationary twin. However, motion is relative; either twin can contend that the other was the traveller and so each can contend that the other should be younger, which is obviously impossible this is the essence of the paradox. The twins' situations appear to be symmetrical and interchangeable. Are they so in fact?

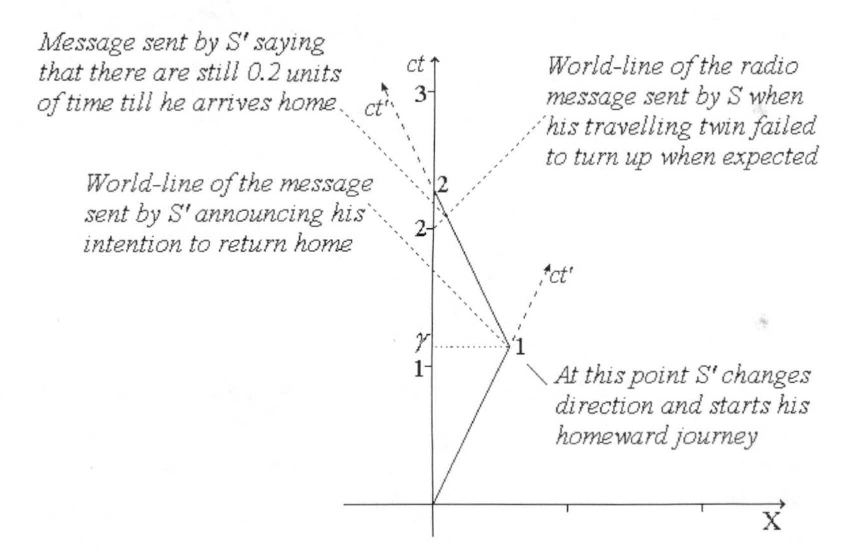

Fig. 1.35. The respective world-lines, ct and ct', of the homebound twin, S, and travelling twin, S′, and of their radio messages. The travelling twin moves at a speed of 0.5 c relative to the stationary twin. The slope of the world-lines of each radio message is 45°. After travelling for a single unit of time as measured by his clock, the travelling twin changes course and makes for home. He arrives home after $t' = 2$ units of time according to his clock but after $t = 2.3$ units of time according to the stationary twin's clock.

Let the travelling twin's velocity $u = 0.5c$. Each twin's world-line defines the time axis in his Minkowski diagram; that of the stationary twin, S, is the ct-axis, that of the travelling twin, S′, is the ct'-axis. However, as the diagram in Fig. 1.35 shows, their respective world-lines are not interchangeable. The orientation of the travelling twin's

[71] A. Einstein (1911) Die Relativitäts-Theorie. *Naturforschende Gesellschaft, Zürich, Vierteljahresschrift* 56, 1–14. Translated by R. Resnick (1968) *Introduction to Special Relatively*, New York: John Wiley & Sons, p. 201.

time axis must change somewhere in space–time if he is to return home; this change is equivalent to an acceleration. Whereas the homebound twin remains throughout in an inertial frame, in order to return home the travelling twin must undergo an acceleration. Thus, their situations are not symmetrical.

Suppose, that at the moment he changes course and makes for home – 1 time unit into his journey according to his clock – the travelling twin, S′, transmits a light signal (a radio message) announcing his intention of returning and that he estimates he will be arriving home after having been away for 2 units of time. This signal is received by the stationary twin, S, 1.73 units of time, according to his clock, after the travelling twin's departure.

When 2 units of time have elapsed, according to the stationary twin's clock, the traveller has still not yet returned. Worried as to his peripatetic twin's fate, the stationary twin sends him another message. This message is received by the travelling twin at a moment, according to his clock, that is only about 1.8 units of time after his departure. So he sends back another message asking why all the panic: 'There are still 0.2 units of time left before I expect to arrive home.'

The message is received just before S′ gets back. When the twins meet, it becomes apparent that whereas the travelling twin's clock has recorded the passage of 2 units of time since he left home, the stationary twin's identical clock – his genetically identical body – has recorded the passage of 2.3 units of time; S has aged more than S′. Each twin's proper time, the time according to which he himself ages, is measured by his individual time axis.

1.4.8 Relativistic Mechanics — Dynamics

The impact of Einstein's revolutionary conception of motion on the study of dynamics was no less profound than it was on the science of kinematics. Like the relativistic kinematics, the new *relativistic dynamics* has stood every experimental test to date. Perhaps its most celebrated consequence was the notion of the underlying equivalence of mass and energy expressed by the equation $E = mc^2$.

In the following pages, we will first examine the consequences of this new conception of motion on our understanding of four familiar physical quantities – linear momentum, mass, force and kinetic energy – and how the notion of mass-energy equivalence arises from it.

1. Linear momentum

In classical mechanics, the linear momentum, $\mathbf{p}$, of a particle of mass m moving with velocity $\mathbf{u}$ relative to a given reference frame is defined as:

$$\mathbf{p} = m\mathbf{u}. \tag{1.79}$$

In experiments conducted at low velocities, linear momentum, thus defined, is generally found to be conserved.[72] However, in experiments with high-energy

[72] See Example 1.4.

particles moving at speeds approaching the speed of light, it is found that linear momentum, as defined classically, is not conserved.

In explaining why this is so, we will restrict ourselves to uniform motion parallel to the X-axis. In the classical conception, a particle's velocity, u, is defined as the ratio between its displacement, Δx, relative to an inertial observer, and the period of time measured by the same observer, Δt, during which this displacement occurs:

$$u = \frac{\Delta x}{\Delta t}. \tag{1.80}$$

This gives for the linear momentum of the particle,

$$p_{classical} = m\frac{\Delta x}{\Delta t}. \tag{1.81}$$

However, this definition of linear momentum ignores the relativity of time. It assigns an unjustified pre-eminence to the time interval, Δt, measured by the observer. As opposed to this, the relativistic definition of linear momentum expresses the particle's linear momentum, $p_{relativistic}$, in terms of the proper time interval, Δt_0, measured in the reference frame of the particle itself:

$$p_{relativistic} = m\frac{\Delta x}{\Delta t_0}. \tag{1.82}$$

Replacing the proper time interval, $\Delta t_0 = \frac{\Delta t}{\gamma}$, in this equation by the observer's time interval gives

$$p_{relativistic} = m\gamma\frac{\Delta x}{\Delta t}. \tag{1.83}$$

Combining Eqs (1.81) and (1.83) produces the relationship

$$p_{relativistic} = \gamma p_{classical}. \tag{1.84}$$

Thus, in relativistic mechanics, the definition of linear momentum is

$$p_{relativistic} = \gamma m u. \tag{1.85}$$

When inertial observers employ this definition of linear momentum, they find that linear momentum is conserved whatever the speed of the particles.

2. Mass

The relativistic definition of linear momentum can also be written in the form

$$p_{relativistic} = m_{relativistic}\, u \tag{1.86}$$

where

$$m_{relativistic} = \gamma m_{classical}. \tag{1.87}$$

Since $\gamma \geq 1$, $m_{relativistic} \geq m_{classical}$.

* A particle's *classical mass*, $m_{classical}$, is its mass as measured from its own reference frame, i.e. from a reference frame at rest relative to the particle. It is the mass assigned to it by classical mechanics. In relativistic mechanics, it is usually designated m_0 and termed the *rest-mass* or *proper mass* of the particle.
* A particle's *relativistic mass*, $m_{relativistic}$, is its mass measured from a reference frame moving at a velocity u relative to the particle; it is usually designated simply m in relativistic mechanics.

A particle's relativistic mass depends on the factor γ, which is itself a function of the observer's motion. Thus, in relativistic mechanics, mass is not an invariant as it is in classical mechanics; it is a relative quantity whose value depends on the relative motion of the observer and the particle. At speeds below $0.5\,c$, the value of γ is so close to unity that the increase in mass is usually undetectable and can be ignored. However, as an object's speed approaches the speed of light, its mass rises asymptotically to infinity and classical mechanics is no longer competent to describe its behaviour.

The dependence of an object's mass on its speed relative to the observer has been confirmed experimentally many times.[73] In a series of experiments carried out by Bucherer[74] in 1909, the ratio, q/m, between the charge and the mass of the electron was found to depend on the particle's velocity; lower values were found at high velocities than at low velocities (Table 1.1). This was attributed to the relativistic increase in the mass of the particles that resulted from their motion.

Table 1.1. The dependence of the charge to mass ratio of electrons on their velocity.

u	γ	Observed charge to mass ratio (10^{11} C/kg)	Proper charge to rest-mass ratio (10^{11} C/kg)[75]
$0.321c$	1.056	1.66	1.75
$0.379c$	1.081	1.63	1.76
$0.430c$	1.108	1.59	1.76
$0.521c$	1.171	1.51	1.77
$0.690c$	1.381	1.28	1.77

[73] A comprehensive list of the experimental confirmations of special relativity can be found at the following website: http://math.ucr.edu/home/baez/physics/Relativity/SR/experiments.html.

[74] Alfred Bucherer 1863–1927, German physicist.

[75] The proper ratio, q/m_0, is obtained by multiplying the observed ratios by the factor γ appropriate to the particle's velocity, u.

The variation in the charge to mass ratios could have been attributed to decreases in the electric charge carried by the particles and not to increases in their mass. However, all the evidence suggests that electric charge remains an invariant quantity in relativistic physics just as it was in classical physics.

3. Force

Force is defined in relativistic mechanics, as it is in classical mechanics, as the rate of change of linear momentum:

$$\mathbf{F} = \frac{d\mathbf{p}}{dt} = \frac{d(m\mathbf{u})}{dt} = \frac{d(\gamma m_0 \mathbf{u})}{dt}. \tag{1.88}$$

Both γ and $\mathbf{u}$ are functions of t. Consequently, differentiation gives

$$\mathbf{F} = \gamma m_0 \frac{d\mathbf{u}}{dt} + m_0 \mathbf{u} \frac{d\gamma}{dt}. \tag{1.89}$$

This relativistic equation reduces to Newton's second law of motion, $\mathrm{F} = m\mathbf{a}$, in the special case where $\gamma = 1$. Since, in general, the force has a component, $m_0 \mathbf{u} \frac{d\gamma}{dt}$, in the direction of the velocity $\mathbf{u}$, in relativistic mechanics, a particle's acceleration, $\mathbf{a} = d\mathbf{u}/dt$, is not necessarily parallel to the force acting upon it as it is in classical mechanics.

If the particle moves at a constant speed u, as in uniform circular motion, $d\gamma/dt = 0$. In this case Eq. (1.57) reduces to

$$F_{Centripetal} = \gamma m_0 a_N = \frac{\gamma m_0 u^2}{R}, \tag{1.90}$$

where R is the radius of the circle.

It can be shown[76] that if F_x, F_y and F_z are the components of a force acting on a particle at rest relative to the inertial observer S, then the components of the force acting on the particle relative to the moving inertial observer S′ are:

$$F'_{x'} = F_x \quad F'_{y'} = \frac{F_y}{\gamma} \quad F'_{z'} = \frac{F_z}{\gamma}. \tag{1.91}$$

4. Kinetic energy

The kinetic energy of a charged particle such as an electron can be expressed in terms of mechanical or electrical measures, depending on the circumstances:

[76]See Appendix 1.5.4.

* In terms of mechanical measures such as the particle's mass and velocity. In classical mechanics, this gives the familiar expression $K = m_e u^2/2$ for the kinetic energy of an electron moving with velocity u in which m_e designates the electron's classical mass.
* In terms of electrical measures such as the particle's charge and the potential difference through which it is accelerated. This gives the equally familiar expression $K = eV$ for the kinetic energy of a particle carrying a charge e that is accelerated through a potential difference V.

Though each definition employs different measures, to be consistent, both should give the same numerical value for the particle's kinetic energy. Equating the two definitions, $K = m_e u^2/2$ and $K = eV$, gives the following expression for the square, u^2, of the velocity of an electron accelerated through a potential V

$$u^2 = \frac{2eV}{m_e}. \tag{1.92}$$

If the two definitions are indeed consistent, measured velocities should correspond with those predicted by this equation. This is borne out at potential differences below ~100,000 V. However, at higher potentials (higher energies), the measured velocities are found to be lower than those it predicts. Instead of increasing linearly with the potential, V, as it should according to Eq. (1.92), the square, u^2, of their velocity rises asymptotically to c^2 (Fig. 1.36).

The source of the problem is that the classical expression for kinetic energy, $K = m_e u^2/2$, recognises neither an upper limit on the speed to which the particle can be accelerated nor that its mass increases with its speed. Whereas these considerations can be ignored for all practical purposes at low energies, they cannot be disregarded at high energies. At such energies, relativistic effects must be taken into account and a different expression must be employed for the particle's kinetic energy, one that takes these effects into consideration.

Kinetic energy is defined in relativistic mechanics as it is in classical mechanics, namely, as the work done by an external force in increasing the speed of the particle. Restricting ourselves to forces and motion parallel to the X-axis, this gives the integral

$$K = \int_0^x F_x dx, \tag{1.93}$$

where F_x is the external force acting on the particle. When carrying out this integration in relativistic mechanics, account must be taken of the change in the

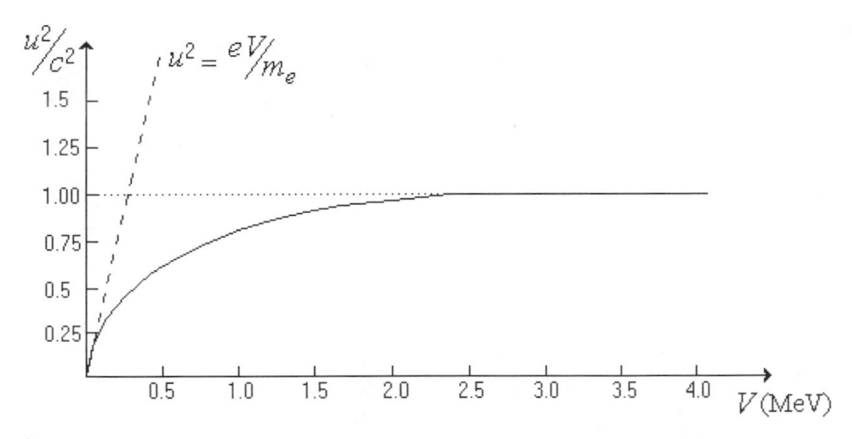

Fig. 1.36. The relationship between the square, u^2, of the velocity of an electron – expressed as a fraction of the square of the speed of light *in vacuo* – and the potential V through which it was accelerated. The broken line shows how the square of the particle's velocity would rise were the classical definition valid at all energies.

particle's mass that occurs as its speed increases. This is done by substituting the relevant relativistic definition of the force, F_x given by Eq. (1.88), in the integral

$$K = \int_0^x F_x dx. \tag{1.94}$$

Performing the integration gives the following expression for a particle's kinetic energy[77]:

$$K = (m - m_0)c^2, \tag{1.95}$$

where m is the particle's relativistic mass and m_0 its rest-mass. Particle velocities forecast by equating this relativistic definition of kinetic energy with the electromagnetic definition are confirmed by experiments at all energies. At low energies, i.e. when $u \ll c$, the relativistic definition of kinetic energy, $K = (m - m_0)c^2$, reduces to the classical definition, $K = m_0u^2/2$.

5. Mass–energy

A corollary of the relativistic definition of kinetic energy is that any change, ΔK, in the kinetic energy of a particle will be accompanied by a corresponding change, Δm, in its relativistic mass, such that

$$\Delta K = \Delta mc^2. \tag{1.96}$$

The additional mass acquired by the particle by virtue of its increased kinetic energy is qualitatively identical to the particle's rest-mass. Consequently, the

[77]The full integration is given in Appendix 1.6.2.

latter can be similarly expressed in terms of an intrinsic energy, E_0, called the particle's *rest-energy*:

$$E_0 = m_0 c^2. \tag{1.97}$$

The total energy, E, of the moving particle is the sum of its own intrinsic energy and its acquired kinetic energy:

$$E = E_0 + K. \tag{1.98}$$

In terms of its rest-mass, m_0, and its linear momentum, p, the total energy of the moving particle is also given by the important and useful formula[78]

$$E^2 = (m_0 c^2)^2 + p^2 c^2. \tag{1.99}$$

In extending this idea, Einstein proposed that a particle's mass is affected not only by changes in its kinetic energy but by changes in any of the forms of energy it may possess: chemical, potential, electrical, etc. On this view, mass is equivalent to energy in whatever form it takes, such that the general relationship between the total energy, E, of a particle and its total relativistic mass, m, is given by the equation

$$E = mc^2. \tag{1.100}$$

It follows that the emission of energy from any system, in any form whatsoever, is done at the expense of the system's overall mass, and that the absorption of energy in any form increases the system's overall mass. Consequently, the separate classical laws of mass and energy conservation are flawed and must be abandoned. Neither mass nor energy alone are conserved in natural processes. The quantity actually conserved – *mass-energy* – combines mass and energy, taking their equivalence into account.

In SI units, the constant of proportionality, c^2, in the equation $E = mc^2$ is approximately equal to 10^{17}. Accordingly, one kilogram of any substance – iron, cream cheese, sea water, etc. – is intrinsically equivalent to about 10^{17} joules of energy, and *vice versa*. Thus, the annihilation of even a minute amount of matter releases considerable amounts of energy: the substantial amounts of energy released by the disintegration of radioactive materials are produced by just minute reductions in their masses.

By the same reasoning, an enormous quantity of energy would have to be absorbed or emitted by a system for any detectable change in its mass to occur: the amounts of energy involved in ordinary chemical reactions are far

[78]See Appendix 1.6.3.

too small to produce any significant or even measurable change in the mass of the reactants.

The tremendous quantities of energy released in stellar processes involve the destruction of matter on a cosmic scale; it is estimated that to maintain the rate at which it emits energy, the sun loses about five million tons of its matter each second (out of its total mass of some 10^{27} tons).

Example 1.10: Calculations in Relativistic Mechanics

A free electron has kinetic energy of 2 MeV. Calculate its

1. total energy;
2. linear momentum;
3. relativistic mass;
4. velocity.

Calculation:

1. The rest-mass of an electron – its classical mass – is $9.11 \cdot 10^{-31}$ kg. Thus, its rest-energy is:

$$m_0 c^2 = 9.11 \cdot 10^{-31} \cdot 9 \cdot 10^{16} = 8.199 \cdot 10^{-14} \mathrm{J}.$$

This quantity can be expressed more conveniently in units of electron-volts.

$$8.199 \cdot 10^{-14} \mathrm{J} \equiv \frac{8.199 \cdot 10^{-14}}{1.6 \cdot 10^{-19}} \mathrm{eV} = 0.512 \, \mathrm{MeV}.$$

By combining Eqs (1.97) and (1.98), the formula for the total energy of a particle can be written in the form $E = m_0 c^2 + K$. Thus, the total energy of a free electron with kinetic energy 2 MeV is:

$$E = m_0 c^2 + K = 0.512 + 2 = 2.512 \, \mathrm{MeV}.$$

2. Substituting the values obtained in (a) for the electron's rest-energy and total energy in Eq. (1.99) gives:

$$E^2 = (m_0 c^2)^2 + (pc)^2$$
$$(2.512)^2 = (0.512)^2 + (pc)^2.$$

From which

$$pc = 2.459 \, \mathrm{MeV},$$

or alternatively

$$p = 2.459 \frac{\mathrm{MeV}}{c}.$$

The unit MeV/c is frequently used to express the linear momentum of particles. Its equivalent value in SI units is:

$$\frac{1 \, \mathrm{MeV}}{c} \equiv \frac{1.6 \cdot 10^{-19} \cdot 10^6}{3 \cdot 10^8} = 5.3 \cdot 10^{-22} \mathrm{kg} \cdot \mathrm{m} \cdot \mathrm{s}^{-1}.$$

3. From the results obtained above, the ratio between the total energy, E, and the rest-energy, E_0, of the electron is:

$$\frac{E}{E_0} = \frac{mc^2}{m_0c^2} = \frac{2.512}{0.512} = 4.906,$$

which gives

$$m = 4.906m_0$$

for the relativistic mass, m, of the electron in terms of its rest-mass (its classical mass).

4. The electron's velocity, u, can be found by substituting the appropriate value for the factor γ in Eq. (1.61) and rearranging.
From Eq. (1.87), $m = \gamma m_0$, and so the appropriate value of γ can be obtained from the ratio between the relativistic mass and the rest-mass of the electron, $m/m_0 = 4.906$ found in the previous question. Substituting this value in Eq. (1.61) gives $u = 0.979c$.

Example 1.11: Transformation of Momentum and Energy

A particle's linear momentum, $\mathbf{p}$, and its total energy, E, are relative quantities, measurements of which depend on the inertial reference frame from which they are made. In classical mechanics the different measurements are reconciled by means of the Galilean transformation. In relativistic mechanics the Lorentz transformation must be employed. What form do the equations of the Lorentz transformation take in the case of linear momentum and total energy?

Linear momentum is a vector quantity; the components of a particle's linear momentum relative to the inertial observer S are p_x, p_y and p_z. The magnitude, p, of the particle's momentum is given by

$$p^2 = p_x^2 + p_y^2 + p_z^2. \quad (1.101)$$

Substitution in the relativistic expression for the particle's total energy relative to the same observer, $E^2 = (m_0c^2)^2 + p^2c^2$, followed by rearrangement of the equation gives

$$\frac{E^2}{c^2} - (p_x^2 + p_y^2 + p_z^2) = m_0^2c^2. \quad (1.102)$$

Both m_0, the particle rest-mass, and c, the speed of light *in vacuo*, are invariants. Hence, the corresponding expression relative to the inertial observer S′ must be

$$\frac{E'^2}{c^2} - (p_{x'}'^2 + p_{y'}'^2 + p_{z'}'^2) = m_0^2c^2, \quad (1.103)$$

where $p'_{x'}$, $p'_{y'}$ and $p'_{z'}$ are the components of the particle's linear momentum relative to S′ and E' is its total energy. Equating Eqs. (1.102) and (1.103) gives

$$\frac{E'^2}{c^2} - (p_{x'}'^2 + p_{y'}'^2 + p_{z'}'^2) = \frac{E^2}{c^2} - (p_x^2 + p_y^2 + p_z^2). \quad (1.104)$$

Making the correspondence

$$p_x \to x \quad p_y \to y \quad p_z \to z \quad \frac{E}{c} \to ct$$

we see that the structure of Eq. (1.104) is identical to that of Eq. (1.76),

$$(ct)^2 - (x^2 + y^2 + z^2) = (ct')^2 - (x'^2 + y'^2 + z'^2),$$

which relates the respective space and time coordinates of the two observers, S and S′, and which was derived by simple algebra directly from the equations of the Lorentz transformation. Hence, by analogy with Eqs (1.53), (1.55), (1.57) and (1.59) of the Lorentz transformation, the transformation equations of the components of the particle's linear momentum and total energy must be:

$$p'_{x'} = \gamma\left(p_x - \frac{uE}{c^2}\right) \quad p'_{y'} = p_y \quad p'_{z'} = p_z \quad E' = \gamma(E - up_x). \tag{1.105}$$

Just as the four space–time coordinates, x, y, z and t, can be depicted in a four-dimensional representative space, so too can the four associated quantities p_x, p_y, p_z and E. Sets of quantities which transform together in this way are called *four-vectors.* Expressing physical laws in terms of four-vectors ensures their covariance under special relativity.

1.4.9 Magnetism — A Relativistic Effect

The application of special relativity to the field of electrodynamics results in an entirely new way of dealing with the magnetic phenomena associated with moving charges, one which is implicitly covariant and avoids the *ad hoc* postulates upon which Maxwell had based his theory. To illustrate how this is done, we return to the case study of Coulomb's law and magnetic forces we first discussed in Section 1.4.1 (p. 39).

As we noted *in situ*, whereas an observer at rest relative to a pair of identical positive charges finds the magnitude of the repulsive force acting between them to correspond to that given by Coulomb's law, $F_E = kq^2/r^2$, a second inertial observer, relative to whom both the first observer and the positive charges are in uniform linear motion, finds that a weaker force, $F' < F_E$, acts between the charges. Seemingly, Coulomb's law is not valid when the observer and the charges are in relative motion. Thus, it should serve as a means for distinguishing between stationary and moving observers.

Classical electromagnetism resolves the discrepancy by the *ad hoc* postulate that each moving charge sets up a magnetic field, B, which, in turn, acts on the other such that an attractive force, $F_M = F_E - F' = Bqu$, acts between them. However, as we noted at the time, this leaves the issue of the apparent non-equivalence of the inertial observers under Coulomb's law unresolved, for the magnetic postulate appears to be no more than an arbitrary contrivance designed to circumvent this. Proper observer equivalence requires that the difference between the electric forces they record be accounted for in terms that have the same import for both, such as their relative motion. Under the classical conception of motion, in which actual forces are invariants, this cannot be done. We will now show how it is achieved under special relativity.

Consider a point charge, q, and a uniform line of charge, Q, of charge density λ (a long thin charged conductor) situated at a distance r from it. In general, the coulombic force that acts between the charge and the line of charge is given by

$$F = \frac{\lambda q}{2\pi\varepsilon_0 r} \tag{1.106}$$

where ε_0 is the vacuum permittivity.

The two inertial observers, S and S′, measure the force that acts between the point charge and the uniform line of charge under three different circumstances:

a. The point charge and the line of charge are at rest relative to the observer S.
b. The point charge is at rest but the line of charge moves with a constant velocity u parallel to the X-axis relative to the observer S.
c. Both the point charge and the line of charge move with a constant velocity u relative to the observer S.

a. In this case (Fig. 1.37a), the charge density that observer S measures is the proper density, $\lambda_0 = Q/l_0$, and the force he observes acting parallel to the Y-axis is the proper force, $F_0 = (\lambda_0 q/2\pi\varepsilon_0 r)$. In special relativity, the corresponding force, F', measured by the moving inertial observer S′ is obtained from the force transformation – Eq. (1.59): $F'_y = F_y/\gamma$. This gives $F' = F_0/\gamma$.

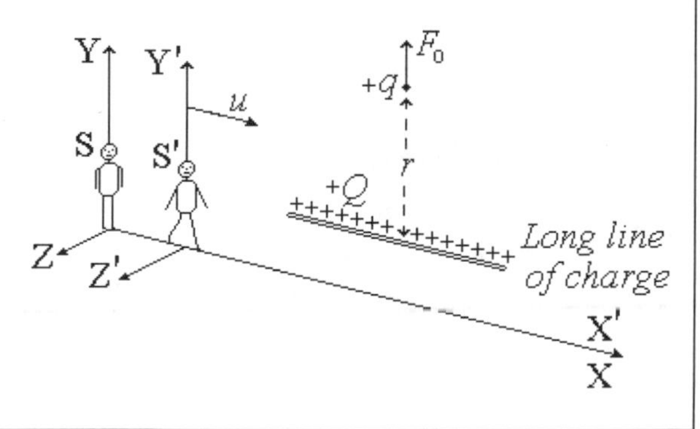

Fig. 1.37a. Relative to the observer S, the force acting on the point charge $+q$ is the proper force F_0. However, because of his motion relative to the point charge and line of charge, the observer S′ measures a smaller force, $F' = F_0/\gamma$.

b. In this instance, the observer S sees the line of charge moving parallel to the X-axis at a velocity u (Fig. 1.37b). Consequently, relative to him, the line of charges contracts by a factor γ. This increases the charge density by the same factor making $\lambda = \gamma\lambda_0$. Hence, the force he observes acting on the point charge is $F = \gamma F_0$.

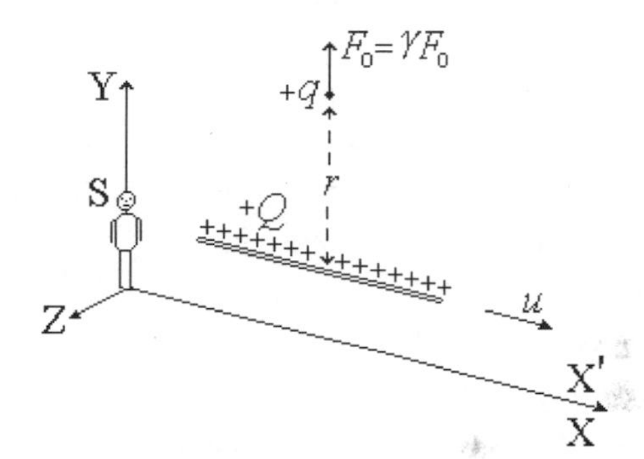

Fig. 1.37b. Relative to S, the length of the moving line of charge is foreshortened to $l = l_0/\gamma$.

The motion of the uniform line of charge is equivalent to a flow of electric charge, i.e., to an electric current $I = \lambda u$.

c. As Fig. 1.37c shows, except for the reversal in the direction of the relative motion, the stationary observer's view of the line of charge and of the point charge in this case is identical to that of the moving inertial observer, S′, in case (a). Thus, the force measured by the stationary observer S in this case will be the same as that measured by S′ in case (a), *viz.* F_0/γ.

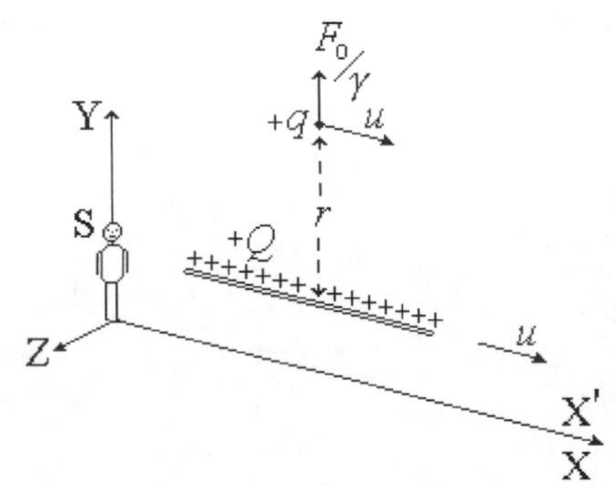

Fig. 1.37c. Both the point charge and the line of charge are in motion relative to S.

This case is equivalent to the familiar example of the force exerted by an electric current on a moving charge.

The difference, $(\frac{F_0}{\gamma} - \gamma F_0) < 0$, between the force acting on the moving charge in (c) and on the stationary charge in (b) is equivalent to an attractive force acting towards and at right angles to the line of charge. Under special relativity, this force, which is attributed to magnetic effects in classical theory, is simply a direct result of the motion of the point charge in (c) relative to S. No further *ad hoc* postulates are needed in order to account for the difference between the forces: Coulomb's law and special relativity are all that is needed.

In this analysis, the reference frames of the two inertial observers, S and S′, are symmetrical and interchangeable. Thus, if Coulomb's law and special relativity are all that is required to account for the forces observed by S, the same must be true for S′. Hence, S and S′ are equivalent and Coulomb's law is covariant.

In general, any difference in the electrical forces inertial observers measure results only from their relative motion; there is no need for the magnetic force postulate. Under special relativity, what are usually called *magnetic forces* arise naturally from the motion of electrical charges; they are not a distinct or separate phenomena but simply the velocity dependent component of the electrical interaction between the charges.

Why then does electromagnetic theory work so well? Why can classical electromagnetism with its *ad hoc* postulates still be used?

The answer is that making the Maxwellian magnetic postulate is, in fact, tantamount to adopting the Einsteinian conception of motion when employing Coulomb's law. To put this another way: the relativistic force of attraction, $\frac{F_0}{\gamma} - \gamma F_0$, that acts on the point charge is actually identical to the magnetic force, $F_M = Bqu$, postulated by classical electromagnetic theory. That this is so, is readily demonstrated by the following series of substitutions.

Putting $F_M = (\frac{F_0}{\gamma} - \gamma F_0)$ and rearranging gives

$$F_M = \gamma F_0 \left(\frac{1}{\gamma^2} - 1 \right). \tag{1.107}$$

Substituting $F_0 = (\lambda_0 q / 2\pi\varepsilon_0 r)$, followed by $\lambda = \gamma\lambda_0$ and $\gamma = \sqrt{\frac{1}{1-\mu^2/c^2}}$ yields

$$\begin{aligned} F_M &= \frac{\gamma\lambda_0 q}{2\pi\varepsilon_0 r} \left(\frac{1}{\gamma^2} - 1 \right) \\ &= \frac{-\lambda u^2 q}{2\pi\varepsilon_0 r c^2}. \end{aligned} \tag{1.108}$$

The negative sign indicates that the force is one of attraction.

Finally, incorporating the electromagnetic definition of the speed of light, $c = \sqrt{\frac{1}{\varepsilon_0 \mu_0}}$ in Eq. (1.108) produces

$$|F_M| = \frac{\mu_0 \lambda u}{2\pi r} \cdot qu. \tag{1.109}$$

Relative to the observer S, the motion of the line of charge in both (b) and (c) is equivalent to a current, $I = \lambda u$. Making this further substitution in Eq. (1.109), we obtain the familiar electromagnetic expression

$$|F_M| = \frac{\mu_0 I}{2\pi r} \cdot qu = Bqu \tag{1.110}$$

for the force acting on a point charge, q, moving with velocity u parallel to a straight conductor carrying a current I, at a distance r from it.

To summarise. Whether we introduce the magnetic field by making the magnetic postulate, i.e. that a moving or changing electric field sets up a magnetic field, or by combining Coulomb's law and the Einsteinian conception of motion, is really just a matter of convenience. One approach is no more fundamental than the other. The same reality underlies Faraday's fields, Maxwell's equations and special relativity. Which way we choose to approach it is a matter of individual choice.

Chapter 1.5
The General Theory of Relativity

The jewel in the crown of classical mechanics was *Newton's theory of gravitation*. Taken together with his laws of motion, it afforded explanations and predictions of unprecedented accuracy, that were repeatedly confirmed by experiment and observation. The classic example was Newton's own derivation and theoretical elucidation of Kepler's three empirical laws.

However, for all its success, classical gravitational theory gave no clear explanation of the source of the force that attracted one body to another. Nor did it explain how this force acted across space: instantaneously or through some medium. Newton himself was circumspect on all these points. Though he recognised that a body's weight is directly proportional to its inertia, its immutable *vis inertiae*, he rejected the notion that 'gravity is essential to bodies'. As he wrote:

> '*It is inconceivable, that inanimate brute matter, should, without the mediation of something else, which is not material, operate upon and affect other matter without mutual contact, as it must be, if gravitation be essential and inherent in it... That gravity should be innate, inherent and essential in matter, so that one body may act on another at a distance through a vacuum, without the mediation of anything else, is to me so great an absurdity, that I believe no man, who has a competent faculty of thinking, can ever fall into it.*'[79]

But if gravity is not an inherent property of matter, then of what is it a property? Newton refused to offer any answer to this question,

> '*Hitherto we have explained the phenomena of the heavens and of our sea by the power of gravity, but have not yet assigned the cause of this power... and I frame no hypotheses.*'[80]

Others were less reticent. Notwithstanding Newton's clear rejection of the notion, even during his lifetime, his theories were interpreted as implying that gravity is indeed an innate inherent property of matter. In the Preface he was asked to write in 1713 to the second edition of *Principia*, the mathematician

---

[79] *Principia*. Translated by Andrew Motte, revised by F. Cajori (1962). Berkeley: University of California Press, p. 634.
[80] *Ibid.* p. 547.

Cotes asserted that 'the attribute of gravity is found in all bodies'. A corollary of this interpretation was that *actions at a distance*, such as gravitational interactions, propagate instantaneously across space. However, with the establishment of the theory of special relativity, which prohibits instantaneous interactions of this type, this became untenable.

Ironically, it was the very accuracy of its predictions and the improvements in instrumentation that they motivated, that gave rise to the first suspicions that something was not quite right in Newton's theory of gravitation. Towards the end of the 19th century, it was found that even after allowing for all the perturbations in their orbits that result from their mutual attractions and from those of their moons, the orbits of the planets deviate very slightly from those predicted by Newton's law of gravity. The effect is most pronounced in the case of Mercury, the planet closest to the sun.

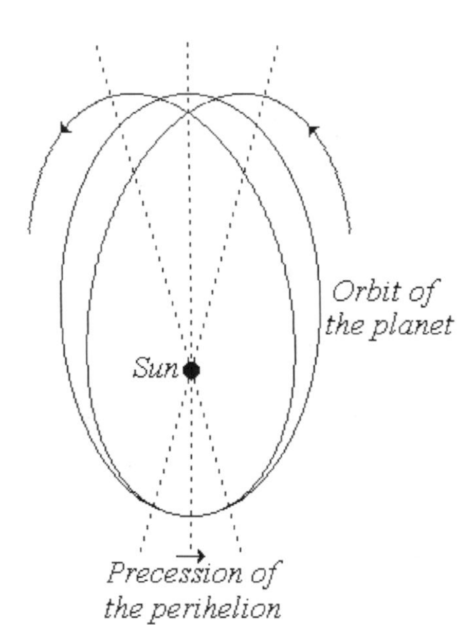

Fig. 1.38. The precession of the perihelion of a planet. After each orbit the perihelion advances in the direction of the planet's motion to a new point in space. The direction of the major axis of Mercury's orbit changes by about 42 seconds of arc per century.

According to Kepler's laws, at the end of each planetary year, a planet should return to the same point in space it occupied at the beginning of the year. However, the perihelion of the planet Mercury – the point closest to the sun in the planet's orbit – is not a fixed point in space; the perihelion precesses. In other words, at the end of each planetary year, Mercury overshoots the point in space it occupied at the beginning of the year. The full extent of this precessional motion is inexplicable in Newton's theory of gravitation (Fig. 1.38).

1.5.1 The Postulates of the General Theory of Relativity

The precession of the perihelion of Mercury's orbit was first explained by Einstein when he broadened his special theory of relativity, which had dealt only with inertial observers, into a general theory that pertained to all observers, irrespective of their coordinate system. Accordingly, the new *principle of general relativity* that he formulated was:

> **The laws of physics must be expressed in such a way that they are form invariant in all reference frames, whatever their motion.**

The exercise of this principle requires that there be a transformation under which all the laws of physics are covariant, whether or not the reference frames from which they are formulated are inertial. This transformation must do for all observers, irrespective of their motion – accelerated or otherwise – what the Lorentz transformation does for inertial observers. In particular, it must reconcile any differences between the observations made in the various reference frames, such that no observer should be able to determine, by means of any experiment, whether he is at rest, in uniform motion in a straight line or in an accelerated motion.

At first sight, it would appear that our daily experiences belie this corollary of the principle of general relativity. Based on the inertial forces we feel at the time, we believe we can always tell whether or not we are accelerating. The palpable forces we experience when travelling in a car or a bus at every turn, at every bump and at every time it stops or starts, tell us that we are in an accelerating non-inertial reference frame. Such forces are not felt by passengers in a stationary vehicle nor by those in a vehicle that is moving along a perfectly straight and smooth horizontal road at uniform velocity. In classical mechanics these *inertial* or *fictitious* forces serve to distinguish, absolutely, between inertial and non-inertial frames of reference.

Newton regarded these inertial forces, which do not derive from interactions between bodies and which appear to be a universal phenomenon, as providing the evidence for the existence of *absolute space.* He wrote:

> '*For instance, if two globes kept at a given distance one from the other, by means of a cord that connects them, were revolved about their common centre of gravity; we might, from the tension of the cord, discover the endeavour [tendency] of the globes to recede from the axis of their motion and from this we might compute the quantity of their circular motion ... even in an immense vacuum, where there was nothing external or sensible with which the globes could be compared.*'[81]

[81] *Principia.* Translated by Andrew Motte, revised by F. Cajori (1962) Berkeley: University of California Press, p. 12.

Newton argued that the tension in the cord linking the two rotating globes was an intrinsic attribute of the system of rotating spheres. Its origin was in their rotation and in this alone. Thus, it constituted absolute proof of their rotation without the need for any external point of reference. Space itself was the frame for their rotation.

It is only on this basis that the Copernican heliocentric system can be accorded preference over the Ptolemaic geocentric system. The description of planetary motions given by the Copernican system is not inherently superior to that given by the Ptolemaic. For it makes no difference to our view of the heavens whether the earth spins on its axis once every 24 hours as is postulated by the Copernican system or the heavenly vault orbits a stationary earth once every 24 hours as the Ptolemaic system posits.[82] Under the Newtonian conception, the real difference between the two models is that if the earth was at rest, free bodies would move across its surface in straight lines. But if it is spinning, their trajectories should be curved – the *Coriolis effect.* Satellite photographs such as those seen every evening in television weather forecasts show masses of atmospheric air moving across the earth's surface in curved paths. Under the Newtonian conception, this proves that the earth is not at rest; that inertial forces arise in the atmosphere and determine the movement of the cyclones and anti-cyclones associated with terrestrial weather patterns. These forces would not arise if the earth was at rest. Classically, it is the weather that proves the earth is spinning.

However, we may ask whether this is the only explanation that can be given for these terrestrial effects? Are changes in motion – accelerations – the only possible way of accounting for the phenomena? Does the presence or absence of *inertial forces* unequivocally distinguish between non-inertial and inertial reference frames?

Suppose we are in a vehicle that runs along a perfectly smooth horizontal track which holds it upright at all times and that its windows are all covered so that we cannot see out. Initially, the vehicle cruises at a uniform speed (there is a speedometer inside) along what we feel to be a straight stretch of track. Suddenly, we and all the other passengers and objects in the vehicle are flung sideways, to our left, say (Fig. 1.39). According to the reasoning above, we would conclude that the vehicle had come to a sharp bend in the track, where, without slowing down, it made a right turn. As a result of this change in direction,

[82] As Laplace wrote: 'It suffices, then, in order to change this hypothesis [the Ptolemaic] into the true system of the world [the Copernican], to assign the apparent movement of the sun to the earth, but in an opposite direction.' (Reprinted 1951) *A Philosophical Essay on Probabilities*, New York: Dover Publications, p. 184.

an inertial force – a *centrifugal force* – had acted upon us, throwing us to the left.

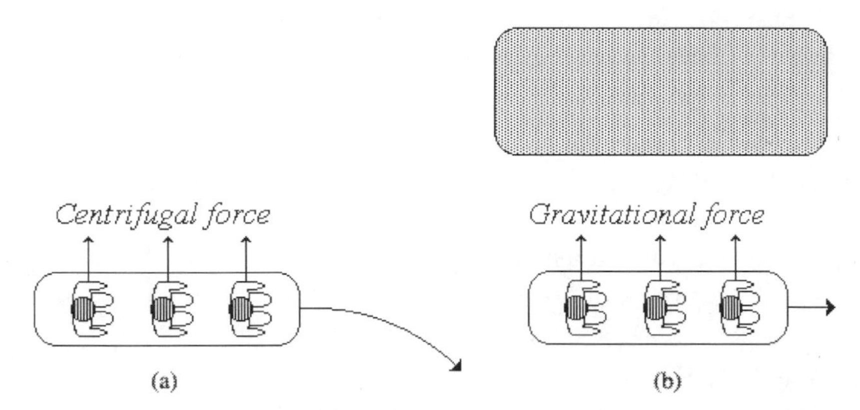

Fig. 1.39. The passengers in the vehicle cannot distinguish between the two possibilities: (a) The vehicle made a sharp right turn and as a result a centrifugal (inertial) force acted on them throwing them to their left; (b) A massive body, whose gravitational pull drew the passengers towards it, suddenly appeared to the left of the vehicle.

However, there is an alternative explanation to the sensations we experience, one that does not involve any change in the our motion. It is that a massive body, which exerts a substantial gravitational pull, suddenly appeared for a few moments outside and to the left of the vehicle, and that we had all 'fallen' sideways towards it because of our own gravitational mass. The vehicle had, in fact, continued to move in a straight line; there was no bend in the track. Though we may consider this second explanation to be somewhat far-fetched, it cannot be dismissed out of hand.[83]

Einstein argued that the two explanations are equally valid and, moreover, that they are indistinguishable. Consequently, there is no way we can know, absolutely, why we were flung sideways. It might have been because of the inertial forces generated by a change in the direction of our motion; equally, it might have been the result of a gravitational force as we continued to travel in a straight line. In the former case, there would have been an acceleration; in the second, there would not. Thus, contrary to our intuition, we cannot always tell whether or not we are accelerating; it may be gravity.

[83]The reader might agree that the notion of the sudden appearance of a massive object is perhaps no less remarkable than the Newtonian notion of 'made-to-order' fictitious inertial forces that are mysteriously created as required by absolute space. The choice of admissible causes defines a culture's attitude to the world. For all its rationalism, Newtonian science remains deeply mystical.

Einstein summarised this notion in a postulate known as the *principle of equivalence*:

> **An observer cannot determine, in any way whatsoever, whether the laboratory he occupies is in a uniform gravitational field or is in a reference frame that is accelerating relative to an inertial frame.**

According to this principle, all the phenomena experienced by a body in a gravitational field can be duplicated by means of an appropriate acceleration; and *vice versa*, all those experienced by an accelerated body can be duplicated in an appropriate gravitational field (Fig. 1.40). Consequently, by choosing the appropriate acceleration for our reference system, we can duplicate the effects of a constant gravitational field, reduce or increase the strength of a given field, or even neutralise it completely.

The notion underlying the principle of equivalence is the sameness of gravitational mass and inertial mass. According to the general theory of relativity, gravitational and inertial mass are not different properties of matter but two aspects of a fundamental and single property of matter. The equality of the magnitude of an object's gravitational mass with that of its inertial mass, which, in classical physics, explains why all bodies fall to the earth's surface with the same acceleration, g, whatever their size or composition, is not fortuitous; it reflects an underlying fundamental unity.

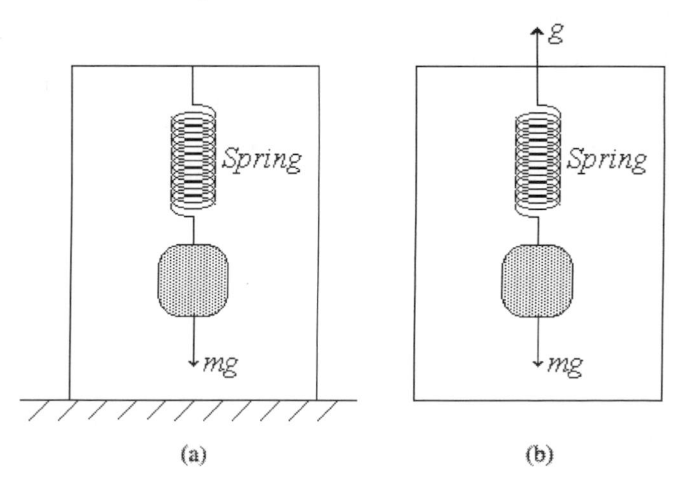

Fig. 1.40. Two equivalent elevators: (a) an elevator at rest on the earth's surface; (b) an elevator in deep space rising with an acceleration g. Identical objects hang from identical springs, each of which is stretched to the same degree. Each duplicates the physical properties of the other. Observers inside the elevators cannot distinguish between the two possibilities.

It must be emphasised that the principle of equivalence is valid only for uniform gravitational fields or for infinitesimal volume elements of non-uniform fields. The reason for this is that an observer in a non-uniform gravitational field could detect its presence by releasing two objects. Due to the non-uniformity of the field the objects would either move towards each other or away from one another. We say that *tidal forces* act in a non-uniform gravitational field.

There are no tidal forces in a uniform gravitational field nor can they be detected in an infinitesimal volume element of a non-uniform field. Thus, the field in an infinitesimal volume element of any gravitational field can be considered, at least for short periods of time, to be uniform. Consequently, we can always posit an accelerated reference frame, relative to which, in a small region of space–time, the gravitational field is cancelled out.

1.5.2 Gravitation and the General Theory of Relativity

Between the years 1907 and 1916, Einstein succeeded in developing a theory of gravitation that is valid for all observers whatever their motion; a truly covariant theory. To a first approximation, Newton's law of gravity ensues from it as the case of the gravitational field around a point mass. In the case of weak gravitational fields, Einstein's theory and that of Newton reach the same conclusions.

A full elucidation of Einstein's theory of gravity requires the solution of the field equations he developed, a task that demands complex mathematical processes and is far beyond the intent of this book. However, some of the physical consequences of the theory, such as the effect of gravitational fields on the measurement of physical quantities, can be deduced from a combination of the principle of equivalence and the special theory of relativity. This is the approach Einstein himself took when he first predicted some of the more surprising consequences of his theory. To ensure the validity of the use of special relativity, we must restrict ourselves to the short distances and short periods of time in which the motions can be considered to be uniform.

As an example of how gravity can be neutralised, consider the following. Suppose that the cable holding a windowless elevator at rest close to the earth's surface snaps. As a result, the elevator and all its contents fall freely in the earth's gravitational field. An observer inside the lift would experience the sensation of *weightlessness*; he would find no evidence of the earth's gravitational field. The acceleration of its free-fall neutralises the effect of the gravitational field inside the elevator. Unable to look out, the observer might well believe that he had been somehow transported to a *gravityless* inertial reference system, deep in space far from any gravitational field. We may summarise this as follows:

Experimentally, a small free-falling laboratory is equivalent to an inertial reference system in which there is no gravitational field.

Hence, the effect of a gravitational field on the measurement of a physical quantity can be investigated by comparing its measurement by an observer in a small free-falling laboratory, with that made by a stationary observer at the same point in the gravitational field.

Suppose that a stationary observer is situated at a point above the earth's surface at a distance r from its the centre, and that a second observer, who is falling freely in a small laboratory from rest at infinity, is momentarily at the same distance from the earth's centre. At that moment, the stationary observer measures the time between events occurring in his neighbourhood and the lengths, parallel to the direction of the gravitational field, of objects at rest around him; likewise, the free-falling observer measures the time between the same events as the stationary observer and the lengths, in the direction of his motion, of the same objects (Fig. 1.41).

In terms of the special theory of relativity, the time intervals and lengths measured by the stationary observer, Δt_0 and L_0, will be proper time intervals and proper lengths respectively, for he is at rest relative to the measurements. Thus, the corresponding time intervals, Δt, measured by the free-falling observer will be longer than those measured by the stationary observer and the corresponding lengths, L, that he measures will be shorter than those measured by the stationary observer. The proper quantities measured by the stationary observer are related to the corresponding quantities measured by the free-falling observer by the equations $L_0 = \gamma L$ and $\Delta t = \gamma \Delta t_0$, respectively. The factor $\gamma = \sqrt{\frac{1}{1-u^2/c^2}}$ where u is the instantaneous relative velocity of the two observers.

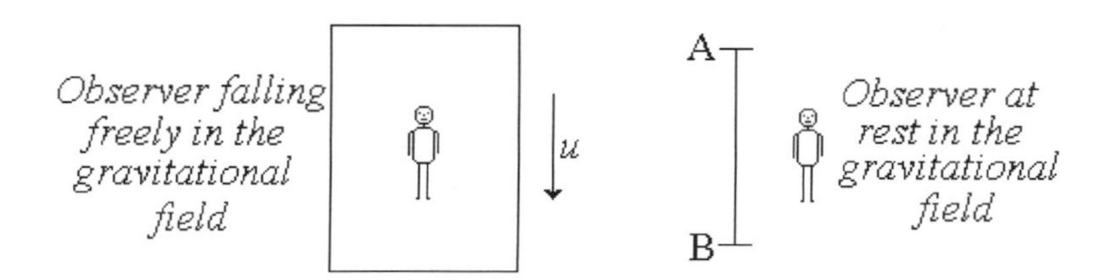

Fig. 1.41. Two observers in a gravitational field, one stationary and the other free-falling. Whereas the stationary observer feels the effects of the field, the one in the free-falling laboratory experiences weightlessness; his situation is equivalent to that of an observer in a region where there is no gravitational field. Both observers measure the length, AB, of an object at rest relative to the stationary observer.

The instantaneous relative velocity of the two observers, u, is the speed of descent of the free-falling laboratory. At infinity, the total potential and kinetic energy of the small laboratory was zero; at the point above the earth's surface where it passes by the stationary observer, its total potential and kinetic energy is $-GMm/r + mu^2/2$. Conservation of energy gives $-GMm/r + mu^2 = 0$ from which

$$u = \sqrt{\frac{2GM}{r}}, \tag{1.111}$$

where m is the mass of the laboratory, M is the mass of the earth and G is Newton's universal gravitational constant.

Substituting $u = \sqrt{2GM/r}$ in $\gamma = \sqrt{\frac{1}{1-u^2/c^2}}$, gives

$$\gamma = \sqrt{\frac{1}{1 - 2GM/rc^2}}. \tag{1.112}$$

Hence, the closer the observers are to the earth's centre, the larger the factor γ and the greater will be the differences between the measurements they make. In general, the stronger the gravitational field or, equivalently, the greater the acceleration of the moving observer, the greater the difference between the measurements made by the stationary and free-falling observers.

The minimum value that r can take, $r_S = 2GM/c^2$, namely, that for which $\gamma = \infty$, is called the *Schwarzschild radius.* It is the critical radius into which a sphere of mass M must be compressed to form a non-rotating black hole. It specifies the boundary of the black hole – the *event horizon* – inside which all matter and radiation are drawn into the singularity at its centre where the known laws of physics break down.

A corollary of the effect of gravity on time measurements is that an observer in a weak gravitational field should measure a lower vibrational frequency than that measured by an observer in a strong gravitational field. For instance, because of the non-uniformity of the earth's gravitational field, spectral lines should be *red-shifted* (displaced towards the lower frequencies) in a laboratory on the top floor of a building as compared to one on the ground floor. The difference is minute but it was detected in 1960 with the help of the Mössbauer effect.[84]

Apart from its effect on measurements of space and time, the gravitational field also affects mass; an object's mass increases, relative to an observer in a gravityless laboratory, the stronger the gravitational field in which it is situated. Consequently, mass cannot be defined without reference to gravity.

The fact that measurements of spatial displacements and of time intervals are affected by the gravitational field in which they are made, provides the first hint of the most radical change brought about by Einstein's theory of general relativity: the notion that gravity is something fundamentally geometrical.

[84]The experiment is described below in Chapter 6.2.

1.5.3 Gravity and Geometry

In its elucidation of the motion of bodies, classical mechanics is totally reliant on the concept of force: whether by virtue of their absence as in inertial reference frames; their presence as when an actual physical interaction takes place; or their mystical summoning-up from absolute space, as though by fiat, *ex nihilo*, in non-inertial systems. Whatever the object's state of motion – at rest, in uniform motion in a straight line or an accelerated motion – forces are invoked in accounting for it. Consider the following three instances:

1. A weightless observer throws an object across his gravityless laboratory, from one side to the other. The object moves in a straight trajectory, no external forces being exerted on it during its flight. The laboratory is an inertial system in which Newton's laws of motion are obeyed (Fig. 1.42a).
2. An observer in an accelerated laboratory, deep in space far from any gravitational field, also throws an object across his laboratory, launching it at right angles to the direction of his acceleration. Despite the fact that here, too, no actual external force was impressed upon it, the object follows a parabolic path. In classical mechanics, the phenomenon is explained by the contrivance of an inertial force – a fictitious force – that is somehow exerted on the object by virtue of it being situated in an accelerated laboratory; it's all a bit metaphysical (Fig. 1.42b).
3. A third observer, at rest in a uniform gravitational field, also throws an object across his laboratory, launching it at right angles to the direction of the field. This object also moves in a parabolic trajectory. To explain this within the framework of his laws of motion, Newton invented a new interaction: gravity (Fig. 1.42c). However, according to the principle of equivalence, gravitational effects are indistinguishable from those of an acceleration. Are we,

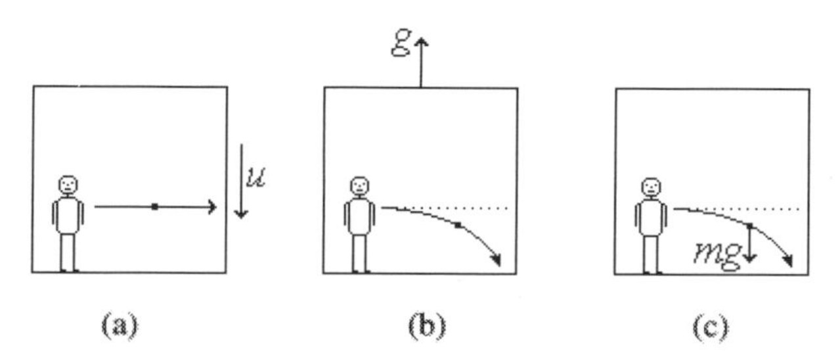

Fig. 1.42. Three observers performing the same experiment. (a) A weightless observer throws an object across his free-falling laboratory; its instantaneous velocity is u. (b) An observer throws an object across his laboratory as it rises with an acceleration g. (c) An observer throws an object across his stationary laboratory in a uniform gravitational field of strength g.

therefore, to conclude, by analogy, that the force of gravity is a metaphysical contrivance too?

General relativity requires that all observers, whether they be in inertial or accelerated reference frames or in gravitational fields, be indistinguishable. However, in the instances described above, it would appear that the first observer can be distinguished from the other two by virtue of the shape of the object's trajectory. Whereas he finds the trajectory to be a straight line, the others find it to be a parabola.

In order for the three observers to be indistinguishable, the straight trajectory observed in the inertial frame must be of the same import for that observer as the parabolic trajectory is for the observers in the non-inertial frames. Einstein realised that, as long as the laws of motion were expressed in terms of forces, this could never be realised; to be truly covariant these laws had to involve no more than just the relative positions and motions of the bodies.

Newton had identified an object's *mass* with the 'quantity of matter' that lies within it: its inertia (*vis inertiæ*) or tendency to persist in a state of rest or uniform motion in a straight line. This innate 'force of matter' is an absolute; an invariant quantity, that is manifested to the same degree whenever an external force is impressed upon the object. It was the concept upon which Newton's first law of motion was predicated. Einstein rejected this and restated the law without recourse to mass or force as follows:

Every object, left alone, moves in a straight line.

At first sight, this might seem to solve nothing, for the trajectories observed by the non-inertial observers still appear to us to be parabolas; they do not look like straight lines. Einstein countered this by noting that we had all been taught to believe, from a young age, that the flat geometry associated with the name of Euclid and employed by Newton in his *system of the world*, correctly describes physical reality. However, during the 19th century, mathematicians had questioned some of the axioms underlying Euclidean geometry and by amending one or more had developed exotic new curved geometries.[85] Moreover, it had been shown that these geometries could be expressed in terms of a mathematical tool called a *tensor* whose relationships are invariant under coordinate transformations; tensors are defined without reference to any particular coordinate system. Vectors are, in fact, simple tensors.

[85]These new geometries replaced Euclid's fifth postulate that through a given point there is only one line parallel to a given line with the postulate that there are an infinite number of lines through the given point parallel to the given line (hyperbolic geometry) or, alternatively, with the postulate that there is no line through the given point parallel to the given line (elliptical geometry).

Einstein suggested that if the tensor forms of these new non-Euclidean geometries were used to depict physical reality, it would be possible to resolve the *parabolas* of the non-inertial observers with the *straight line* of the inertial observer as required by the principle of general relativity and to describe motion without resort to forces, actual, inertial or even gravitational.

A simple way of distinguishing between the various geometries is by reference to the value they assign to the ratio between the circumference of a circle and its radius (Fig. 1.43).

* For circles drawn on the conventional flat surfaces of Euclidean geometry, this ratio equals 2π.
* For circles drawn on the curved spherical surfaces of the so-called *elliptical geometry*, the ratio is less than 2π.
* For circles drawn on the saddle shaped surfaces of *hyperbolic geometry*, the ratio is greater than 2π.

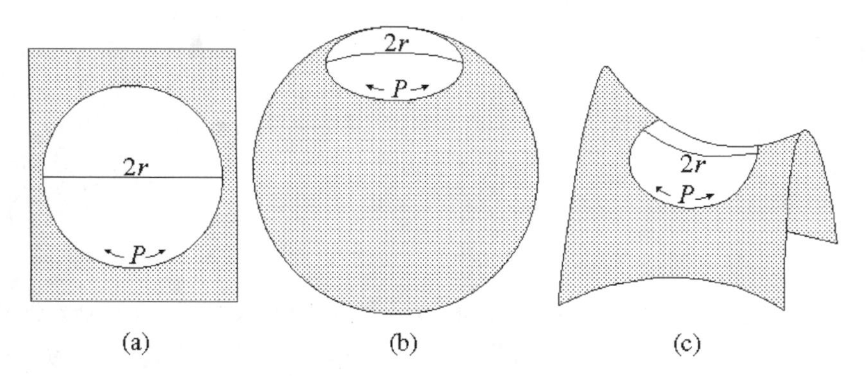

Fig. 1.43. Three different geometries. (a) Euclidean: the ratio, P/r, between the circumference, P, and the radius, r, of a circle drawn on a flat surface is 2π. (b) Elliptical: for a circle drawn on a sphere $P/r < 2\pi$. (c) Hyperbolic: for a circle drawn on a saddle surface $P/r > 2\pi$.

In all geometries, the *straight line* is the line of shortest length that connects two points; mathematicians call these lines *geodesics* or *geodetic lines.* Their shape depends on the particular geometry involved. For example, the geodesics of Euclidean geometry are our familiar straight lines, whereas those of the elliptical geometry are the great circles on a spherical surface.

Geodesics are described by invariant formulae from which all the other invariants of the surface can be derived. Though its shape may change, the line that is a geodesic in one coordinate system will also be one in another. Consequently, by describing motion exclusively in terms of geodesics, Einstein could formulate a new covariant mechanics and theory of gravity.

According to this new mechanics, the trajectories of all bodies through space–time are geodesics. This axiom is consistent with the formulation of Newton's first law of motion given above and enables us to express the laws of mechanics in terms of a real physical quantity – the shortest distance between two points – rather than the spurious concepts of inertial and gravitational forces. Similarly, gravity is associated with the geometry of space–time. A body's presence affects the shape of the space–time in its neighbourhood and that of its geodesics. This, in turn, affects the trajectories of other bodies moving through this space–time. The classical theory of gravitation is applicable in small spaces and in weak fields, where to a close approximation, Euclidean geometry is valid.

Suppose that a circle is drawn around the centre of the earth and two observers set out to measure its circumference and radius. One observer is in a free-falling laboratory – a gravityless laboratory – and the other in a stationary laboratory in the earth's gravitational field (Fig. 1.44). Since the gravitational field is perpendicular to the circle at all points, the two observers agree as to its circumference, P. However, the proper radius, r_0, measured by the stationary observer is greater than the radius, r, measured by the free-falling observer. The free-falling laboratory is equivalent to an inertial reference system in which Newton's laws and Euclidean geometry are valid. Hence, the ratio, P/r, between the circumference of the circle and the radius measured by the free-falling observer is equal to 2π. On the other hand, since $r_0 > r$, the

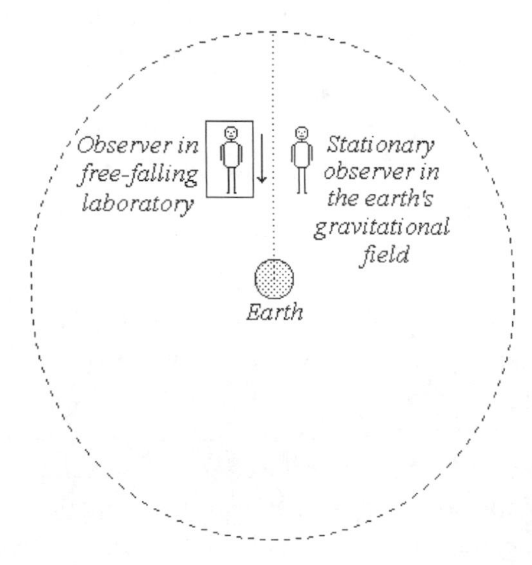

Fig. 1.44. A circle drawn round the centre of the earth. Relative to the falling observer, the circle's radius, r, is shorter than the proper radius, r_0, measured by the stationary observer. The two observers agree about the circle's circumference.

ratio, P/r_0, between the circumference of the circle and the radius measured by the stationary observer is less than 2π. Thus, the geometry appropriate to the stationary observer in the earth's gravitational field is the elliptic geometry. The gravitational field experienced by the stationary observer is a manifestation of the difference between his geometry and that of the free-falling observer.

It can be shown that an observer rotating on a carousel finds that the ratio between the circumference of the circle and its radius is greater than 2π; the geometry applicable to an observer in a rotating laboratory is, therefore, the hyperbolic geometry. All the phenomena classically attributed to centrifugal forces and the Coriolis effect can be elucidated by describing the motions observed from the carousel in terms of geodesics in this non-Euclidean geometry.

We can now return to the precession of the perihelion of the planet Mercury. Kepler's laws, like Newton's theory of gravitation, presume that Euclidean geometry is the right one for describing a planet's motion. However, the planet Mercury is the closest of all to the sun and so it moves in an appreciable gravitational field. The geometry that should be used to describe its motion is the elliptic geometry in which the circumference of a circle (or ellipse) is less than that of a circle (or ellipse) with the same radius in Euclidean space. The periodic time (sidereal period), T, calculated from Kepler's laws is the time it takes the planet to return to a given point in its orbit relative to the sun. In Euclidean geometry this is also the time it takes for the planet to return to a given point in space. But this is not so in non-Euclidean geometry. Because of the elliptic geometry of the space in which it moves, the planet Mercury has to travel a shorter distance than the length of its Keplerian orbit in order to return to a given point in space. Consequently, when, after a time T, the planet returns to its perihelion – the point in its orbit closest to the sun – it has overshot the point in space that was the previous location of its perihelion. As a result of this, the perihelion precesses through space in the direction of the motion of the planet.

Newtonian gravity obeys the principle of superposition, i.e. the field, g, that it predicts for any point is the sum of the fields produced there by all the masses in space: $g = g_1 + g_2 \ldots + g_n$. But in the Einsteinian theory, the shape of space–time which determines the gravitational field at a point is not that predicted by simply adding together the effects of the different masses in space on the geometry at the point. In general relativity, not only do bodies affect the shape of space–time, but their separate effects also act on one another and contribute to the final field. This interaction affects the energy of the gravitational field and is the source of the physical difference between the classical and the relativistic fields.

In the Einsteinian universe, changes in the shape of space–time propagate with the speed of light; waving our hands does not instantaneously move the galaxies. The changes in the shape of space–time are accompanied by energy changes which, it has been suggested, should be observable as *gravity waves.* However, gravitational changes take place over æons and so such waves will be of very low intensity, which makes their detection very difficult.

The world-lines of light rays are geodesics too, whose shape depends on the gravitational field through which the light propagates. Hence, in the narrow Euclidean sense, light does not propagate in a straight line in a gravitational field; light rays cannot be used to draw orthogonal coordinate systems in which Euclidean geometry is valid. Time intervals and lengths, too, depend on the gravitational fields in which they are measured.

Consequently, it will come as no surprise to learn that according to general relativity, the velocity of light *in vacuo* depends on the strength of the gravitational field through which it propagates. The speed of light *in vacuo* is not the supreme law of nature that special relativity supposes it to be. As Einstein wrote:

> '*The principle of the constancy of the velocity of light in vacuo, which is one of the two postulates of the special theory of relativity, does not have an unlimited validity. Light rays can only bend if the velocity of light changes from place to place... The results of special relativity are valid only so long as the effects of gravitational fields on the phenomena can be ignored.*'[86]

[86] A. Einstein (1920) *Relativity.* Translated by R.W. Lawson, London: Methuen & Co., p. 76.

Appendices to Part One

1.6.1 Velocity Addition in Special Relativity

Consider an object moving relative to the two observers, S and S′. The components of its velocity relative to the observer S are

$$v_x = \frac{dx}{dt} \quad v_y = \frac{dy}{dt} \quad v_z = \frac{dz}{dt}, \tag{1.113}$$

and relative to the observer S′

$$v'_{x'} = \frac{dx'}{dt'} \quad v'_{y'} = \frac{dy'}{dt'} \quad v'_{z'} = \frac{dz'}{dt'}. \tag{1.114}$$

In general, the velocity components measured by S can be related to the corresponding components measured by S′, by making appropriate substitutions in the above definitions. For instance, to express the S components in terms of the S′ components, expressions for the infinitesimals dx, dy, dz and dt in terms of dx', dy', dz' and dt' are substituted in their place in the S definitions. The relevant expressions are obtained from the derivatives of the equations of the Lorentz transformation:

$$\begin{aligned} dx' &= \gamma(dx - udt) & dx &= \gamma(dx' + udt') \\ dy' &= dy & dy &= dy' \\ dz' &= dz & dz &= dz' \\ dt' &= \gamma\left(dt - \frac{udx}{c^2}\right) & dt &= \gamma\left(dt' + \frac{udx'}{c^2}\right) \end{aligned} \quad . \tag{1.115}$$

For example, to obtain an expression for the component $v'_{x'}$ in terms of v_x:

$$\begin{aligned} v'_{x'} &= \frac{dx'}{dt'} \\ &= \frac{\gamma(dx - udt)}{\gamma\left(dt - {}^{udx}\!/_{c^2}\right)}, \end{aligned} \tag{1.116}$$

and substituting $dx = v_x dt$ from Eq. (1.113) gives

$$v'_{x'} = \frac{\gamma(v_x dt - udt)}{\gamma\left(dt - {}^{uv_x dt}\!/_{c^2}\right)}$$

$$= \frac{v_x - u}{1 - {}^{uv_x}\!/_{c^2}}. \tag{1.117}$$

Similarly, for the other two velocity components,

$$v'_{y'} = \frac{dy'}{dt'}$$

$$= \frac{dy}{\gamma\left(dt - {}^{uv_x}\!/_{c^2}\right)} \tag{1.118}$$

$$= \frac{v_y}{\gamma\left(1 - {}^{uv_x}\!/_{c^2}\right)}$$

and

$$v'_{z'} = \frac{v_z}{\gamma\left(1 - {}^{uv_x}\!/_{c^2}\right)}. \tag{1.119}$$

The corresponding equations which give the velocity components relative to the observer S in terms of those of S′ are

$$v_x = \frac{v'_{x'} + u}{1 + {}^{uv'_{x'}}\!/_{c^2}} \tag{1.120}$$

$$v_y = \frac{v'_{y'}}{\gamma\left(1 + {}^{uv'_{x'}}\!/_{c^2}\right)} \tag{1.121}$$

$$v_z = \frac{v'_{z'}}{\gamma\left(1 + {}^{uv'_{x'}}\!/_{c^2}\right)}. \tag{1.122}$$

Equations (1.118) to (1.122) are the velocity addition rules of the special theory of relativity. When the relative velocity, u, of the two observers is much less than the speed of light *in vacuo*, c, these equations reduce to the classical velocity addition rules – Eqs (1.45), (1.46) and (1.47).

In the case where the 'object', whose velocity is measured by the observer S′ is a light ray, i.e. $v'_{x'} = c$, the velocity that the observer S will measure is

$$v_x = \frac{v'_{x'} + u}{1 + {}^{uv'_{x'}}\!/_{c^2}} = \frac{c + u}{1 + {}^{uc}\!/_{c^2}} = c.$$

This result confirms that these velocity addition rules conform to the principle of the constancy of the speed of light.

Similarly, the rules conform to the requirement that no object or signal can move at a speed greater than c. Thus, whereas in classical mechanics, a velocity of $0.9c + 0.5c = 1.4c$ is needed to overtake an object moving at $0.9c$ with a relative velocity of $0.5c$, according to the relativistic Eq. (1.120), if $u = 0.9c$ and $v'_{x'} = 0.5c$, the required velocity is

$$v_x = \frac{v'_{x'} + u}{1 + \left(uv'_{x'}/c^2\right)} = \frac{0.5c + 0.9c}{1 + \left(0.9c \cdot 0.5c/c^2\right)} = 0.9655c,$$

i.e. the required velocity is less than c.

1.6.2 The Kinetic Energy of a Particle in Special Relativity

The kinetic energy, K, of a particle moving parallel to the X-axis is equal to the work done in accelerating the particle from rest by the force, F_x, acting on the particle parallel to the X-axis:

$$K = \int_0^x F_x \cdot dx. \tag{1.123}$$

Substituting the relevant relativistic definition of force, $F_x = \frac{dp_x}{dt} = \frac{m_0 d(\gamma u)}{dt}$, from Eq. (1.88) gives

$$K = m_0 \int_0^x \frac{d(\gamma u)}{dt} dx. \tag{1.124}$$

The accelerated particle acquires a velocity $u = dx/dt$ parallel to the X-axis. In terms of this velocity, the particle's kinetic energy is given by

$$K = m_0 \int_0^u u d(\gamma u). \tag{1.125}$$

Applying the rule, $\int xdy = xy - \int ydx$, for integrating by parts, we obtain

$$K = m_0 \gamma u^2 - m_0 \int_0^u \gamma u du. \tag{1.126}$$

Because u is always less than c, we can use the following substitution in order to evaluate the integral:

$$u = c \sin \alpha, \tag{1.127}$$

where c is the speed of light *in vacuo*. This, in turn, produces the following additional relationships:

$$du = c \cos \alpha . d\alpha \quad \gamma = \frac{1}{\cos \alpha}. \tag{1.128}$$

Substituting these relationships in Eq. (1.112) gives

$$K = \frac{m_0 c^2 \sin^2\alpha}{\cos\alpha} - m_0 \int_0^{\alpha} \frac{c \sin\alpha(-c\cos\alpha)}{\cos\alpha} \cdot d\alpha$$

$$= \frac{m_0 c^2 \sin^2\alpha}{\cos\alpha} + m_0 c^2 \int_0^{\alpha} \sin\alpha \cdot d\alpha. \qquad (1.129)$$

Evaluating the integral gives

$$K = \frac{m_0 c^2 \sin^2\alpha}{\cos\alpha} + m_0 c^2 [\cos\alpha]_0^{\alpha}$$

$$= \frac{m_0 c^2 \sin^2\alpha}{\cos\alpha} + m_0 c^2 (\cos\alpha - 1)$$

$$= m_0 c^2 \left(\frac{\sin^2\alpha}{\cos\alpha} + \cos\alpha - 1 \right). \qquad (1.130)$$

Simplifying this result gives

$$K = m_0 c^2 \left(\frac{\sin^2\alpha + \cos^2\alpha - \cos\alpha}{\cos\alpha} \right)$$

$$= m_0 c^2 \left(\frac{1}{\cos\alpha} - 1 \right). \qquad (1.131)$$

Returning to the original format by making the substitution $1/\cos\alpha = \gamma$ gives the following relativistic expressions for the particle's kinetic energy:

$$K = m_0 c^2 (\gamma - 1)$$

$$= c^2 (\gamma m_0 - m_0)$$

$$= c^2 (m - m_0). \qquad (1.132)$$

1.6.3 The Total Energy of a Particle

The total energy, E, of a particle can be expressed in terms of its rest-mass, m_0, and its linear momentum, p, through a synthesis of the following four fundamental relationships:

$$\gamma = \sqrt{\frac{1}{1 - u^2/c^2}}, \qquad (1.61)$$

$$m = \gamma m_0, \qquad (1.87)$$

$$E = mc^2 \text{ and} \qquad (1.100)$$

$$p = mu. \qquad (1.86)$$

Algebraic elimination of the particle's relativistic mass, m, its velocity, u, and the factor γ from these equations produces the relationship

$$E^2 = m_0^2 c^4 + p^2 c^2. \tag{1.133}$$

1.6.4 The Transformation of Force

The components of a force acting on a particle relative to the inertial observers S and S′ are

$$\begin{aligned} F_x &= \frac{dp_x}{dt} \quad F_y = \frac{dp_y}{dt} \quad F_z = \frac{dp_z}{dt} \\ F'_{x'} &= \frac{dp'_{x'}}{dt'} \quad F'_{y'} = \frac{dp'_{y'}}{dt'} \quad F'_{z'} = \frac{dp'_{z'}}{dt'}. \end{aligned} \tag{1.134}$$

In the special case in which the particle is momentarily at rest relative to the observer S, the force he measures will be the *proper force.*

In general, the S components of the force can be converted in those of S′, or *vice versa,* by making the appropriate substitutions in the above definitions. For example, to convert the S′ definitions into those of S, the appropriate expressions for the infinitesimals $dp'_{x'}$, $dp'_{y'}$, $dp'_{z'}$ and dt' in terms of dp_x, dp_y, dp_z and dt are substituted in the definitions of the S components. These expressions are obtained by taking the derivatives of the following momentum and time transformation equations,

$$p'_{x'} = \gamma\left(p_x - \frac{uE}{c^2}\right) \quad p'_{y'} = p_y \quad p'_{z'} = p_z \quad t' = \gamma\left(t - \frac{ux}{c^2}\right).$$

In the special case where the particle is momentarily at rest relative to the observer S, i.e. when $dx = 0$ and $dE = 0$, the relevant substitutions are

$$dp'_{x'} = \gamma\, dp_x \quad dp'_{y'} = dp_y \quad dp'_{z'} = dp_z \quad dt' = \gamma\, dt. \tag{1.135}$$

Making these substitutions in the definitions of the S′ components gives

$$\begin{aligned} F'_{x'} &= \frac{dp'_{x'}}{dt'} & F'_{y'} &= \frac{dp'_{y'}}{dt'} & F'_{z'} &= \frac{dp'_{z'}}{dt'} \\ &= \frac{\gamma\, dp_x}{\gamma\, dt} & &= \frac{dp_y}{\gamma\, dt} & &= \frac{dp_z}{\gamma\, dt} \\ &= F_x & &= \frac{F_y}{\gamma} & &= \frac{F_z}{\gamma}. \end{aligned} \tag{1.136}$$

Part One: Questions, Exercises and Problems

Questions

1. Why must the equations which describe the Doppler effect in light be different from those which describe the effect in sound?
2. The light spot formed on a television screen by the stream of electrons aimed at it from inside the tube, can sweep across the screen at speeds greater than the speed of light *in vacuo*. Does this contradict the principles of special relativity?
3. Show that if the speed of light was infinite, the Lorentz transformation would reduce to the Galilean transformation.
4. Since the speed of light is finite, it is evident that 'the world as we see it at this moment' and 'the world as it is at this moment' are not the same thing. Clarify this and elucidate the differences between the concept of 'the present' in classical and relativistic mechanics.
5. Show that when the velocity of a particle is much less than the speed of light *in vacuo*, the relativistic expression for its kinetic energy, $K = (m - m_0)c^2$, reduces to the classical relationship $K = (mu^2/2)$.
6. A hot object cools on the pan of a delicate balance; does the balance show any change in the object's weight? When a stretched spring is released and returns to its natural length is its mass conserved? Is mass conserved in chemical reactions? When a condenser is charged?
7. 'I don't say that matter and space are the same thing. I only say, there is not space, where there is no matter; and that space itself is not an absolute reality.' Gottfried Leibniz (1646–1716). Contrast this approach to that of Newton. With which of them might Einstein have agreed?
8. It is related that in his youth Einstein asked, 'What would I see in a mirror that I held in front of me if I ran forward at the speed of light?' How would you answer this question?
9. How does the shape of a sphere change as its linear velocity increases?
10. A rigid body is defined as one which shows no deformation when a force is impressed on it. Is the existence of such a body consistent with special relativity?
11. Light travels through water at a velocity of about $2.25 \cdot 10^8\,\mathrm{m/s}$. Could a particle move through water at a velocity greater than this?
12. 'From Einstein's point of view Ptolemy and Copernicus are equally right.' Max Born. Comment on this statement.
13. 'Let us then take a leap off a precipice so that we may contemplate Nature undisturbed.' Arthur Eddington. What did this English mathematician and physicist mean by this?
14. Relative to an observer on the earth, the tangential velocity of the fixed stars in their nightly passage across the sky, is given by the equation $v = 2\pi r/T$, where r is the distance to the star and T is the length of an earth day. This velocity will be greater than that of the speed of light *in vacuo*, c, for all the stars at a distance from the earth of more than $cT/2\pi = 365/2\pi \approx 60$ light years. The vast majority of the stars are further away than this. How can this be reconciled with the axiom of special relativity that no object can move with a velocity greater than c relative to any observer?

Exercises

1. In laboratory S′, which is moving with a velocity of $0.75c$ relative to laboratory S, a lamp flashes with a frequency of 10 Hz. What will be its frequency according to an observer in laboratory S?
2. The proper periodic time of a pendulum is 2 s. What will be its period relative to an observer moving at a velocity of $0.95c$ relative to it?
3. The average proper life-time of muons – unstable high-energy particles – is 2.2 μs. Cosmic rays contain muons and measurements made from the earth indicate an average life-time for these muons of 16 μs. Calculate the velocity of the cosmic ray muons relative to the earth.
4. The proper life-time of a particle is 100 ns.

   a. How long will the particle exist if it is moving at a velocity of $0.9c$?
   b. How far does the particle move relative to the observer at this velocity before disintegrating?
   c. How far did the particle move in its own reference frame?

5. Observer S determines that an event occurred at a point with coordinates $x =$ 100 km and $t = 200\,\mu$s. Observer S′ moves with a velocity of $0.95c$ relative to observer S. Assuming that $x = x'$ when $t = t' = 0$, find the coordinates of the event according to the observer S′.
6. The length of a spaceship on earth is 2,500 m. However, when it is in motion in space its length relative to an observer on earth is 2,350 m. Calculate the spaceship's velocity.
7. A spaceship passes the earth on its way to the moon at a velocity of $0.7c$. The proper distance from the earth to the moon is 380,000 km.

   a. According to clocks on the earth, how long does it take for the spaceship to reach the moon?
   b. What is the distance from the earth to the moon according to an observer on the spaceship?
   c. According to the clocks on the spaceship, how long did it take to reach the moon?

8. Burning 1 kg of petroleum releases about 10^8 J of energy. How far could a vehicle which conventionally consumes 1 kg of petroleum every 20 km travel were it able to exploit the total energy content of the petroleum?
9. For an electron moving at a velocity of $0.9c$ calculate:

   a. its total energy;
   b. its kinetic energy;
   c. its linear momentum.

10. Calculate the velocity of a particle which has a kinetic energy of 10 MeV if the particle is:

    a. an electron;
    b. a proton.

11. At what velocity does the kinetic energy of a particle equal its rest energy?

12. Two protons in a particle accelerator approach one another head-on. Relative to a stationary observer the velocity of the first proton is $0.75c$ and of the second $0.85c$. What is the velocity of the first proton relative to the second?
13. A body, moving relative to a stationary observer with a velocity of $0.8c$, emits a particle in the direction of its motion at a velocity of $0.7c$. Calculate the particle's velocity relative to the stationary observer.

Problems

1. At what velocity is the linear momentum of a particle equal to m_0c? What are the total and kinetic energies of the particle under these circumstances?
2. How much work must be expended to accelerate an electron from

   a. $0.28c$ to $0.29c$?
   b. $0.98c$ to $0.99c$?

3. Two identical particles, each with a rest-mass of $3\,\mathrm{MeV/c^2}$, collide elastically head-on. Before the collision the linear momentum of each particle was $1.25\,\mathrm{Mev/c}$. A new particle is formed in the collision. Calculate the rest-mass of this particle.
4. A stationary particle divides spontaneously into two smaller particles with rest-masses m_1 and m_2. The first particle moves away with velocity $v_1 = 0.75c$. If the ratio between the rest-masses is $m_1/m_2 = 0.75$, calculate the velocity of the second particle.
5. The radius of the circular path of an electron moving with a velocity v at right angles to a magnetic field B is given classically by:

$$r = \frac{mv}{eB}.$$

   This equation is valid so long as the electron's velocity is substantially less than c.

   a. How can the equation be used in the case of high-speed electrons?
   b. Calculate the radius of the path of a 10 MeV electron moving at right angles to a magnetic field of strength 2 T.

6. Find the electric potential difference required to accelerate electrons to the speed of light *in vacuo* according to classical mechanics. To what velocity would the electrons actually be accelerated by this potential?
7. When a certain particle, which carries unit electric charge, is accelerated by a potential difference of $5 \cdot 10^8\,\mathrm{V}$ and enters a magnetic field of strength 1 T, it moves in a circular path of radius 3.62 m. Calculate the particle's rest-mass.
8. Show that the rest-mass of a particle is related to its kinetic energy, K, and linear momentum, p, by the equation

$$m_0 = \frac{p^2c^2 - K^2}{2Kc^2}.$$

9. A stationary observer sees two spaceships, A and B, moving towards him from opposite directions at speeds of $0.8c$ and $0.9c$ respectively. Calculate:

   a. According to an observer on A, at what velocity is he approaching the stationary observer?

b. According to an observer on A, at what velocity is he approaching the spaceship B?
c. According to an observer on B, at what velocity is he approaching the stationary observer?
d. According to an observer on B, at what velocity is he approaching the spaceship A?

10. An event occurs at a point whose coordinates are (x_1, t_1) in the S system and (x_1', t_1') in the S′ system. This event causes the occurrence of a second event at the point (x_2, t_2) in the S system. Show that no matter what the motion of the S′ system, an observer in it can never see the second event (the consequence) before the first (the cause), i.e. that whatever the motion $t_2' - t_1' > 0$.

Part Two Quantum Theory

Despite its revolutionary notions concerning the nature of space and time, Einstein's special theory of relativity did not require any changes in Maxwell's electromagnetic theory. Even the realisation that there is no æther, astounding as it was at the time, could be taken in its stride. Notwithstanding Maxwell's own conviction that an æther did exist, it was not a crucial element of his theory.

However, the non-existence of the æther did raise profound questions about the actual physical nature of electromagnetic waves. For if there is no æther, what medium carries the energy transported by an electromagnetic wave? How can empty space carry energy? How does sunlight reach us across 150,000,000 km of empty space? And what reality can we ascribe to a wave that propagates through empty space? There was nothing in special or general relativity that could answer these questions.

The existence or otherwise of the æther was not the only troubling issue faced by electromagnetic theory towards the end of the 19th century. Systematic quantitative studies of the emission and absorption of light and heat carried out at the time had uncovered phenomena that were totally inexplicable in the framework of Maxwell's theories. There appeared to be a fundamental flaw in the wave theory of light.

The new theory – *quantum theory* – that was developed in the first decades of the 20th century to explain these phenomena, provided an answer to the question of the physical nature of electromagnetic waves. Subsequently, this same theory furnished the conceptual framework for a new understanding of the structure of matter and of its chemical and physical properties.

However, like the theory of relativity before it, quantum theory also requires fundamental changes in our intuitive 'common-sense' grasp of reality. This time, it is not the equivalence or otherwise of our observations that is questioned but their very nature and that of their underlying reality.

What do we actually mean when we say: 'the mass of the electron is $9.31 \cdot 10^{-31}$ kg'? That a particle's mass is a property that adheres to it; something it actually *has*? Or is it the result of a measurement, which exists only through the observer's consciousness?

In general, does a measurement reveal the particular result it does because that is the value the quantity actually *had* at the moment it was measured? Or did the quantity only exist *potentially* before being brought into being by the act of measurement? And what if the potentiality allows for more than one possible outcome of a measurement? How are the various possibilities disentangled when a measurement is actually made and just a single outcome is obtained?

We shall be concerned with these and many similar esoteric questions during our investigation of quantum theory.

Chapter 2.1

The Quantum Hypothesis

2.1.1 Radiators and Radiation

The energy emitted and absorbed by material objects in the form of electromagnetic waves is generally termed *electromagnetic radiation* or simply *radiation.*

Radiation is emitted at the expense of the energy of the radiating object; absorbed radiation adds to the energy of the absorbing object.

Any specific distribution of electromagnetic radiation, such as the colours of the rainbow, is termed a *spectrum.* If the distribution comprises a continuous region of frequencies or wavelengths, it is a *continuous spectrum*; if it comprises a series or group of discrete frequencies, it is a *line spectrum.*

The distribution of the electromagnetic radiation emitted by a body that acts as a source of radiation is its *emission spectrum* (Fig. 2.1a). The rainbow is the sun's emission spectrum in the visible region.

The distribution of the electromagnetic radiation transmitted by an absorbing medium placed in the path of radiation that exhibits a continuous spectrum, is called an *absorption spectrum* (Fig. 2.1b). When viewing a rainbow through a coloured filter, certain colours are seen to be missing; they are the ones absorbed by the filter. The spectrum of the transmitted light is the filter's absorption spectrum.

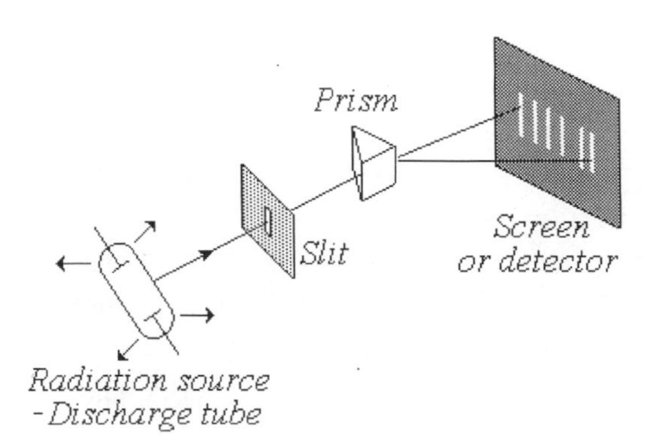

Fig. 2.1a. Viewing an emission spectrum in the visible range. The light emitted from the source is dispersed into its components by a prism or diffraction grating. These components (colours) illuminate the screen or detector and constitute the source's emission spectrum.

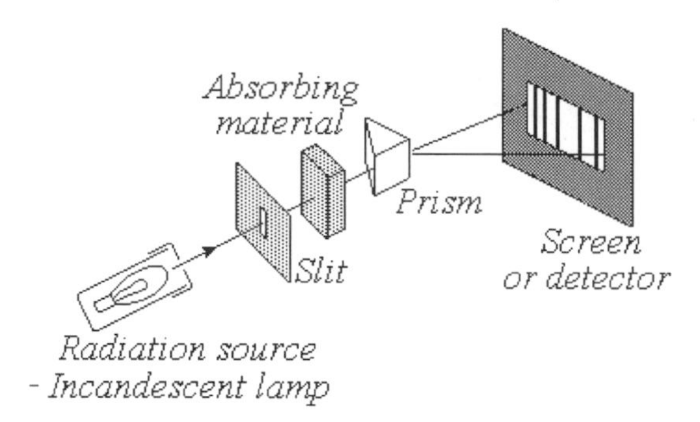

Fig. 2.1b. Viewing an absorption spectrum in the visible range. Light, such as that emitted by an incandescent lamp which exhibits a continuum of frequencies, is passed through a sample of the material whose absorption spectrum is being investigated. The frequencies absorbed by the sample appear as dark lines or bands on the continuous spectrum projected onto the screen.

In general, the electromagnetic radiation emitted and absorbed by a body depends on two factors:

1. its temperature;
2. the nature of the body and of the material from which it is made.

Quantum theory evolved from the attempts to find the correct functional relationship between these factors and the frequencies and intensities of the electromagnetic radiation that material bodies emit and absorb.

2.1.2 Thermal Radiation

The electromagnetic radiation that bodies emit and absorb by virtue of their temperature is called *thermal or heat radiation.* The spectrum of heat radiation is a continuous one that comprises all the wavelengths (or frequencies) in a specific region of the electromagnetic spectrum. A plot of the intensity of the heat radiation emitted by an object as a function of the radiation's wavelength – the *intensity spectrum* – resembles a statistical distribution (Fig. 2.2).

The wavelengths of the heat radiation emitted from a body at a temperature below 1,000 K are in the invisible infra-red region of the electromagnetic spectrum: $\lambda > 1.2\,\mu$m. As the temperature rises, radiation of lower wavelength appears in the spectrum. At about 1,200 K the spectrum includes wavelengths in the visible red, $\lambda_{Red} \approx 0.7\,\mu$m. Raising the temperature further introduces still shorter wavelengths and the colour of a hot glowing object changes accordingly, first to yellow and finally to white. The temperature of the element in an incandescent lamp is almost 3,000 K.

In 1792, Prévost[1] proposed that all bodies radiate heat continually at a rate that depends on their temperature and, likewise, absorb heat at a rate that depends on the temperature of their surroundings. Hence, an object placed

[1]Pierre Prévost 1751–1839, Swiss physicist.

in surroundings whose temperature is higher than its own is irradiated by a greater flux of heat energy than it emits. The object absorbs the difference and its temperature rises accordingly until the flow of heat energy it emits exactly balances that it absorbs from its surroundings (convection and conduction having ceased). From that point on, the temperature of the object remains constant, equal to that of its surroundings. The object is in *thermal equilibrium*, absorbing heat from its surroundings at the same rate as it loses it by radiation.

Heat radiation is formed inside a body by the random thermal motions of its atoms and molecules. On its way to the surface, this radiation is repeatedly absorbed and re-emitted by the body's particles. In this process, the original character of the radiation is totally obliterated. Hence, irrespective of the identity of the object's atoms and molecules, as a result of this smoothing out process, the intensity spectrum of the heat radiation inside every object at a given temperature is the same.

However, not all the heat radiation reaching the object's surface from the inside is emitted, just as not all of that incident on it from outside is absorbed. The flux of heat radiation actually emitted or absorbed by an object depends on the nature of its surface. Different kinds of surfaces absorb and emit heat radiation to a different extent. Black surfaces are good absorbers, whereas polished or white surfaces are not.

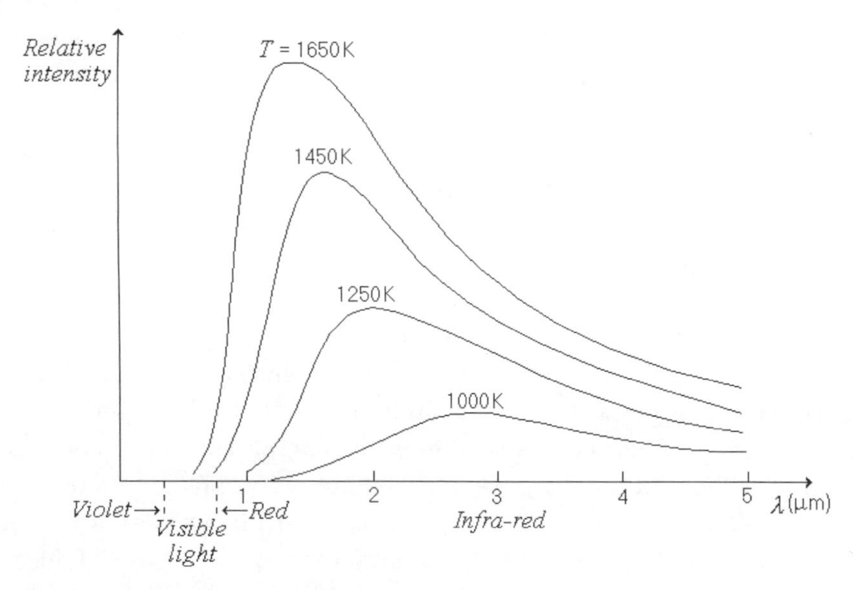

Fig. 2.2. The intensity spectrum of the heat radiation emitted by an object at different temperatures. As the temperature rises the spectrum includes shorter wavelengths. Likewise, the wavelength of maximum intensity – the wavelength of the dominant colour of the radiation – decreases with rising temperature.

From Prévost's *theory of exchanges* it follows that good absorbers must be equally good emitters, and poor absorbers must be equally poor emitters. Were this not so, an object that was a good absorber of heat radiation but only a poor emitter, could not maintain thermal equilibrium with its surroundings. It would absorb too much heat radiation, emit too little and so spontaneously warm up to a temperature above that of its surroundings. But this would be contrary to our common experience. After reaching thermal equilibrium, objects do not spontaneously raise their temperature above that of their surroundings; nor do they spontaneously lower it. Evidently, at thermal equilibrium, all objects emit heat radiation that is equivalent in its quality and quantity to that which they absorb.

2.1.3 Black-body Radiation

Common materials and objects do not absorb all the radiation incident upon them; they are not perfect absorbers of radiation. Nevertheless, we can imagine a perfect absorber, an ideal body which does absorb all the electromagnetic radiation that strikes it, whatever its wavelength or intensity. Because such a body would appear black in whatever light it is illuminated, it is called a *black-body*.

Since, by definition, a black-body is a perfect absorber, it must also be a perfect emitter, i.e. it must be able to emit radiation of every wavelength at any intensity. The heat radiation emitted by a black-body is called *black-body radiation*.

No substance is a perfect black-body, though soot, which absorbs some 95% of the visible and infra-red radiation incident upon it, closely approximates to one in this range. However, by making a small hole in the wall of a hollow object we can construct a device which, to all intents and purposes, absorbs all the radiation incident upon it. The device works somewhat like a fly-trap. Any radiation coming upon the hole from outside will pass through it, enter the cavity and be trapped inside (Fig. 2.3). The hole is thus a perfect absorber of radiation; it is a black-body. By definition, it will also be a perfect emitter of radiation. Any radiation generated in the cavity that happens upon the hole will escape without hindrance, irrespective of its wavelength[2] or intensity.

Suppose that a hollow object with a small aperture in one of its walls is kept at a constant temperature. Every point on the inside surface of the cavity

[2]Strictly speaking, on condition that the hole is substantially larger than the wavelength of the radiation. Microwaves cannot pass through the perforations in the metal plate covering the window of a microwave oven because of their relatively long wavelength.

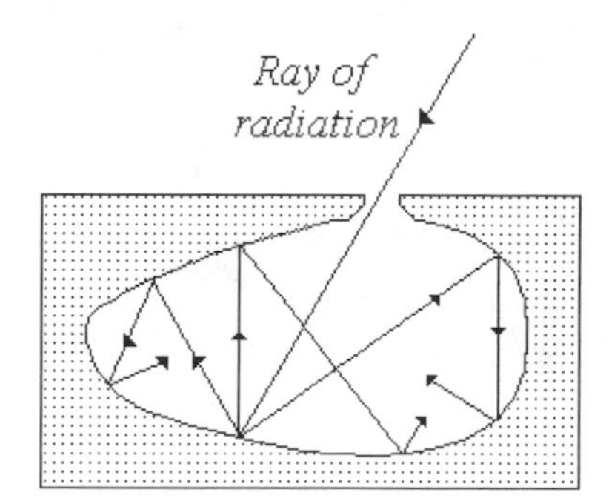

Fig. 2.3. A hollow object with a small aperture in one of its walls. Any radiation that enters through the hole will be trapped inside.

will be in thermal equilibrium with all the other points on the surface and heat radiation of the same quality and quantity will be emitted and absorbed by each point, irrespective of the material from which the inside surface of the cavity is made. The cavity will be filled by electromagnetic radiation that comprises all the wavelengths of the heat radiation characteristic of the object's temperature, each at its appropriate intensity. Any of this *cavity radiation* that happens upon the aperture from inside will escape through it unhindered. Viewed from outside, this representative sample of the cavity radiation will be the black-body radiation characteristic of the particular temperature.

Experimental investigations of black-body radiation conducted towards the end of the 19th century revealed the following three important features:

1. The intensity of the heat radiation emitted from a black-body is greater than that emitted by any other body at the same temperature (Fig. 2.4).
2. The energy, E, radiated each second from each unit area of the surface of a black-body is directly proportional to the fourth power of the body's absolute temperature, T:

$$E = \sigma T^4 \tag{2.1}$$

   where the proportionality constant $\sigma = 5.672 \cdot 10^{-8}\mathrm{W} \cdot \mathrm{m}^{-2} \cdot \mathrm{K}^{-4}$. This rule is known as the *Stefan*[3]*–Boltzmann*[4] *law.*
3. The absolute temperature, T, of a black-body is inversely proportional to the wavelength *in vacuo*, $\lambda_{\max}$, of the radiation it emits with the greatest intensity (Fig. 2.5):

$$T\lambda_{\max} = 0.29 \cdot 10^{-2}\mathrm{m} \cdot \mathrm{K}. \tag{2.2}$$

This rule is known as *Wien's*[5] *law.*

[3]Josef Stefan 1835–1893, Austrian physicist.
[4]Ludwig Boltzmann 1844–1906, Austrian physicist.
[5]Wilhelm Wien 1864–1928, German physicist who was awarded the Nobel prize in 1911 for his discovery of the law of heat radiation.

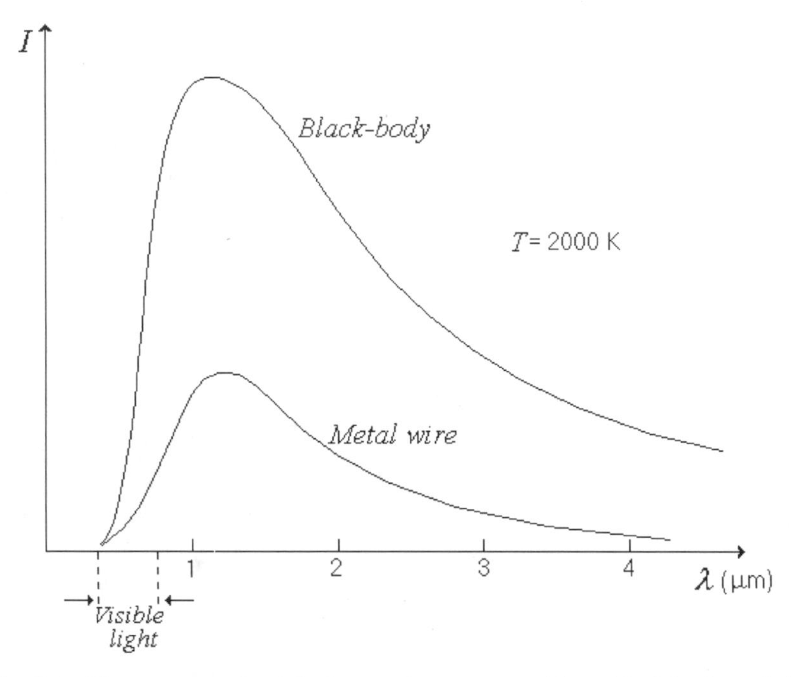

Fig. 2.4. The intensity, I, of the heat radiation emitted by a black-body and the filament in an incandescent lamp at a temperature of 2,000 K as a function of the wavelength, λ, of the radiation.

The best physical evidence for the 'big bang' hypothesis in cosmology is the radio frequency radiation that appears to be black-body radiation corresponding to a temperature of 3 K, discovered coming from outer space. The intensity of the radiation is almost completely uniform in all directions, which suggests that the universe itself is its source. According to the hypothesis, the universe arose out of a point singularity about 15 to 20 billion years ago. At the universe's beginning, when its temperature was extremely high, the radiation and elementary particles were in a kind of dynamic equilibrium, in which particles were unceasingly changing into radiation and the radiation back into particles. Under these conditions, neutral atoms and molecules could not form. With the passage of time, the universe expanded and cooled and when its temperature had dropped to 3,000 K, normal matter was first formed. At this stage the synergetic link between radiation and matter was broken and the radiation, which was black-body radiation corresponding to a temperature of 3,000 K, began to propagate through the expanses of space. It is hypothesised that the ubiquitous 3 K black-body radiation is the residue of the radiation released when matter and radiation divorced, the temperature of the radiation having fallen a thousand-fold since then as a result of the subsequent further expansion and cooling of the universe.

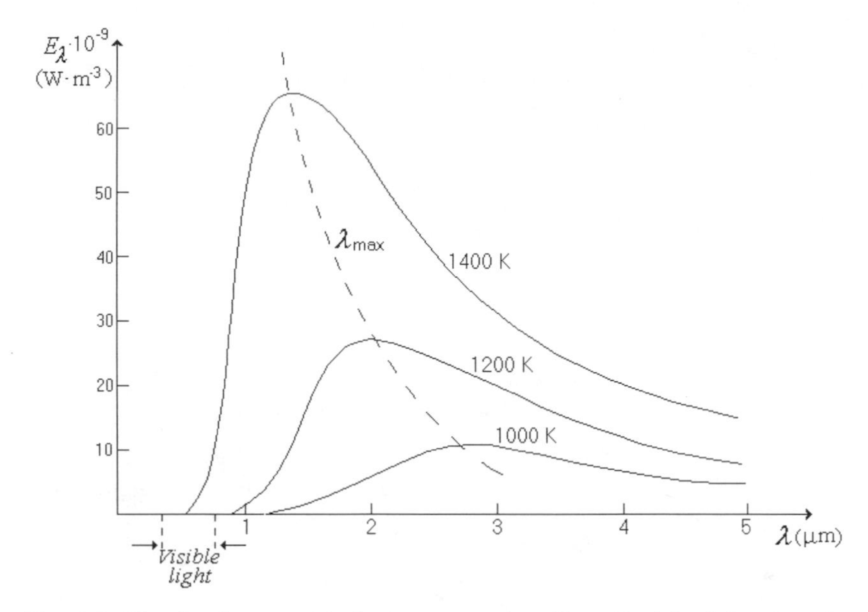

Fig. 2.5. The distribution of the heat energy radiated by a black-body between the various wavelengths at a number of temperatures. E_λ is defined such that $E_\lambda \cdot \Delta\lambda$ is the energy radiated each second at the wavelengths between λ and $(\lambda + \Delta\lambda)$ from a unit area of the body's surface. The area under each curve gives the total energy radiated each second from each unit area at the particular temperature.

Example 2.1: Wien's Law

Assuming they are black-bodies, find the wavelength of the heat radiation of maximum intensity emitted by:

1. an object at room temperature, say 17°C;
2. a hot iron, say 150°C;
3. a hot star with a temperature 50,000°C.

Calculation: By Wien's law – Eq. (2.2) – the wavelength of the heat radiation of maximum intensity emitted by a black-body is:

$$\lambda_{\text{max}} = \frac{0.29 \cdot 10^{-2}}{T}\ \text{m}.$$

This gives for the object at room temperature, $T = 290\,\text{K}$:

$$\lambda_{\text{max}} = \frac{0.29 \cdot 10^{-2}}{290} = 10^{-5}\ \text{m}$$

and for the hot iron, $T = 423\,\text{K}$:

$$\lambda_{\text{max}} = \frac{0.29 \cdot 10^{-2}}{423} = 6.8 \cdot 10^{-6}\ \text{m}.$$

Both of these wavelengths are in the infra-red region.

For the hot star $T = 50,273\,\text{K}$:

$$\lambda_{\text{max}} = \frac{0.29 \cdot 10^{-2}}{50,273} = 5.8 \cdot 10^{-8}\,\text{m}.$$

This wavelength is in the far ultra-violet.

Example 2.2: Black-body Radiation and Astronomy

An impressive example of the importance of black-body radiation is its application to astronomy (and cosmology). For example, the temperature of the surface of a star can be calculated on the assumption that the light it emits is thermal radiation and that the star is a black-body; this is correct to a close approximation. We shall illustrate this by calculating the temperature of the sun's surface in two different ways.

First calculation: In this calculation, the only data we will require are the average temperature of the earth's surface, approximately 290 K, and the angle that the sun's disc subtends at the earth, $\theta = 32'$ (Fig. 2.6).

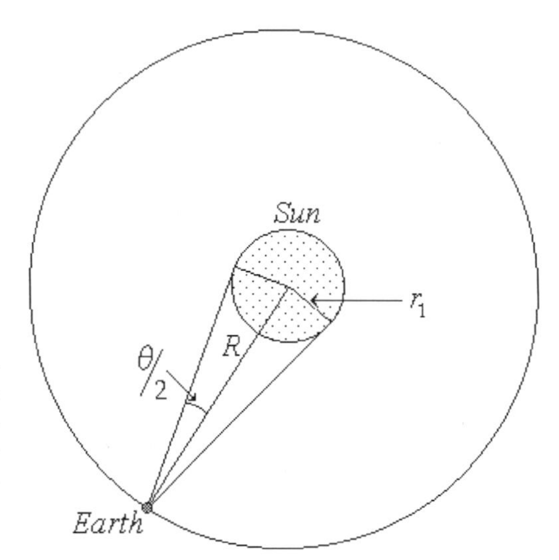

Fig. 2.6. The earth's orbit round the sun (not to scale). The sun's disc subtends an angle $\theta = 32'$ at the earth. Expressed in radians, the angle θ is the ratio of the diameter of the sun's disc, $2r_1$, to the distance from the earth to the sun, R: $\theta = 2r_1/R = 32/60 \cdot 2\pi/360 = 9.31 \cdot 10^{-3}$.

From Stefan–Boltzmann's law, Eq. (2.2), the total heat energy, Q, radiated every second from the surface of a black-body at an absolute temperature T is

$$Q = \sigma T^4 \cdot \text{(surface of the black-body)}.$$

The surface area of a sphere of radius r is $4\pi r^2$. Therefore, assuming both the earth and the sun are black-bodies, the total heat energy radiated each second by the sun is

$$Q_{Sun} = \sigma T_{Sun}^4 \cdot 4\pi r_1^2, \tag{2.3}$$

where T_{Sun} is the temperature of the sun's surface, and the total heat energy radiated each second and by the earth is

$$Q_{Earth} = \sigma 290^4 \cdot 4\pi r_2^2, \tag{2.4}$$

where r_2 is the radius of the earth.

The fraction of the sun's total heat radiation that can be absorbed by the earth is given by the ratio $\pi r_2^2/4\pi R^2$, between the area of the earth's diameter plane, πr_2^2, and the surface area of the sphere of radius R (the distance from the earth to the sun) that surrounds the sun, $4\pi R^2$ (Fig. 2.7).

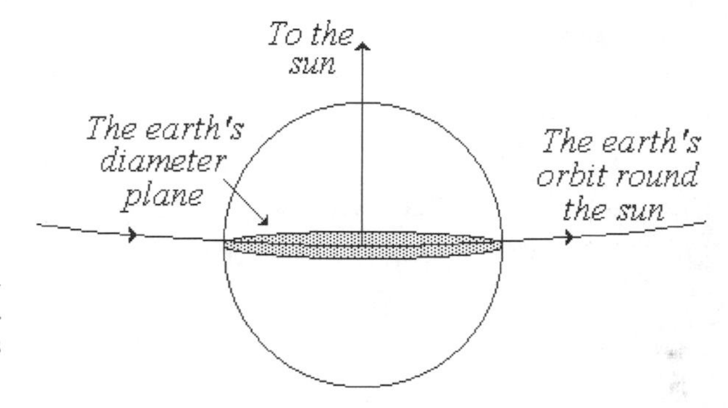

Fig. 2.7. The earth at a point in its orbit round the sun. The area of the earth's diameter plane is πr_2^2.

Accordingly, the amount of heat energy actually absorbed by the earth is

$$Q_{Sun} \cdot \frac{\pi r_2^2}{4\pi R^2} = \sigma T_{Sun}^4 \cdot 4\pi r_1^2 \cdot \frac{\pi r_2^2}{4\pi R^2}. \tag{2.5}$$

At equilibrium, the rate at which the earth absorbs heat energy from the sun is balanced by that at which it radiates heat energy:

$$\begin{aligned} Q_{Sun} \cdot \frac{\pi r_2^2}{4\pi R^2} &= Q_{Earth} \\ 4\pi r_1^2 \cdot \sigma T_{Sun}^4 \cdot \frac{\pi r_2^2}{4\pi R^2} &= \sigma 290^4 \cdot 4\pi r_2^2 \end{aligned} \tag{2.6}$$

Simple algebra and the substitution $\theta = 2r_1/R$ gives

$$T_{Sun}^4 = 290^4 \frac{4R^2}{r_1^2} = 290^4 \cdot \frac{16}{\theta^2}. \tag{2.7}$$

From which

$$T_{Sun} = 6010\,\text{K}. \tag{2.8}$$

Second calculation: An alternative method of determining the temperature of the sun's surface is by means of Wien's law – Eq. (2.2). The wavelength of maximum intensity in sunlight is $\lambda_{\max} = 4.9 \cdot 10^{-7}$m. Substituting this value in Wien's law gives

$$T_{Sun} = \frac{0.29 \cdot 10^{-2}}{4.9 \cdot 10^{-7}} = 5920\,\text{K}. \tag{2.9}$$

2.1.4 Difficulties in the Classical Theory of Radiation

Finding an explanation for the intensity spectrum of the thermal radiation emitted by a black-body was one of the most pressing unsolved problems of physics in the last decade of the 19th century. All the attempts to derive a mathematical function that correctly predicts the intensity spectra in Fig. 2.5 from classical postulates failed. Without going into all the details, we can explain the basic reasons for this failure.

According to Newtonian and Maxwellian physics, the average energy of an oscillator, $\overline{E}_{Oscillator}$, that is in thermal equilibrium with its surroundings, depends only on its absolute temperature, T:

$$\overline{E}_{Oscillator} = kT. \tag{2.10}$$

The constant of proportionality in this equation, $k = 1.38 \cdot 10^{-23}\,\mathrm{J} \cdot \mathrm{K}^{-1}$, is called *Boltzmann's constant.*

This relationship is derived from the *principle of the equipartition of energy*, according to which all the oscillators in a given system will possess the same average energy at equilibrium, irrespective of their frequencies. For example, in a system of 100 oscillators, 20 with a natural frequency of 1,000 Hz and 80 with a natural frequency of 5,000 Hz, on average 20% of the system's energy should be associated with the low frequency oscillators and 80% with the higher frequency ones.

The electric oscillators that generate the heat radiation in a material object are the particles – the atoms and molecules – of which it is composed. The frequencies of the heat radiation emitted by the object are those at which these particles can vibrate. It follows that the more frequencies there are in a particular range, the greater will be the intensity of the radiation in that range. If, for example, a system's oscillators can vibrate at ten different frequencies in the range between 500 Hz and 1,500 Hz, but at 100 different frequencies in the range between 7,500 Hz and 8,500 Hz, then the energy emitted in the latter range will be ten times greater than that emitted in the former. In general, the more permitted frequencies there are in a given region of the spectrum, the greater will be the emission intensity in that region. Thus, the problem of explaining the intensity spectrum of black-body radiation, reduces to determining the number of frequencies allowed to the black-body's electrical oscillators in each region of the spectrum.

It can be shown[6] that the electrical oscillators in a system such as the cavity of a constant-temperature hollow body, possess many more high-frequency vibrational modes than low frequency ones. Consequently, if the energy of a cavity that is in thermal equilibrium is allocated, on average, equally between all the allowed vibrational modes, as required by the equipartition principle, most of this energy will be associated with the higher frequencies. Thus, cavity radiation of any temperature should be predominantly very high-frequency radiation, i.e. it should comprise just a little infra-red (low frequency radiation), lots of ultra-violet (high-frequency radiation) and even more X-rays (very high-frequency radiation). When first presented, this construal of heat radiation was termed the *ultra-violet catastrophe*, since it implied that every physical system should spontaneously explode in a burst of high-frequency radiation.

However, this is contrary to experience. Black-body radiation always exhibits its maximum intensity near the middle of its range of frequencies and not at the high-frequency end. In practice, the highest frequencies are not operative; the system's energy is not, in fact, equipartitioned between all its frequencies (Fig. 2.8).

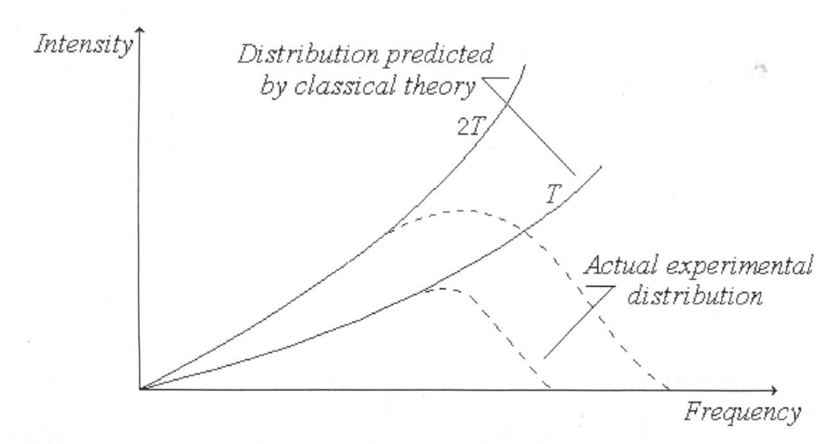

Fig. 2.8. The distribution of the intensity of black-body radiation as a function of its frequency at two temperatures, T and $2T$. The distribution predicted by classical physics – the continuous line – coincides with that found experimentally – the broken line – only at low frequencies. The area under each curve represents the total energy of the black-body radiation. Whereas the areas under the experimental curves are finite, those predicted by classical physics are potentially infinite, which is obviously an impossibility.

[6]See Example 2.3.

Example 2.3: The Frequencies in Cavity Radiation

Cavity radiation can be regarded as a system of standing waves produced from the interference of the electromagnetic waves reflected to and fro between the cavity walls. Such a system is analogous to the standing waves produced in a wire stretched between two fixed points. Using this analogy, we can show how the number, Δn, of possible vibrational modes (standing wave frequencies) in a wavelength band of a given width, $\Delta\lambda$, increases with the frequency of the vibrations.

Calculation:

Standing waves produced in a stretched wire of length L by waves have wavelengths

$$\lambda = 2L/n \quad (n = 1, 2, 3, \ldots).$$

The number of such wavelengths in the band between λ and $(\lambda + \Delta\lambda)$ is

$$\Delta n = \frac{2L}{\lambda} - \frac{2L}{\lambda + \Delta\lambda}.$$

In the case where $\Delta\lambda \ll \lambda$, this gives

$$\Delta n = \left(\frac{2L}{\lambda^2}\right) \cdot \Delta\lambda.$$

Thus, the number of possible wavelengths in a range of width $\Delta\lambda$ increases as the square of wavelength decreases, i.e. as the square, f^2, of the frequency increases. Thus, in a band of a given width, more vibrational modes are permitted at high frequencies than at low frequencies.

It can be shown that in the case of a cavity of volume V, the number of possible wavelengths in a range of width $\Delta\lambda$ is

$$\Delta n = \left(\frac{8\pi V}{\lambda^4}\right) \cdot \Delta\lambda, \tag{2.11}$$

from which the preponderance of the short wavelengths (higher frequencies) clearly follows.

2.1.5 Planck's Quantum Hypothesis

By the end of the 19th century, the atomic nature of matter had been well established. It was recognised that matter could only exist in integer multiples of a natural unit – one hundred atoms, a mole of atoms, etc. – and could only be allocated in discrete amounts. However, as regards energy, it was still believed that it could be absorbed or emitted continuously in any arbitrary amount. This was a necessary corollary of the equipartition principle, which required the equal and unhindered distribution of a system's energy between all its vibrational modes, irrespective of their frequencies.

In 1900, by abandoning this tenet, Planck[7] first succeeded in deriving a function that correctly described black-body radiation. Suggesting that the energy of an oscillator is not 'a continuous, infinitely divisible quantity, but a discrete quantity composed of an integral number of finite equal parts',[8] Planck showed that Wien's law required that the energy of each of these finite equal parts be directly proportional to the oscillator's frequency, f. On this basis, he argued that the oscillator's energy can only take the specific values

$$E_{Oscillator} = 0, hf, 2hf, 3hf, \ldots, \tag{2.12}$$

all others being forbidden. The constant of proportionality, $h = 6.63 \cdot 10^{-34}\text{J} \cdot \text{s}$, is the universal constant now called *Planck's constant.*

It follows that an oscillator's energy can only change by certain discrete amounts, i.e. that it can emit or absorb energy only in small 'packets' – what are now called *quanta* – the energy of such a packet or *quantum* depending on the oscillator's frequency, such that

$$E_{Quantum} = hf. \tag{2.13}$$

It was on the basis of this hypothesis – the *quantum hypothesis* – that Planck succeeded in deriving a mathematical function which correctly predicted the intensity spectrum of black-body radiation. In doing so, Planck showed that the average energy of an oscillator in thermal equilibrium at an absolute temperature T is correctly given by the equation

$$\overline{E}_{Oscillator} = \frac{hf}{e^x - 1}, \tag{2.14}$$

where the factor x is given by the equation

$$x = \frac{hf}{kT} \tag{2.15}$$

and not by the classical relationship $\overline{E}_{Oscillator} = kT$. Thus, energy is not allocated equally between a system's oscillators; each oscillator's average energy depends not only on its temperature but also on its frequency.

We can now explain, qualitatively, why the ultra-violet catastrophe does not happen. At normal temperatures, the energy, hf, of the quanta associated with radiation of high frequency is much greater than the average kinetic energy, kT, of the atoms comprising the source of the radiation, i.e. $hf \gg kT$. Under these

[7]Max Planck 1858–1947, German physicist who was awarded the Nobel prize in 1918 for the quantum hypothesis.

[8](1901) *Annalen der Physik*, **4**, 553, translation in M.H. Shamos (ed.) (1959) *Great Experiments in Physics*, New York: Dover Publications, pp. 301–314.

conditions, the probability of any atom (oscillator) acquiring sufficient energy to create such a large energy quantum is negligible. Thus, the oscillators' high frequencies are not operative and so the heat radiation emitted does not contain high-frequency radiation.

At low frequencies, the energy of the quanta is much smaller than the average kinetic energy of the atoms, i.e. $hf \ll kT$. In this case, the factor x in Eq. (2.14) will be correspondingly small and to a close approximation we can say that

$$e^x = 1 + x = 1 + \frac{hf}{kT}. \tag{2.16}$$

Substituting this in Eq. (2.14) gives

$$\overline{E}_{Oscillator} = \frac{hf}{\left(1 + \frac{hf}{kT}\right) - 1} = kT, \tag{2.17}$$

which is the classical Eq. (2.10) for the average energy of an oscillator at equilibrium. The energy quanta associated with changes in the energies of low frequency oscillators are so minute that changes in the oscillators' energies appear to be continuous. This is why the classical approach correctly predicts the actual experimental distribution of intensities at low frequencies.

Despite the success of the quantum hypothesis in explaining black-body radiation, Planck himself continued to believe that the radiation itself was a pure wave phenomenon. It was difficult to accept the notion that a natural process was discontinuous. This idea stood in total contradiction to one of the most venerable postulates of western thinking – *natura non facit saltum* – nature does not make jumps. Only after the quantum hypothesis had repeatedly demonstrated its competence in explaining different physical phenomena did it gain acceptance as a fundamental doctrine of physics.

A guideline for the coexistence of the classical physics of Newton and Maxwell with the new quantum physics was first formulated by Niels Bohr.[9] Known as the *correspondence principle*, it states that:

> '*The results predicted by quantum physics must be identical to those predicted by classical physics in those situations where classical physics corresponds to the experimental facts.*'

[9]Niels Bohr 1885–1962, Danish physicist who was awarded the Nobel prize in 1922 for his research on the structure of the atom and the radiation emitted from them.

This principle reconfirms the rule that in the physical sciences it is the experimental facts that are the final arbiter. Whatever else changes, a theory remains useful and valid so long as it satisfies this requirement and correctly predicts the facts.

2.1.6 Atomic Spectra

In contrast with the typical continuous spectrum of thermal radiation, the spectrum of the light emitted from a low pressure discharge tube containing an atomic gas such as atomic hydrogen, helium or sodium vapour is a *line spectrum*, each line corresponding to a specific frequency or wavelength (Fig. 2.9). Atomic emission spectra can also be observed by analysing the light emitted from the gas when heated to a high temperature.

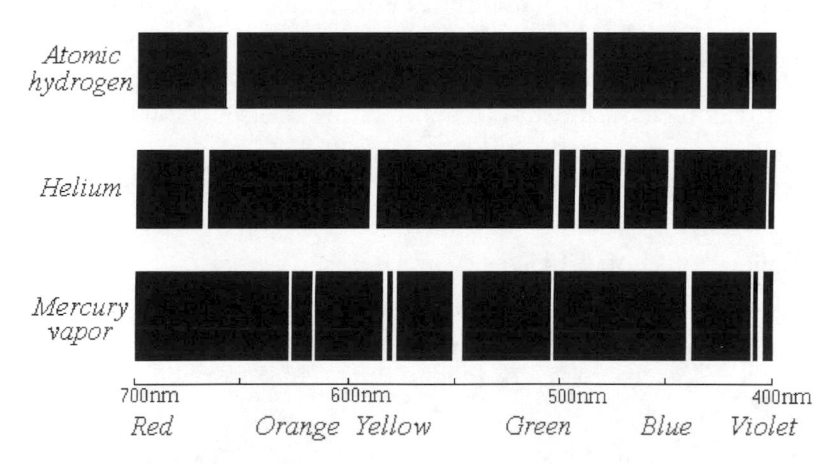

Fig. 2.9. Emission spectra of a number of atomic gases. The light emitted by the excited gases comprises discrete frequencies which appear as lines in the spectrum.

The absorption spectra of atomic gases are also line spectra. When white light is passed through an atomic gas, some of the components of the light are absorbed. The absorbed components are absent in the transmitted light. Their absence shows up in the spectrum of the emergent light as dark lines on the illuminated coloured background of its continuous spectrum (Fig. 2.10). The frequencies of the absorption lines generally correspond to those of lines in the emission spectrum. However, there are generally fewer lines in the absorption spectrum of an element than in its emission spectrum. Light absorption is an example of resonance; the atoms in the gas absorb energy from the incident white light at frequencies that match one of their own natural frequencies.

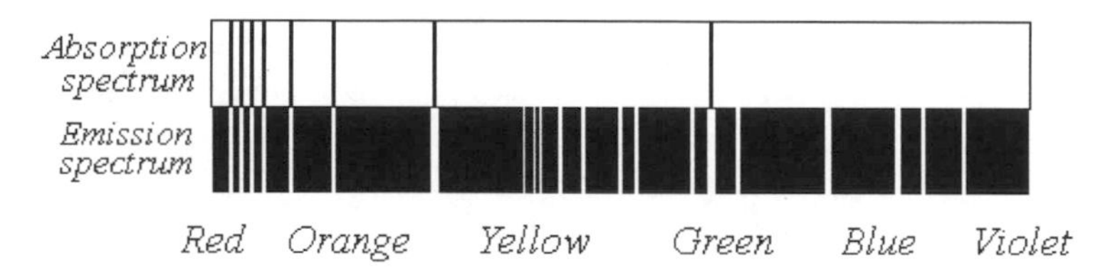

Fig. 2.10. The absorption and emission spectra of sodium vapour. Note that there are less lines in the absorption spectrum than in the emission spectrum and that all the absorption frequencies correspond to those of emission lines. Note also the differences in the intensities of the various lines.

In the middle of the 19th century, Kirchoff[10] demonstrated that every chemical element exhibits a distinctive emission and absorption spectrum; an element's spectra are its unique finger-prints. Thus, chemical elements can be identified by means of their spectra. On the basis of this insight, Kirchoff was able to elucidate a phenomenon first observed some 50 years earlier by Fraunhofer[11] in his investigations of the spectrum of sunlight. Since sunlight is the heat radiation characteristic of a temperature of about 6,000 K, its spectrum should be continuous. However, when Fraunhofer examined the spectrum of sunlight closely, he found it crossed by some 700 dark lines. Kirchoff explained that these are absorption lines, introduced into the sunlight as it passes though the cool outer layers of the sun's chromosphere.

The presence of some 70 chemical elements in the sun's chromosphere has been established by the study of the spectrum of sunlight. The element helium was observed spectroscopically in the sun's chromosphere some 30 years before it was isolated on earth. All our information about the composition of the stars and the galaxies is drawn from the spectra of starlight. And it is from shifts in the wavelengths of the spectral lines in starlight which are attributed to a Doppler effect, that much of our knowledge and ideas about the origin and expansion of the universe have been gleaned.

Since the emission of light was thought to result from some sort of vibration, harmonic relationships like those found in vibrating mechanical systems were sought between the frequencies of the spectral lines, but none such were discovered. However, early in the 20th century, after more data had been collected, Ritz[12] noticed that in many instances the sum of the frequencies of two lines in an atom's spectrum gives the frequency of a third line, and that the sum of the frequencies of these three lines gives the frequency of a fourth line, and so on. This is called the *Ritz combination rule* (Fig. 2.11).

[10]Gustav Kirchoff 1824–1887, German physicist.
[11]Josef von Fraunhofer 1787–1826, German physicist and optician.
[12]Walther Ritz 1878–1909, German physicist.

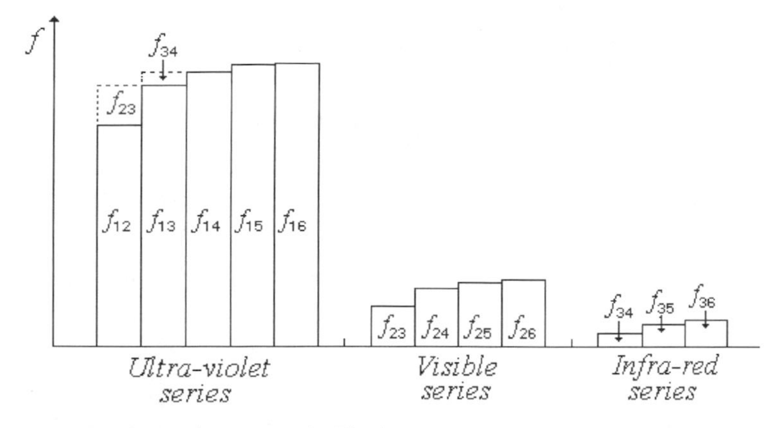

Fig. 2.11. Ritz's combination rule. The lines in an imaginary atom's spectrum. The spectral lines are identified by the two subscript numbers of each line's frequency. The first number indicates the series or group to which the line belongs. The diagram shows the following relationships: $f_{12} + f_{23} = f_{13}$; $f_{12} + f_{23} + f_{34} = f_{14}$.

Ritz's empirical rule was first given a theoretical foundation in 1913, when Bohr, in the framework of his research on the structure of the atom, employed Planck's quantum hypothesis to explain the origin of atomic line spectra.[13] According to the quantum hypothesis, an electric oscillator, such as an atom, absorbs and emits radiation in discrete quanta, each of energy hf where f is the frequency of the radiation. The apparently continuous absorption and emission of radiation by a body is, in fact, the averaged out result of a myriad of discrete events, of countless quanta being absorbed or emitted by innumerable atomic oscillators.

Fig. 2.12. A hypothetical system of stationary states showing the transitions between the energy levels that accompany the absorption or emission of energy by the system. The system rises to a higher level when it absorbs energy and drops to a lower one when it emits energy.

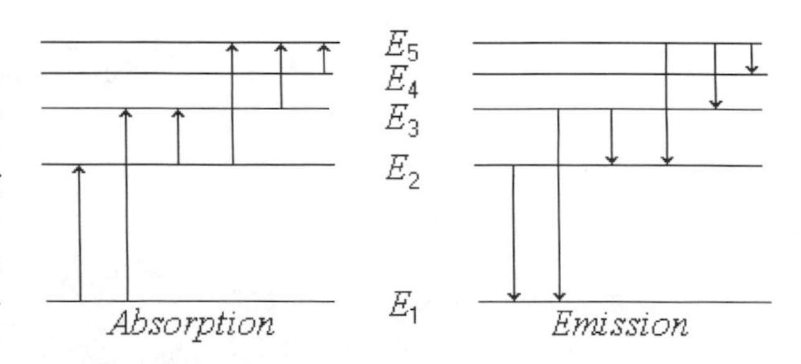

Bohr postulated that between these discrete events, when the atoms are not actually emitting or absorbing a quantum of energy, they are in non-radiating *stationary states*, each of which is characterised by a specific amount of energy, $E_1, E_2, E_3, \ldots$, called its *energy level* (Fig. 2.12). The lowest energy level in a

[13]See Chapter 3.2.

system is its *ground state* and is usually designated E_1. The system's higher energy levels are called *excited states.*

In passing from one stationary state to another, atoms (oscillators) emit or absorb a quantum of radiation. The difference, $E_i - E_j$, between the energy level, E_i, of the initial stationary state, and that of the final stationary state, E_j, is balanced by the energy, hf_{ij}, of the quantum of radiation associated with transitions between the states (Fig. 2.13). Bohr posited that the frequency, f_{ij}, of the radiation emitted or absorbed in a transition is given by

$$f_{ij} = \frac{E_i - E_j}{h}. \tag{2.18}$$

Equation (2.18) is known as *Bohr's equation.*

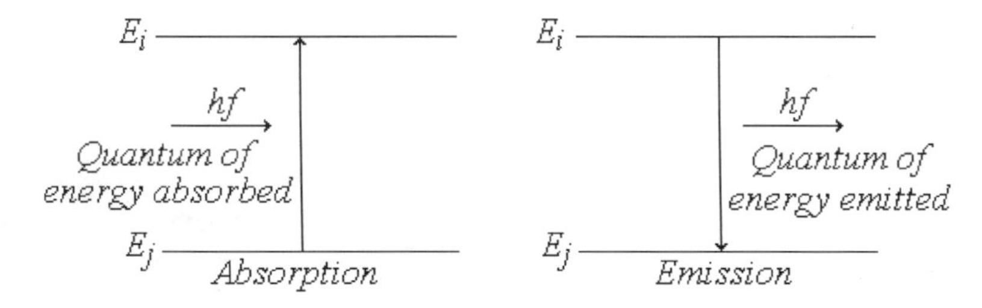

Fig. 2.13. The absorption and emission of a quantum of energy. When absorbing a photon the atom rises ('jumps') from the lower energy level to the higher one. An atom that emits a photon 'falls' to a lower energy level.

To appreciate how Bohr's postulate accounts for the Ritz combination rule, consider the relationship $f_{12} + f_{23} = f_{13}$ between the frequencies of the spectral lines of the imaginary atom in Fig. 2.11. Energy conservation requires that there be the same energy change whether the transition from the E_1 level to the E_3 level takes place directly or by stages, i.e.

$$(E_1 - E_3) = (E_1 - E_2) + (E_2 - E_3). \tag{2.19a}$$

Dividing this expression by h gives

$$\frac{(E_1 - E_3)}{h} = \frac{(E_1 - E_2)}{h} + \frac{(E_2 - E_3)}{h}. \tag{2.19b}$$

From Bohr's equation, the frequencies of the spectral lines f_{13}, f_{12} and f_{23}, are

$$f_{13} = \frac{(E_1 - E_3)}{h} \quad f_{12} = \frac{(E_1 - E_2)}{h} \quad f_{23} = \frac{(E_2 - E_3)}{h}$$

respectively. Substituting these values in Eq. (2.19b) gives

$$f_{13} = f_{12} + f_{23}.$$

Thus, Ritz's combination rule follows from energy conservation and Bohr's postulate.

With the help of some statistical theory, Bohr's postulate also explains why there are generally fewer lines in the absorption spectrum of an element than in its emission spectrum, i.e. why gaseous atoms do not usually absorb all the frequencies they emit. Obviously, an atom cannot absorb the quantum of radiation that corresponds to a transition from a particular state to a higher excited state unless it is already occupying the particular state; it cannot rise from E_3 to E_4 if it is not already in E_3. In a sample of gaseous atoms, the ratio between the number of atoms occupying an excited state, N_i, to the number, N_1, occupying the ground state is given by the Maxwell–Boltzmann distribution:

$$\frac{N_i}{N_1} = e^{-(E_i - E_1)/kT}, \tag{2.20}$$

where k is the Boltzmann constant and T is the sample's absolute temperature.[14]

At room temperature (25°C) the value of the factor kT is

$$kT = 1.36 \cdot 10^{-23} \cdot 298 = 4 \cdot 10^{-21}\,\text{J}. \tag{2.21}$$

The energy difference between a pair of adjacent atomic energy levels is typically about 2 eV:

$$E_i - E_1 \approx 2\,eV = 3.2 \cdot 10^{-19}\,\text{J}. \tag{2.22}$$

Whence, the ratio between the number of atoms occupying the two energy levels is

$$\frac{N_i}{N_1} = e^{-2/2.57 \cdot 10^{-2}} \approx e^{-80}. \tag{2.23}$$

It follows that, at room temperature, the population of the excited state is negligible and the probability of the occurrence of a transition that starts from this state is close to zero. As a result, under normal conditions, the lines corresponding to transitions from excited states to higher states do not usually appear in the atom's absorption spectrum; the only lines that do appear are those corresponding to transitions that start from the ground state. With rising temperature, the ratio N_i/N_1 increases, and the population, N_i, of the excited state grows accordingly. Consequently, the probability of transitions starting

[14]Equation (2.20) is obtained from the statistical analysis of a system of identical but distinguishable particles in which there are no restrictions on the way they are deployed. A sample of gaseous atoms, in which the atoms can be distinguished by their individual trajectories, satisfies these requirements. At thermal equilibrium, the probability that a particle in such a system has a specified velocity or position is proportional to the Boltzmann factor, $e^{-E/kT}$, regardless of whether E is kinetic or potential energy.

from this excited state increases, and so more lines begin to appear in the absorption spectrum.

Ritz's empirical rule was also the starting point for the *matrix mechanics* developed by Heisenberg[15] in 1925. Asserting that in a logically consistent system no entities should be introduced except such as are physically observable, he listed the observed spectral frequencies emitted by an atom in a square array as follows:

$$\begin{array}{cccc} f_{11}=0 & f_{12} & f_{13} & \cdots \\ f_{21} & f_{22}=0 & f_{23} & \cdots \\ f_{31} & f_{32} & f_{33}=0 & \cdots \\ \vdots & \vdots & \vdots & \ddots \end{array}$$

Each element in the array corresponds to a transition between a pair of energy levels.

Heisenberg went on to suggest that the values of any observable quantity associated with the transitions could be listed in a corresponding square array. On this view, physical quantities such as a particle's position and momentum are not ordinary functions of time, but represent matrices, each of whose elements is associated with a transition from one energy level to another. Calculations can be performed with these infinite matrices by the usual rules of matrix theory, subject to certain distinctive commutation rules, which here take the place of the quantum hypothesis.

2.1.7 The Franck–Hertz Experiment

In 1914, convincing experimental evidence for the existence of the stationary atomic states postulated by Bohr was obtained by Franck[16] and Hertz[17] in the course of their investigations of collisions between electrons and gaseous mercury atoms.

In order to study these collisions, the electrodes of a glass triode, containing low pressure mercury vapour, were connected to variable potentials, V_1 and

[15] Werner Heisenberg 1901–1976, German physicist who was awarded the Nobel prize in 1932 for the invention of quantum mechanics.

[16] James Franck 1882–1964, German–American physicist who was awarded the Nobel prize in 1925 for discovering the laws governing collisions between electrons and atoms.

[17] Gustav Hertz 1887–1961, German physicist who was awarded the Nobel prize in 1925 for discovering the laws governing collisions between electrons and atoms.

Fig. 2.14. Electrons released from the cathode are accelerated by the variable potential V_1 towards the grid. Only those electrons emerging from the grid with an energy greater than eV_2 can reach the anode. Between the cathode and the grid the electrons collide with atoms of the mercury vapour.

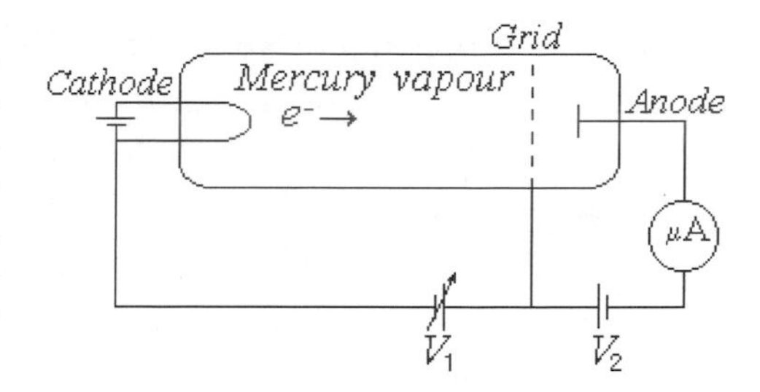

V_2, as shown in Fig. 2.14. Electrons released from the heated cathode C are accelerated towards the grid G by the potential V_1. The anode A was kept at a potential, V_2, lower than the grid. Consequently, electrons passing through the grid could reach the anode only if their kinetic energy was equal to or greater than eV_2. The current, I, shown by the microammeter, indicated the rate at which electrons arrived at the anode.

In the space between the cathode and the grid, the accelerated electrons collide with atoms of the mercury vapour. The electrons are very much lighter than the mercury atoms, $m_e \approx m_{Hg}/370,000$, and so in elastic collisions with the mercury atoms the electrons lose only a minute amount of their energy; the electrons recoil as though they had struck a massive wall. Under these circumstances, electrons reaching the grid should have sufficient energy to carry on to the anode and a significant anode current, I, should be recorded by the microammeter. Furthermore, the higher the potential through which the electrons are accelerated, the more of them will reach the anode and the greater will be the anode current.

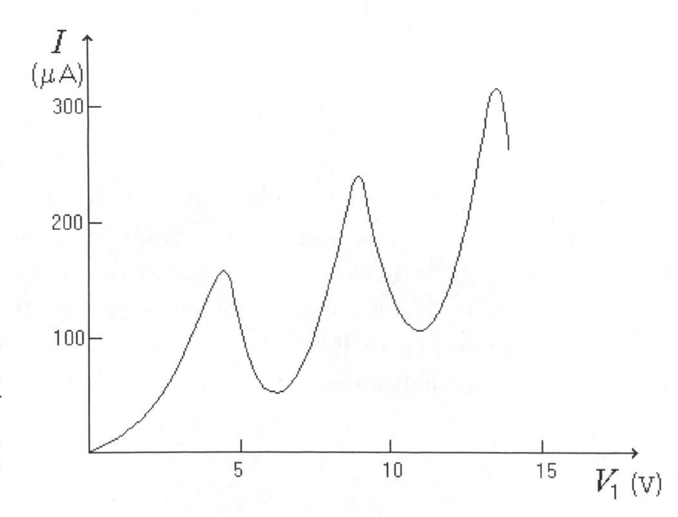

Fig. 2.15. Franck–Hertz experiment. The anode current, I, as a function of the potential difference, V_1, between the cathode and the grid.

In their experiment, Franck and Hertz recorded the changes in the anode current, I, as they varied the potential, V_1, between the cathode and the grid. The results they obtained are shown in Fig. 2.15. These show that at first the anode current increases as the potential, V_1, through which the electrons are accelerated is raised; this is consistent with elastic collisions between the electrons and the mercury atoms. However, at a potential of 4.9 V, the current falls sharply, almost to zero. Raising the potential beyond this value restores the current, and it continues to increase as the potential rises until suddenly it drops a second time at a potential of 9.8 V. Raising the potential further, once again restores the current only for it to fall a third time at a potential of about 14.7 V. The sudden changes in the anode current indicate that at those specific potentials, the number of electrons reaching the anode drops drastically. It appears as though at these potentials the electrons are 'robbed' of their energy as they pass through the mercury vapour and cannot, therefore, overcome the opposing potential difference, V_2, between the grid and the anode.

The simplest explanation for the loss of the electrons' energy is that at these specific potentials, their collisions with the mercury atoms are not elastic. The inference is that, whereas electrons with other kinetic energies collide elastically with the mercury atoms, electrons with kinetic energies of 4.9 eV, 9.8 eV or 14.7 eV collide inelastically with mercury atoms, losing all their energy in the collisions.

From the point of view of the mercury atoms in the vapour, it appears that, under these conditions, they can only absorb mechanical energy from the electrons in amounts that are an integer multiple of 4.9 eV. These inelastic collisions can be summarised by the symbolic representation:

$$\text{Hg} + \text{fast electron} \rightarrow \text{Hg}^* + \text{slow electron},$$

where Hg^* represents a mercury atom in an excited state. Thus, an electron of energy 4.9 eV loses all its energy in a single inelastic collision; an electron of energy 9.8 eV in two inelastic collisions; an electron of energy 14.7 eV after three inelastic collisions. It would appear that in each of these inelastic collisions, a mercury atom rises from its ground state, E_1, to an excited state, E_2, the difference between their energy levels being 4.9 eV. Unless the energy of the electron is equal to the difference between two energy levels in the mercury atom, or equals an integer multiple of that difference, the atom does not accept any energy from the electron and the collision between them is elastic.

According to Bohr's postulate, Eq. (2.18), the excited mercury atom, Hg^*, can return to its ground state, Hg, by emitting a quantum of radiation of frequency

$$f = \frac{E^*_{\text{Hg}} - E_{\text{Hg}}}{h} = \frac{4.9}{4.14 \cdot 10^{-15}} = 1.18 \cdot 10^{15}\,\text{Hz}.$$

Franck and Hertz detected the emission of ultra-violet radiation of this frequency from the glass triode.

This relatively simple experiment demonstrated both the existence of stationary states in the mercury atoms and the validity of Bohr's equation for the relationship between their energy levels and the atom's emission spectra. In further experiments, it was found that mercury atoms can absorb other specific amounts of energy such as 6.7 eV and 8.8 eV, indicating the existence of other energy levels in the atom.

Chapter 2.2

The Photoelectric Effect

Though the success of Planck's quantum hypothesis may have established the fact that the absorption and emission of radiation were discrete processes, it was still generally believed that radiation actually propagated as a wave. This all began to change in 1905 when Einstein published his revolutionary explanation of the photoelectric effect.

2.2.1 The Photoelectric Effect — The Problem

When Hertz first succeeded in producing electromagnetic waves from oscillating electrical circuits, he noted that 'it is essential that the pole surfaces of the spark gap [from which the waves were emitted] should be frequently re-polished'.[18] He also observed that sparks were more readily induced between the poles in the secondary (detector) circuit when they were illuminated with ultra-violet radiation. He called this phenomenon, for which he had no explanation, the *photoelectric effect*.

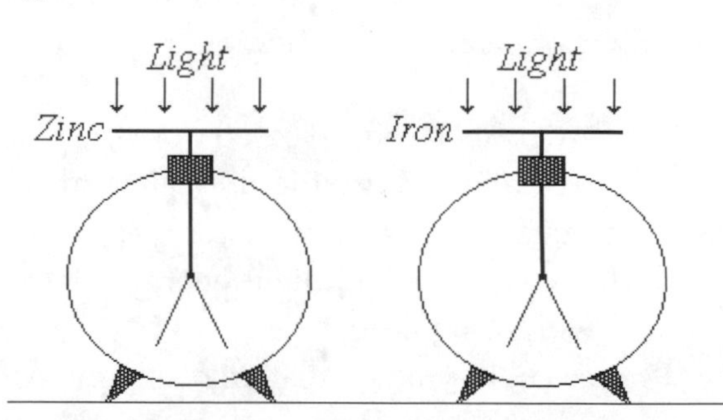

Fig. 2.16. Hallwach's experiment. Two gold-leaf electroscopes, one with an iron plate, the other with a zinc plate, are charged, first positively and then negatively. In each case, light of different colours is beamed on to the plates.

A detailed study of the photoelectric effect was subsequently made by Hallwachs.[19] He studied the effect of the following three factors on the response of a charged gold-leaf electroscope to visible light and ultra-violet radiation (Fig. 2.16).

[18] Heinrich Hertz (1892), *Electric Waves*, translated by D.E. Jones (1893), New York: Macmillan.
[19] Wilhelm Hallwachs 1859–1922, German physicist.

1. The sign of the charge carried by the electroscope, positive or negative.
2. The metal from which the electroscope plate was made, iron or zinc.
3. The intensity of the light and ultra-violet radiation.

Hallwachs made the following observations:

* Illuminating the plate of a positively charged electroscope with visible light or ultra-violet radiation produced no response whatsoever, whether the plate was made of iron or zinc. Even if a very intense beam of radiation was used, no charge was lost by the positively charged electroscope.
* A negatively charged iron plate electroscope showed no response when its plate was illuminated by an intense beam of either visible light or of ultra-violet radiation.
* Visible light also had no effect on a negatively charged zinc plate electroscope. However, the moment the plate of the negatively charged zinc plate electroscope was illuminated with even a weak beam of ultra-violet radiation, the electroscope's leaves closed, indicating that it had lost its negative charge.

The discharge of the negatively charged zinc plate electroscope is readily explained by supposing that the ultra-violet radiation supplies the energy needed by the electrons constituting its negative charge to break the bond holding them to the zinc plate and to escape from it. However, this still leaves a number of unanswered questions:

1. Why are no electrons released from the iron plate electroscope by the ultra-violet radiation? Why the difference in the behaviour of the two metals? All metals emit electrons when heated – the *thermionic effect.* Why does irradiating metals with light energy affect them differently from warming them with heat energy?
2. Whereas ultra-violet radiation of any intensity, no matter how weak, instantaneously releases electrons from the negatively charged zinc plate, electrons are not released from either plate by visible light, no matter how intense the beam. Why is this? The intensity of an electromagnetic wave, namely, the rate at which it delivers energy, is a function of its amplitude, not of its frequency. Thus, a radiation's colour (its frequency) should not be a factor in its ability to deliver the energy needed for the release of the electrons from the metal plates.

   Furthermore, the greater the intensity of the ultra-violet radiation, the greater should be the kinetic energy of the liberated electrons – the *photoelectrons*. However, other experiments showed that although the intensity of the ultra-violet radiation does affect the number of electrons emitted from the zinc plate – the more intense the beam the more electrons are released – it has no effect on their individual energy.

3. The energy carried by a wave is distributed equally along its front. Thus, the energy of the incident electromagnetic wave should be delivered at the same rate to every electron in the surface of the metal plate and a quite considerable length of time should pass before they will have accumulated sufficient energy to be released, whereupon they should all be emitted in a single explosive burst. Yet the zinc plate electroscope loses its negative charge the very moment its plate is illuminated with ultra-violet radiation, even at low beam intensities. How does this happen?

Classical electromagnetic theory had no answer to these questions. Evidently, the photoelectric effect requires a totally new theory of the transport of energy by radiation.

2.2.2 Einstein's Equation

The photoelectric effect was first elucidated by Einstein in 1905. In doing so, he adopted and extended Planck's quantum hypothesis, postulating that 'the incident light is composed of quanta of energy'. Einstein proposed that radiation is not only absorbed and emitted in quanta but that it also propagates as such. Thus, a light beam can be considered to be a stream of particles, *light quanta* or *photons*, each of energy

$$E_{Photon} = hf \tag{2.24}$$

where f is the frequency of the light. The photons move through space in straight lines and at the speed of light. Light intensity falls as we move away from a light source because the photons it emits are spread out over a larger surface area. However, the energy of each individual photon remains unchanged.

Having ascribed a particle nature to light, Einstein went on to treat the photoelectric effect as a simple mechanical problem of energy conservation.

> '*The quanta penetrate the surface of the material and their respective energies are at least in part changed into the kinetic energy of the electrons ... The simplest process conceivable is that a quantum of light gives up all its energy to a single electron ... Upon reaching the surface, an electron originally inside the body will have lost a part of its kinetic energy. Furthermore, one may assume that each electron in leaving the body does an amount of work which is characteristic of the material. Those*

electrons that are ejected normal to and from the immediate surface will have the greatest velocities.'[20]

A photon striking a metal surface surrenders all of its energy to a single electron in the metal. The electron first uses this energy to break the bond holding it to the metal; any surplus energy appears as the kinetic energy, $K_{Electron}$, of the liberated electron. Einstein summarised these ideas in the following simple relationship called *Einstein's equation*:

$$\begin{aligned} K_{Electron} &= E_{Photon} - W \\ &= hf - W. \end{aligned} \tag{2.25}$$

W is the work done in releasing the electron from the metal plate. The minimum amount of work, W_{min}, required to release an electron from a metal plate is called the metal's *work function*; it is a characteristic property of the metal.

Einstein's simple hypothesis explains the results of the experiments with the charged electroscopes as follows:

1. Ultra-violet radiation does not liberate electrons from an iron plate because the metal's work function is greater than the energy of the photons, i.e. $W_{\min} > hf_{UV}$. In order to release electrons from iron, radiation of higher frequency must be used; radiation whose photons have a greater individual energy.
2. Visible light comprises a range of frequencies – those of all the colours of the rainbow. Thus, it contains photons in a range of energies, from about 2 to 3 electron-volts. However, the work function of zinc (and certainly that of iron) is greater than the energy of even the most energetic of these photons. Hence, visible light cannot release electrons from a zinc plate.
3. Electrons are released the moment the ultra-violet radiation strikes the zinc plate because the energy concentrated in the photons is transferred to them virtually instantaneously. There is no accumulation of energy.
4. The photons in an intense beam of radiation possess no more individual energy than those in a weak beam of the same radiation; the more intense beam just contains more photons. Thus, an intense beam can liberate more photoelectrons from the metal plate. However, not all the photoelectrons are released with the same kinetic energy; those extracted from deep inside the metal plate have to expend some of their energy in rising to the surface. The photoelectrons with the greatest kinetic energy, $K_{\max}$, are those released from the plate's surface, since these have to expend no more than the metal's

[20] (1905) *Annalen der Physik*, **17**, 144, translation in M.H. Shamos (ed.) (1959) *Great Experiments in Physics*, New York: Dover Publications, p. 235.

work function, W_{min}, in breaking loose,

$$K_{\text{max}} = hf - W_{\text{min}}. \tag{2.26}$$

The colour (frequency) dependence of the response of certain materials to incident light had in fact been known since the invention of photography, some 60 years before the discovery of the photoelectric effect. It was well known that most photographic plates were not sensitive to red light; red and black objects looked the same in photographs. Furthermore, the plates could be safely exposed and developed in a room illuminated with red light. On the other hand, even very weak blue light ruined the plates.

The imprinting of a photographic image on a photographic plate or film involves a chemical reaction, the energy for which is supplied by light; such reactions are called *photochemical reactions.* The reaction involved in photography is essentially the reduction of silver ions to metallic silver:

$$\text{Ag}^+ + e^- \xrightarrow{\text{photon}} \text{Ag}.$$

Red light, which has a relatively low frequency, did not blacken the plates used by the first photographers because its photons have insufficient energy to promote the reduction of the silver ions. By the addition of suitable dye-forming chemicals to the emulsion containing the silver ions, modern photographic films are made sensitive to all the colours of the rainbow and even to infra-red radiation.

Einstein's proposal that light propagates as photons stood in total opposition to electromagnetic wave theory. The fundamental concept implicit in Einstein's revolutionary *quantum theory of light* was that photons are real particles, a notion totally alien to the classical wave theory of light. There was no way of arriving at the concept of photons from Maxwell's equations. Moreover, it was subsequently shown that the mathematical function derived by Planck for the energy distribution between the wavelengths in black-body radiation (the intensity spectrum) could also be derived by regarding cavity radiation as a sort of *photon gas* between whose particles the radiation's energy was suitably distributed (Section 3.4.5). The actual nature of light was once again an open question.

Example 2.4: Photons and Wavelengths

Calculate the energies of the photons associated with the following radiation:

1. violet light: $\lambda = 413\,\text{nm}$;
2. X-rays: $\lambda = 0.1\,\text{nm}$;
3. radio waves: $\lambda = 10\,\text{m}$.

Calculation: From Eq. (2.24), $E_{Photon} = hf$, and the relationship $c = f\lambda$, we obtain for the energy of the photons in violet light:

$$E_{Photon} = \frac{hc}{\lambda} = \frac{6.63 \cdot 10^{-34} \cdot 3 \cdot 10^{8}}{413 \cdot 10^{-9}} = 4.8 \cdot 10^{-19}\,\text{J}. \tag{2.27}$$

This energy is more conveniently expressed in electron-volts:

$$E_{Photon} = \frac{4.8 \cdot 10^{-19}}{1.6 \cdot 10^{-19}} = 3\,\text{eV}.$$

Similar calculations give for the energy of the X-ray photons 12,400 eV, and for that of the radio wave photons $1.24 \cdot 10^{-7}$eV.

Example 2.5: Counting Photons

Calculate the number of photons that strike a unit area of a surface each second, when it is illuminated by a beam of violet light of wavelength 413 nm and intensity 5 W/m^2.

Calculation: The intensity of a beam of radiation is defined as the energy transported by the beam each second across a plane of unit area. In terms of the photons which comprise it, the intensity, I, of a beam of radiation is given by the formula

$$I = NE_{Photon}, \tag{2.28}$$

where N is the number of photons transported by the beam across a plane of unit area each second.

The energy carried by each photon in the beam of violet light is $E_{Photon} = 4.8 \cdot 10^{-19}$J (see Example 2.4). Hence, the number of photons striking the unit area each second is

$$N = \frac{I}{E_{Photon}} = \frac{5}{4.8 \cdot 10^{-19}} = 1.04 \cdot 10^{19}\,\text{m}^{-2}\,\text{s}^{-1}.$$

2.2.3 Planck's Constant

Planck's constant, h, is one of the four fundamental constants of physics. The other three are the velocity of light *in vacuo*, c, the universal gravitational constant, G, and the elementary charge, e. Planck's constant is the underlying constant of quantum physics. Because of its very small value, $h = 6.63 \cdot 10^{-34}\,\text{J} \cdot \text{s}$, quantum effects are not usually detected on the macroscopic scale. However, if its value was zero, quantum effects would not exist at all.[21]

In 1916, Millikan designed and carried out an accurate experiment that both confirmed the validity of Einstein's equation and fixed the precise value

[21]The three fundamental constants, h, c, and e, can be combined with the permittivity of empty space ε_0, to give a dimensionless number, $\alpha = e^2/2\varepsilon_0 hc \approx 1/137$, called the *Fine Structure Constant.* This number is central to the theory of quantum electrodynamics – QED, that unites quantum theory and the theory of relativity in explaining the interaction of photons and electrons. See Chapter 4.4.

of Planck's constant. The key component of the experiment was a glass vacuum tube called a *photoelectric tube* or *photoelectric diode*, into which two electrodes were fixed: one, a metal plate that emitted electrons when illuminated by visible light; the other smaller, intended for the collection of the photoelectrons. The electrodes were connected to a potentiometer, such that the illuminated metal plate was given a positive potential and the collector a negative potential (Fig. 2.17). The physical quantity that Millikan measured in the experiment was the kinetic energy of the most energetic photoelectrons released from a metal by incident light of a given frequency.

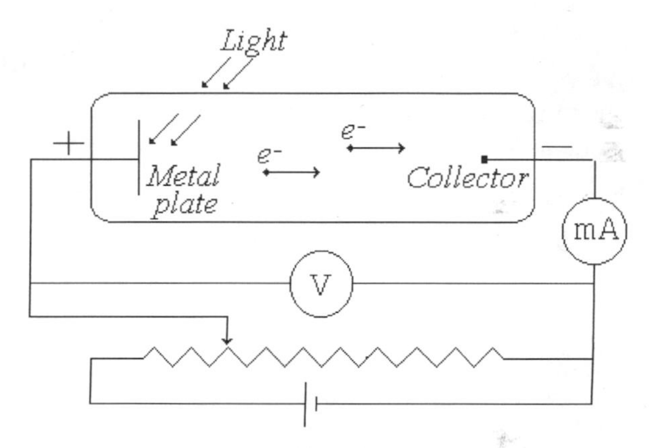

Fig. 2.17. Millikan's apparatus used to investigate the photoelectric effect. Light falls on the metal plate and the photoelectrons emitted from it are collected by the electrode. The potential difference, V, between the electrodes can be adjusted by the potentiometer.

In order to reach the collector electrode, an electron emitted from the illuminated metal plate has to move against the potential difference, V, between the two electrodes. It does so at the expense of its kinetic energy. As long as photoelectrons manage to overcome the potential difference and reach the collecting electrode, an electric current, whose magnitude is shown by the ammeter reading, flows through the circuit. In order to measure the kinetic energy of the most energetic photoelectrons released by the incident light, Millikan determined the minimum electrical potential – the *stopping potential* – needed to reduce the current to zero.

The metal plate in the photoelectric tube was illuminated with monochromatic light of known frequency, f. The potential between the electrodes was adjusted by means of the potentiometer until no current passed through the ammeter. The stopping potential, V_0, was then recorded. The experiment was repeated with light of different frequencies and with plates made of different metals.

The change in an electron's kinetic energy, ΔK, as it moves through a potential difference, V, is given by

$$\Delta K = eV. \tag{2.29}$$

Thus, the stopping potential, V_0, is related to the maximum kinetic energy, K_{max}, possessed by the photoelectrons by the equation

$$K_{max} = eV_0. \tag{2.30}$$

Substituting this expression in Einstein's equation – Eq. (2.26) – gives

$$eV_0 = hf - W_{min}, \tag{2.31}$$

or, alternatively

$$V_0 = \frac{hf}{e} - \frac{W_{min}}{e}. \tag{2.32}$$

This function, which was derived directly from Einstein's equation, predicts a linear relationship between the stopping potential, V_0, and the frequency, f, of the light. This was confirmed by Millikan's experiment.

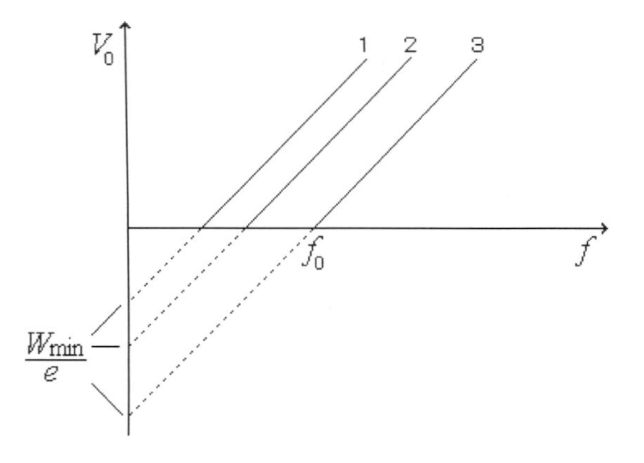

Fig. 2.18. The stopping potential, V_0, as a function of the frequency, f, of the light with which they are irradiated for three different metals. The metals' work functions, W_{min}, can be calculated from the points where the extrapolated straight lines cut the vertical axis.

Typical experimental results for three different metals are shown in Fig. 2.18. Planck's constant can be calculated from the slope, h/e, of the lines. The point at which each line cuts the horizontal axis gives the minimum frequency – the *threshold frequency*, f_0, which releases electrons from the metal. From the line's intersect with the vertical axis, the value of work function, W_{min}, of each metal can be obtained.

Example 2.6: Photoelectrons

Calculate

1. the maximum kinetic energy;
2. the stopping potential of the electrons emitted from a tungsten plate by light whose wavelength *in vacuo* is 200 nm. The work function of tungsten is 4.6 eV.

Calculation: Substituting the formula $f = c/\lambda$ in Einstein's equation gives

$$K_{\max} = hf - W_{\min}$$
$$= \frac{hc}{\lambda} - W_{\min}. \quad (2.33)$$

Substituting the relevant numerical values into this equation gives the maximum kinetic energy expressed in units of electron-volts:

$$K_{\max} = \frac{6.63 \cdot 10^{-34} \cdot 3 \cdot 10^{8}}{200 \cdot 10^{-9} \cdot 1.6 \cdot 10^{-19}} - 4.6 = 1.6\,\text{eV}.$$

The stopping potential is therefore 1.6 V.

2.2.4 X-rays

In the photoelectric effect, photons are annihilated when they strike the metal plate, each one surrendering its energy to an electron. Is the reverse process possible? Can the kinetic energy of electrons be used to produce photons?

This reverse process had been discovered, but incorrectly interpreted, a number of years before Planck put forward his quantum hypothesis. In 1895, Röntgen[22] had found that an invisible radiation of great penetrating power was produced when the 'electrical discharge from a fairly large induction coil' was passed through a cathode ray tube.

> '*The most striking feature of this phenomenon is the fact that an active agent [the radiation] here passes through a black cardboard envelope, which is opaque to the visible and ultra-violet rays of the sun... an agent which has the power of producing active fluorescence.*
> *We soon discover that all bodies are transparent to this agent, though in very different degrees... If the hand be held between the discharge tube and the screen, the darker shadow of the bones is seen within the slightly dark shadow image of the hand itself.*'[23]

Röntgen called this radiation *X-rays.* He found that X-rays propagate in straight lines from which they are deflected neither by electric nor magnetic fields. They ionise the matter through which they pass, cause fluorescent materials to glow and darken photographic plates.

---

[22] Wilhelm Röntgen 1845–1923, German physicist who was awarded the Nobel prize in 1901 for the discovery of X-rays.

[23] (1895) *Sitzungsberichte der Wurzburger Physikalischen – Medicinischen Gesellschaft*, December. Translated by G.F. Barker (1899) *Röntgen Rays*, New York: Harper & Brothers, pp. 3–4.

X-rays are now known to be high-frequency electromagnetic radiation, 10^{17} Hz to 10^{20} Hz. They show all the characteristic wave properties of electromagnetic radiation such as diffraction and polarisation. However, their refractive index is almost exactly equal to unity in all materials and so they cannot be focused by lenses.

Figure 2.19 shows the basic design of a typical device for producing X-rays. Electrons, accelerated by potential differences of tens of thousands of volts, are aimed at a metal target in a vacuum tube. Upon striking the target, the electrons are abruptly stopped and X-rays and heat are generated. Just 1% of the kinetic energy of the electrons is converted into X-rays; the rest appears as heat. Consequently, the target must be cooled when the tube is operated.

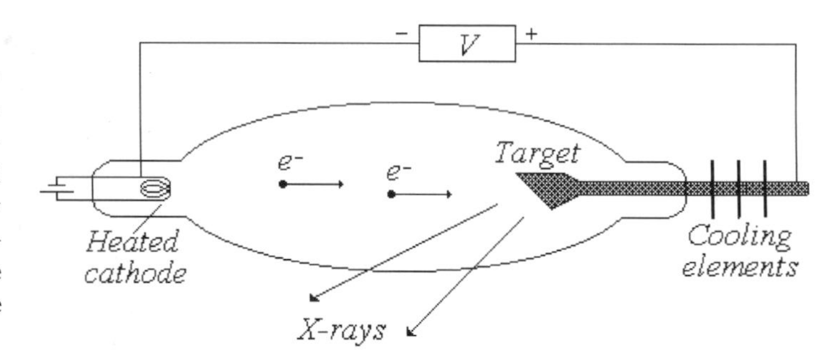

Fig. 2.19. An X-ray tube. Electrons emitted from the hot cathode are accelerated towards the target by the potential difference V. X-rays are produced from the target.

Typical intensity spectra of the X-ray radiation produced at various potential differences by such devices are shown in Fig. 2.20. All the spectra exhibit a sharp cut-off at a specific wavelength below which radiation is not produced. This minimum wavelength, $\lambda_{\min}$, is inversely proportional to the potential difference, V, at which the device is operated. At low and medium potential differences, 5 kV to 20 kV, the spectrum above the cut-off is continuous. Changing the material from which the target in the tube is made has no effect on the spectrum of this *continuous radiation*. At high potential differences, >20 kV, sharp intensity peaks appear in the spectrum at wavelengths that are a characteristic of the material from which the anode target is made.[24]

Initially, the origin of the continuous radiation was accounted for in classical terms. According to classical electromagnetic theory, all accelerated or decelerated charges radiate electromagnetic waves. The radiation produced by decelerated charges was called *bremstrahlung*. It was suggested that, on striking the target in the X-ray tube, the electrons are decelerated and that this

[24]The source of the *characteristic radiation*, i.e. the constituent that depends on the material from which the target is made, is elucidated in Chapter 3.3.

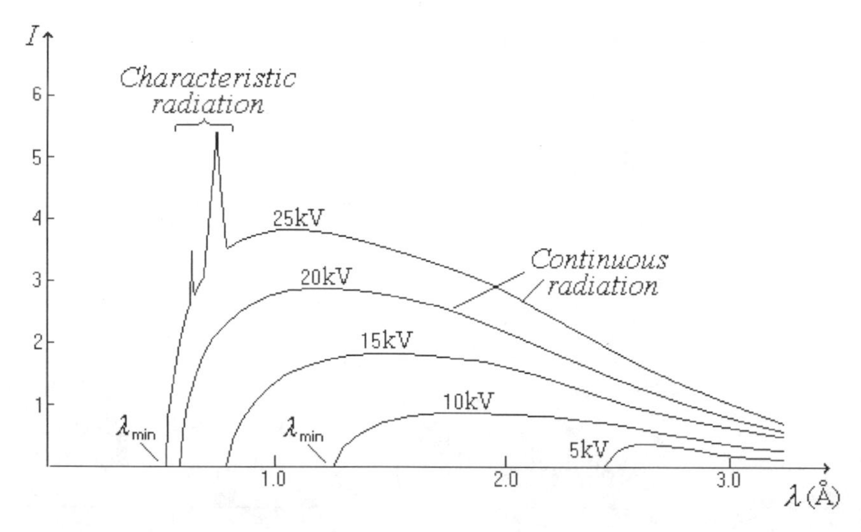

Fig. 2.20. The intensity, I, of the radiation produced from a typical X-ray tube (molybdenum target) as a function of its wavelength, λ, for different values of the potential difference at which the tube is operated. Note the characteristic radiation which appears as sharp peaks of intensity on the spectrum of the continuous radiation at high potential differences.

deceleration was the source of the continuous bremstrahlung radiation. However, no explanation could be found for why radiation is only produced down to a certain minimum wavelength nor why this minimum wavelength depends directly on the potential at which the X-ray tube is operated.

By contrast, these features are readily explained by the quantum theory of radiation. Although most of the electrons striking the target are decelerated in a series of multiple collisions with the target's metal atoms, a few may lose all or almost all of their energy in a single collision. The electrons that undergo multiple collisions generate the heat produced when the tube is operated; those stopped by a single collision generate the X-rays. Thus, the production of the X-rays is a sort of reverse photoelectric effect. In the simplest case, an electron loses all its energy in a single event, generating a single X-ray photon with a corresponding energy.

The kinetic energy of an electron accelerated from rest by a potential difference V is $K_{Electron} = eV$. The most energetic photon that can be produced from such an electron will possess all this energy:

$$(E_{Photon})_{\max} = eV = hf_{\max}. \tag{2.34}$$

The wavelength of the radiation associated with these photons will be

$$\lambda_{\min} = \frac{c}{f_{\max}} = \frac{ch}{eV}, \tag{2.35}$$

i.e. the minimum wavelength produced by the electrons is inversely proportional to the accelerating potential, V, as is found experimentally.

Example 2.7: Minimum X-ray Wavelength

An X-ray tube is connected to a 50,000 volt source. Calculate the shortest wavelength found in the spectrum of the radiation produced.

Calculation: Substituting the appropriate numerical values in Eq. (2.35), gives

$$\lambda_{\min} = \frac{ch}{eV} = \frac{3.0 \cdot 10^{8} \cdot 6.63 \cdot 10^{-34}}{50,000 \cdot 1.6 \cdot 10^{-19}} = 2.5 \cdot 10^{-11}\ \text{m}$$

for the minimum wavelength found in the continuous radiation.

2.2.5 X-rays and Crystallography

The nature of X-rays was the subject of much controversy during the years immediately following their discovery. Were they high energy cathode rays or a wave phenomenon? And if they were waves, what sort were they? At one point, Röntgen even suggested they might be *longitudinal* æther waves.

The wave nature of X-rays was ultimately established in 1912, when von Laue[25] suggested that crystals could be used as diffraction gratings for X-rays. Diffraction occurs whenever the progress of a wave-front is restricted by an object; the extent and nature of the diffraction depending on the ratio between the wavelength and the size of the object. When the spacing between a grating's elements is of the order of magnitude of the wavelength of the radiation incident upon it, the grating diffracts the radiation in certain preferred directions. Von Laue anticipated that the spacing between the atoms in crystals would be of the order of magnitude expected for the wavelength of X-rays. The idea was tried and diffraction patterns were indeed obtained. The result was viewed as confirmation of the wave nature of X-rays.

In 1913, William Bragg and his son Lawrence,[26] proposed a simple method of measuring the wavelengths of X-rays by means of their 'reflection' from crystal planes. Crystals comprise atoms (or ions) that are arranged in a spatial

[25] Max von Laue 1879–1960, German physicist who was awarded the Nobel prize in 1914 for discovering X-ray diffraction.

[26] William Bragg 1862–1942 and Lawrence Bragg 1890–1971, English physicists who were awarded the Nobel prize in 1915 for the analysis of crystal structure by X-rays.

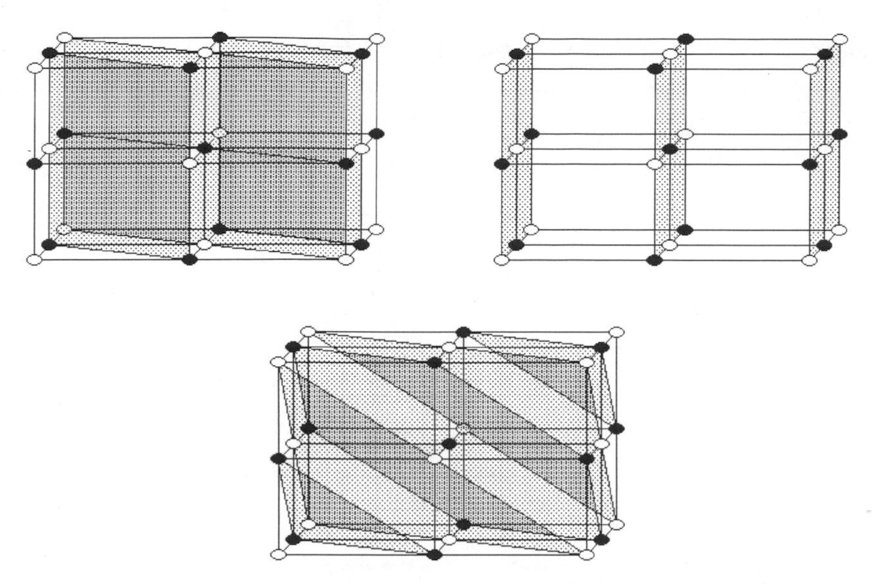

Fig. 2.21. 'Families' of parallel planes in the same crystal. In each family the distance between the planes is different.

lattice, such that they define 'families' of parallel planes (Fig. 2.21). The Braggs proposed that beams of X-rays would be reflected from these planes as though they were plane mirrors (Fig. 2.22).

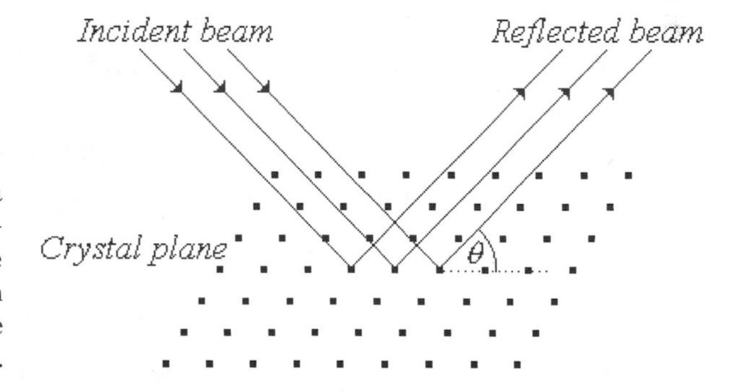

Fig. 2.22. The 'reflection' of a beam of electromagnetic radiation from a crystal plane. The glancing angle θ is that between the crystal plane and both the incident and the reflected beam.

The reflection of X-rays from a crystal plane can be understood in terms of Huygens' principle, in the same way as it explains the reflection of a beam of light from a plane mirror. According to this principle, when a wave-front strikes a reflecting surface, each point on the surface acts as a source of secondary light waves. The secondary waves interfere constructively only where the angle of reflection of the beam equals its angle of incidence such that the new wave-front is the common tangent to these secondary waves (Fig. 2.23). Similarly, when a beam of X-rays is aimed at a crystal, its atoms act as sources of secondary waves and the beam is reflected at an angle equal to its angle of incidence.

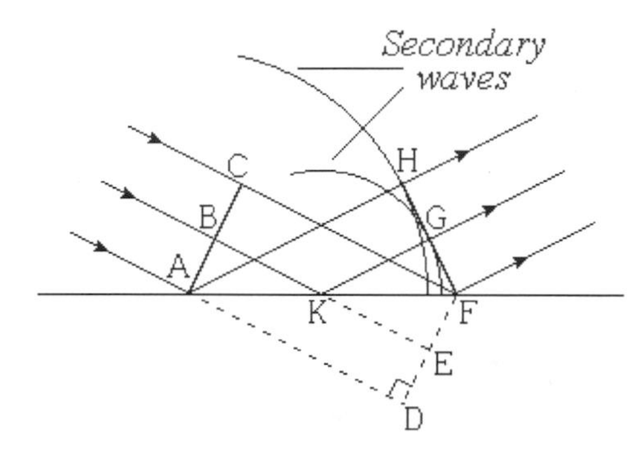

Fig. 2.23. The plane wave-front ABC represents a parallel beam of light which strikes a plane mirror obliquely. In the absence of the mirror, the wave-front would have travelled to DEF. Instead, it is reflected and occupies the position FGH. According to Huygens' principle, the new wave-front is tangential to the secondary waves whose centres are at the points A and K. The right angled triangles KGF, KBA and KEF are congruent; so too are the triangles AHF, ACF and ADF. Hence, the incident and reflected wave-fronts are equally inclined to the mirror.

The X-rays reflected from a family of parallel planes all emerge from the crystal at the same glancing angle θ relative to the planes. However, the distances the rays have travelled inside the crystal may differ, depending on the depth of the plane from which they were reflected. Whether these rays interfere constructively or destructively when they are recombined in the reflected beam depends on their path difference.

Consider rays reflected from two adjacent planes spaced at a distance d (Fig. 2.24). The path difference between the rays is

$$\mathrm{AB} + \mathrm{BC} = 2d \sin\theta. \tag{2.36}$$

Destructive interference occurs when the path difference equals an odd number of half the radiation's wavelength:

$$(2n+1)\frac{\lambda}{2} = 2d \sin\theta_n \quad (n = 1, 2, 3, \ldots). \tag{2.37}$$

At these angles, the reflected beam will then be of minimum intensity. The reflected beam of radiation will be of minimum intensity at glancing angles given by $\theta_n = \sin^{-1}\left(\frac{(2n+1)\lambda}{4d}\right)$.

Constructive interference occurs when the path difference equals an integer number of the radiation's wavelength:

$$n\lambda = 2d \sin\theta_n \quad (n = 1, 2, 3, \ldots). \tag{2.38}$$

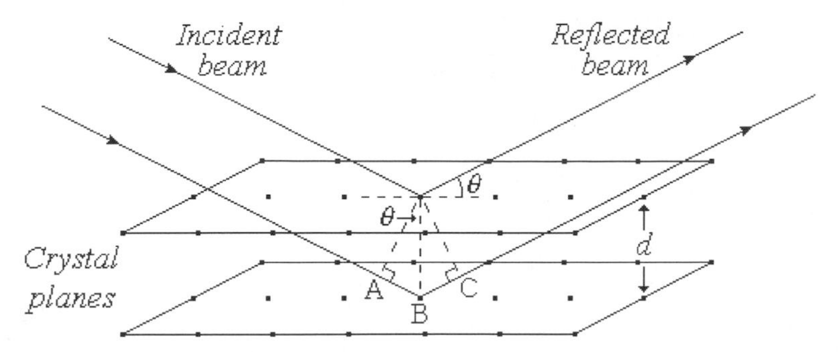

Fig. 2.24. The reflection of X-rays from two adjacent parallel crystal planes. The optical path of the ray reflected from the lower plane is longer than that reflected from the upper plane.

The reflected beam will then be of maximum intensity; n is termed the *order* of the reflected beam. The reflected beam of radiation will be of maximum intensity at glancing angles given by $\theta_n = \sin^{-1}\left(\frac{n\lambda}{2d}\right)$.

Thus, when a crystal is rotated in a beam of X-rays, the intensity of the beam reflected from it rises and falls as the angle at which the beam strikes the crystal changes. Equation (2.38) is known as *Bragg's equation.*

If the wavelength of the radiation is known, X-ray crystallography can be employed to determine the structure of crystals and to measure interatomic distances in molecules. However, its use is limited to regularly repeating structures, as only these can give rise to constructive interference in the reflected or diffracted beam. The structure of many proteins has been elucidated by this method; the helical structure of DNA was first indicated by the X-ray diffraction pattern to which it gave rise.

Example 2.8: X-ray Crystallography

The Braggs found that when the characteristic radiation produced by an X-ray tube with a palladium target is aimed at a rock-salt crystal (sodium chloride), the reflected beam has a maximum intensity at the following glancing angles: $5°59'$, $12°3'$, $18°14'$. Calculate the wavelength of the characteristic radiation.

Calculation: Assuming the glancing angles given above correspond to the first, second and third orders respectively, the wavelength of the radiation could be calculated from Bragg's equation, $\lambda = \frac{2d\sin\theta_n}{n}$, if the value of d, the spacing between the crystal planes was known.

The crystal form of rock salt is a cubic lattice – Fig. 2.25. Each cube in the lattice has an ion at each corner, making eight in all. Thus, in a large crystal, each ion, except for those on the crystal's surface, will be shared by eight cubes; on average, one ion per cube.

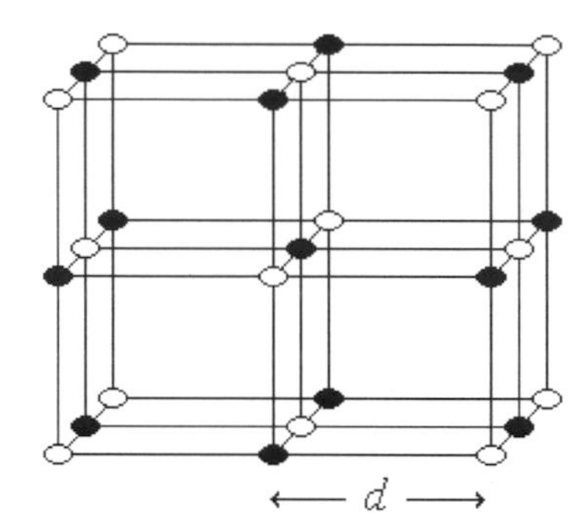

Fig. 2.25. A rock-salt (Na^+Cl^-) crystal. Note that the central ion is shared by eight cubes of the cubic lattice. The volume of each cube is d^3.

Every positive sodium ion in the rock-salt crystal is matched by a negative chlorine ion. Thus, a crystal containing a mole of Na^+ ions of mass 0.023 kg, will also contain a mole of Cl^- ions of mass 0.0355 kg. The density of rock-salt is 2,180 kg/m^3, and so in each cubic meter of rock salt there are $N = \frac{2 \cdot 2180}{(0.023+0.0355)}$ moles of ions.

Since, on average, each ion occupies a single cube of the crystal lattice, there will be $6.02N \cdot 10^{23}$ cubes in each cubic meter. Hence

$$\frac{2 \cdot 2180}{(0.023 + 0.0355)} \cdot 6.02 \cdot 10^{23} \cdot d^3 = 1 \tag{2.39}$$

where d^3 is the volume of each cube. From which we obtain for the spacing between the crystal planes:

$$d = 2.8 \cdot 10^{-10}\,\text{m}. \tag{2.40}$$

Substituting this value in Bragg's equation gives for the radiation's wavelength $\lambda = 0.59 \cdot 10^{-10}\,\text{m}$.

Chapter 2.3

Photons

Einstein presented his quantum theory of light and its elucidation of the photoelectric effect in a scientific paper that appeared in the spring of 1905. He followed it, just a few weeks later, with a ground-breaking paper on the mathematics of Brownian motion. His special theory of relativity was published in two papers during the summer of that *annus mirabilis* and before the year's end he had also completed his doctoral thesis on molecular dimensions. In the space of just a few months, Einstein had set much of the agenda of physics for the next 50 years.

In this chapter we will examine some of the properties of the light quanta – the *photons* – whose existence Einstein first conceived.

2.3.1 Photon Mass

The quantum theory of light postulates that electromagnetic radiation propagates as particles – photons – that travel at the speed of light. According to this theory, the energy possessed by a photon depends only on the frequency, f, of the radiation with which it is associated:

$$E_{photon} = hf. \tag{2.41}$$

However, as we saw in Chapter 1.4, according to the special theory of relativity, the total energy, E, of any particle is given by

$$E^2 = m_0^2 c^4 + c^2 p^2, \tag{1.67}$$

where m_0 is the rest-mass of the particle, p is its linear momentum and c is the velocity of light *in vacuo*. How can these two definitions of the particle's energy be reconciled in the case of the photon?

According to the principle of the constancy of the speed of light, no inertial observer can ever catch up with a photon in free space and move together with it such that it appears to be at rest; photons are never at rest. Consequently, the concept of the 'rest-mass of a photon' has no meaning; the rest-mass of a photon can be taken to be zero.

Substituting the photon's zero rest-mass, $m_0 = 0$, in Eq. (1.67) gives

$$E^2_{photon} = c^2 p^2_{photon}, \tag{2.42}$$

from which we obtain

$$p_{photon} = \frac{E_{photon}}{c}. \tag{2.43}$$

Thus, even though a photon has no rest-mass, a linear momentum, p_{photon}, can be attributed to it. Substituting $E_{photon} = hf$ and $c = f\lambda$, in Eq. 2.43 gives

$$p_{photon} = \frac{hf}{c} = \frac{h}{\lambda}, \tag{2.44}$$

where λ is the radiation's wavelength *in vacuo*. This simple relationship between the photon's linear momentum, p_{photon}, and the wavelength, λ, or frequency, f, of the radiation with which it is associated, is the fundamental link between the quantum theory of light and the wave theory of light.

Although photons have no rest-mass, it is often useful to attribute a moving mass to them. Equating the relativistic expression for the total energy of a particle, $E = mc^2$, with that for the energy of a photon, $E = hf$, gives

$$m_{photon} = \frac{hf}{c^2}. \tag{2.45}$$

In an article published in 1911, Einstein predicted that as a consequence of photon mass, a ray of light passing through the sun's gravitational field close to its surface would be deflected through an angle of 0.00025° (0.9 s of arc). Similar predictions had previously been made using Newtonian mechanics by calculating the deflection of an ordinary particle that passed close to the sun at the speed of light. Neither of these calculations took account of the curvature of space – its non-Euclidean nature predicted by the theory of general relativity. In 1915, Einstein repeated his calculations to allow for space curvature and predicted a deviation of 0.0005° (1.8 s of arc). This prediction was confirmed during the total solar eclipse in 1919 when it was found that a star whose line of sight passed close to the sun's surface appeared to be displaced from its true astronomical position (Fig. 2.26). The observed deviation was 0.00045° ± 25%. This result was received as evidence of the gravitational mass of the photon and as a confirmation of the theory of general relativity.

Photons rising from the earth's surface lose energy as a result of the work that must be done against the earth's gravitational pull. This loss does not appear as a reduction in their speed – photons always travel at the speed of light – but as a reduction in the frequency of the radiation. The phenomenon is called the *gravitational red-shift* because, in the case of visible light, the frequency of the rising photons is shifted in the direction of the red end of the spectrum. This reduction in frequency was observed in experiments conducted in 1960, employing the Mössbauer effect.[27]

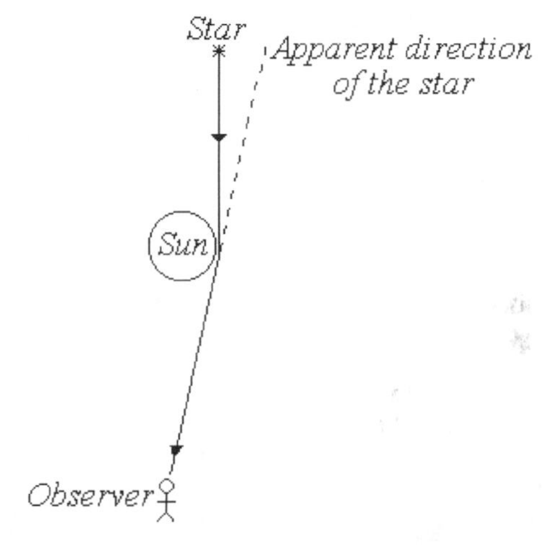

Fig. 2.26. The deflection of a ray of starlight by the sun's gravitational field. During a total solar eclipse, when the sun's glare does not blind the observers, the star can be viewed and appears to be displaced from its true position. The deviation is attributed to the bending of the light ray as it glances the sun's surface.

Like any stream of particles, a beam of photons exerts a force on any surface it strikes. The phenomenon, called *radiation pressure*, was discovered even before the development of the quantum theory. It was found that a small light paddle-wheel could be made to rotate by aiming a beam of light at it. The phenomenon was explained at the time in terms of classical electromagnetism. It was reasoned that the force exerted by the radiation resulted from the interaction of the electrical and magnetic fields in the incident electromagnetic waves with the electrons in the surface of the paddle-wheel. It was shown that the linear momentum transferred by the incident radiation to the surface, $p_{radiation}$, is related to the energy it delivers, $E_{radiation}$, by

$$p_{radiation} = \frac{E_{radiation}}{c}. \tag{2.46}$$

The similarity between this relationship derived from classical electromagnetic considerations and that derived from relativistic considerations, Eq. (2.43), is evident and is suggestive of a consistency underlying electromagnetic theory, relativity and the quantum theory of radiation.

[27]See Chapter 6.2.

Example 2.9: Radiation Pressure

Calculate, from classical considerations, the force exerted on a perfectly reflecting mirror by a laser beam of cross-section $10^{-6}\,\text{m}^2$ and intensity $10^6\,\text{W/m}^2$, that strikes the mirror normal to its surface.

Solution: In general, force is the rate of change of linear momentum: $F = \Delta p/\Delta t$. Since the laser beam strikes the mirror normally and is perfectly reflected from it, the rate of change of the linear momentum of the light incident and reflected from it per second, Δp_{light}, i.e. the force exerted on the mirror, is equal to twice the momentum, p_{light}, delivered by the light beam every second:

$$f = \Delta p_{light} = p_{light} - (-p_{light}) = 2p_{light}.$$

Substituting $p_{light} = E_{light}/c$ gives

$$F = \frac{2E_{light}}{c},$$

where E_{light} is the radiation energy supplied by the laser beam each second: $E_{light} = 10^{-6} \cdot 10^6 = 1\,\text{J/s}$. Hence,

$$F = \frac{2 \cdot 1}{3 \cdot 10^8} = 6.7 \cdot 10^{-9}\,\text{N}.$$

2.3.2 The Compton Effect

Confirmation of the quantum theory of light and, in particular, of its prediction that a linear momentum can be attributed to photons, was obtained in a landmark experiment carried out by Compton[28] in 1923. The experiment set out to investigate the scattering of high energy photons – X-ray photons – by free electrons.[29]

According to classical electromagnetic theory, radiation incident on free electrons should be scattered in all directions, without undergoing any change in its frequency. The incident waves set up vibrations in the electrons at frequencies equal to their own, and these, in turn, generate the secondary waves of the scattered radiation, also at the same frequencies. Overall, the free electrons should not take up any energy from the incident radiation.

Relativistic considerations appear to support this conclusion. Conservation of linear momentum requires that a free stationary electron that absorbs an

[28]Arthur Compton 1892–1962, American physicist who was awarded the Nobel prize in 1927 for the discovery of the effect named after him.

[29]Scattering by bound electrons is discussed in Chapter 4.1 below.

amount of energy, $E_{radiation} = pc$, from an incident beam of electromagnetic radiation, must also acquire a corresponding amount of linear momentum, $p = E_{radiation}/c$. The energy absorbed by the electron appears as kinetic energy, K, such that

$$K = E_{radiation} = pc. \tag{2.47}$$

However, according to special relativity, the kinetic energy of a moving particle is given by

$$\begin{aligned} K &= mc^2 - m_0c^2 \\ &= \sqrt{m_0^2c^4 + p^2c^2} - m_0c^2. \end{aligned} \tag{2.48}$$

Clearly, this relativistic expression is incompatible with the conservation requirement given by Eq. (2.47) that $K = pc$.

Thus, it would appear that for free electrons to absorb the energy of incident electromagnetic radiation would require a violation of the principles of energy and linear momentum conservation. We may note, in passing, that this problem does not arise in the photoelectric effect, where the electrons taking up energy from the radiation are bound to the atoms in the metal plate. In this case, the atoms participate together with the electrons in absorbing the energy and momentum of the incident radiation. The heavy particles absorb the marginal amounts of momentum or energy not absorbed by the electrons, thereby ensuring momentum and energy conservation.

However, notwithstanding all these considerations, free electrons do, in fact, absorb energy from high-frequency electromagnetic radiation. When a beam of X-rays of frequency f passes through a medium containing free electrons, X-rays of lower frequencies $f' < f$, are detected in the emergent beam. In terms of the quantum theory of light, the lower frequencies indicate the presence of lower energy photons in the emergent beam. This implies that photons in the incident beam must have surrendered energy during the passage of the radiation through the scattering medium. Apparently, an interaction does occur between the incident X-ray photons and the free electrons in which energy is transferred from the former to the latter. Compton suggested that this interaction could be elucidated by means of an amalgam of classical mechanics, the quantum hypothesis and special relativity.

According to Compton, photons of the incident X-rays collide elastically with the free electrons. After colliding, the photons and electrons move off in different directions (Fig. 2.27).

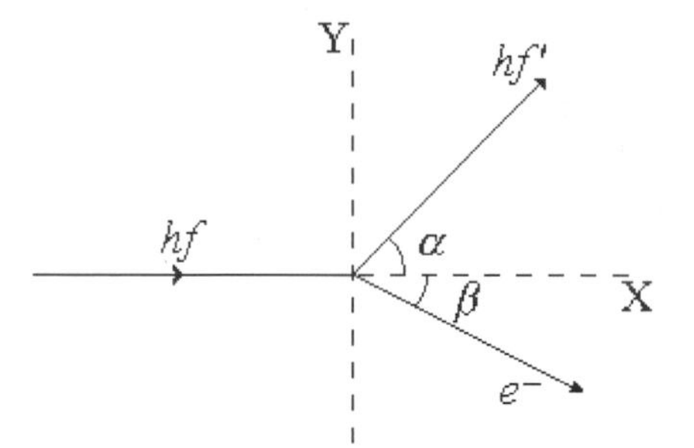

Fig. 2.27. An elastic collision between a photon of frequency f and a stationary free electron. The electron recoils at an angle β relative to the direction of the incident photon.

The energy of the photon after the collision, hf', is less than that of the incident photon, hf; the difference, $h(f - f')$ appears as the kinetic energy, K, of the electron:

$$K = h(f - f'). \tag{2.49}$$

The total (relativistic) energy, E, of the electron after the collision is equal to the sum of its rest-energy, m_0c^2, and its kinetic energy, K:

$$\begin{aligned} E &= m_0c^2 + K \\ &= m_0c^2 + h(f - f'). \end{aligned} \tag{2.50}$$

According to the quantum theory of radiation, the relationship between the linear momentum of a photon and the frequency of its associated radiation is given by $hf = cp_{photon}$. Substitution in Eq. (2.50) gives

$$E = m_0c^2 + c(p_{photon} - p'_{photon}), \tag{2.51}$$

where p_{photon} is the magnitude of the photon's linear momentum before the collision and p'_{photon} is its magnitude afterwards.

A second expression for the electron's total energy after the collision can be obtained from the equation for the total energy of a particle given by special relativity:

$$E^2 = m_0^2c^4 + c^2p_{electron}^2, \tag{2.52}$$

where m_0 is the rest-mass of the electron and $p_{electron}$ is the magnitude of its linear momentum after the collision. Combining the two expressions for the electron's energy – Eqs (2.51) and (2.52) – gives

$$m_0^2c^4 + c^2p_{electron}^2 = [m_0c^2 + c(p_{photon} - p'_{photon})]^2. \tag{2.53}$$

After the collision, the original linear momentum, $\mathbf{p}_{photon}$, of the incident photon will be divided between that, $\mathbf{p}_{electron}$, of the recoiling electron and that,

$\mathbf{p}'_{photon}$, of the scattered photon:

$$\mathbf{p}_{photon} = \mathbf{p}'_{photon} + \mathbf{p}_{electron}. \tag{2.54}$$

Separating this vector equation into components gives

$$p_{photon} = p'_{photon} \cos\alpha + p_{electron} \cos\beta \quad \text{(x-axis)} \tag{2.55}$$

$$0 = p'_{photon} \sin\alpha + p_{electron} \sin\beta \quad \text{(y-axis)} \tag{2.56}$$

where α and β are, respectively, the scattering angles of the photon and electron.

By applying the general trigonometric relationship $\sin^2\theta + \cos^2\theta = 1$ and eliminating β from these two equations we obtain

$$p^2_{electron} = p^2 - 2p_{photon}p'_{photon} \cos\alpha + p'^2_{photon} \tag{2.57}$$

Substituting $p^2_{electron}$ from Eq. (2.57) in Eq. (2.53) and re-arranging gives

$$\frac{1}{p'_{photon}} - \frac{1}{p_{photon}} = \frac{(1-\cos\alpha)}{m_0 c}. \tag{2.58}$$

This relationship can be re-written in terms of the difference, $\lambda' - \lambda$, between the respective wavelengths of the emergent and incident radiation by the substitutions $1/p'_{photon} = \lambda'/h$ and $1/p_{photon} = \lambda/h$ from Eq. (2.44):

$$\lambda' - \lambda = \frac{h}{m_0 c}(1 - \cos\alpha). \tag{2.59}$$

Thus, Compton predicted that the reduction in the radiation's wavelength should depend only on the scattering angle, α; all the other factors, h, m_0 and c, that appear in Eq. (2.59) being constants.

In the experiment he conducted to test this prediction, Compton directed a beam of X-rays onto a graphite target. Since the binding energy of the outermost electrons in the carbon atoms that comprised the target is negligible compared to that of the high energy X-ray photons, these electrons could be considered to be essentially free. Using the Bragg technique, Compton compared the wavelength of the incident X-rays with that of the scattered X-rays at specific angles, $45°, 90°$ and $135°$. In each case, he found that the difference between the wavelength of the scattered X-rays and that of the incident X-rays, $\lambda' - \lambda$, corresponded to that predicted by Eq. (2.59). The phenomenon is called the *Compton effect*. Shortly afterwards, Wilson[30] and Bothe[31] succeeded in tracking the recoiled electrons by the trails they left in a cloud-chamber.

[30] Charles Wilson 1869–1959, Scottish physicist who was awarded the Nobel prize in 1927 for the invention of the cloud-chamber.

[31] Walther Bothe 1891–1957, German physicist who was awarded the Nobel prize in 1954 for his invention of experimental methods for counting very fast particles and for measuring very short time intervals.

Since the increase in the wavelength of the scattered radiation, $\Delta\lambda = \lambda' - \lambda$, is independent of the wavelength of the incident radiation, the fractional change, $\Delta\lambda/\lambda$, becomes negligible at long wavelengths. Thus, the Compton effect is only exhibited by short-wave radiation: X-rays exhibit a maximum fractional change of about 10^{-2}; γ-rays of the order of one.

Furthermore, because the mass of the scattering particle, m_0, appears in the denominator of Eq. (2.59), increases in the wavelength of even high-energy radiation are only detected when the scattering entities are as small as electrons. The change in wavelength caused by a scattering particle as large as an atom is totally negligible. Thus, no detectable change occurs in the wavelength of the X-rays scattered by the atoms and ions comprising the crystal lattice in the Bragg technique.

In terms of quantum theory, the mechanism of the Compton effect can be regarded as a sequence of two electron–photon interactions:

1. In the first, the free electron absorbs energy $E = hf$ and linear momentum of magnitude $p = h/\lambda$ from the incident photon.
2. In the second, a photon with energy $E' = hf'$ and linear momentum of magnitude $p' = h/\lambda'$ is emitted from the electron.

Overall, the electron acquires amounts of energy and linear momentum of magnitudes $\Delta E = E - E'$ and $\Delta p = p - p'$, respectively, such that they satisfy the relativistic requirement $\Delta E = \sqrt{m_0^2 c^4 + (\Delta p)^2 c^2} - m_0 c^2$.

The Compton effect is not the only phenomenon amenable to interpretation as a sequence of particle and photon interactions. All the interactions between charged particles described by classical theory in terms of oscillators and electromagnetic fields, can be elucidated in terms of a sequence of particle–photon interactions that are most conveniently illustrated by the pictorial representations, known as *Feynman*[32] *diagrams*, suggested by the space–time diagrams of the processes first developed by the physicist of that name (see Chapter 4.4 below).

For example, consider the exchange of momentum and energy that accompanies the repulsive interaction between two positively charged particles – Fig. 2.28.

Suppose, that at a particular moment, one of the charged particles is at point A and the other at point B. The first particle – particle (1) – emits a photon which moves to B where it is absorbed by a second particle – particle (2).

[32]Richard Feynman 1918–1988, American physicist who was awarded the Nobel prize in 1965 for his work on quantum electrodynamics.

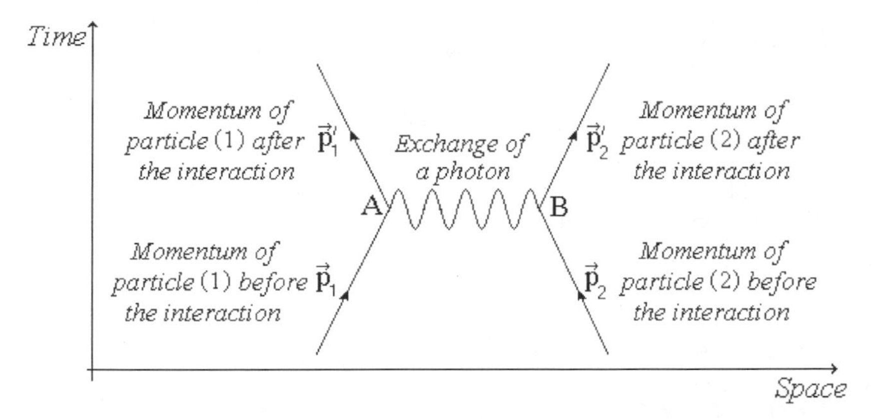

Fig. 2.28. A Feynman-type space–time diagram. The electromagnetic interaction between two charged particles, (1) and (2), at the moment they are situated at the points A and B, respectively, shown as an exchange of a virtual photon between them.

As a result, particle (2) absorbs energy and linear momentum and its motion is changed accordingly. However, the motion of particle (1) must also change, for there has been a matching reduction in its energy and linear momentum; it lost the photon.

Of course, the reverse process is equally possible. Particle (1) can absorb a photon emitted from particle (2), resulting in similar changes in their motions. Thus, the mechanism of the interaction can be viewed as an exchange of photons between the particles. It can be shown that the force that results from the exchange of zero rest-mass photons between two charged particles varies as the inverse square of the distance between them, i.e. in accordance with Coulomb's law.

Because they are never directly observed, the photons that 'carry' the force from one particle to another are called *virtual photons*; they appear in neither the initial nor final conditions of an experiment. In order for the photons generated in such interactions to become actual and observable, a supply of energy must be provided.

Example 2.10: The Compton Wavelength

A photon strikes a free electron and is scattered through an angle of 90° relative to its initial direction of motion. Calculate the change in the wavelength of the photon.

Solution: From Eq. (2.59), the change in the photon's wavelength is given by

$$\Delta\lambda = \frac{h}{m_0 c}(1 - \cos\alpha).$$

In the case where $\alpha = 90°, \cos\alpha = 0$, and so

$$\Delta\lambda = \frac{h}{m_0 c} = 2.43 \cdot 10^{-12}\,\mathrm{m}, \tag{2.60}$$

$h/m_0 c$ is called the *Compton wavelength*; it defines the change in the wavelength of any photon that is scattered through an angle of 90° by a free electron.

2.3.3 Photons — Light Particles

The success of the quantum theory of radiation in elucidating the photoelectric and Compton effects reopened the debate on the nature of light. Although the interference, diffraction and polarisation of light had been convincingly explained by the wave theory, the interactions between radiation and matter appeared to offer equally convincing evidence that light is comprised of particles – photons. So do we need both models, the wave and the particle, for a complete understanding of the phenomenon of light? Or can the interference, diffraction and polarisation of light also be explained by the particle theory of radiation? In other words, is the quantum theory of light a partial or a complete theory of light? We will start our investigation of this question by reference to Young's famous double-slit experiment.

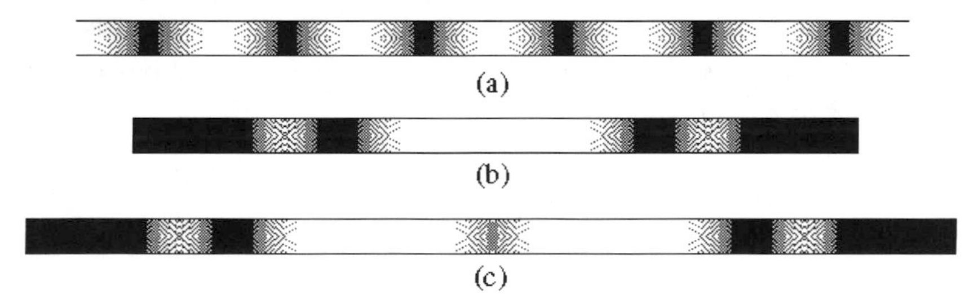

Fig. 2.29. The patterns that appear in photographs of: (a) The typical interference pattern of alternating bright and dark fringes obtained in Young's experiment when monochromatic light is projected through two adjacent narrow slits; (b) The diffraction pattern typically obtained when monochromatic light is projected through a single slit: a broad central bright band with weaker side bands; (c) The two overlapping diffraction patterns produced when each slit is opened for half the time but never both at the same time.

As we described in Chapter 1.2, Young observed that a sequence of bright and dark fringes or bands – an *interference pattern* – appears when light from a single source is projected onto a screen through two adjacent narrow slits. If the screen is replaced by a photographic plate, a permanent image of the interference pattern can be obtained by the usual photographic processes (Fig. 2.29a).

An interference pattern is only obtained if the light passes through both slits concurrently, i.e. if both slits are open at the same time. If just one is open, a different pattern appears. This pattern, characteristic of single slit diffraction, is usually referred to as a *diffraction pattern* (Fig. 2.29b). If the slits are opened sequentially, first the one and then the other, but never both at the same time, the final image that appears in the photograph is of two adjacent and overlapping diffraction patterns (Fig. 2.29c).

How is the formation of the interference pattern produced in Young's experiment accounted for by the wave theory of light and the quantum theory of light?

1. Young's experiment according to the wave theory of light

According to Huygens' principle and the wave theory of light, the role of the two slits in the production of the interference pattern in Young's experiment is clear and well defined. They are coherent sources of secondary light waves, i.e. they are 'orchestrated and in time with one another' such that the phase difference between them is constant. The interference pattern results from the superposition of these secondary waves on the screen. Bright bands – intensity maxima – appear where the secondary waves interfere constructively; dark bands – intensity minima – where they interfere destructively.

In general, the amplitude, A, of the compound wave formed by the superposition of the secondary waves at a point on the screen in Young's experiment is given by

$$A^2 = A_1^2 + A_2^2 + 2A_1A_2\cos\phi, \qquad (2.61)$$

where A_1 and A_2 are the amplitudes of the secondary waves and the angle ϕ is the phase difference between them at the particular point.[33] The intensity, I, of the light at the particular point is, in turn, proportional to the square of the amplitude of the compound wave formed by the superposition of the secondary waves: $I \propto A^2$. Hence:

* Intensity maxima (constructive interference) appear where $\cos\phi = +1$, i.e. at those places on the screen where the phase difference satisfies the condition $\phi = 2n\pi$ $(n = 0, \pm1, \pm2\ldots)$.
* Intensity minima (destructive interference) appear where $\cos\phi = -1$, i.e. where $\phi = (2n+1)\pi$ $(n = 0, \pm1, \pm2\ldots)$.

[33]The derivation of this equation is given in Supplementary Topic A which is devoted to the mathematical description of wave motion.

The phase difference, ϕ, between the secondary waves arriving at a particular point on the screen depends only on the difference, δ, in the lengths of the optical paths to the point (Fig. 2.30):

$$\phi = \frac{2\pi\delta}{\lambda}, \tag{2.62}$$

where λ is the wavelength. Thus, according to the wave theory of light, the appearance of the interference pattern obtained in Young's experiment is determined by the difference in the lengths of the optical paths taken by the light from the two coherent sources.

* Intensity maxima correspond to an optical path difference $\delta = n\lambda$.
* Intensity minima correspond to an optical path difference $\delta = (2n+1)\lambda/2$.

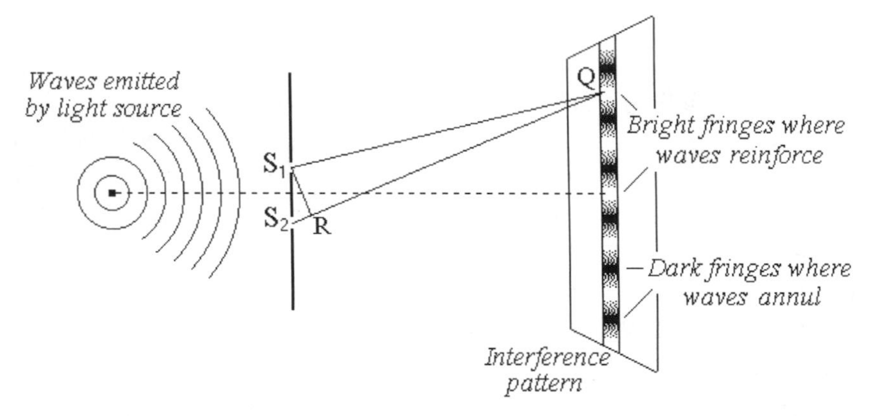

Fig. 2.30. Young's experiment. S_1 and S_2 are two narrow horizontal slits through which light passes from the source to the screen. The slits constitute coherent sources of secondary waves. Whether the secondary waves reinforce or annul at a particular point on the screen depends on the difference, $\delta = S_2R$, between the respective lengths of the optical paths, S_1Q and S_2Q, from the slits to the point.

If two independent light sources are used in place of the two slits, no interference pattern is observed. A stationary and unchanging interference pattern is only obtained if the sources are coherent, such that there is a constant phase difference between them. Typical sources such as flames, discharge tubes and incandescent lamps emit light intermittently, as a multitude of uncoordinated wave trains each approximately 1 m long. Consequently, two such independent and unrelated sources cannot maintain a constant phase difference between them; they cannot be coherent. The interference patterns they produce change so rapidly, about a billion times a second, that the screen's illumination appears uniform. However, since the sources of the secondary waves in Young's experiment – the two narrow slits – share the same original light source, they automatically and simultaneously suffer the same phase changes and so they are, of necessity, related and coherent.

2. Young's experiment according to the quantum theory of light

According to quantum theory, the light intensity at a point on the screen depends on the number of photons striking there. The darkening of an exposed photographic plate results from a chemical reaction initiated by the photons in the incident light. Thus, the photograph of the interference pattern (and likewise that of a diffraction pattern) records the distribution of the photons striking the plate. The bright bands in the photographic image of the interference pattern correspond to areas upon which many photons fell and the dark bands to areas upon which almost no photons fell.

However, the role of the slits in the quantum interpretation of interference is not immediately clear. How exactly do they affect the distribution of the photons on the photographic plate? And why must both slits be kept open for an interference pattern to appear? We will investigate these questions by considering the behaviour of individual photons as they pass from the light source to the photographic plate.

In a simple experiment carried out in 1908, Taylor[34] investigated how an interference pattern develops in Young's experiment when extremely weak light is used, so weak that the probability of there being more than one photon at any time in the space around the slits was negligible.[35] For this purpose, he assembled a light-proof box containing a weak light source, a series of filters, two slits and a photographic plate, arranged as required for Young's double-slit experiment (Fig. 2.31).

The assembled apparatus was left to stand for a few days after which the photographic plate was removed and processed. After a fresh photographic plate had been inserted, the apparatus was again left to stand, this time for longer. The photographic plate was then again removed and processed. This was repeated several times, on each occasion allowing a longer time for the interference pattern to develop, up to a maximum of a few months.

When left for just a few days, no clear or distinct pattern appeared in the developed photograph; just a random sprinkling of white dots. However, after longer periods of time, the shape and form of a typical interference pattern began to emerge in the distribution of the dots; and after still longer periods – about three months – a distinct interference pattern appeared (Fig. 2.32).

[34] Sir Geoffrey Ingram Taylor 1886–1975, English physicist.

[35] See Example 2.10 for a calculation of this probability.

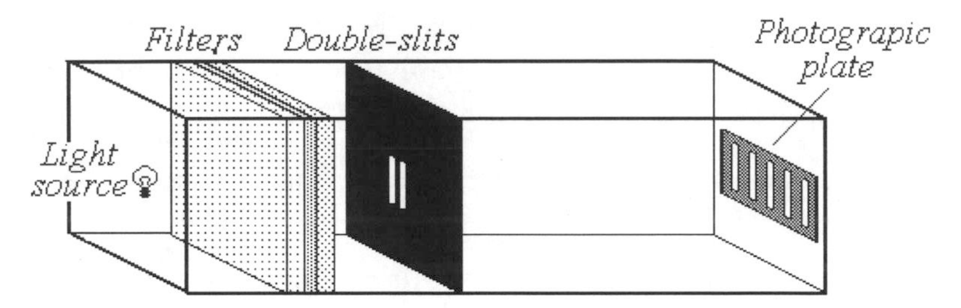

Fig. 2.31. Taylor's experiment. A light-proof container enclosing the components of Young's double-slit experiment. After passing through the filters, the light was so weak that the probability of there being more than one photon at any time in the space around the slits was negligible. The interference pattern appears on the photographic plate. In alternative assemblies, the photographic plate is replaced by a sensitive electronic light detector that emits a signal every time a photon strikes it.

The appearance of the dots in the first images confirms that the interference pattern is indeed the result of the distribution of a multitude of photons on the surface of the photographic plate, each dot corresponding to the incidence of a single photon. However, since there was never more than one photon in the space around the slits at any time, the interference pattern that emerged in the later images could not be the result of an interaction between the photons; it could not be the outcome of the constructive or destructive combination of photons.

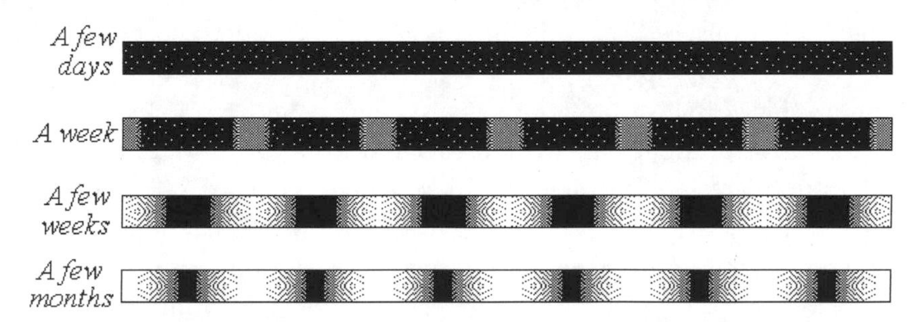

Fig. 2.32. Taylor's experiment. The interference patterns that appear in the photographic images after different periods of time, from a few days (the top pattern) to a few months (the bottom one).

Evidently, the interference pattern observed in Young's experiment is an inherent property of each photon; it is as though each photon 'interferes' with itself as it passes through the apparatus from the source to the screen. Even when the light is of greater intensity and many photons are simultaneously passing through the apparatus, each interferes only with itself and not with the others.

Does this mean that each photon splits in two as it passes through the slits? If the photographic plate is replaced by an electronic light detector that emits a signal every time a photon strikes it, or by a fluorescent screen which flashes when struck by a photon, it is found that the signals or flashes are all of the same intensity. Similarly, if photon detectors are installed next to each slit, they never simultaneously record the passage of a photon. There is, therefore, no evidence for a splitting of the photons during the experiment, so this cannot be the purpose for the slits.

Is it then the case that some of the photons pass through one slit and others through the other, such that when added together the joint effect of their distribution on the photographic plate is to produce the interference pattern? To examine this possibility, the Taylor experiment is repeated but with just one slit open. After allowing time for a pattern to develop on the photographic plate, this slit is closed and the second is then opened for the same period of time as the first. Under these circumstances no interference pattern is produced. Instead, two adjacent overlapping single-slit diffraction patterns appear (Fig. 2.29c). Clearly, the interference pattern is not the sum of the distribution of the photons that pass through one slit plus that of the photons passing through the other.

Let us now suppose that, instead of being opened and closed one at a time, both slits are left open all the time, but a detector designed to reveal through which slit each photon actually passes is installed in the apparatus. In this case too, two adjacent overlapping single-slit diffraction patterns are produced and not the typical double-slit interference pattern. Moreover, it makes no difference whether the detector is placed closer to the slits than is the screen, or further away, i.e. whether the photon's choice can be detected before or after it strikes the screen; the very presence of the detector destroys the interference pattern. If the detector is turned off or removed from the apparatus, the interference pattern reappears.[36] Hard as it is for us to understand, the photons appear to behave as though they were 'aware' of the intent of the experiment.

Evidently, an interference pattern only appears when the experiment provides no means of determining through which of the slits each photon passes

[36]The question of when precisely each photon decides which way to go has no answer.
'The common-sense idea that there is an objective reality "out there all the time" is a fallacy. When reality and knowledge are entangled, the question of *when* something *becomes* real cannot be answered in a straightforward manner.' Paul Davies (1995) *About Time*, New York: Simon & Schuster.
'The quantum principle shows that there is a sense in which what the observer will do in the future defines what happens in the past – even in a past so remote that life did not then exist.' John A. Wheeler (1977) *Foundational Problems in the Special Sciences*, Dordrecht: Reidel, p. 3.

on its way from the source to the screen, i.e. only when both slits are open concurrently and are 'unwatched'.

The patterns appearing on the screen or the photographic plate are undoubtedly the result of the distribution of the photons striking their surfaces. However, there is no way of anticipating exactly where a particular photon will strike. Likewise, it is impossible to tell which path a particular photon took, through one slit or the other; a photon cannot be assigned a specific trajectory. We cannot even suppose that half the photons went through one slit and half through the other. We can only predict how a very large number of photons will be distributed on the surface of the screen or photographic plate.

On this view, the interference and diffraction patterns are statistical phenomena, associated with the probability that, in a given experimental set-up, a photon will strike a particular point on the screen. The greater the probability, the more photons strike the point and the stronger the light intensity at the point. Hence, the bright fringes correspond to a high probability; the dark fringes to a low probability.

The essential features of the explanations given by the wave and quantum theories of light to Young's experiment are summarised in the following matrix:

	Wave Theory	Quantum Theory
Pattern of fringes:	The intensity bands arise from the variations in the value of the square of the amplitude, A^2, of the resultant wave at each point on the screen.	The intensity bands arise from the variations in the probability, P, of a photon striking different points on the screen
Role of the slits:	To act as two coherent sources of the secondary waves that interfere on the screen.	To present two potential routes by which a photon can pass from the source to the screen

Is light then a wave or a particle phenomenon? Can the two explanations, that of the wave theory in terms of wave amplitudes and that of the photon theory in terms of counting photons, be reconciled? Or are the wave and particle hypotheses of light in fact as irreconcilable as classical physics had taught?

To answer these questions, let us reconsider the experiment conducted by Compton on the scattering of X-rays. In order to measure the wavelength of the X-rays, Compton employed the technique developed by the Braggs which imputes a wave character to the X-rays. However, the theory and analysis upon which Compton predicted the change he expected to find in the wavelength of the radiation was based on the assumption that X-rays are composed of photons, i.e. particles that collide with free electrons according to the laws of mechanics. Thus, overall, we may say that Compton's experiment demonstrated that X-rays have both particle and wave characteristics, each attribute manifesting itself independently in a different part of the experiment in line with the experimenter's expectations.

In modern physics, the wave and particle hypotheses of light complement one another; they are not mutually exclusive. In our understanding of light intensity we link these two qualities by saying that the number of photons, N, striking a particular point on a photographic plate is directly proportional to the intensity, I, of the light wave formed at the point, i.e. to the square, A^2, of its amplitude:

$$N \propto A^2. \tag{2.63}$$

Thus, in Young's experiment, the probability, P, of a photon striking close to a particular point on the screen will be directly proportional to the square of the amplitude of the compound wave formed at the point by the superposition of the secondary waves:

$$P \propto A^2 = A_1^2 + A_2^2 + 2A_1A_2 \cos\phi. \tag{2.64}$$

At those points where the interference is constructive, the square of the amplitude is at a maximum and so the probability of a photon striking the point is high. On the other hand, at those points where the interference is destructive the amplitude is at a minimum and the probability of a photon striking the point is negligible.

The single-slit diffraction pattern is also a statistical phenomenon; it too reflects the probability of a photon striking different points on the photographic plate. If just one slit is open, the probability, P_1, that a photon will strike a particular point depends on the square of the amplitude, A_1^2, at the particular point of the secondary waves it emits. Overall, this probability corresponds to a single-slit diffraction pattern. Likewise, when just the other slit is open, the probability, P_2, that a photon will strike any point depends on A_2^2. Overall, this probability corresponds to a single-slit diffraction pattern too. When the slits are opened one at a time, one after the other, the total probability, P_{12}, that a photon will strike a particular point on the photographic plate is simply the sum of the two separate probabilities:

$$P_{12} = P_1 + P_2,$$

or in terms of the wave amplitudes:

$$P_{12} \propto A_1^2 + A_2^2.$$

This probability, which corresponds to two single-slit diffraction patterns, is different from the probability, $P \propto= A_1^2 + A_2^2 + 2A_1A_2 \cos\phi$, associated with the formation of an interference pattern; the interference factor, $2A_1A_2 \cos\phi$, is missing.

Installing a photon detector to determine through which slit each photon passed removes any uncertainty we may have as to the path taken by the photons. As such, it is essentially equivalent to opening the slits one at a time and so it produces the same distribution pattern.

2.3.4 The Locality Paradox

When first proposed, the elusiveness attributed to photons by the quantum theory of light troubled many physicists. Known as the *Copenhagen interpretation*, by virtue of its advocacy by the Danish physicist Bohr, this construal of quantum theory was vigorously opposed by no less a figure than Einstein himself: 'God does not play dice', he declared. If photons actually exist, their motion should be determinable and not 'spooky' as the Copenhagen interpretation asserted. In 1935, Einstein proposed an experiment that he believed would demonstrate the inadequacy of the Copenhagen interpretation.

The quantum theory of light accounts for light polarisation by assigning an intrinsic angular momentum – *spin* – to the photons. In principle, photon spin can take two forms: it can either be parallel or anti-parallel to the direction of the photon's linear momentum. The various modes and orientations light polarisation actually takes can all be accounted for in terms of these two basic polarisation states.

Suppose, Einstein argued, that a stationary particle emits two photons that move off back-to-back in opposite directions. The conservation of angular momentum requires that the photons' spins be in opposite orientations, such that their sum be zero. Thus, if the spin of one is subsequently measured, that of the other will automatically become known too.

In the context of classical physics this 100% correlation between the spins presents no difficulties. The classical conception asserts that the particles actually possessed their respective spins from the moment they were emitted; the measurement just reveals this fact.

However, the Copenhagen interpretation sees things differently. It asserts that each photon has an equal probability of being in the parallel or anti-parallel orientation. They exist in both possible states simultaneously; this circumstance is termed a *quantum superposition.* A photon does not 'know in advance' how it will be spinning when the measurement is made, just as we cannot tell through which slit a photon passes in Young's experiment; each orientation or pathway has the same probability. Accordingly, we cannot tell whether a particular photon was emitted with its spin in one orientation or another. All we can say is that each photon has a 50% chance of being polarised in one or other of the two possible orientations. And so things remain, until the spin of one of the photons is actually measured. This act instantly fixes the spin orientations of both photons. For if the measured photon is found to have one orientation, the other photon must, of necessity, have the opposite orientation. Should its spin be subsequently measured, it will be found to be in the opposite orientation to that found for the first photon.[37]

But what if the two photons had travelled some distance from one another when the first measurement was made? How does the second photon 'learn' that this measurement has been made and that the orientation of its spin has, of necessity, been fixed from that very moment? What mechanism ensures that the second measurement accords with the first, as required by the laws of conservation?[38] The controversy surrounding this issue became known as the *locality paradox* or *entanglement.*

[37] Light is unpolarised if we cannot determine its polarisation. However, this does not mean that unpolarised light comprises an equal number of spin parallel and spin anti-parallel photons. Being unpolarised is a property of each photon in a beam of unpolarised light. Each has an equal chance of being in the parallel or anti-parallel spin orientation. Hence, when passing through a birefringent calcite crystal, each photon of unpolarised light has a 50% chance of emerging in one or other of the polarised beams. Until an actual measurement is made, there is no way of knowing how each particle is spinning. Schrödinger's cat, which may be alive or dead in the unopened box, is another example of quantum superposition.

[38] If the measurements were made by two different observers, it would only be if and when they met and compared notes that the second observer would discover that the result he found had been pre-determined by the prior observation made by the first observer. But what if when they met, it transpired that the second observer had, in fact, made his measurement before the first? Whose measurement then was the determining one? And what if they never met? The Copenhagen interpretation only deals with how a single brain – that of a self-aware individual or the 'collective brain' of a group of individuals – acquires information about a quantum system. It is the incorporation of this information into a single 'brain' that fixes the system's state. That this may not be the only 'brain' around is disregarded by the Copenhagen interpretation.

Einstein concluded that these difficulties could only be reconciled in one of the following two ways:

1. If, at the very moment the first measurement is made, the result communicates itself instantaneously to the second particle and sets its spin state accordingly.[39]
2. The quantum theory of light is incomplete and does not fully specify the actual 'goings-on'; there are *hidden variables* which find no expression in conventional quantum theory but without which a full and *real* explanation cannot be given.

The first alternative clearly violates special relativity, under which the instantaneous transmission of information is prohibited. Not surprisingly, it was not Einstein's choice; he opted for the hidden variables. However, subsequent studies have shown that they too exhibit a non-locality every bit as paradoxical as that he sought to avoid. It transpires, that 'any realist interpretation of quantum theory in which it is deemed meaningful to say that an individual system possesses the values of its physical quantities in a way that is analogous to classical physics, will exhibit non-locality.'[40]

A number of experiments of the type suggested by Einstein have been carried out since he put forward his critique. Their design ruled out the possibility of any link between the production of the photons and the measurements made upon them or that the detectors could have affected the findings. The overwhelming conclusion, 70 years after it was first put forward, is that the 'spooky' predictions of the Copenhagen interpretation are vindicated. Accordingly, it is this pragmatic construction of quantum theory that we will adopt throughout this book.

Example 2.11: Photons and Interference Patterns

The interference pattern formed by a train of single photons can easily be seen at night with the following simple arrangement of two slits. With a sharp blade, two narrow parallel slits, each about 5 mm long and 0.2 mm wide, are cut close to each other in a piece of card. The observer holds the card close to his eye and looks through the slits in the direction of a street-lamp (preferably of the gas-discharge type) that is about 100 m from him. An interference pattern forms on the retina of the eye and is clearly seen.

To show that the pattern is formed by single photons, we will calculate how many photons there are at any moment, on average, between the slits and the retina of the

[39]This 'retro-active causation' can be resolved by postulating faster-than-light particles – *tachyons.*

[40]Chris J. Isham (1995) *Lectures on Quantum Theory*, London: Imperial College Press, p. 215.

observer's eye, in the typical case where the electrical power rating of the lamp is 100 W and just 1% of it is radiated as visible light.

Solution: The frequency, f, of the light in a typical street-lamp (sodium discharge) is about $5 \cdot 10^{14}$ Hz and so the energy E of each photon in this light is

$$E = hf = 5 \cdot 10^{14} \cdot 6.6 \cdot 10^{-34} = 3.3 \cdot 10^{-19}\,\text{J}.$$

Light energy is emitted by the lamp at a rate of 100·0.01 W. The number of photons, N, emitted from the lamp in each second is, therefore,

$$N = \frac{100 \cdot 0.01}{3.3 \cdot 10^{-19}} = 3 \cdot 10^{18}\,\text{s}^{-1}.$$

Let the number of photons entering the observer's eye through the slits each second be n. If the distance from the observer to the lamp is 100 m and the light is radiated isotropically, then the fraction, n/N, of the photons striking the slits each second is

$$n/N = \frac{\text{area of the two slits}}{\text{area of a sphere of radius 100 m}} = \frac{2 \cdot 5 \cdot 0.2 \cdot 10^{-6}}{4\pi 100^2} = 1.6 \cdot 10^{-11}.$$

Hence,

$$n = 1.6 \cdot 10^{-11} \cdot 3 \cdot 10^{18} = 4.8 \cdot 10^{7}.$$

The velocity of light is $3 \cdot 10^8$ m/s, and so, on the assumption that the photons move in a uniform file along a straight line, the distance, d, between successive photons entering the eye is

$$d = \frac{3 \cdot 10^8}{4.8 \cdot 10^7} = 6.2\,\text{m}.$$

The distance from the slits to the observer's retina is only a few centimeters at most, and so on average no more than one photon will be found at any time between the slits and the retina.

Chapter 2.4
The Mechanics of Minute Particles

The interpretation given by the quantum theory of radiation to the interference and diffraction of light rests on the principle that an explicit trajectory cannot be assigned to the motion of an individual photon. We can only state that the photon was at point A at one moment and appeared at point B some time later. Along which of the potential routes linking A and B it actually went can never be known. All that we can ever determine is the probability, P, that it took a particular path. Furthermore, we saw that this is equivalent to ascribing a wave amplitude, A, to each potential path whose square, A^2, is proportional to the probability that the photon took that path.

We now investigate whether this elusiveness is a feature only of photons or is a general property of all minute particles.

2.4.1 De Broglie's Hypothesis

According to the evidence of our senses, material objects have well-defined shapes and sizes, i.e. they are localised in space. However, the theory of special relativity asserts that the most distinctive attribute of material objects – their mass, is essentially equivalent to the most distinctive attribute of the fields associated with the interactions between them – their energy. The theory states that the two – mass and energy – are interchangeable forms of the same fundamental physical quantity. Hence, we might reasonably expect that what pertains to the one should pertain to the other.

This is the essence of the hypothesis put forward by De Broglie[41] in his doctoral thesis in 1924:

> '*Just as the energy and momentum associated with electromagnetic waves can be interpreted in terms of particles – photons, the dynamical properties of material particles, such as their energy and momentum, can be interpreted in terms of waves – "matter waves".*'

[41] Louis De Broglie 1892–1987, French physicist who was awarded the Nobel prize in 1929 for the discovery of the wave nature of the electron.

Quantum theory interprets electromagnetic fields in terms of particles – photons. To be consistent, we should be able, in the reverse manner, to interpret material particles in terms of fields – *matter fields.*

De Broglie proposed that the two fundamental mathematical formulae, $E = hf$ and $p = h/\lambda$, previously thought to bear only on photons, were applicable to all particles, whatever their nature. The inference is that every material particle has a matter wave associated with it of frequency $f = E/h$, where E is the total energy of the particle. The wavelength, λ, of the particle's matter wave is given by

$$\lambda = \frac{h}{p} = \frac{h}{mu}, \tag{2.65}$$

where $p = mu$ is the particle's relativistic linear momentum. Equation (2.65) is known as *De Broglie's equation.*

The import of De Broglie's hypothesis is that every particle flow can be viewed as a propagating disturbance in a field, i.e. as a wave motion. Just as light, which was traditionally thought of as a wave, can be understood as a stream of particles – photons, so any stream of material particles can be understood as a wave.

De Broglie's revolutionary hypothesis was confirmed in 1927 when Davisson[42] and Germer[43] in the United States and George Thomson[44] in the United Kingdom independently discovered that electrons can be diffracted. The American team was studying the scattering of low-energy electrons from metal surfaces. The surfaces had to be scrupulously clean. When a narrow beam of electrons was aimed at a nickel crystal that had been heated in hydrogen to remove any trace of oxides from its surface, exceptionally high intensities of scattered electrons were found at specific angles relative to the incident beam (Fig. 2.33).

In terms of waves, this experiment is similar to that in which the diffraction of X-rays by crystal planes was first demonstrated by the Braggs, except that in this case it is the matter wave associated with the electrons that is diffracted and not a high-frequency electromagnetic wave. The electrons are reflected at the atomic planes in the metal crystal, and their reflected matter waves interfere constructively under the conditions for constructive interference given by the

[42]Clinton Davisson 1881–1958, American physicist who was awarded the Nobel prize in 1937 for the discovery of electron diffraction.

[43]Lester Germer 1896–1971, American physicist.

[44]Sir George Thomson 1892–1975, English physicist who was awarded the Nobel prize in 1937 for the discovery of electron diffraction. His father, J.J. Thomson, had demonstrated that the electron was a particle; the son showed it was a wave.

Bragg equation:

$$n\lambda_{electron} = d \sin \theta_n. \tag{2.66}$$

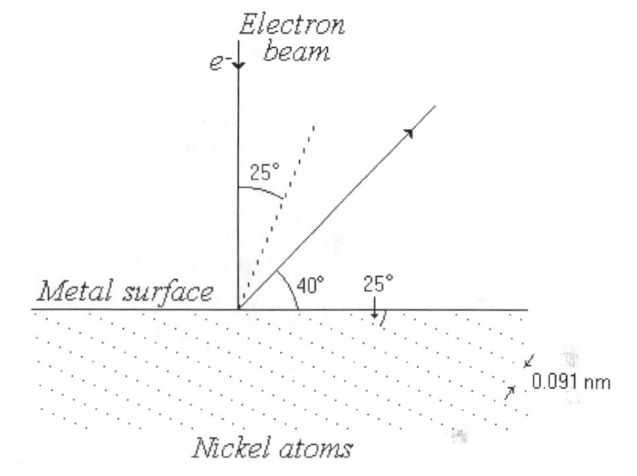

Fig. 2.33. The Davisson–Germer experiment: the scattering of electrons from a nickel surface. Electrons accelerated through a potential difference of 54 V were strongly scattered at an angle of 50° to the incident beam. The angle between the reflected beam and the plane of the atoms is 65°.

At the time the experiment was conducted, the spacing between the atomic planes in nickel crystals had already been determined from X-ray diffraction experiments and was known to be $d = 0.091\,\text{nm}$. Substituting this in the Bragg equation gives for the wavelength of the electron's matter wave ($n = 1$)

$$\lambda = 2d \sin \theta = 2 \cdot 0.091 \cdot \sin 65° = 0.165\,\text{nm}. \tag{2.67}$$

As the following calculation shows, this wavelength almost exactly equalled that predicted by the De Broglie hypothesis for the matter waves associated with electrons accelerated from rest through a potential difference of 54 V.

The linear momentum of a particle, in terms of its mass, m, and its kinetic energy, K, is given by[45]

$$p = \sqrt{2mK}. \tag{2.103}$$

Substitution in the De Broglie equation gives

$$\lambda = \frac{h}{p} = \frac{h}{\sqrt{2mK}}. \tag{2.68}$$

In general, the kinetic energy of the electrons accelerated from rest by a potential difference V is $K = eV$. This gives for the wavelength of the

[45] See Appendix 2.5.1.

matter wave of an electron accelerated from rest by a potential of 54 V (a 'slow' electron whose velocity is about $4.3 \cdot 10^6$ m/s)

$$\lambda = \frac{h}{\sqrt{2meV}} = \frac{6.63 \cdot 10^{-34}}{\sqrt{2 \cdot 9.11 \cdot 10^{-31} \cdot 54 \cdot 1.6 \cdot 10^{-19}}} = 0.167\,\text{nm}. \quad (2.69)$$

The close correspondence between the experimental value and the value computed from De Broglie's equation was seen as confirmation of the hypothesis.

The diffraction of other minute particles, such as neutrons, protons and atomic nuclei, was subsequently demonstrated in other experiments. In each case, the measured wavelengths of the matter waves corresponded to those predicted by De Broglie's hypothesis. Furthermore, interference patterns of alternating intensity maxima and minima were obtained in double-slit experiments, similar to that carried out by Young, but using an electron source in place of the light source. These experiments established the wave character of material particles beyond any doubt. Moreover, just as in the case of photons, individual material particles defied all attempts to determine precisely how they moved from one point to another. It proved as impossible to assign a definite trajectory to individual material particles as it had been for photons. Elusiveness is a property of all minute particles.

The elusiveness and particle-wave complimentarity exhibited by minute particles is a difficult concept to accept. Born[46] has suggested the following elucidation:

> *'The ultimate origin of the difficulty lies in the fact (or philosophical principle) that we are compelled to use the words of common language when we wish to describe a phenomenon... Classical physics has restricted itself to the use of concepts of this kind; by analysing visible motions it has developed two ways of representing them by elementary processes: moving particles and waves. There is no other way of giving a pictorial description of motions – we have to apply it even in the region of atomic processes, where classical physics breaks down.*
>
> *Every process can be interpreted either in terms of corpuscles or in terms of waves, but on the other hand it is beyond our power to produce proof that it is actually corpuscles or waves with which we are dealing, for we cannot simultaneously determine all the other properties which are distinctive of a corpuscle or a wave, as the case may be. We can therefore say that the wave and corpuscular descriptions are only to be regarded as complementary ways of viewing one and the same objective process... The corpuscular description means at bottom that we carry out measurements with the object of getting*

[46]Max Born 1882–1970, German–British physicist who was awarded the Nobel prize in 1954 for the probabilistic interpretation of the wave function.

exact information about momentum and energy relations (e.g. in the Compton effect), while experiments which amount to determinations of place and time we can always picture to ourselves in terms of the wave representation (e.g. passage of electrons through thin foils and observations of the deflected beam).'[47]

In general, the wave nature of a phenomenon is only evinced when the dimensions of the obstacles or apertures that the wave encounters are no more than just a few fold larger than its wavelength. The wavelength of electrons with energies of a few hundred electron-volts, such as those which generate the images on oscilloscope and television screens, are of the order of magnitude of an atom, $\sim 10^{-10}$ m. Consequently, when considering the motion of such electrons through oscilloscopes or television tubes – systems of macroscopic proportions – we can ignore the electrons' wave character and describe them as classical particles. However, when describing the motion of the electrons in an atom, their wave character cannot be ignored; indeed, without taking it into consideration, many of the properties of the atom defy comprehension, as we will see in Part Three of this book.

The wavelengths of the matter waves associated with macroscopic particles are so small that their wave properties are never discerned. Thus, for example, the De Broglie wavelength of a grain of sand of mass 10^{-6} kg (1 mg) moving with a velocity of 10 m/s, is

$$\lambda = \frac{h}{mu} = \frac{6.63 \cdot 10^{-34}}{10^{-6} \cdot 10} = 6.63 \cdot 10^{-29}\,\text{m}. \tag{2.70}$$

Obstacles or apertures of this order of magnitude do not exist and so under normal conditions the motion of macroscopic particles can be described by classical Newtonian mechanics.

The resolving power of a microscope depends on the wavelength of the waves with which it 'sees'; the smaller the wavelength the greater its power of resolution. A light microscope can reach a maximum useful magnification of about $\times 500$. Electrons accelerated through a potential of 100,000 V have a wavelength of 0.0037 nm, which is about 100,000 smaller than that of visible light. Consequently, a microscope operating with the matter waves associated with such electrons, should have a power of resolution many times greater than that of a light microscope. Such a microscope is called an *electron microscope*.

Since electrons carry an electric charge, their motion is affected by electromagnetic fields; a Lorentz force acts on them. This makes it possible to focus the electrons in the electron microscope by means of *magnetic lenses*. In practice, the useful magnification obtained is less than the theoretical because of

[47] Max Born (1969) *Atomic Physics*, New York: Dover Publications, p. 97.

limits set by the magnets. Nevertheless, magnifications of $\times 1{,}000{,}000$ can be achieved, making possible the examination of the most minute constituents of a system.

Two types of electron microscopes have been developed: transmission and scanning. The design of the transmission electron microscope closely resembles that of an optical microscope (Fig. 2.34). To avoid excessive attenuation of the electron beam the sample being examined must be sliced very thin. As a result, only two-dimensional images are obtained with the transmission microscope.

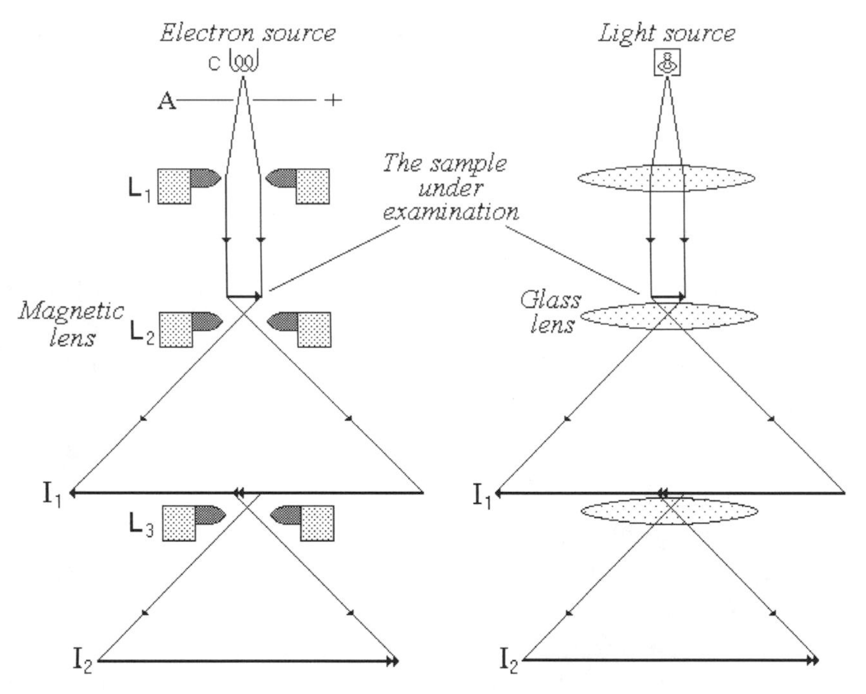

Fig. 2.34. A light microscope and a transmission electron microscope. The components of an electron microscope match those of the light microscopes. The sample being examined is irradiated with electrons instead of with light. The electrons are accelerated from the cathode C to the anode A. The magnetic lens L_1 collimates the beam incident on the sample. The lens L_2 produces the first image I_1, and the lens L_3 projects the final image I_2 on a photographic plate or television screen.

The scanning electron microscope produces a seemingly three-dimensional image of the sample being examined. The samples magnified in these microscopes must be electrically conducting. Samples of non-conducting material such as biological specimens have a thin metal layer evaporated onto their surface. The electron beam is focused to a fine spot ($\sim$10 nm) that scans the surface of the sample releasing secondary electrons from it. The number of secondary electrons depends on the nature of the surface and the image is formed by collecting these secondary electrons.

A more recent development is a scanning electron microscope that employs the tunnel effect. In this device the tip of a very fine-pointed metal probe moves over and across the surface of the conducting sample. A potential difference is maintained between the probe and the sample's surface. Although the tip does not actually touch the surface, electrons can tunnel across the narrow gap between them. By plotting the tip-surface current or the movement of the tip required to maintain a constant current, the sample's surface can be mapped. Resolutions of less than one tenth of a nano-metre are routinely achieved with this microscope.

Example 2.12: Electron Diffraction — Thomson's Experiment

The wave character of the electron was also demonstrated independently in 1927 by George Thomson, the son of the discoverer of the electron. Thomson found that circular diffraction patterns were produced when a narrow beam of high energy electrons was aimed through a thin metal foil (Fig. 2.35).

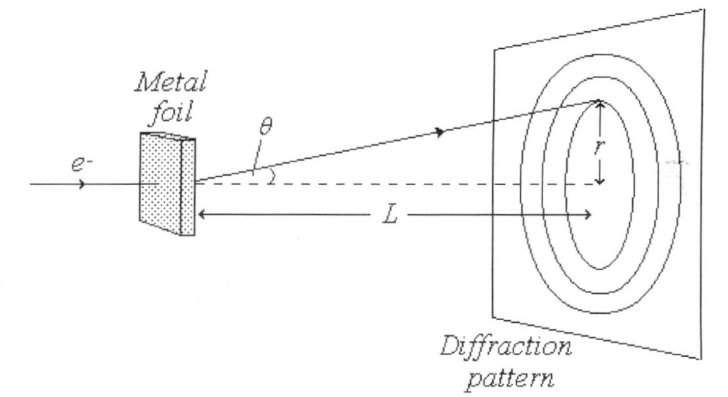

Fig. 2.35. The Thomson experiment. A beam of electrons is diffracted by passing them through a thin metal foil.

When using an aluminium foil, Thomson obtained the following values for the radius, r, of the first order circle in the diffraction pattern and the corresponding potential, V, through which the electrons had been accelerated (Table 2.1).
Do these values confirm De Broglie's hypothesis?

Calculation: A metal foil comprises a myriad of perfectly ordered micro-crystals that are randomly scattered throughout it. The rows of atoms in the micro-crystals form a lattice, and so each micro-crystal can act as a diffraction grating for waves of appropriate wavelength, λ. The lattices are identical and so the diffraction patterns they produce will be identical. However, because they are randomly scattered throughout the foil, these patterns will likewise be oriented randomly. The overall result is a circular diffraction pattern, the foil acting as though it was a single grating, rotating about an axis normal to its plane.

Table 2.1. The radius, r, of the first order diffraction circle observed by Thomson and the corresponding potential, V, through which the electrons had been accelerated.

V (volts)	r (mm)	V (volts)	r (mm)
17,500	3.10	44,000	2.08
30,500	2.45	48,600	1.98
31,800	2.34	56,500	1.80
40,000	2.12		

As with any diffraction grating, constructive interference is obtained at the angles given:

$$d \sin \theta_n = n\lambda \quad (n = 1, 2, 3 \ldots),$$

where d is the spacing of the grating.

The radius of the first order circle in the diffraction pattern is much smaller than the distance L from the foil to the screen and so to a close approximation:

$$\sin \theta_1 = \tan \theta_1 = r/L.$$

Thus, the wavelength corresponding to the diffraction pattern is

$$\lambda = \frac{dr}{L}.$$

According to De Broglie's hypothesis, the wavelength of electrons accelerated through a potential V is given by

$$\lambda = \frac{h}{\sqrt{2m_e eV}}.$$

If the experimental value for the wavelength equals that predicted by the De Broglie hypothesis

$$\frac{dr}{L} = \frac{h}{\sqrt{2m_e eV}},$$

or, alternatively,

$$r\sqrt{V} = \frac{Lh}{d\sqrt{2m_e e}}.$$

For a particular experimental set-up, all the component quantities of the product $\frac{Lh}{d\sqrt{2m_e e}}$ are constants. De Broglie's hypothesis will be confirmed, therefore, if the product $r\sqrt{V}$ is constant for a series of measurements with the particular set-up. Examination of the data in the table above shows that in each case the product falls within the range

$$r\sqrt{V} = (425 \pm 15) \cdot 10^{-3}\,\mathrm{m} \cdot \mathrm{V}^{\frac{1}{2}};$$

a result that confirms the hypothesis within the experimental error.

2.4.2 Heisenberg's Uncertainty Principle

Both classical and relativistic mechanics assert that, in principle, there is no constraint on what can be known about the physical parameters of a system; any uncertainty in the actual value of these quantities results only from human frailty or ignorance. Thus, one of the postulates of classical mechanics is that the position and linear momentum of a particle can be exactly determined at any moment in time.[48] A sequence of such determination defines the particle's trajectory through space–time.

Furthermore, classical and relativistic mechanics are fundamentally deterministic, i.e. they assert that from a knowledge of the state of a mechanical system at any moment and of the laws governing it, everything about its past and its future can be ascertained. In practice, exact predictions may be hard or even impossible to achieve due to practical or technical inadequacies in the accuracy of our measurements, but the possibility of perfectly accurate measurements is not *a priori* precluded.

The confidence expressed in these assertions mirrors the 18th century European belief that nature is *reliable* and that there is no natural limit to what man can know about the world; that all knowledge is just 'out there' waiting to be acquired. Nothing is, in principle, inexplicable; uncertainty is just a feebleness of the mind. Given time, all the mechanisms that govern the universe would be discovered. The great expectations of this mechanistic philosophy are illustrated by the following extract from the writings of the great French physicist Laplace.[49]

> *'All events, even those which on account of their insignificance do not seem to follow the great laws of nature, are a result of it just as necessarily as the revolutions of the sun. In ignorance of the ties which unite such events to the entire system of the universe, they have been made to depend upon final causes or upon hazard [chance]... But these imaginary causes have gradually receded with the widening bounds of knowledge and disappear entirely before sound philosophy, which sees in them only the expression of our ignorance of the true causes.*
>
> *Given for one instant an intelligence which could comprehend all the forces by which nature is animated and the respective situation of the beings who compose it – an intelligence sufficiently vast to submit these data to analysis – it would embrace in the same formula the movements of the greatest bodies of the universe and those of the*

[48] The motion of a conservative classical system is completely described by expressing its energy as a function of its position (its space coordinates) and its linear momentum; the so-called *Hamiltonian function.* Position and momentum are designated *canonically conjugate variables.*

[49] Pierre Simon de Laplace 1749–1827, French mathematician and physicist.

lightest atoms; for it, nothing would be uncertain and the future, as the past, would be present to its eyes. The human mind offers . . . a feeble idea of this intelligence.

The curve described by a simple molecule of air or vapour is regulated in a manner just as certain as the planetary orbits; the only difference between them is that which comes from our ignorance.'[50]

In the deterministic view of Laplace, we could even predict how a dice would fall, were it not for our benightedness. Precise information as to its starting position, together with a knowledge of the laws of motion, is all that we would need.

But the inherent elusiveness of those minute particles that manifest a wave character raises the question of whether, *a priori*, their position and linear momentum can be simultaneously determined. In principle, is such an observation possible? We will examine this question by means of the following thought experiment.

The power of resolution of a microscope depends on the wavelength, λ, of the waves with which it 'sees'. Because of diffraction effects, well-defined images of objects, whose size is comparable to the wavelength, cannot be obtained; the position of such objects cannot be exactly determined. The uncertainty, Δx, in the object's x coordinate, is of the order of magnitude of the wavelength

$$\Delta x \approx \lambda. \tag{2.71}$$

Clearly, this uncertainty can be reduced by employing shorter wavelengths. However, at the moment it is 'seen' by the waves, the motion of the object changes. For, in order to see the object, the waves must impinge upon it and, in doing so, some of their own linear momentum is transferred to it. From De Broglie's equation, $p = h/\lambda$, the shorter the wavelength employed, the greater the linear momentum, p, associated with the waves and the greater will be the concomitant change in the object's motion. In the collision between the waves and the object, the maximum possible change, Δp_x, in the object's linear momentum parallel to the X-axis will be of the order of magnitude of the linear momentum of the incident waves, i.e.

$$\Delta p_x \approx \frac{h}{\lambda}. \tag{2.72}$$

Substituting λ from Eq. (2.68) in Eq. (2.69) gives

$$\Delta x \cdot \Delta p_x \approx h. \tag{2.73}$$

[50](1951) *A Philosophical Essay on Probabilities*, translated by F.W. Truscott & F.L. Emory, New York: Dover Publications, pp. 3ff.

The significance of this relationship is that the smaller the uncertainty, Δx, in the object's position the greater the uncertainty, Δp_x, in its linear momentum, and *vice versa.* Employing shorter wavelengths would indeed improve the measurement of the object's position, but this would be achieved at the expense of a loss of accuracy in the measurement of its momentum. Hence, in principle, exact simultaneous measurements of an object's position and linear momentum cannot be made.

A rigorous mathematical analysis shows that the actual relationships between the inherent uncertainties in a particle's space coordinates and the components of its linear momentum are

$$\begin{aligned} \Delta x \cdot \Delta p_x &\geq \frac{h}{4\pi} \\ \Delta y \cdot \Delta p_y &\geq \frac{h}{4\pi} \\ \Delta z \cdot \Delta p_z &\geq \frac{h}{4\pi}. \end{aligned} \tag{2.74}$$

Equation (2.73) is clearly consistent with these relationships.[51] This limitation on the simultaneous measurement of complementary physical quantities is known as *Heisenberg's uncertainty principle.* The quantity $h/2\pi$, which occurs frequently in quantum physics, is usually designated by the symbol $\hbar$ called 'h-bar'.

The uncertainty principle can also be expressed in terms of the inherent uncertainty, ΔE, in a particle's energy and the inherent uncertainty, Δt, in a time interval:

$$\Delta E \cdot \Delta t \approx \eta. \tag{2.75}$$

For example, this relationship relates the uncertainty, ΔE, in the energy of the photon emitted when an atom falls from an excited stationary state to the ground state to the uncertainty, Δt, in the lifetime of the excited state. The uncertainty in the lifetime of the excited state is of the order of its lifetime. It follows, that the more stable the state, the greater is Δt and the smaller is ΔE.

In general, if the energy of an excited stationary state is accurately determined, as it usually can be by spectroscopic measurements of the frequency of the photons emitted in transitions from it, no accurate information can be obtained as to the moment the individual photons were actually emitted. Similarly, a particle can 'borrow' an amount of energy, ΔE, from apparently empty

[51]These relationships are also written sometimes with h or $h/2\pi$ on the right-hand side or with $\approx$ rather than $\geq$ showing the equality. This difference is not very important since the relationships give only estimates.

space without violating energy conservation, so long as it returns the loan within a period of time $\Delta t \approx \hbar/\Delta E$. The phenomenon is known as 'vacuum fluctuations'. The creation, apparently *ex nihilo*, of the virtual photons which mediate the rapid interactions between particles described by Feynman diagrams, is explained in this way.

It must be emphasised that the uncertainties expressed in the expressions such as $\Delta x \cdot \Delta p_x \approx \hbar$ and $\Delta E \cdot \Delta t \approx \hbar$ are not the result of technical or practical difficulties in making the measurements; they are inherent in the wave character of material objects. They are a consequence of the fact that every wave phenomenon has some *a priori* uncertainty associated with it. For example, measuring the frequency of a musical note requires a wavetrain of a certain length (Fig. 2.36). In order to obtain the frequency, we divide the length of time that the note is heard by the number of periods counted during that time, i.e. by the number of periods in the wavetrain. The longer the note, the more accurate will be the measurement of its frequency. On the other hand, limiting the time allowed for the measurement reduces its accuracy. In the extreme case, if the wavetrain is less than one period long, the measurement of the note's frequency becomes meaningless.

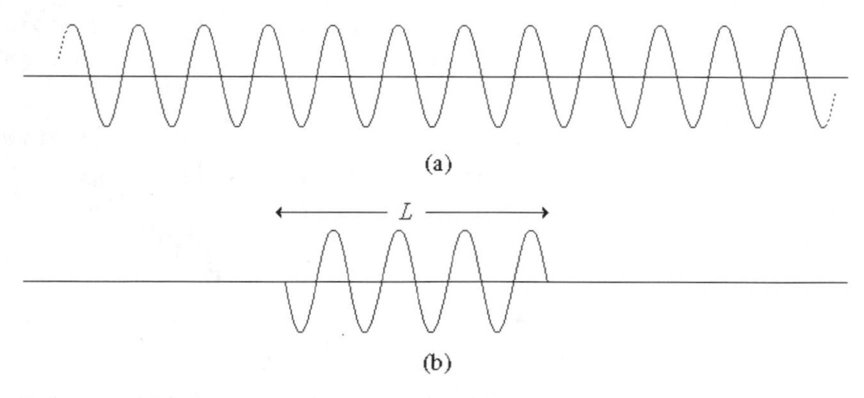

Fig. 2.36. (a) A section of an infinite sinusoidal wave. The frequency or wavelength can be measured with almost complete accuracy if a sufficiently long section is taken. (b) A finite wavetrain of length L. Although the frequency and wavelength can be measured with considerable accuracy, there is always an uncertainty as to exactly where the wavetrain starts and ends.

Ascribing a precise trajectory to a moving particle implies absolute certainty as to its position at any moment. Such trajectories are typically delineated in classical and relativistic mechanics by continuous functions such as $E(x, t)$ or $p(x, t)$, that describe the energy or linear momentum of a particle moving parallel to the X-axis in terms of its space and time coordinates. However, the uncertainty principle precludes the validity of such functions *a priori.* Why, then, do classical and relativistic mechanics work so well?

The reason is the diminutive size of Planck's constant, h. The fundamental uncertainty prescribed by Heisenberg's uncertainty principle in a macroscopic object's position, $\Delta x \approx \hbar/\Delta p_x$, is totally negligible compared to its dimensions. The motion of macroscopic objects can be properly described by the deterministic mathematical functions of classical and relativistic mechanics because the inherent uncertainties associated with them are too small to be detectable. This is a further example of the correspondence principle.[52]

However, the fundamental uncertainties in the position or momentum of minute particles are not negligible relative to their dimensions and so their motion cannot be described by deterministic mathematical functions. Consequently, when setting out to examine the behaviour of minute particles, we must first decide which physical quantity is the more important to us: its position or its momentum; its energy or the stability of its present state. Information about the one is always acquired at the expense of information about the other. Precise knowledge of a particle's momentum will leave us with little more than probabilities about its position. Similarly, accurate measurement of the particle's energy in a particular state leaves us ignorant of how long the particle remains in the state.

Any particle property or physical variable that can be measured is termed an *observable.* In classical and relativistic physics, the measurement of observables is not subject to any inherent uncertainty, only to technical or random errors. However, in quantum physics, the degree to which a physical variable can be precisely measured is subject to the uncertainty principle. Repeated errorless measurements of the variable will not always give precisely the same value. The spread (variance) in the errorless measurements of a variable reflects the inherent uncertainty in the particular particle property.[53] If the spread is small, the variable is said to be a *sharp* variable; if it is large, it is said to be a *fuzzy* variable. The values of a quantum system's sharp variables are called their *characteristic, proper* or *eigen* values. With a few exceptions, the sharp observables are the same quantities as those conserved in classical physics.

---

[52] It has been realised in recent years that the behaviour of macroscopic systems that obey deterministic laws can also be unpredictable. This happens in systems that are affected by an excessive number of variables or are sensitive to minuscule changes in their initial conditions or parameters. An example of this *deterministic chaos* is the so-called *butterfly effect.* It is suggested that meteorological dynamics are so sensitive that the flap of a butterfly's wings at one place may result in violent changes in the weather thousands of kilometres away.

[53] The spread, ΔQ, in a physical quantity Q is given by the familiar statistical parameter (the standard deviation), $\Delta Q = \sqrt{\overline{Q^2} - (\overline{Q})^2}$.

Example 2.13: Heisenberg's Uncertainty Principle

Calculate the fundamental uncertainty in the x coordinate of

1. an electron;
2. a ball of mass 10 gm, both moving with a velocity of 100.00 ± 0.01 m/s parallel to the X-axis.

Calculation: According to Heisenberg's principle, the uncertainty in the x coordinate of a particle is $\Delta x \geq h/4\pi\Delta p_x$. The uncertainty in the linear momentum of a particle of mass m moving with a velocity $u \pm \Delta u$ is

$$\Delta p_x = m\Delta u.$$

Thus, the fundamental uncertainty in the position of the electron is

$$\Delta x_{electron} \geq \frac{h}{4\pi \cdot 0.01\, m_{electron}} \geq \frac{6.63 \cdot 10^{-34}}{4\pi \cdot 0.01 \cdot 9.11 \cdot 10^{-31}} \approx 0.01\,\text{m}.$$

Clearly, the concept of the electron as a particle has no validity under these conditions. The fundamental uncertainty in the position of the ball is

$$\Delta x_{ball} \geq \frac{h}{4\pi \cdot 0.01\, m_{ball}} \geq \frac{6.63 \cdot 10^{-34}}{4\pi \cdot 0.01 \cdot 10^{-2}} \approx 10^{-30}\,\text{m}.$$

This uncertainty in the position of the ball is totally negligible. Hence, the position and linear momentum of macroscopic objects can be determined, to all intents and purposes, absolutely.

2.4.3 Matter Waves

The concept of a matter wave defies simple intuitive visualisation. Matter waves do not correspond to vibrations in an underlying physical medium nor are they the result of particle vibrations. A matter wave is the moving particle itself, conceived of as a disturbance in a field – the matter field.

Unlike mechanical, electrical and magnetic fields, matter fields can be neither detected nor mapped experimentally. The quantities that oscillate in sound and electromagnetic waves, respectively, the air pressure and the magnitude of the electric and magnetic fields, have a clear physical reality and can be measured and interpreted experimentally. The former can be measured with a pressure gauge, the latter with a test charge. However, the analogous quantity – the 'thing' – that oscillates in matter waves can neither be measured nor interpreted in terms of any experiment.

The 'thing' that oscillates in matter waves is usually referred to as the *wave function* or the *probability amplitude.* Wave functions (or probability amplitudes) are usually designated by the Greek letter psi, Ψ. In general, the wave function, Ψ, is a complex number[54] whose value is a function of the space coordinates x, y and z and of the time coordinate t.

That part of the wave function that depends only on the space coordinates is known as the time-independent wave function and is usually designated by the lower case letter ψ.

Although no direct physical meaning can be ascribed to the wave function, matter waves exhibit all the typical wave properties. Thus, for example, the superposition of two or more wave functions produces another wave function. Moreover, although it has no direct physical meaning itself, the wave function contains all the information that can be known about the particle. This information can be extracted from the wave function by performing certain mathematical operations on it and can be used to make calculations and predictions about the outcome of experiments.

Thus, for example, the square of the absolute value of the wave function, $|\Psi|^2$, is a measurable quantity and can be interpreted experimentally. It represents the intensity of the matter wave and indicates the probability, P, of finding the particle around a given point:

$$P \propto |\Psi|^2. \tag{2.76}$$

$|\Psi|^2$ is called the *probability density*. This interpretation of the square of the wave function was first proposed by Born in 1926. It can be drawn, by analogy, from the link between the square, A^2, of the amplitude of an electromagnetic wave at a given point (its intensity at the point) and the probability, P, of finding a photon there.

A corollary of this interpretation is that, in the one-dimensional case, the probability, $P(x)\,dx$, of a particle being found between x and $x + dx$ at time t is

$$P(x)\,dx = |\Psi(x, t)|^2\,dx, \tag{2.77}$$

and the probability, P_{ab}, of finding the particle between $x = a$ and $x = b$ is

$$P_{ab} = \int_a^b |\Psi(x, t)|^2\,dx. \tag{2.78}$$

[54] For example, $\Psi = a + ib$, where $i^2 = -1$. The complex number $\Psi* = a - ib$ is called the *complex conjugate* of Ψ. In exponential form $\Psi = Ae^{i\theta}$ and $\Psi* = Ae^{-i\theta}$. Their product $\Psi\Psi* = |\Psi|^2$ is a real number. The reader is referred to Supplementary Topic A for a more complete explanation of the use of complex numbers in the description of waves in general and matter waves in particular .

Clearly, the particle must be somewhere and so the sum of the probabilities for all the possible values of x must equal unity:

$$\int_{-\infty}^{\infty} |\Psi(x,t)|^2 dx = 1. \tag{2.79}$$

Any wave function satisfying Eq. (2.79) is said to be *normalised.*

Although repeated measurements of a physical variable may vary because of its inherent uncertainty, an average or *expectation* value can always be assigned to it. For example, repeated measurements of a particle's x coordinate may reveal it to be in a different position each time. However, if we know the particle's wave function, we can find the expectation (average) value, $\langle x \rangle$, of its position from the equation

$$\langle x \rangle = \int_{-\infty}^{\infty} xP(x)\,dx = \int_{-\infty}^{\infty} x|\Psi(x,t)|^2 dx. \tag{2.80}$$

In the same way, expectation values can be found for all the particle's other physical variables. For example, the expectation value of its potential energy, U, is given by

$$\langle U \rangle = \int_{-\infty}^{\infty} U(x)\,P(x)\,dx = \int_{-\infty}^{\infty} U(x)\,|\Psi(x,t)|^2 dx. \tag{2.80a}$$

The formal condition that a physical variable must fulfil if its value is to be sharply observable is that the effect upon the wave function of a mathematical operation specific to the variable, e.g. multiplication by a given function or taking a certain derivative, must be simply to return a multiple of the wave function. In mathematical notation this is expressed as

$$[\mathrm{E}]\Psi = E_n\Psi, \tag{2.81}$$

where E is the particular variable, [E] is the specific mathematical operation and E_n is a value of the variable. It transpires that this is only possible for certain wave functions called the *eigenfunctions* and only for certain values of the variable – its eigenvalues.

2.4.4 Wave Functions and Probability Amplitudes

Wave functions can also be used to describe real events such as the displacement of a particle from one point to another. This is the basis of the new mechanics – *wave mechanics* – developed by Born, Heisenberg and Schrödinger[55]

[55] Erwin Schrödinger 1887–1961, Austrian physicist who was awarded the Nobel prize in 1933 for innovations in atomic theory.

during 1925 and 1926. In place of the equations of motion of classical mechanics, from which the particle's exact position in space at every moment can be computed, this new mechanics provides wave functions that contain all that can be known about the particle subject to the uncertainty principle.

In principle, if not always in practice, the interpretation of actual events in terms of wave functions is a simple technique. It involves just the following three rules.

Rule 1

When an event, all of whose initial and final conditions have been fully specified, can take place in a number of alternate ways, it being impossible to determine which one was actually taken, the wave function for the event, Ψ, is given by the sum of the wave functions, $\Psi_1, \Psi_2, \Psi_3 \ldots$, that represent, respectively, each of the different alternative ways:

$$\Psi = \Psi_1 + \Psi_2 + \Psi_3 + \cdots . \tag{2.82}$$

This superposition of the wave functions results in interference, and the overall probability of the event, P, is given by

$$P \propto |\Psi|^2 = |\Psi_1 + \Psi_2 + \Psi_3 + \ldots|^2. \tag{2.83}$$

Constructive interference corresponds to a high probability of the event's occurrence; destructive interference to a low probability.

Rule 2

If the way the event took place can be established, there is no superposition of the wave functions. In this case, the overall probability of the event, $P_{123\ldots}$, is given by the sum of the separate probabilities for each alternative way:

$$P_{123\ldots} = P_1 + P_2 + P_3 \cdots \propto |\Psi_1|^2 + |\Psi_2|^2 + |\Psi_3|^2 + \cdots . \tag{2.84}$$

Rule 3

When the event takes place as a series of steps or concomitant happenings, its wave function is obtained from the product of the wave functions of the steps/happenings:

$$\Psi = \Psi_1 \Psi_2 \Psi_3 \cdots . \tag{2.85}$$

In order to demonstrate how wave functions are used to describe real physical phenomena, we will consider a double-slit experiment similar to Young's, except that it employs an electron source in place of the light source and a suitable detector to record the pattern of electron intensities at a given distance from the slits (Fig. 2.37).

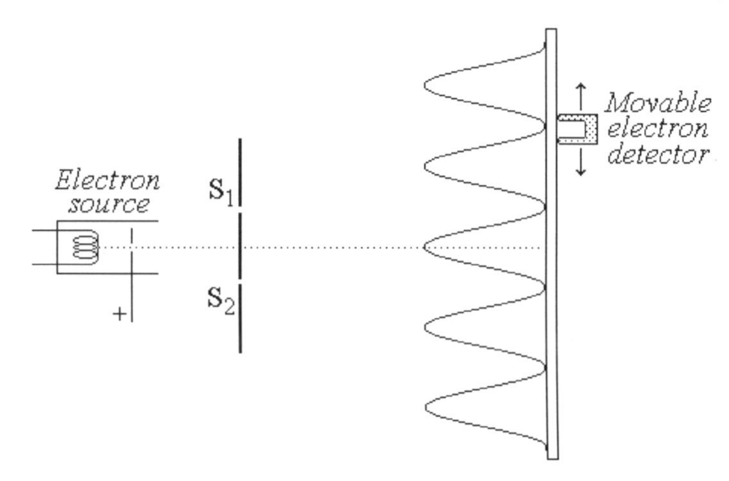

Fig. 2.37. A double-slit experiment with electrons. When the two slits, S_1 and S_2, are kept open at the same time, an interference pattern of alternating maxima and minima is recorded by the movable electron detector. With just one slit open or if a detector is installed in the apparatus which determines through which slit each electron passes, a single broad diffraction band is recorded.

The two slits are alternative routes by which electrons can pass from the source to the movable electron detector. If just one slit is open at any one time or if both are kept open all the time but a detector that records through which slit each electron passes is also installed in the apparatus, the route taken by each electron can be unequivocally ascertained. In this case, the probability, P_1, of an electron passing through the first slit is given by $|\Psi_1|^2$, and the probability, P_2, of it passing through the second slit by $|\Psi_2|^2$. Overall, the probability, P_{12}, of an electron reaching the movable detector at a particular point is

$$P_{12} = P_1 + P_2 = |\Psi_1|^2 + |\Psi_2|^2 \tag{2.86}$$

and there is no interference. The distribution of electrons recorded by the movable detector corresponds to two overlapping single-slit diffraction patterns (Fig. 2.29c).

However, if both slits are kept open at the same time and no additional detectors are installed in the apparatus, there is no way of ascertaining through which slit each electron actually passes. In this case, the wave function that describes the passage of an electron from the source to the movable detector is $\Psi = \Psi_1 + \Psi_2$, and the probability, P, that an electron reaches the movable detector at a particular point is

$$P = |\Psi|^2 = |\Psi_1 + \Psi_2|^2 = |\Psi_1|^2 + |\Psi_2|^2 + 2|\Psi_1||\Psi_2|\cos\phi \qquad (2.87)$$

where ϕ is the phase difference between the wave functions at the point.[56] The term $2|\Psi_1||\Psi_2|\cos\phi$ is an interference term whose value varies with that of $\cos\phi$. Many electrons reach those points where $\cos\phi = +1$ and which correspond to constructive interference of the wave functions. In contrast, the chance of an electron reaching those points that correspond to destructive interference where $\cos\phi = -1$ is negligible. Thus, a typical interference pattern of alternating intensity maxima and minima is recorded by the detector (Fig. 2.29a).

In general, the alternate routes or stages by which a physical event can take place are delineated by the structure of the apparatus or system in which the event occurs. Accordingly, every change in the system's dimensions, components or design will affect the event's overall wave function. This accounts for the 'awareness' of the experimental set-up that we attributed to the photons in the description of the experiment in the previous chapter. It is as though each of the particles involved in an event ascertains all the ways in which it can occur and thus becomes aware of the experimental set-up. The overall wave function is formed from the appropriate addition of the wave functions of all the alternate routes available to the particles.

Despite the difficulty many people have in accepting these strange and non-intuitive ideas, the technique of ascertaining the outcome of events by means of wave functions has passed every single test and experimental assay to which it has been subjected, without exception. It does work.

> '*Throughout much of history, one of the major positions in the methodology of science has been what is sometimes referred to as the "save the phenomenon" position ... The basic tenet of this position is that judgements of the value of a scientific theory should be made in terms of only two criteria: the ability of the theory to save (account for or predict) the phenomena and the theory's simplicity. Among the criteria viewed as inappropriate is the realist's claim that a theory must be true. For an advocate of the save the phenomena position, a theory is nothing more than a useful fiction or instrument to account for phenomena.*'[57]

[56] Mathematically, the superposition of the complex wave functions is similar to the addition of the amplitudes of coherent waves. Both can be represented by phasor (rotating vector) diagrams. The reader is referred to Supplementary Topic A for an explanation of their use.

[57] Michael J. Crowe (1990) *Theories of the World from Antiquity to the Copernican Revolution*, New York: Dover Publications, p. 69.

Although quantum mechanics satisfies the first criterion, it does indeed 'save' the phenomena, its simplicity is not self-evident.

> *'As man's investigation of the universe progressed to the almost infinitely large distances of interstellar space or to the almost infinitesimal magnitudes of atomic structures, it began to be realised that these other worlds could not be adequately described in terms of the bricks and mortar and plumbing of terrestrial architecture ... Intensive research into the ultimate nature of our universe is thus gradually changing the meaning we attach to such words as "explanation" or "understanding". Originally they signified a representation of the strange in terms of the commonplace; nowadays, scientific explanation tends more to be a description of the relatively familiar in terms of the unfamiliar, light in terms of photons; matter in terms of waves.'*[58]

2.4.5 The Wave Function of a Free Particle

A free particle is one that is not subject to any force; it is free to roam anywhere. As such, there are no restrictions on its total energy; it can take any value. The energy eigenvalues of a free particle – its energy levels – merge into an unbroken continuum.

The wave function of a free particle cannot be such that it assigns the same probability to all points in space–time. The particle must actually be somewhere and its matter wave must give the greatest probability at that place. Hence, its wave function cannot be a uniform infinite sinusoidal wave, for such a wave assigns the same probability to every point in space.

The fact that the wave function of a free particle cannot be a uniform infinite sinusoidal wave is also confirmed by considerations of its phase velocity. The phase velocity, v, of a sinusoidal wave is equal to the product of its frequency and wavelength. According to De Broglie's hypothesis, the frequency of a particle's matter wave is given by $f = E/h$, where $E = mc^2$ is the particle's total energy. Likewise, its wavelength is given by $\lambda = h/mu$, where u is the particle's velocity. Combining these three expressions gives

$$v = f\lambda = \frac{mc^2}{h} \cdot \frac{h}{mu} = \frac{c^2}{u}. \tag{2.88}$$

Since, by reason of special relativity, the particle's velocity $u < c$, it follows from Eq. (2.88) that if the matter wave was uniform and sinusoidal, its phase velocity $v > c$, i.e. the wave would propagate at a speed greater than that of light. This of itself presents no problem, for no physical quality is actually moving; matter waves, as such, can neither be measured nor interpreted in

[58] Walter J. Moore (1957) *Physical Chemistry*, Harlow: Longmans, Green & Co, p. 1.

terms of any experiment. But it does mean that a single uniform sinusoidal matter wave cannot represent a free particle, for its phase velocity, v, cannot be the velocity of a real object.

A free particle's matter wave is, in fact, a wave group (wavetrain or wave packet), that is built up from the superposition of an infinity of sinusoidal matter waves of different amplitudes and wavelengths (Fig. 2.38). The wave group is the result of the superposition of the component waves. The waves interfere destructively everywhere except in the region where the particle actually is. The velocity of the particle is not that of the component sinusoidal matter waves but the *group velocity*, i.e. the velocity with which the group of waves propagates. Thus, a free moving particle is analogous to a signal transmitted by a matter wave.

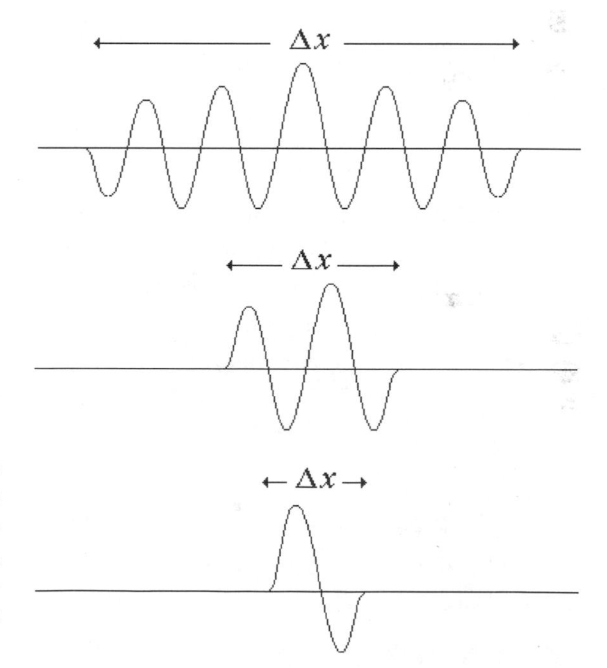

Fig. 2.38. The matter waves of free particles. The more confined the wave group the smaller the uncertainty, Δx, in the particle's position, but the greater will be the uncertainty, $\Delta\lambda$, in the wavelength, i.e. in the particle's linear momentum.

The inherent uncertainty in a particle's position depends on the width of its wave group, Δx. The smaller the wave group the lower the uncertainty in the particle's position, and *vice versa*. However, the smaller the wave group, the less well defined is its wavelength and the greater, therefore, its uncertainty, $\Delta\lambda$. This, in turn, implies a greater uncertainty in the particle's linear momentum. Hence, exactness in fixing the particle's position is achieved at the expense of the accuracy with which its momentum can be measured, which is specifically what Heisenberg's uncertainty principle, $\Delta x \cdot \Delta p_x \approx \hbar$, asserts.

The time a wave packet takes to move a distance equal to its width is

$$\Delta t = \frac{\Delta x}{u}. \tag{2.89}$$

This is the time required for the particle to move a noticeable distance, i.e. for an appreciable change to occur in the system. The uncertainty in the particle's energy, ΔE, is given by the energy–time uncertainty relationship, $\Delta E \approx \hbar/\Delta t$. Hence, in general, the longer it takes for an appreciable change to occur in a system, the more sharply observable is its energy, and *vice versa.*

2.4.6 Quantum Mechanics — Schrödinger's Equation

Classical mechanics cannot be employed in systems in which the wave character of particles is manifested. The attempt to assign trajectories to particles that show pronounced wave properties does not yield a correct description of their actual behaviour. A new mechanics is needed for such systems; this new mechanics is called *quantum mechanics* or *wave mechanics.* In content, this new mechanics is identical to the matrix mechanics developed by Heisenberg.

The equations of motion of classical mechanics ensue from the three postulates Newton called the 'Axioms or Laws of Motion'. In the one-dimensional case, the second of these axioms – Newton's second law – can be expressed as the familiar differential equation

$$F = m\frac{d^2x}{dt^2}. \tag{2.90}$$

This equation has no *a priori* foundation. However, its solutions are found to be consistent with experiment in many physical situations. Its value and validity spring from this accord.

In place of the equations of motion of classical mechanics from which the particle's exact position in space at every moment can be computed, quantum mechanics provides wave functions that contain all that can be known about the particles subject to the uncertainty principle. The wave functions of quantum mechanics can be derived from a fundamental differential equation known generally as *Schrödinger's equation,* that has the same status in quantum mechanics as that held by Newton's laws in classical mechanics. It too is a postulate that has no *a priori* standing and is only as good as the consistency between its solutions and experiment.

Whatever other properties it might have, we would expect the Schrödinger equation to incorporate the following fundamental principles:

1. The conservation of energy: this principle is so basic that its exclusion is unthinkable.
2. De Broglie's hypothesis: quantum mechanics is specifically intended for particles that show distinctive wave properties.
3. The principle of linear superposition: the interference of matter waves must be inherent in the equation.

We shall now show how a differential equation that embraces these fundamental principles can be constructed.

The conservation of a particle's energy is defined by the equation

$$K = E - U, \tag{2.91}$$

where K is the particle's kinetic energy, E its total mechanical energy and U its potential energy.

In terms of its mass, m, and its linear momentum, p, the kinetic energy of a particle of is given by (Appendix 2.5.1)

$$K = \frac{p^2}{2m}. \tag{2.92}$$

Substituting the De Broglie equation, $p = h/\lambda$, in this relationship gives

$$K = \frac{h^2}{2\lambda^2 m} = E - U. \tag{2.93}$$

Let us assume, for simplicity's sake, that the time independent part of the particle's one-dimensional wave function, ψ, has the form

$$\psi = A \sin kx, \tag{2.94}$$

where $k = 2\pi/\lambda$. The second derivative of this function is

$$\begin{aligned} \frac{d^2\psi}{dx^2} &= -k^2 \sin kx \\ &= -k^2\psi \\ &= -\frac{4\pi^2}{\lambda^2}\psi. \end{aligned} \tag{2.95}$$

Eliminating λ^2 from Eqs (2.93) and (2.95) gives

$$\frac{d^2\psi}{dx^2} = -\frac{8\pi^2 m}{h^2}(E - U)\psi. \tag{2.96}$$

This equation is the one-dimensional form of the general time independent Schrödinger equation.[59] It is not the only equation we could have constructed;

[59] See Appendix 2.5.2.

others could have been put together that incorporate the same three principles. However, it is the solutions of this equation and of its three-dimensional analogues that are found to be consistent with experimental results.

In principle, the procedure for solving a system's Schrödinger equation is quite straightforward. The mathematical function that describes the system's potential energy in terms of its space and time coordinates is substituted in the equation in place of U and the resulting differential equation is then solved subject to the normalisation requirement that the particle must be somewhere.

However, though the procedure sounds simple, in most cases, the mathematics required for the solution of this differential equation are anything but straightforward.

2.4.7 Quantum Mechanics — Potential Wells

The simplest example of a quantum system is that of a particle for which $U = 0$. In the one-dimensional time independent case this gives for the particle's Schrödinger equation

$$\frac{d^2\psi}{dx^2} = -\frac{8\pi^2 m}{h^2} E\psi. \tag{2.97}$$

We will illustrate the solution of this equation by considering the case of a particle trapped in a one-dimensional infinite potential well of length L, a box with infinitely high impenetrable walls (Fig. 2.39).[60]

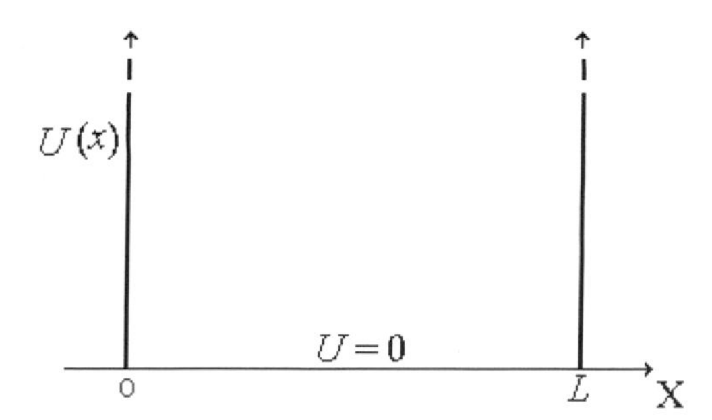

Fig. 2.39. A one-dimensional infinite potential well of width L. A particle situated inside the well can never escape, the height of the walls being infinite. The particle can only fly to and fro in the space between the walls.

We could solve Eq. (2.97) by the standard mathematical techniques,[61] but instead, let us use our physical intuition. Inside the well, the potential energy

[60] Other examples are discussed below as they arise in context.

[61] This is done in Appendix 2.5.2.

of the particle is everywhere zero; at the walls of the well it rises to infinity. Lacking the energy required to mount the infinite energy barrier that the walls of the well represent, the particle can never escape. It is trapped in the well. This being so, its matter wave must reduce to zero at the walls of the well. This *boundary condition* must be reflected in the solution of the particle's Schrödinger equation. It suggests that the trapped particle's matter wave has the form of a standing wave, of the type encountered in a vibrating wire stretched between two fixed points in which there are nodes at each end (Fig. 2.40).

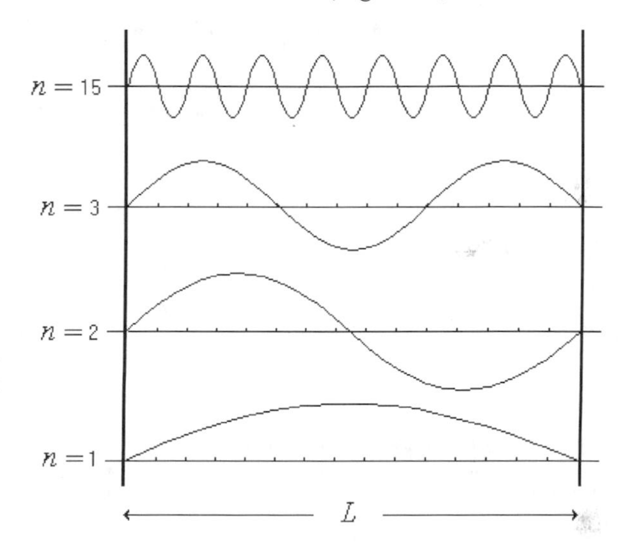

Fig. 2.40. The standing waves produced in a stretched wire of length L for different values of the integer n. The wave functions of a particle trapped in an infinite potential well have a similar form.

The wavelengths, λ_n, that can produce such standing waves in a stretched wire of length L are given by

$$\lambda_n = \frac{2L}{n} \quad (n = 1, 2, 3, \ldots), \tag{2.98}$$

i.e. the allowed wavelengths are $2L/1,\ 2L/2,\ 2L/3,\ \ldots$.

The integer n, which characterises the pattern of the trapped particle's matter wave, is called a *quantum number*. Each value of n corresponds to a different vibrational mode of the matter wave.

Since the particle's potential energy is zero, its total energy is its kinetic energy: $E = K$. Substituting the trapped particle's allowed wavelengths, $\lambda_n = 2L/n$, in the quantum mechanical expression for its kinetic energy, $K = h^2/2\lambda^2 m$, gives the following expression for its total energy, E_n:

$$E_n = \frac{n^2 h^2}{8mL^2} \quad (n = 1, 2, 3, \ldots). \tag{2.99}$$

These are the energy eigenvalues that define the *energy levels* of the particle trapped in the one-dimensional infinite potential well (Fig. 2.41). Delineation of the energy of a particle in a three-dimensional well would require three quantum numbers.

The eigenvalues of the trapped particle's linear momentum are obtained by substituting the expression for its energy eigenvalues, $E_n = n^2h^2/8mL^2$, in the formula $p = \pm\sqrt{2mE_n}$:

$$p_n = \pm\frac{n\pi\hbar}{L} \quad (n = 1, 2, 3, \ldots). \tag{2.100}$$

The particle is moving to and fro and so its average linear momentum is zero.

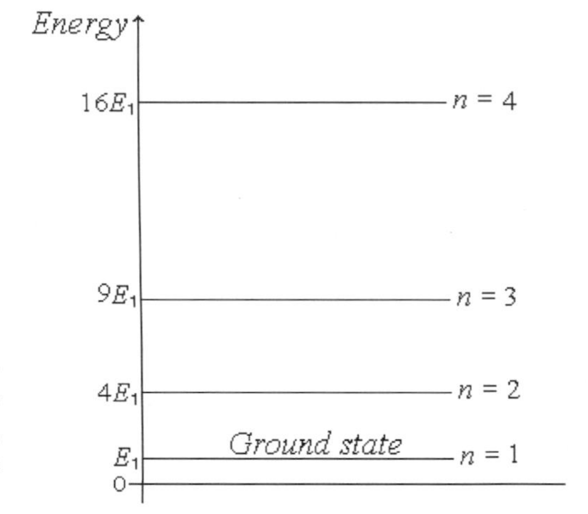

Fig. 2.41. The energy levels of a particle trapped in a one-dimensional infinite potential well. Even in the ground state, $n = 1$, the particle's energy is greater than zero.

In complete contrast to the situation in classical or relativistic mechanics, where the energy of a trapped or bounded particle can change continuously, in quantum mechanics the energy of a confined particle is quantised, i.e. the particle's energy can only change by certain discrete amounts. Thus, the energy, E_n, of a particle trapped in a one-dimensional infinite potential well always equals an integer multiple of the basic quantum, $E_1 = h^2/8mL^2$. Furthermore, the particle's energy can never be less than a single basic quantum; its quantum number, n, can never be less than unity. Thus, the particle can never be completely at rest even at the absolute zero of temperature.

In the case of a small macroscopic particle such as a grain of sand of mass $\sim 10^{-6}$ kg trapped in a well of length 1 m, the size of its basic quantum of energy is approximately 10^{-62} J. This is such a minute quantity of energy, that it is in practice undetectable, and so changes in the particle's energy are, to all intents and purposes, continuous and can be properly described by classical mechanics. This is a further example of the correspondence principle.

On the other hand, the basic quantum of energy of a particle the size of an electron trapped in a potential well of length 10^{-10} m (the approximate size of an atom) is approximately 30 eV. This is a significant quantity of energy; it is the energy acquired by an electron accelerated through a potential of 30 V. Consequently, the quantisation of the energy of an atom's electrons is a central feature in the structure of the atom and it is in such systems that quantum mechanics finds most of its applications.

In general, the total energy of a particle that is confined to a specific region of space, such as the particles constituting an atom, can only take certain discrete values. The particles are restricted to certain stationary states in which their energy is *quantised.* The energy eigenvalues of these stationary states, i.e. their energy levels, are sharply observable.

The complete solution to the particle's Schrödinger equation is obtained by its integration subject to the following two conditions:

1. the boundary condition that the wave function must equal zero at the well's walls, i.e. $\psi(x) = 0$ when $x = 0$ or L;
2. the normalisation condition, $\int_0^L |\psi^2(x)|dx = 1$.

This gives for the particle's wave function

$$\psi(x)_n = \sqrt{\frac{2}{L}} \cdot \sin\left(\frac{n\pi x}{L}\right) \quad (n = 1, 2, 3, \ldots). \tag{2.101}$$

The standing waves illustrated in Fig. 2.40 correspond to this equation. The corresponding probability densities, $|\psi(x)_n|^2$, are shown in Fig. 2.42.

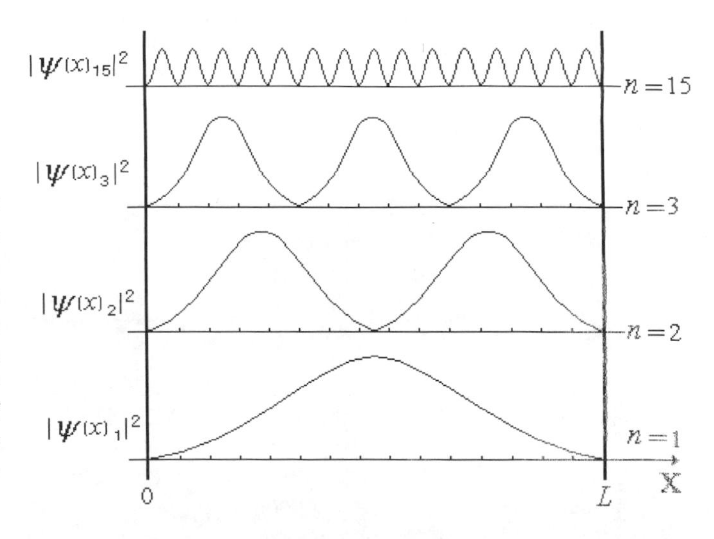

Fig. 2.42. The probability densities, $|\psi(x)_n|^2$, for various values of the quantum number n. The probability of finding the particle close to a point is proportional to the probability density at the point. Note that in every case the probability of finding the particle at the walls of the infinite potential well is zero.

The *eigenfunctions* that describe the stationary states the particle can occupy have the form

$$\Psi_S(x, y, z, t) = \psi(x, y, z)e^{-iEt/\hbar}, \tag{2.102}$$

where E is the energy eigenvalue of the stationary state, i.e. its energy level. The probability of finding the particle in a particular stationary state does not change with time. Hence, the probability density of a stationary state, $|\psi(x, y, z)e^{-(iEt/\hbar)}|^2$, is constant. This is the meaning quantum mechanics gives to the concept of a stationary state.

2.4.8 The Tunnel Effect

Infinite potential wells do not exist in reality; real energy barriers are of a finite height (Fig. 2.43). In classical physics, a particle can only escape from a finite potential well if it acquires sufficient energy to climb over its walls; there is no way it can get out with less energy. A ball rolling down one side of a bowl can rise to the same height on the opposite side, but no higher. However, this is not the case in quantum mechanics. Because of the inherent uncertainties in its position, the wave function of a particle does not equal zero at the walls of the well, and so there is a non-zero probability of finding the particle there. Consequently, the possibility exists, slight though it may be, that the particle will penetrate or 'tunnel' through the walls, i.e. through the energy (potential) barrier keeping it in the well. Thus, it can escape from the potential well even

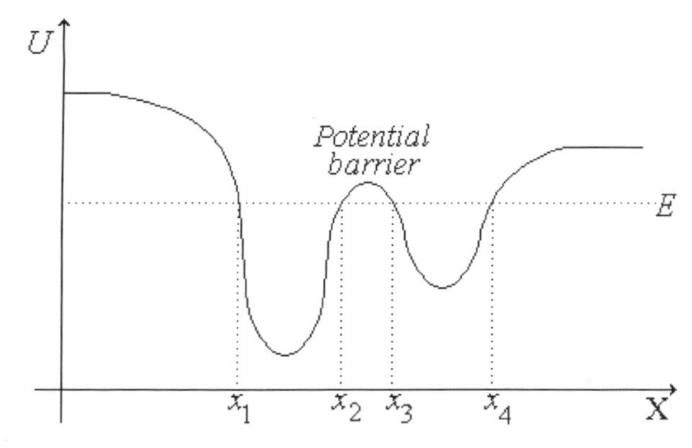

Fig. 2.43. A hypothetical potential energy diagram. Classically, a particle with energy E can oscillate in either of the two potential wells, i.e. between x_1 and x_2 or between x_3 and x_4. However, it has insufficient energy to pass over the potential barrier between x_2 and x_3; it is trapped in one or other of the wells. Quantum mechanics asserts that there is a non-zero probability that the particle will 'tunnel' through the barrier.

when it has insufficient energy to climb over its walls. This phenomenon is called the *tunnel effect* and is a characteristic of quantum behaviour.

Most of the particles striking the wall of a potential well will be reflected from it. Those that do get across to the other side are never observed inside the wall, where they would have a negative kinetic energy: $K_{wall} = E - U < 0$ (Fig. 2.44).

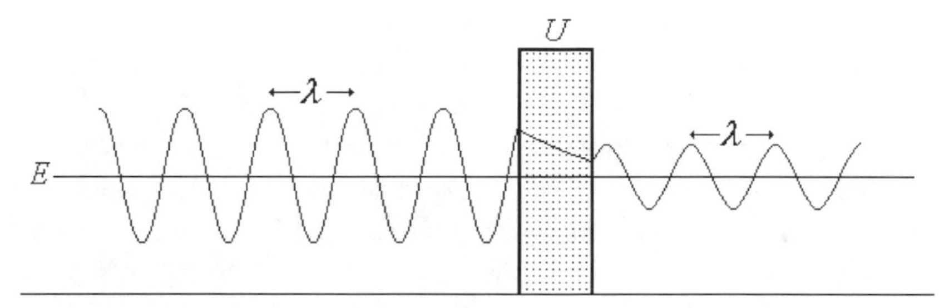

Fig. 2.44. The tunnel effect. The wave function of a particle which strikes the barrier from the left. Its energy, E, is less than the height of the barrier, U. Nevertheless, the wave function to the right of the barrier is not zero, though its amplitude is less than on the left. As a result, there is a non-zero probability of finding the particle to the right of the barrier.

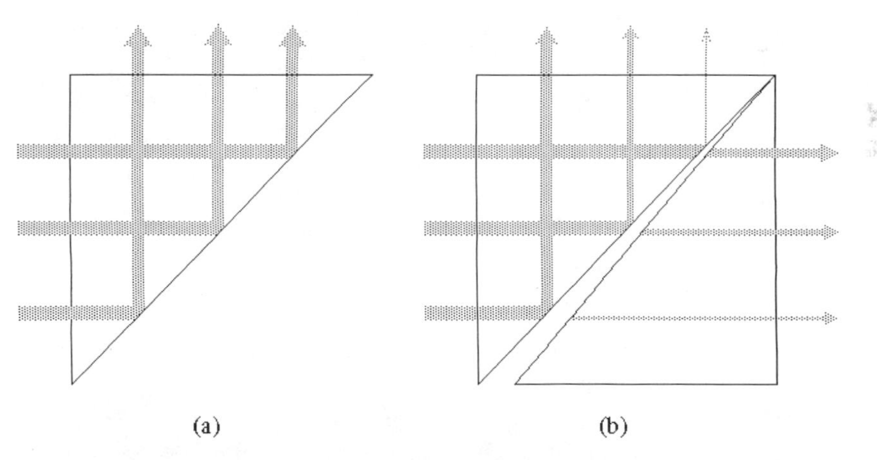

Fig. 2.45. (a) Total internal reflection. (b) The spacing between the prisms is of the order of a few wavelengths. Light 'jumps' across the gap and appears in the second prism. No light is detected in the gap.

Many phenomena are explained by the tunnel effect. For example, when two copper wires are in contact with one another, an electric current flows between them even though each is covered by a thin layer of non-conducting copper oxide. The conducting electrons simply 'tunnel' through the oxide barrier. The emission of α particles (Helium nuclei) from the nuclei of radioactive atoms is another example of *quantum tunnelling*. Even though the α particles have insufficient energy to overcome the barrier that holds nucleons inside the atom's nucleus, there is a non-zero probability that they will appear outside the nucleus.

The phenomenon of barrier penetration also occurs in macroscopic systems traditionally described in terms of waves. For example, how do the waves inside a glass prism 'know' that there is air on the outside and that they should be totally reflected internally (Fig. 2.45a)? It must be that on reaching the edge of the prism, the waves continue some small distance into the air in order to 'feel' what is outside. If a second prism is placed close to, but not actually touching the first, light reaches and enters the second prism (Fig. 2.45b). However, no light is detected in the space between them. The phenomenon can also be demonstrated with water and sound waves.

Appendices to Part Two

2.5.1 The Kinetic Energy and Linear Momentum of a Particle

Eliminating the particle's velocity, u, by simple algebra, from the familiar classical equations $K = mu^2/2$ and $p = mu$, for the kinetic energy and linear momentum of a particle of mass m respectively, gives

$$p = \sqrt{2mK}. \tag{2.103}$$

2.5.2 The Wave Function of a Trapped Particle

The general time-independent Schrödinger equation for a particle of mass m and potential energy $U(x, y, z)$ is

$$\frac{-h^2}{8\pi^2 m}\left(\frac{\partial^2 \psi}{\partial x^2} + \frac{\partial^2 \psi}{\partial y^2} + \frac{\partial^2 \psi}{\partial z^2}\right) = (E - U)\psi. \tag{2.104}$$

The particle's wave function, ψ, and energy levels, E, are obtained from the solution of this equation.

In the one-dimensional case, Eq. (2.104) reduces to

$$\frac{d^2\psi}{dx^2} = \frac{-8\pi^2 m}{h^2}(E - U)\psi. \tag{2.105}$$

The potential energy of a particle trapped in an infinite potential well is zero: $U = 0$. At the walls of the well the system's potential energy rises to infinity: $U = \infty$. Substituting $U = 0$ in Eq. (2.105) gives:

$$\frac{d^2\psi}{dx^2} = \frac{-8\pi^2 mE}{h^2}\psi. \tag{2.106}$$

Making the substitution

$$k^2 = \frac{8\pi^2 mE}{h^2} \tag{2.107}$$

reduces Eq. (2.106) to the familiar harmonic form

$$\frac{d^2\psi}{dx^2} = -k^2\psi. \tag{2.108}$$

The general solution of this equation is

$$\psi(x) = A\sin kx + B\cos kx. \tag{2.109}$$

The particle is trapped inside a potential well of width L, and so its wave function must equal zero at the well's walls. This requirement defines the following boundary conditions for the wave function: that $\psi(x) = 0$ when (i) $x = 0$; (ii) $x = L$. The first condition gives

$$\psi(0) = A\sin(0) + B\cos(0) = 0. \tag{2.110}$$

Satisfying this equation requires that $B\cos(0) = 0$, i.e. that $B = 0$. This reduces Eq. (2.109) to the form

$$\psi(x) = A\sin kx. \tag{2.111}$$

The second boundary condition gives

$$\psi(L) = A\sin kL = 0. \tag{2.112}$$

Satisfying this equation requires that $\sin kL = 0$, i.e. that

$$kL = n\pi \tag{2.113}$$

or

$$k^2 = \frac{n^2\pi^2}{L^2}, \tag{2.114}$$

where $n = 1, 2, 3, \ldots$. Substituting for k^2 in Eq. (2.107) gives for the energy of the trapped particle:

$$E = \frac{n^2h^2}{8mL^2}. \tag{2.115}$$

In order to complete the solution of the wave equation, we adopt Born's interpretation of the wave function, i.e. that the probability P of finding the particle in the region between x and $x + \Delta x$ is given by the square of the absolute value of the wave function:

$$P = |\psi(x)|^2 \cdot \Delta x. \tag{2.116}$$

The particle must be somewhere inside the potential well, i.e. the probability of finding it somewhere in there must equal unity. This defines the normalisation requirement

$$\int_0^L |\psi(x)|^2 dx = 1. \tag{2.117}$$

Substituting $\psi(x) = A \sin kx$ from Eq. (2.111) gives

$$A^2 \int_0^L \sin^2 kx \cdot dx = A^2 \int_0^L \frac{1 - \cos 2kx}{2} dx$$
$$= A^2 \left(\frac{L}{2} - \frac{\sin 2kL}{4k} \right)$$
$$= 1. \tag{2.118}$$

However, according to Eq. (2.112), $\sin 2kL = 0$. Hence,

$$A = \sqrt{\frac{2}{L}}. \tag{2.119}$$

Finally, using Eq. (2.113) and substituting for A in Eq. (2.111) gives

$$\psi_n(x) = \sqrt{\frac{2}{L}} \sin \left(\frac{n\pi x}{L} \right). \tag{2.120}$$

The waves illustrated in Fig. 2.40 match this equation.

Part Two: Questions, Exercises and Problems

Questions

1. Explain the following:

   a. No indication of the features on the inside surface of a red-hot cavity can be obtained by looking through a small hole in its wall.
   b. Cavities between the glowing coals of a very hot barbecue are indistinct in outline and brighter than the coals themselves.
   c. The temperature of hot glowing objects can be determined by their colour.
   d. If a piece of porcelain, having a dark pattern on a white background, is raised to white heat, the pattern appears reversed; the dark patterned parts become brighter than the white parts.
   e. A 'thermos' flask, which consists of a glass vessel with thin double-walls, the space between them being evacuated and the surfaces facing each other across the vacuum being silvered, can be used for storing hot or cold liquids for long periods without significant temperature change.

2. Show that the energy of a photon in electron-volts, E, is related to its wavelength in nano-metres, λ, by the equation $E = 1240/\lambda$.
3. A metal plate emits electrons when it is illuminated with green light but not when the light's color is yellow. Will it emit electrons when illuminated with (a) red light; (b) blue light?
4. A photoelectric cell is illuminated with green light. How will moving the light source farther away affect (a) the stopping potential; (b) the current through the cell?
5. Explain the following:

   a. An insulated metal plate that is illuminated with ultra-violet light emits electrons for just a short time; the emission then ceases.
   b. Distant stars and galaxies could not be photographed were light not of particle character.

6. An infra-red lamp and an ultra-violet lamp have the same power rating. Which emits photons at the greater rate?
7. Why is the Compton effect not observed with visible light?
8. Explain the tunnel effect in terms of the uncertainty principle.
9. How does the size of a potential well affect the energy levels of a particle inside it?
10. How would the graph of the probability density of a particle trapped in an infinite well look when (a) $n = 100$; (b) $n \to \infty$, i.e. under classical conditions?
11. Explain and distinguish between:

    a. The uncertainty principle;
    b. The correspondence principle;
    c. The complementary principle – particle/wave duality.

12. The frequency of the beats generated by waves of frequency f and f' is $|f - f'|$. In order to be certain that $f = f'$ how long must we wait to see whether beats

are generated or not? And after a period of time Δt, with what accuracy could we declare that the two frequencies are equal?

Exercises

1. The human eye is most sensitive to light of wavelength 550 nm. At what temperature will the radiation emitted by a black-body be most visible to the eye?
2. Calculate the rate at which a body emits heat at a temperature of 500°C if its surface area is $0.5\,\mathrm{m}^2$.
3. Small differences in the surface temperature of a body can be detected with a thermograph. Healthy skin has a temperature of 34°C; skin covering a malignancy has a slightly higher temperature. Calculate the ratio between the rate at which heat is emitted from healthy skin and that at which it is emitted from skin that is one degree warmer.
4. The threshold frequency for the emission of photoelectrons from copper is $1.1 \cdot 10^{15}$ Hz. Calculate the maximum energy – in Joules and in electron-volts – of the photoelectrons emitted from a copper plate by radiation of frequency $1.5 \cdot 10^{15}$ Hz.
5. The work function of sodium is 2.3 eV.

   a. What is the longest wavelength of light that can release photoelectrons from sodium?
   b. What will be the maximum kinetic energy of the electrons emitted from sodium by light of wavelength 2000 Å ($1\,\text{Å} = 10^{-10}\,\mathrm{m}$)

6. The maximum kinetic energy of the photoelectrons emitted from a tungsten plate illuminated by light of wavelength 248 nm is $8.6 \cdot 10^{-20}$ J. Calculate the work function of tungsten.
7. The work function of sodium is $3.12 \cdot 10^{-19}$ J. A sodium lamp emits yellow light with a wavelength of 590 nm. Calculate the stopping potential of the electrons emitted from a sodium plate illuminated by a sodium lamp.
8. Calculate the energy of a photon of wavelength 700 nm.
9. Calculate the wavelength of a photon of energy $5 \cdot 10^{-19}$ J.
10. A sodium lamp emits yellow light with a wavelength of 590 nm. How many photons does a 100 W sodium lamp emit each second?
11. The radiation emitted by an X-ray tube has a minimum wavelength of 0.1 Å. At what potential is the tube being operated?
12. Calculate the frequency of a photon which has a linear momentum of $1.1 \cdot 10^{-23}$ kg·m/s.
13. A surface of area two square metres is uniformly illuminated with light of intensity $10\,\mathrm{W/m^2}$ and wavelength 600 nm. How many photons strike the surface each second?
14. A beam of 0.1 nm X-rays is reflected from a crystal at a glancing angle of 10.3°. Calculate the spacing between the reflecting crystal planes.
15. The spacing between adjacent parallel planes in a crystal is 0.3 nm. Calculate the smallest glancing angle at which Bragg diffraction will be observed with radiation of wavelength 0.3 Å.

16. Calculate the fractional change in the photon energy as a result of Compton scattering through an angle of 90° for photons of:

    a. X-rays with a wavelength of 0.025 nm;
    b. γ rays with an energy of 1 MeV.

17. A beam of X-rays is scattered by free electrons. At a scattering angle of 45° the wavelength of the scattered beam is $2.2 \cdot 10^{-12}$ m. Calculate the wavelength of the incident beam.
18. Calculate the wavelength of the matter wave associated with electrons of energy 15 keV.
19. The electrons in an electron microscope are accelerated through a potential of 40,000 V. On the assumption that a microscope's power of resolution is determined by the wavelength with which it operates, calculate the power of resolution of this electron microscope.
20. A beam of 600 eV electrons strikes the surface of a crystal. The spacing between the atoms in the crystal is $2.8 \cdot 10^{-10}$ m. What is the smallest glancing angle at which a strong reflected beam of electrons will be obtained?
21. The position of a 1 keV electron is determined to an accuracy of 1 Å. Estimate the accuracy to which the electron's linear momentum can simultaneously be known.
22. The mean kinetic energy of thermal neutrons is given by $3kT/2$, where k is Boltzmann's constant and T is their absolute temperature. Calculate the mean De Broglie wavelength of neutrons at room temperature (27°C).
23. Calculate the lowest possible energy of a neutron that is trapped in a potential well of width 10^{-14} m (the approximate size of a nucleus).
24. Calculate the energy of (a) an electron and (b) a photon, whose wavelength *in vacuo* is 1.0 nm.
25. An electron is trapped in an infinite potential well of width 0.25 nm. How much energy must the electron acquire in order to 'jump' from its ground state ($n = 1$) to the $n = 4$ state?

Problems

1. Assuming that the human body has a surface area of two square metres and radiates like a black-body at a temperature of 35°C, calculate the rate at which it loses heat in surroundings that have a temperature of 15°C.
2. Radiant energy from the sun strikes the earth with a maximum mean intensity of 1.4 kW/m^2 – a quantity called the *solar constant* that is sufficient to melt a layer of ice 1.6 cm thick every hour. On the assumption that sunlight is black-body radiation and that the ratio between the radius of the earth's orbit to the radius of the sun is 216, calculate the temperature of the sun's surface.
3. A 100 mW argon laser emits a beam of light of wavelength 488 nm *in vacuo*.

    a. How many photons does the laser emit each second?
    b. What current will flow through a photoelectric cell whose cathode (emitting plate) is illuminated by the laser's light beam if 10% of the photons release an electron from it?

c. What potential will be required between the cell's electrodes in order to stop the current, if the threshold frequency of the cathode is $5.2 \cdot 10^{14}$ Hz?

4. When the velocity of a certain particle is three times that of an electron, the ratio between their De Broglie wavelength's is $1.813 \cdot 10^{-4}$. Identify the particle. (Assume that the velocities are significantly less than c and so non-relativistic equations can be used.)
5. Show that in the ground-state, the uncertainty in the energy of a particle trapped in a potential well is approximately equal to its lowest allowed energy.
6. a. Calculate the difference, $\Delta E_n = E_{n+1} - E_n$, between the energy levels of a particle of mass m that is trapped in an infinite potential well of length L in the case where $n \gg 1$.
   b. Compare the values of ΔE_n and ΔE_{n+1} in the case where $n \gg 1$. Comment on the result.
   c. Find an expression for the ratio $\Delta E_n / E_n$. How does the value of this ratio change as n grows larger?

7. A photoelectric cell (photodiode) is connected to a potentiometer as shown in the diagram. Note that the emitting plate is connected to the negative terminal and the collector to the positive one.

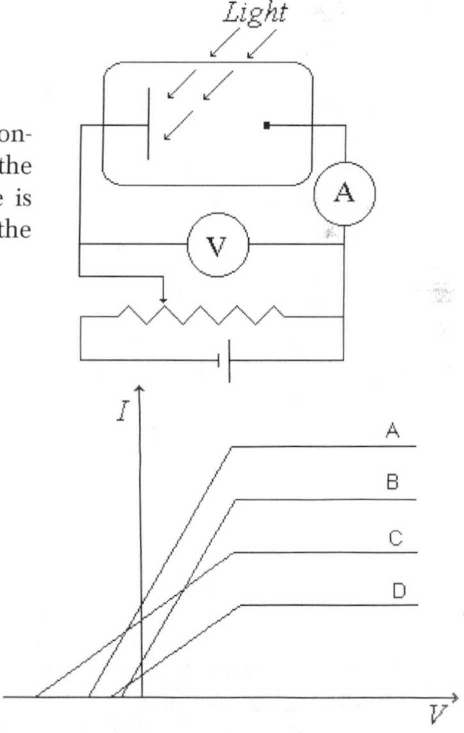

The emitting plate was illuminated with light of different wavelengths and intensities. In each case, the relationship between the current, I, flowing through the diode and the potential, V, across it (the diode characteristic) was investigated. The results obtained are summarised in the four graphs A to D.

a. What does the horizontal part of each graph indicate?
b. What was happening in the sloping part of the graphs?
c. Why didn't the current cease altogether when the potential between the plates was zero?

d. Which graph corresponds to light:

   i. of greatest intensity;
   ii. of highest frequency;
   iii. of longest wavelength?

8. With what potential must the electrons be accelerated in an electron microscope if its theoretical power of resolution is to equal that of microscope operating with X-rays of energy 100 keV?
9. The following values were obtained for the stopping potential, V, when a metal plate in a photoelectric tube was illuminated with light of different wavelengths, λ.

   Draw an appropriate graph and obtain from it the values of:

   a. the metal's threshold frequency;
   b. the metal's work function;
   c. Planck's constant.

λ (nm)	V (volts)
366	1.48
405	1.15
436	0.93
492	0.62
546	0.36
579	0.24

10. In the upper atmosphere, oxygen molecules dissociate into oxygen atoms as a result of a photochemical reaction initiated by photons in the incident sunlight. Only photons with a wavelength less than $1.75 \cdot 10^{-7}$ m *in vacuo* can bring about the dissociation. On this basis, calculate the bond energy of the oxygen molecule.
11. The frequency, f, of a harmonic oscillator of mass m and elasticity constant k is given by the equation $f = \frac{1}{2\pi}\sqrt{k/m}$. The energy of the oscillator is given by $E = \frac{p^2}{2m} + \frac{kx^2}{2}$, where p is the system's linear momentum and x is the displacement from its equilibrium position. Use the uncertainty principle, $\Delta x \cdot \Delta p \approx \hbar/2$, to express the oscillator's energy E in terms of x and show, by taking the derivative of this function and setting $dE/dx = 0$, that the minimum energy of the oscillator (its ground state energy) is $E_{\text{min}} = hf/2$.
12. The electrons in a television tube are accelerated through a potential of about 10,000 V. Evaluate the danger to those watching due to the brehmstrahlung radiation produced when these electrons strike the screen and produce the image.
13. The following lines appear in the emission spectrum of an element:

$$\lambda_1 = 122\,\text{nm} \quad \lambda_2 = 102.5\,\text{nm} \quad \lambda_3 = 97\,\text{nm} \quad \lambda_4 = 95\,\text{nm}$$
$$\lambda_5 = 656\,\text{nm} \quad \lambda_6 = 486\,\text{nm} \quad \lambda_7 = 434\,\text{nm}$$
$$\lambda_8 = 1880\,\text{nm} \quad \lambda_9 = 1280\,\text{nm} \quad \lambda_{10} = 4050\,\text{nm}$$

   a. Does the spectrum exhibit the Ritz combination rule?
   b. Calculate the energy of the photons associated with each spectral line.
   c. Sketch an energy level diagram that is consistent with the element's spectral lines.

14. Under usual conditions of temperature and density, the behaviour of a gas can be interpreted in terms of classical physics, for example, by Maxwell's kinetic theory. However, when the De Broglie wavelength of the atoms is of the order of the mean distance, d, between them, the classical approach fails.

   a. Show that the classical approach fails when $d \approx \sqrt{h^2/mkT}$.

b. Assuming that the mean distance between the atoms is 1 nm, at what temperature will quantum effects begin to appear in the gas helium?

15. In a demonstration of the diffraction of electrons by a thin aluminium foil (Example 2.12) it was found that when the electrons were accelerated through a potential of 1600 V, the radius of the first circle in the diffraction pattern was 11.4 mm. The distance from the foil to the screen was 15.0 cm. The spacing between the atoms in aluminum metal is 4.04 Å. Calculate the wavelength of the electrons:

    a. according to De Broglie's hypothesis;
    b. on the basis of the dimensions of the diffraction pattern.

16. When an electron penetrates the surface of a metal, its energy changes as a result of the metal's work function.

    a. Does an electron's energy increase or decrease when it penetrates a metal's surface?
    b. In the Davisson and Germer experiment, the electrons were accelerated through low potentials; 54 V was a typical value. Calculate the velocity of the electrons before they entered the nickel crystal and after entering on the assumption that the work function of nickel is 2.6 eV.
    c. Calculate the wavelengths of the electrons before and after they penetrate the surface.

Part Three The Nuclear Atom

The word atom is of Greek origin – $\alpha\tau o\mu o\varsigma$ – meaning *indivisible.* The ancient Greek philosophers Leucippus, Epicurus and Democritus supposed that the properties of materials were determined by the different shapes and forms of their atoms. According to this notion, iron atoms are solid and strong with hooks that lock them into a solid; water atoms are smooth and slippery; salt atoms, because of their taste, sharp and pointed; and air atoms light and whirling, pervading all other materials.

Other ancient Greek philosophers, such as Empedocles, Heraclitus and Aristotle, considered all matter to be composed of the four elements, earth, air, fire and water, in varying proportions. According to this notion, lead differs from gold only in the proportions of the four elements it contains and so the problem of converting lead into gold is just a matter of correctly blending the right combination of elements. The search for this recipe, which, with the addition of a pinch of a very special and rare spice called the *philosopher's stone*, would turn base metals into gold and enrich its finder beyond his wildest dreams, occupied the minds and time of many learned men for more than 1,000 years. Atoms were forgotten in the frenzy of the alchemists' search for the formula and the essence.

Modern ideas about the structure of matter originated in the 17th century. In *The Sceptical Chymist*, published in 1661, Robert Boyle[1] laid down the modern criterion of an element: a substance that cannot be decomposed chemically. This definition set the agenda for much of experimental chemistry for the next 200 years. A counterpart to this was the renascence of the atom brought about by the writings of Descartes,[2] Gassendi[3] and Newton. Anticipating kinetic theory, Hooke[4] even suggested that the source of the 'spring of the air'[5] was the collisions of its atoms with the walls of the container.

Summing up the general belief in atoms, Newton wrote in the final query he appended to his book *Opticks*:

[1] Robert Boyle 1627–1691, English chemist and physicist.
[2] René Descartes 1596–1650, French philosopher and mathematician.
[3] Pierre Gassendi 1592–1655, French philosopher.
[4] Robert Hooke 1635–1703, English physicist.
[5] The air's elasticity.

'All things being considered, it seems probable to me that God in the beginning formed matter in solid, massy, hard, impenetrable, moveable particles, of such sizes and figures, and with such other properties, and in such proportion, as most conduced to the end for which He formed them; and that these primitive particles, being solids, are incomparably harder than any porous bodies compounded of them; even so very hard, as never to wear or break into pieces.'[6]

In his *Essay on Human Understanding*, the philosopher Locke[7] re-examined the problem of how atoms could account for the many and varied properties of material objects. He distinguished between *primary qualities* such as motion, size and shape, that he argued are inherent in the material's atoms, and *secondary properties* such as colour, taste and smell, that exist only in the mind of the observer and result from an interaction between the atoms in the object and those in the sense organs of the observer. It was this dichotomised interpretation of perception, i.e. that though an object's shape is actually seen its colour is only in the mind, that marked the severance of science from philosophy. Whereas philosophy continued to explore the mind and its ideas, experimental science set out to examine the physical world and the atoms from which it is composed.

Boyle's empirical law, that the volume of a gas is inversely proportional to its pressure, was given a sound mathematical basis by Daniel Bernoulli[8] in 1738, when he showed that it could be derived from mechanical considerations of the collisions between the particles of the gas and the walls of its container (Example 1.1). In so doing, Bernoulli founded the classical kinetic theory of matter. Assuming material substances and physical systems to be composed of minute particles, this theory seeks to explain their macroscopic properties in terms of the Newtonian motion of their constituent particles. Direct evidence of the motion of the these particles was first obtained by Brown[9] in 1827, when he observed the random movement of fine grains of pollen in an aqueous suspension using the recently invented achromatic microscope (Ch. 1.3).

In 1803 Dalton[10] published the first comprehensive atomic theory. In *A New System of Chemical Philosophy* he proposed that the elements are comprised of indestructible atoms, all the atoms of a particular element being identical. The notion of the molecule, compounded of atoms, as the basic and characteristic chemical unit of matter was proposed by Avogadro[11] in 1811, and, in 1812,

---

[6] *Opticks* (Reprinted 1952) New York: Dover Publications, p. 400.
[7] John Locke 1632–1704, English philosopher.
[8] Daniel Bernoulli 1700–1782, Swiss mathematician.
[9] Robert Brown 1773–1858, Scottish botanist.
[10] John Dalton 1766–1844, English chemist.
[11] Count Amedeo Avogadro 1776–1856, Italian physicist.

Berzelius[12] first suggested that the forces binding the atoms together were electrical in origin.

Modern atomic theory began with the discovery of the electron. This discovery opened the way for the integration of the mechanical, the chemical and the electrical properties of the atom into a single atomic theory – the nuclear model of the atom.

The notion of an essentially indestructible and indivisible atom remains fundamental to much of modern chemistry and biology, though, as we will see in Part Five of this book, by the middle of the 20th century the techniques of splitting atoms had been mastered by modern physics with devastating results.

[12]Jons Berzelius 1779–1848, Swedish chemist.

Chapter 3.1
The Structure of the Atom

3.1.1 The Thomson Model of the Atom

Experiments carried out following their discovery in 1897 showed that electrons could be obtained, by one means or another, from just about any material. For example, by the action of ultra-violet light on metals, by heating metal wires, by the ionising action of X-rays or from the spontaneous disintegration of radioactive substances. Evidently, electrons were a component of all neutral atoms, the counterbalancing positive charge being carried by one or more of their other constituent parts.

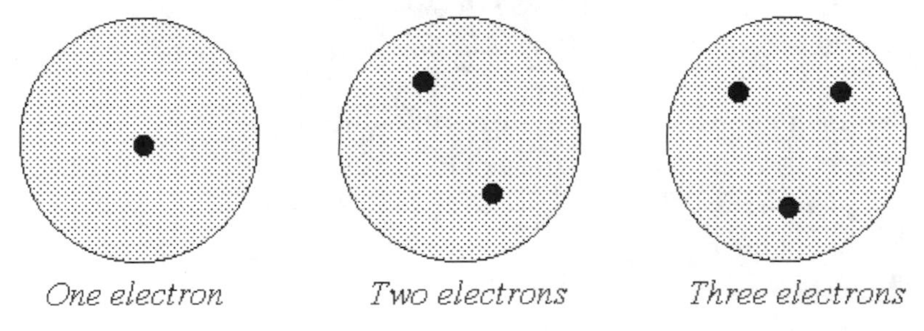

Fig. 3.1. The Thomson model of the atom; the *plum-pudding atom.* In multi-electron atoms, the attractive force on the electrons is balanced by their mutual repulsion and a stable configuration results.

Since the electron's mass is just 0.05% that of the hydrogen atom, itself the smallest of all atoms, the contribution the electrons make to the overall mass of an atom is minuscule. Most of an atom's mass must be associated with its positively charged constituent. In the light of this, J.J. Thomson proposed in 1904 that a neutral atom is a sphere of homogeneous positively charged material, in which negatively charged electrons are scattered like 'the plums in a pudding'. The positive charge attracts the electrons to the centre of the sphere, however, in atoms containing more than a single electron – *multi-electron atoms* – this attraction is balanced by the mutual repulsion between the electrons. At equilibrium, a balanced distribution of the electrons results (Fig. 3.1).

Calculations showed that the electrons in multi-electron atoms would arrange themselves in stable concentric rings and, on certain assumptions, even in rings of eight, reminiscent of the groups of eight in Mendeleév's[13] periodic table of the elements. As the number of electrons in an atom increased, more rings would be formed; the electrons in large multi-electron atoms would be arranged in a series of concentric rings. Thomson went on to suggest that the number of electrons in the outermost ring would determine the atom's electrovalency and was thus related to the element's position in the periodic table. On this basis, the properties of the series of elements

He Li Be B C N O F Ne
Ne Na Mg Al Si P S Cl A

could be explained as follows:

> '*The first and last element in each of these series has no valency [their outermost ring has an unalterable number of electrons], the second is a monovalent electropositive element, the last but one is a monovalent electronegative element, the third is a divalent electropositive element, the last but two a divalent electronegative element, and so on.*'[14]

However, Thomson met with great difficulties when he tried to explain the emission spectrum of atomic hydrogen in terms of his model. It can be shown, that if the electron in a Thomson hydrogen atom is disturbed from its equilibrium position at the centre of the atom as the result of a collision or some other excitation, it will oscillate harmonically at a single frequency. The excited electron constitutes an electric oscillator, and in accordance with electromagnetic theory it should emit electromagnetic waves of a frequency equal to that of its oscillation. Thus, the emission spectrum of hydrogen atoms should comprise just a single line (in the far ultra-violet). However, the emission spectrum of atomic hydrogen actually comprises a large number of lines (Fig. 3.2). The Thomson model had no explanation for this.

To summarise, Thomson's model proposed that:

1. Atoms comprise a sphere of positive material of uniform and 'normal' density ($\sim 10^3\ \mathrm{kg/m^3}$) that accounts for virtually all of their mass and determines their spatial extension, e.g. their diameter.
2. At equilibrium, an atom's electrons are at rest and are distributed throughout the sphere of its positive component.

[13]Dmitri Ivanovich Mendeléev 1834–1907, Russian chemist.
[14]J.J. Thomson (1904) *Phil. Mag.*, **7**, 237.

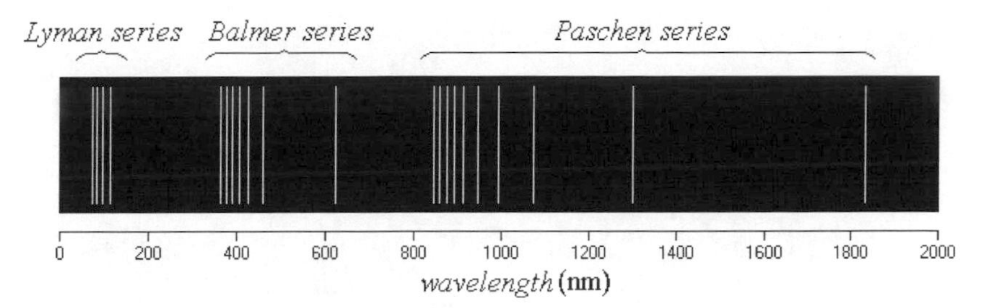

Fig. 3.2. The emission spectrum of atomic hydrogen. The spectral lines appear in distinct series, named after their discoverers. Each series starts from a characteristic minimum wavelength, all the other lines in the series corresponding to longer wavelengths.

These aspects of the model were investigated by Rutherford[15] and his coworkers in 1910.

3.1.2 The Nuclear Atom

The discovery of X-rays, and of the penetrating α and β radiations emitted by radioactive substances, provided investigators at the beginning of the 20th century with powerful new tools for atomic research. In the first experiments performed using these probes, a narrow collimated beam of β-rays was aimed at a metal foil (β-rays are high energy electrons, ~1 MeV). The beam penetrated the foil and was found to be just slightly broadened when it emerged from the other side. This broadening of the particle beam was attributed to the repulsion of the β particles by the electrons in the metal's atoms. The number of electrons in each atom was estimated from the extent of the broadening. Their number was found to depend on the nature of the atom; it varied from about 12 in the case of aluminium to about 100 in the case of gold. These experiments confirmed that the number of electrons in an atom was not great enough for them to make a significant contribution to the atom's total mass.

In 1910, Geiger[16] and Marsden[17] investigated the scattering of α-rays by metal foils. A narrow collimated beam of α particles was aimed at a gold foil of approximately 1 μm thickness, i.e. about 10,000 atoms thick. The particles in α-rays are high energy helium nuclei, ~6 MeV. They carry a positive charge of $+2e$ and have a mass that is about 7,300 times greater than that of the electron.

[15] Lord Ernest Rutherford 1871–1937, English physicist who was awarded the Nobel prize in 1908 for his researches on the structure of the atom and radioactivity.

[16] Hans Geiger 1882–1945, German physicist.

[17] Sir Ernest Marsden 1889–1970, New Zealand physicist.

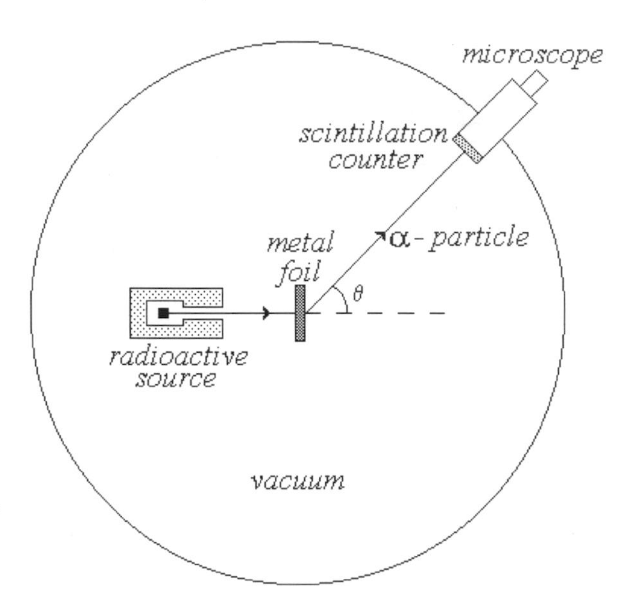

Fig. 3.3. Geiger and Marsden's apparatus for investigating the scattering of α particles by a metal foil. The particles emerging from the foil at a particular angle θ were detected by the faint flashes of light produced when they struck the scintillation counter.

Because of their relatively much greater mass, α particles are not significantly deflected from their paths by the electrons in the metal's atoms.

The α particles emerging from the gold foil were detected by the faint light flashes each produced on striking a glass plate coated with zinc sulphide; the plate was viewed through a low power microscope. This particle detector or *scintillation counter*, as it was later termed, was mounted on an arm that could swivel in a circle around the metal foil. The apparatus was enclosed in a vacuum chamber in order to avoid collisions between the α particles and air molecules (Fig. 3.3).

The number of α particles scattered through a particular angle, θ, during a given period of time, was found by counting the number of scintillations observed through the microscope at that angle. In the first series of experiments, Geiger found that the average deflection of the α particles was 0.87°. This result was in line with those that had been obtained from the scattering of β-rays.

At Rutherford's suggestion, the experiment was repeated, perhaps with greater attention to detail. This time, the experimenters noticed that a few of the α particles, about 1 in 8,000, were deflected through angles greater than 90°, i.e. they recoiled backwards. When he was informed of this result Rutherford is reported to have said: 'It is almost as incredible as if you had fired a fifteen inch shell at a sheet of tissue paper and it came back and hit you.' Such large deflections in the paths of the α particles could only result from a collision with a massive target or from an interaction with an electric or magnetic field of great strength.

That atoms are not impenetrable massive objects had been shown by the relative ease with which β-rays and most of the α-rays traversed the metal foils; this excluded the first possibility. Thus, the deflections had to be electrical or perhaps possibly magnetic in origin. The presence, inside the gold foil, of magnetic fields of the required intensity and orientation stretched the imagination, and so Rutherford concluded that the interaction had to be electrical (Fig. 3.4).

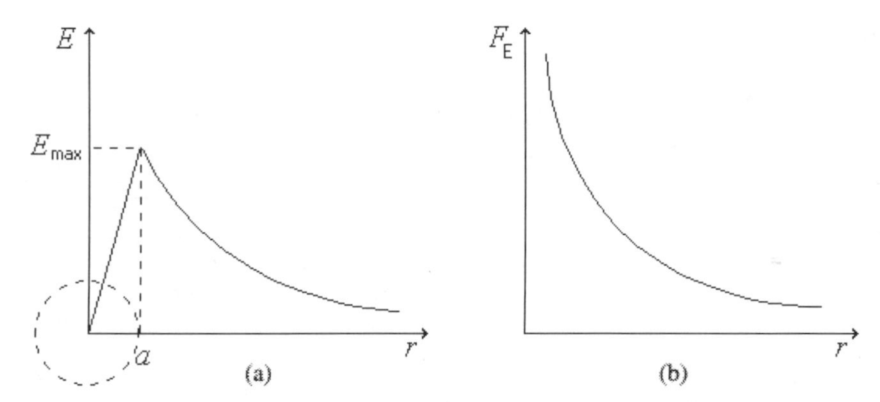

Fig. 3.4. (a) The electric field, E, associated with a homogeneous charged sphere of radius a (a Thomson atom); the maximum field strength is at the sphere's surface. (b) The repulsive force, F_{E}, between an α particle and the positive nucleus in a Rutherford atom as a function of the distance, r, between them; the closer the encounter between the particles, the greater the repulsive force between them becomes.

Confronted with these results, Rutherford abandoned the Thomson model. He pointed out that:

> '*the theory of J.J. Thomson does not admit of a very large deflection in traversing a single atom, unless it is supposed that the diameter of the sphere of positive electrification is minute in comparison with the diameter of the sphere of influence of the atom.*'[18]

After considering the dynamics of the problem, Rutherford suggested that the positive component of the atom is localised in a very small volume, which he called the *atomic nucleus*, whose diameter is of the order 10^{-14} m. Only by adopting this hypothesis could the very large deflections experienced by the α particles be explained (Fig. 3.5).

In Rutherford's atom, the diameter of its sphere of influence is determined by its electrons. These gird the nucleus at a relatively great distance from

[18](1911) *Phil. Mag.*, **21**, 669.

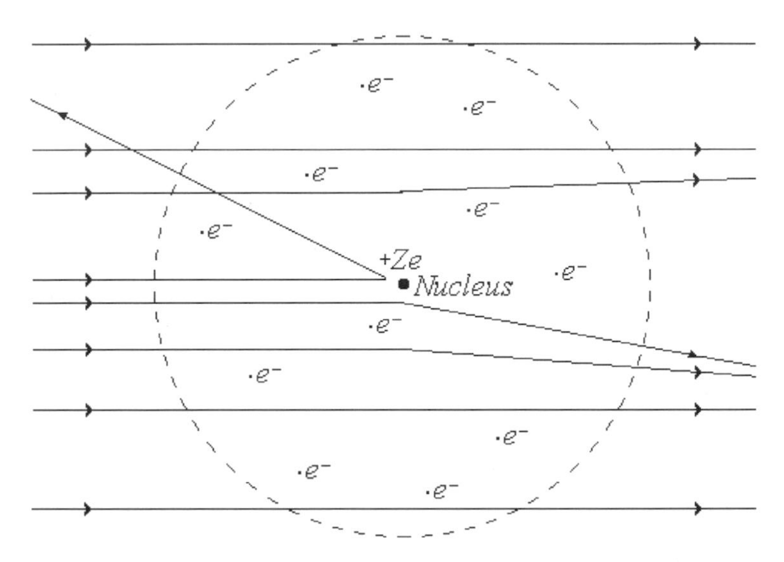

Fig. 3.5. The scattering of a beam of α particles by a nuclear atom (not to scale). The atom's size, i.e. the diameter of its sphere of influence, is determined by its electrons. Most of the α particles pass through the atom without any noticeable deflection. Only those passing close to the nucleus, which carries a charge of $+Ze$, where Z is the number of electrons in a neutral atom, are significantly deflected. A particle moving very close to the nucleus may even be deflected backwards.

it: about 10^{-10} m. Thus, the diameter of the nuclear atom is some four to five orders of magnitude larger than that of its nucleus; the nuclear atom is largely empty space. Since virtually all the atom's mass is concentrated within its minute nucleus, the density of the nuclear material must be immense. Calculations showed it to be about 10^{14} tons per cubic meter; this is one hundred billion times greater than the density of water.

The electric field close to the surface of the nucleus is about 10^{21} V/m, i.e. some 10^8 times greater than that at the surface of the Thomson atom (ignoring the effect of the electrons). Rutherford explained that the scattering of the α particles through large angles resulted from a close encounter between an α particle and an atomic nucleus. The repulsive force, F_E, between the positive α particle and the positive charge, $+Ze$, carried by the nucleus is given by Coulomb's law, and in close encounters this force is very great (Fig. 3.4b). The small angle deflections, which are the vast majority, result from relatively distant encounters between α particles and nuclei, in regions where the electric field and the concomitant repulsive force between the particles is weak.

Rutherford showed that under the influence of a Coulombic force, the path of an α particle passing by a nucleus would be a hyperbola, the nucleus being situated at its external focus. The angle of deflection of the particle, θ, is that between the two asymptotes of the hyperbola (Fig. 3.6). Every α particle that is directed at a point within a certain target area around the nucleus will be deflected through an angle equal to or greater than a particular value of θ. The target area, πb^2, that corresponds to a particular angle of deflection, is called the *cross-section* for that deflection; b is called its *impact parameter.*

In order to incur a large deflection, an α particle must approach the nucleus with an impact parameter of 10^{-13} m or less. Rutherford predicted that the fraction of the incident α particles deflected through any given angle, θ, on traversing the metal foil, would be proportional to $\text{cosec}^4(\theta/2)$. This was confirmed in a further series of experiments conducted by Geiger and Marsden in 1913.

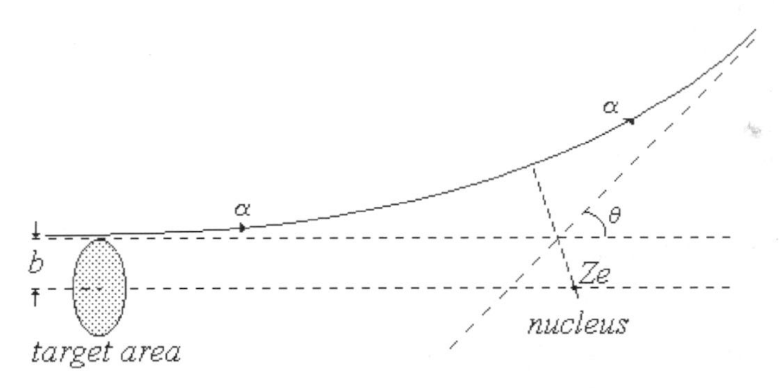

Fig. 3.6. The path of an α particle as it passes close to a nucleus that carries a nuclear charge of $+Ze$. The imaginary target area, πb^2, is that into which the particle must be aimed in order that its deflection be at least through the angle θ.

Rutherford drew attention to the importance of the magnitude of the nuclear charge in the determination of the properties of an atom. However, though the 'primitive atom' of negative electricity – the electron – had been discovered some 15 years previously, the particle that carried the natural unit of positive electricity, and which must be a constituent of every nucleus, had not yet been identified.

The identification of the 'primitive atom' of positive electricity was first made by Rutherford. In a suitable discharge tube, positive (canal) rays can be produced that comprise simply hydrogen atoms that have lost their electrons. No positively charged atom lighter than these had ever been detected. Furthermore, they never carry more than a single natural unit of positive charge, $+e$. Rutherford reasoned that the positive particles in these rays must be the nuclei of hydrogen atoms. He identified these particles – hydrogen nuclei – as the carriers of the natural unit of positive electricity and called them *protons*. Thus, a neutral hydrogen atom comprises a single proton and a single electron.

To summarise Rutherford's model of the atom:

1. Atoms comprise a very small and extremely dense nucleus that contains the entire positive charge, $+Ze$, of the atom and virtually all of its mass.
2. Z electrons, each of a charge e^-, surround the nucleus at a relatively great distance from it, determining its diameter; together with the nucleus they comprise the neutral atom.

Example 3.1: The Atomic Nucleus

What is the closest that an α particle with a kinetic energy of 6 MeV can approach to the nucleus of a gold atom in a gold foil?

Calculation: Gold atoms in a foil are arranged in a regular lattice which restricts their movement. Thus, the gold nucleus can be assumed to be at rest. Since both the α particle and the nucleus carry positive charges, $+q_\alpha$ and $+q_{\mathrm{Au}}$ respectively, a Coulombic force of repulsion will act between them. Initially, the distance between the α particle and the nucleus is so great that the interaction between them is negligible. However, as the α particle approaches the gold nucleus, its kinetic energy, K, is progressively converted into the electrostatic potential energy

$$U(r) = \frac{q_\alpha q_{\mathrm{Au}}}{4\pi\varepsilon_0 r} \tag{3.1}$$

at a distance r between the charged particles (Appendix 3.5.1).

At its closest approach to the nucleus, a distance r_0 from it, the kinetic energy of the α particle equals zero, i.e. all its original energy has been converted into electrostatic potential energy:

$$U(r_0) = \frac{q_\alpha q_{\mathrm{Au}}}{4\pi\varepsilon_0 r_0} = 6\,\mathrm{MeV}. \tag{3.2}$$

The charges carried by the α particle and the gold nucleus are $q_\alpha = +2e$ and $q_{\mathrm{Au}} = +79e$, respectively. We shall assume that the space inside the atom can be regarded as empty space, such that the electrostatic constant of proportionality, k, has its usual value *in vacuo*, i.e. $k = 1/4\pi\varepsilon_0 = 9 \cdot 10^9\,\mathrm{N \cdot m^2 \cdot C^{-2}}$.

Converting the energy of the α particle to units of Joules and substituting in Eq. (3.2) gives

$$U(r_0) = 6 \cdot 10^6 \cdot 1.6 \cdot 10^{-19} = \frac{9 \cdot 10^9(+2e)(+79e)}{r_0}\,\mathrm{J}. \tag{3.3}$$

Thus,

$$r_0 = \frac{9 \cdot 10^9 \cdot 158 \cdot (1.6 \cdot 10^{-19})^2}{6 \cdot 10^6 \cdot 1.6 \cdot 10^{-19}} = 3.8 \cdot 10^{-14}\,\mathrm{m}. \tag{3.4}$$

The result we have obtained is indicative of the order of magnitude of the gold nucleus. The actual diameter is approximately $7 \cdot 10^{-15}$ m.

Chapter 3.2

The Bohr Model of the Atom

Rutherford was not the first to propose a nuclear structure for the atom. The idea had been put forward previously in 1903 by Nagaoka.[19] He had suggested that the atom consists of a massive positive centre surrounded by rings of orbiting electrons, like the rings of tiny particles around the planet Saturn. Such systems had been studied some 30 years earlier by Maxwell, who had shown that, in classical mechanics, a condition for the dynamic stability of such systems of rotating particles is that only forces of attraction act between its components. The only forces acting in the Saturnic system are attractive gravitational forces. However, the orbiting electrons in Nagaoka's atom would repel one another. Nagaoka recognised this but argued that it was the small oscillations of the electrons induced by their instabilities that gave rise to the radiation emitted by atoms. He also suggested that the α and β particles emitted by radioactive atoms were the result of more violent instabilities. His ideas were not found sufficiently convincing at the time.

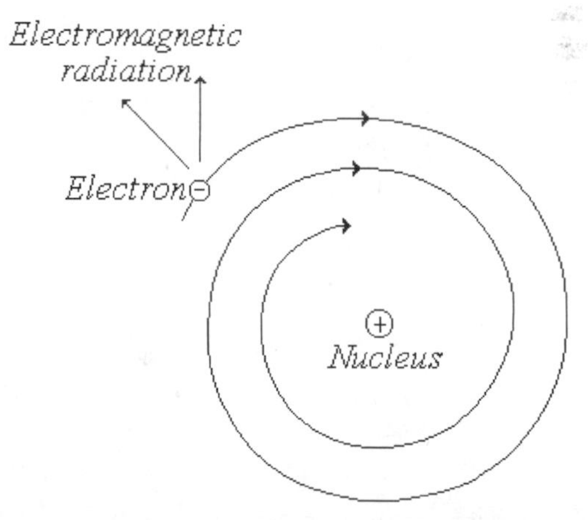

Fig. 3.7. The spiral motion of an electron around the nucleus predicted by classical considerations. The circulating electron radiates electromagnetic waves and loses energy. As a result, it 'falls' towards the positive nucleus. Starting from a distance of 1 nm, it would crash onto the nucleus after just 10^{-12} s.

A further problem arises from the electromagnetic tenet that all accelerated charges radiate energy in the form of electromagnetic waves. The orbiting electrons are accelerated and so should radiate electromagnetic waves, until, having lost all their energy, they collapse onto the nucleus (Fig. 3.7). Thus, even

[19] Hantaro Nagaoka 1865–1950, Japanese physicist.

a hydrogen atom, with its single electron, should be prohibited from having a nuclear configuration.

Rutherford was not unaware of these problems. He understood that the nuclear model he had proposed begged the question of the stability of the atomic electrons; the simplistic analogy with the solar planetary system was no solution. Furthermore, a comprehensive atomic theory could not ignore the incredibly immense density, $\sim 10^{17}$ kg/m^3, that the nuclear model predicts for the nucleus; it too must be explained.

The first solution to be proposed to the problem of the stability of the electrons in a nuclear atom was that put forward by Bohr in 1913. Bohr avoided the difficulties presented by classical electromagnetic theory by simply denying its sole or absolute validity in atomic systems. Instead, he adopted Planck's quantum hypothesis and postulated that an atom emits or absorbs energy in discrete quanta, i.e. that the emission and absorption of energy by an atom is not the continuous process enjoined by classical electromagnetics but is a discrete quantum process.

Bohr then went on to argue that when an atom is not actually emitting or absorbing energy, which is most of the time, it is in a non-radiating stable state which he designated a *stationary state.* Each of these stationary states is characterised by a specific amount of energy called its *energy level.*

Bohr's third premise was that the emission and absorption of radiation involves transitions between the stationary states. The quantum of radiation energy – the photon – that an atom emits or absorbs in a transition between two stationary states being the amount of energy specified by the difference in their energy levels.

In distinguishing between the non-radiating stationary states and the radiative transitions between them, Bohr assumed

> '*that the dynamical equilibrium of the systems [the atoms] in the stationary states can be discussed by help of the ordinary mechanics, while the passing of the systems between different stationary states cannot be treated on that basis . . . the latter process is the one given by Planck's theory.*'[20]

Bohr proposed that whereas transitions between stationary states are governed by quantum considerations, the stationary states could be described classically. Thus, when the atom is in a stationary state, specific trajectories – *stationary*

[20] (1913) 'On the Constitution of Atoms and Molecules', *Phil. Mag.*, xxvi, 1–25.

orbits – whose parameters are determined by classical considerations, can be assigned to the electrons.[21]

Although now superseded by the quantum mechanical model of the atom, Bohr's model marked a major breakthrough in our understanding of atomic structure. It was the first model to offer a comprehensive explanation of how atomic structure determines the nature of the radiation emitted and absorbed by an atom. We shall illustrate this by showing how the model dealt with the following three phenomena:

1. The spectra of the hydrogen atom, H, and of hydrogen-like atoms, such as He^+ and Li^{++}, each of which comprises a nucleus and a single electron. These spectra are particularly well defined in the optical region. This enabled precise measurements to be made which resulted in the discovery of a number of empirical relationships.
2. The Zeeman effect: the splitting of spectral lines in a magnetic field.
3. The characteristic radiation emitted from X-ray tubes: Moseley's experiment.

3.2.1 The Hydrogen Atom

Bohr began his study of hydrogen and hydrogen-like atoms by presupposing that the interaction between the atom's negative electron and its positive nucleus is Coulombic in nature. Thus, the force of attraction, F_E, that acts between an electron and a nucleus carrying a charge $+Ze$ is given by

$$F_E = -\frac{kZe^2}{r^2}, \tag{3.5}$$

where r is the distance between them and k is the electrostatic constant of proportionality.

Classically, when two particles are bound by a central force such as the electrostatic force of attraction that acts between two oppositely charged particles, the smaller particle orbits the larger one in an elliptical path.[22] Applying this principle to the structure of hydrogen or hydrogen-like atoms, Bohr postulated that in their non-radiating stationary states, the tiny electron revolves

[21] The notion of fixed trajectories would have to be abandoned later, after the discovery of the wave nature of the electron and quantum mechanics instead of classical mechanics was applied to the question of atomic structure.

[22] The Keplerian elliptical planetary orbits are the classic example of this rule.

around the massive nucleus in an elliptic or, in the simplest case, a circular orbit.

In the case of a circular orbit of radius R, the electron's potential energy, U, and kinetic energy, K, are given, respectively, by the equations

$$U = -\frac{kZe^2}{R} \tag{3.6}$$

and

$$K = \frac{m_e v^2}{2} = \frac{kZe^2}{2R}, \tag{3.7}$$

where m_e is the mass of the electron and v its orbital velocity.[23] Accordingly, the total energy, E, of the orbiting electron is

$$E = U + K = -\frac{kZe^2}{2R}. \tag{3.8}$$

The orbiting electron's total energy is negative because it is bound to the nucleus, i.e. energy must be invested to release it.

Having established these classical expressions for the electron's energy in a circular stationary orbit, Bohr now made the following *ad hoc* assumptions concerning the manner in which energy might be released during the formation of a hydrogen atom from a free proton and a free electron, i.e. as the electron spirals down from infinity and settles into a stationary orbit around the proton.

> '*Let us assume that, during the binding of the electron a homogeneous [monochromatic] radiation is emitted of frequency φ , equal to half the frequency of revolution of the electron in its final orbit; then, from Planck's theory, we might expect that the amount of energy emitted by the process is equal to nhφ where h is Planck's constant and n an entire [integer] number . . . the assumption concerning the frequency of the radiation suggests itself, since the frequency of revolution of the electron at the beginning of the emission is zero.*'[24]

Bohr supposed that during the atom's formation, n photons of a hypothetical radiation of frequency φ would be emitted, thereby releasing an amount of energy $\Delta E = nh\varphi$. He further suggested that the frequency, φ, of the hypothetical monochromatic radiation could be assumed to be equal to the mean of the initial and final frequencies of the electron's spiral. The electron's initial frequency of revolution at infinity is zero. Hence, if the frequency of the stationary orbit into which it finally settles is f, the frequency of the hypothetical radiation will be $\varphi = \frac{(0+f)}{2} = \frac{f}{2}$.

[23]See Appendix 3.5.1 for the proofs of these formulae.

[24]N. Bohr (1913) On the Constitution of Atoms and Molecules, *Phil. Mag.*, **26**, pp. 1–25. Quotes here and on pages 245–246 appear in M.H. Shamos (ed.) (1959) *Great Experiments in Physics*, New York: Dover Publications.

The frequency of revolution, f, of an electron circulating at a constant velocity, v, in an orbit of radius, R, is $f = v/2\pi R$, from which

$$\varphi = \frac{v}{4\pi R}. \tag{3.9}$$

Thus, the total energy released during the formation of the atom would be

$$\Delta E = nh\varphi = \frac{nhv}{4\pi R}. \tag{3.10}$$

Assuming that the electron started out with zero energy at infinity, its energy in the stationary state will be

$$E = 0 - \Delta E = -\frac{nhv}{4\pi R} \quad (n = 1, 2, 3, \ldots), \tag{3.11}$$

where n is the number of photons of the hypothetical radiation emitted during the atom's formation. Based on his premise that the stationary orbits could be described classically, Bohr equated this expression with that – Eq. (3.8) – given by classical considerations for the electron's total energy:

$$E = -\frac{kZe^2}{2R} = -\frac{nhv}{4\pi R}. \tag{3.12}$$

Finally, a comparison of the classical expression for the orbiting electron's kinetic energy, $K = m_e v^2/2 = kZe^2/2R$, with that for its total energy, $E = -kZe^2/2R$, shows that the electron's kinetic energy in a circular orbit is equal to the magnitude of its total energy, i.e. $K = |E|$. Hence, $m_e v^2/2 = nhv/4\pi R$, from which

$$m_e vR = n \cdot \frac{h}{2\pi} = n\hbar. \tag{3.13}$$

The quantity $m_e vR$, that appears in this equation, is the orbital angular momentum, L, of the electron in the stationary orbit, i.e.

$$L = n\hbar \quad (n = 1, 2, 3, \ldots). \tag{3.14}$$

This simple equation expresses the key concept introduced into physics by Bohr's atomic theory, namely, that an atomic electron's orbital angular momentum is *quantised*; it can only take certain values which are all integer multiples of $\hbar$. The integer n, called the *principal quantum number*, designates the electron's orbit. The stationary orbit corresponding to $n = 1$ is called the *ground state* (Fig. 3.8).

Having derived this simple expression from the quantum hypothesis, Bohr went on to suggest that it was a general rule that applied to all electrons moving in circular stationary orbits, irrespective of the nuclear charge. This concept of angular momentum quantisation subsequently assumed the status of an axiom, underivable from any deeper law, depending for its validity on the consistent agreement between its corollaries and experimental results. It is the usual starting point for derivations of Bohr's atomic model.

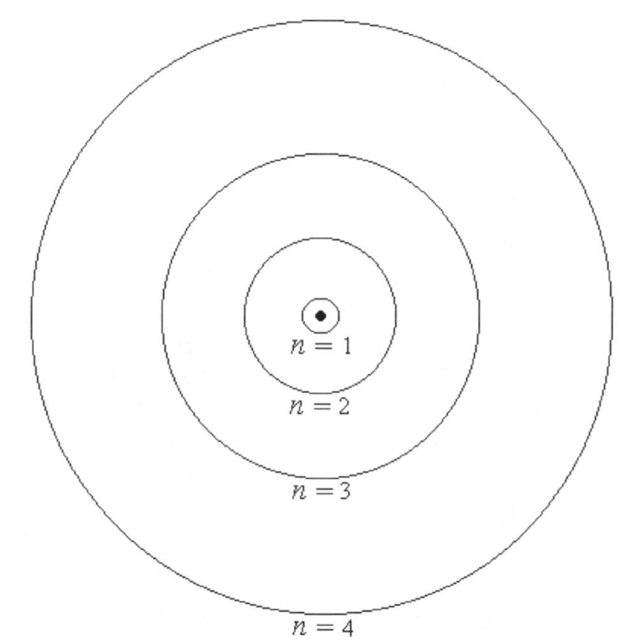

Fig. 3.8. The stationary electronic orbits predicted by Bohr's model of the hydrogen atom. The circles represent the orbits corresponding to the principal quantum numbers 1, 2, 3 and 4.

The radii, R_n, of the stationary circular orbits in hydrogen and hydrogen-like atoms, such as He^+ and Li^{++}, are obtained by substituting the expression for the electron's velocity, $v = n\hbar/m_e R$, obtained from Eq. (3.13b), in the classical expression for its kinetic energy, $K = m_e v^2/2 = kZe^2/2R$. This gives

$$R_n = \frac{n^2\hbar^2}{m_e\, k\, Z\, e^2} \quad (n = 1, 2, 3, \ldots). \tag{3.15}$$

The radius of the ground state ($n = 1$) in the hydrogen atom is usually designated a_0 and termed the *Bohr radius*: $a_0 = 5.29 \cdot 10^{-11}$ m. The radius of the ground state orbit in He^+ ($Z = 2$) is $a_0/2$ and in Li^{++} ($Z = 3$) it is $a_0/3$.

The principle energy levels, E_n, in hydrogen and hydrogen-like atoms are obtained by substituting the radii of the stationary orbits in the expression for the electron's total energy, $E = -kZe^2/2R$. This gives

$$E_n = -\frac{k^2 Z^2 m_e e^4}{2 n^2 \hbar^2} \quad (n = 1, 2, 3, \ldots). \tag{3.16}$$

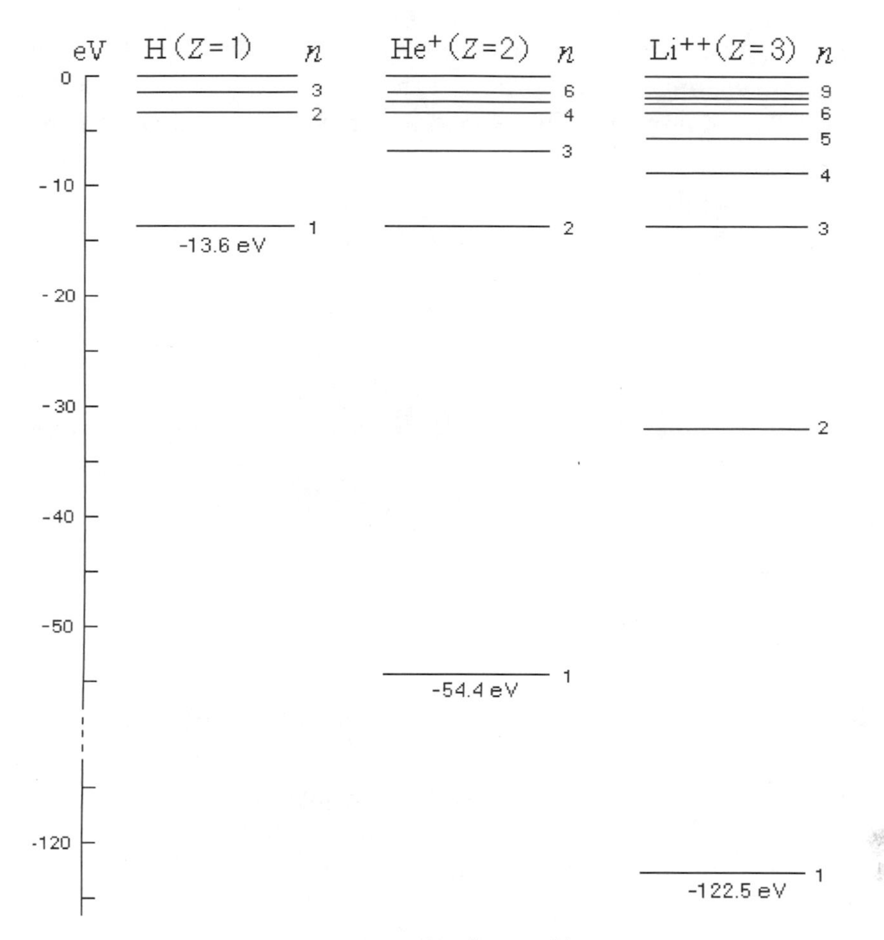

Fig. 3.9 The energy level diagrams of H, He^+ and Li^{++}. As the principal quantum number, n, increases, the spacing between the energy levels decreases and the atomic electron's total energy approaches zero.

The energy levels in H, He^+ and Li^{++} are shown diagramatically in Fig. 3.9.

In terms of the nuclear charge, Z, the energy of the ground state ($n = 1$) in each of the three atoms, H, He^+ and Li^{++}, is given by

$$E_1 = 13.5Z^2 \text{ eV}. \tag{3.17}$$

This defines the minimum amount of energy – the *ionisation energy* – that must be invested in order to remove the orbiting electron from a hydrogen or hydrogen-type atom in its ground state. Thus, Bohr's model predicts that the ionisation energy of hydrogen ($Z = 1$) should be 13.6 eV, that of He^+ ($Z = 2$) $13.6 \times 2^2 = 54.4$ eV, and that of Li^{++} ($Z = 3$) $13.6 \times 3^2 = 122.4$ eV. These figures are all confirmed experimentally.

Having derived a general formula for the energy levels in hydrogen and hydrogen-like atoms, Bohr went on to show how, when taken together with his postulate that transitions between stationary states are accompanied by the emission or absorption of a photon, it could account for the regularities, that had been recognised for some time, between the wavelengths and frequencies of the lines in the hydrogen emission spectrum and for which no satisfactory explanation had yet been found. Chief amongst these empirical protocols were the *Ritz combination rule* and the *Rydberg equation.*

As we showed in Chapter 2.1, the combination rule formulated by Ritz follows directly from energy conservation and Bohr's equation, $f_{ij} = \frac{E_i - E_j}{h}$, for the frequency of the photon emitted or absorbed in a transition between the energy levels E_i and E_j.

The empirical equation Rydberg[25] had formulated for the wavelengths of the lines in the emission spectrum of hydrogen was

$$\frac{1}{\lambda} = R_0 \left(\frac{1}{n_j^2} - \frac{1}{n_i^2} \right), \tag{3.18}$$

where R_0 is a constant known as *Rydberg's constant for hydrogen*, $n_j = 1, 2, 3, \ldots$ and $n_i = n_j + 1, n_j + 2, n_j + 3, \ldots$. The experimental value found for the constant R_0 was $1.09 \cdot 10^7\,\mathrm{m}^{-1}$. A special case of this general relationship had been discovered earlier by Balmer[26] in 1885 when he noted that the wavelengths of the lines in the hydrogen spectrum in the visible region could all be expressed by the simple formula

$$\frac{1}{\lambda} = B \left(\frac{1}{2^2} - \frac{1}{n_i^2} \right) \qquad n_i = 3, 4, 5, \ldots, \tag{3.18a}$$

where B is a constant.

Bohr showed that Rydberg's equation and the value of the constant R_0 could be directly derived from his theoretical model, as follows. From the expression he had derived for the energy level of a stationary state – Eq. (3.16) – the difference between the energy level, E_i, designated by the principal quantum number n_i and the energy level, E_j, designated by n_j is

$$E_i - E_j = -\frac{k^2 m_e e^4}{2\hbar^2} \left(\frac{1}{n_j^2} - \frac{1}{n_i^2} \right). \tag{3.19}$$

[25]Johannes Rydberg 1854–1919, Swedish physicist.

[26]Johann Balmer 1825–1898, Swiss mathematician and physicist.

A transition between these states would be accompanied by the emission or absorption of a photon of frequency f_{ij} as given by Bohr's equation

$$f_{ij} = \frac{E_i - E_j}{h} = \frac{k^2 Z^2 m_e e^4}{2\hbar^2 h}\left(\frac{1}{n_j^2} - \frac{1}{n_i^2}\right). \tag{3.20}$$

Making the substitutions $c = f\lambda$ and $Z = 1$ gives for the hydrogen atom

$$\frac{1}{\lambda_{ij}} = \frac{k^2 m_e e^4}{2\hbar^2 hc}\left(\frac{1}{n_j^2} - \frac{1}{n_i^2}\right). \tag{3.21}$$

The similarity between the form of this equation and that of the empirical expression formulated by Rydberg – Eq. (3.18) – is evident. Furthermore, the agreement is quantitative as well as qualitative. The calculated SI value of the quotient $k^2 m_e e^4 / 2\hbar^2 hc$ outside the brackets in this equation is $1.093 \cdot 10^7\ \mathrm{m}^{-1}$, almost exactly the same as the experimental value of Rydberg's constant for hydrogen.

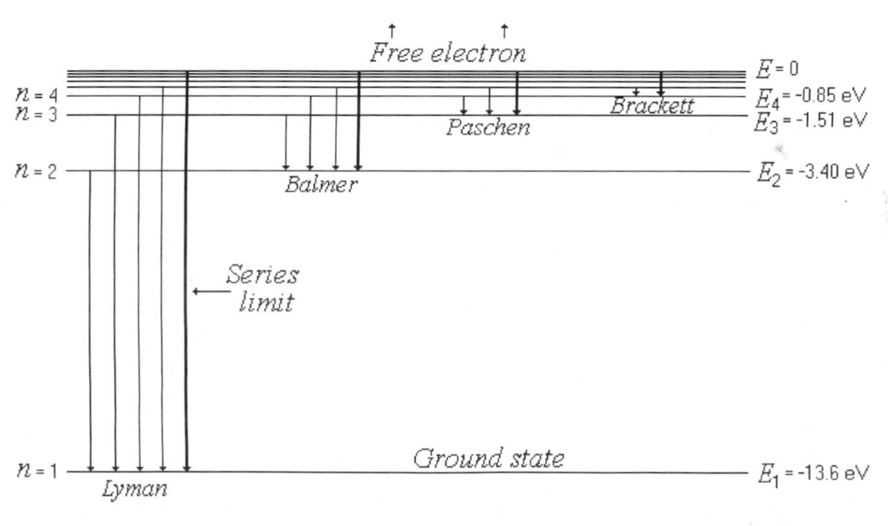

Fig. 3.10 The energy levels and transitions that are responsible for the emission spectrum of a hydrogen atom. The spectral lines in the Lyman series correspond to transitions from higher levels to the ground state: $n_j = 1$. Those in the Balmer series to transitions from higher levels down to the $n_j = 2$ level, those in the Paschen series down to the $n_j = 3$ level and those in the Brackett series down to the $n_j = 4$ level. When $E \geq 0$ the electron is free and no longer in orbit around the nucleus.

Summarising his argument, Bohr wrote:

'*We see that this expression accounts for the law connecting the lines in the spectrum of Hydrogen. If we put* $n_j = 2$ *and let* n_i *vary, we get the ordinary Balmer series. If we* *put $n_j = 3$, we get the series in the ultrared [infra-red] observed by Paschen ... If we put $n_j = 1$ and $n_j = 4, 5, \ldots$, we get series respectively in the extreme ultraviolet and the extreme utrared [infra-red], which are not observed but the existence of which may be expected.'*

The series for which $n_j = 1$ and $n_j = 4$ were discovered soon afterwards. The former is now known as the *Lyman series* and the latter as the *Brackett series.* The transitions associated with the four series are shown in Fig. 3.10.

In his original calculation, Bohr had assumed, as we have done so far, that the mass of the electron orbiting the nucleus was its rest-mass, m_e. This would be correct if the hydrogen nucleus was of infinite mass and were it fixed at the point in space around which the electron rotates in its stationary circular orbits. But the mass of the nucleus, M, is not infinite and so the nucleus and the electron actually revolve around their common centre of mass. It can be shown,[27] that such a system is equivalent to one in which a single particle of *reduced mass*, m', given by the formula

$$m' = \frac{m_e}{1 + \frac{m_e}{M}} \tag{3.22}$$

revolves around the larger mass. In the case of a normal hydrogen atom $m' = 0.99945 m_e$.

Though not large, differences due to this effect are detectable at high resolution. For example, the isotope of hydrogen, *deuterium*, whose mass is almost exactly twice that of a hydrogen atom, was first detected spectroscopically in 1932, as a result of the difference between its reduced mass, $m' = 0.99973 m_e$, and that of normal hydrogen, $m' = 0.99945 m_e$. Because of this difference, the spectral line corresponding to the transition from $n = 3$ to $n = 2$ that occurs at a wavelength of 656.3 nm in the spectrum of hydrogen, occurs at 656.1 nm in the deuterium spectrum. In a sample of naturally occurring hydrogen, the intensity of the deuterium spectral lines is about 1/6000 that of the corresponding normal hydrogen lines, a ratio which reflects their relative abundance.

Following Bohr's initial analysis of the circular stationary orbits, Sommerfeld[28] went on to investigate the elliptical orbits postulated by the model. Their description required the separation of the principal quantum number, n, into a *radial quantum number*, n', and an *azimuthal quantum number*, k, such that $n = n' + k$.[29] It then became apparent that it is actually the azimuthal quantum number, k, and not the principal quantum number, n, that fixes the electron's

[27] See Appendix 4.3.2,

[28] Arnold Johannes Wilhelm Sommerfeld 1868–1951, German physicist.

[29] The ratio of the axes of an elliptical orbit is equal to k/n.

total orbital angular momentum:

$$L = k\hbar \quad (k = 1, 2, 3 \ldots, n). \tag{3.23}$$

In the case of a circular orbit, $k = n$, and so the need for more than one quantum number did not arise in Bohr's original derivation.

At high resolution, the major spectral lines, such as those in the Balmer series, are seen to comprise a number of finer lines. Sommerfeld explained that this *fine structure* within the major spectral lines was the result of relativistic effects, such as the periodic variations in the electron's mass as its velocity varied in its elliptical orbit around the nucleus. He suggested that these effects would cause a precessional motion of the axis of the elliptical orbits similar to that of the planet mercury. This precessional motion has a small energy associated with it, whose effect is to split the principal energy levels, E_n, set by the quantum number n into as many sub-levels. The fine structure is associated with transitions from the sub-levels, $k = 1, 2, \ldots, n_2$, in an upper state, to those, $k = 1, 2, \ldots, n_1$, in a lower state (Fig. 3.11).[30]

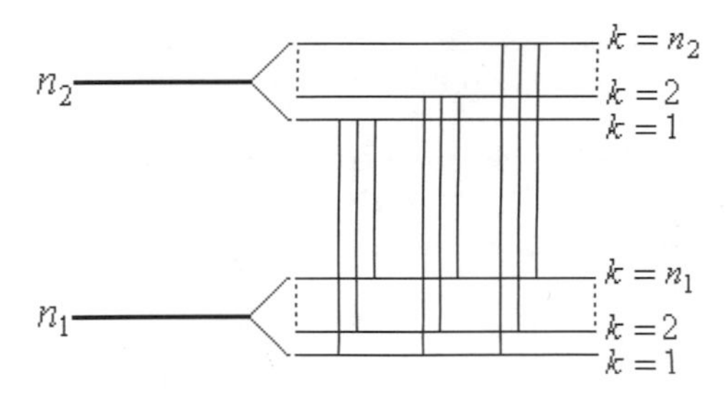

Fig. 3.11 An idealised diagram of the possible transitions from sub-levels in an upper state, n_2, to sub-levels in a lower state, n_1. The existence of the sub-levels is indicated by the fine structure of narrow lines, that is observed within the broader principal spectral line corresponding to the transition from n_2 to n_1. The finer spectral lines correspond to transitions from sub-levels in the upper state to sub-levels in the lower state.

Since in general $k = 1, 2, 3, \ldots, n$, the higher the principal quantum number, n, the more k values there are and the greater the number of different transitions that can be imagined. It might, therefore, be expected that a correspondingly large number of fine structure lines would be observed. However, this is not so. The only transitions that actually show up as spectral lines are those for which $\Delta k = \pm 1$. Formulae such as this, which specify in terms of the quantum numbers which transitions are allowed, are known as *selection rules.*

[30]The function that yields the amount of energy level splitting includes the dimensionless fine structure constant, $\alpha = e^2/2\varepsilon_0 hc \approx 1/137$.

They are of paramount importance in the interpretation of spectra and indicate that factors other than just energy conservation are involved.

Example 3.2: Spectral Transitions

Calculate the frequency and the wavelength of the spectral lines associated with transitions in a hydrogen atom:

1. from the first excited level to the ground state;
2. from the energy level $n = 6$ to the energy level $n = 5$.

Calculation: In case (a), $n_j = 2$ and $n_j = 1$; in case (b), $n_j = 6$ and $n_j = 5$. Substituting these quantum numbers in Eq. (3.20) gives respectively:

$$f_{21} = 2.47 \cdot 10^{15}\,\text{Hz} \quad f_{65} = 4.02 \cdot 10^{13}\,\text{Hz}.$$

The corresponding wavelengths *in vacuo* are:

$$\lambda_{21} = \frac{c}{f_{21}} = 121.5\,\text{nm} \quad \lambda_{65} = \frac{c}{f_{65}} = 7460\,\text{nm}.$$

Example 3.3: The Correspondence Principle

A hypothetical macroscopic hydrogen-type atom of radius 1 cm would be expected to obey classical electrodynamics. Do the properties predicted by Bohr's model for such an atom correspond to those given by classical electrodynamics, as the correspondence principle requires they should?

According to classical electrodynamics, an electron moving in a circular orbit will radiate electromagnetic waves of a frequency equal to that of its frequency of revolution. The frequency of revolution of an electron moving with a velocity, v, in a circular orbit of radius R, is $f_{revolution} = v/2\pi R$. From Eq. (3.6b), the velocity, v, of the electron in the hypothetical hydrogen atom ($Z = 1$) is

$$v = \sqrt{\frac{ke^2}{m_e R}} = \sqrt{\frac{9 \cdot 10^9 \cdot (1.6 \cdot 10^{-19})^2}{9.1 \cdot 10^{-31} \cdot 10^{-2}}} = 1.6 \cdot 10^2\,\text{m/s}, \tag{3.24}$$

which gives for its frequency of revolution

$$f_{revolution} = \frac{v}{2\pi R} = \frac{1.6 \cdot 10^2}{2\pi \cdot 10^{-2}} = 2545\,\text{Hz}. \tag{3.25}$$

According to Eq. (3.15), which was derived from Bohr's model, the principal quantum number of a hydrogen atom stationary state of radius 1 cm is

$$n = \sqrt{\frac{R_n m_e kZe^2}{\hbar^2}} = \sqrt{\frac{0.01 \cdot 9.1 \cdot 10^{-31} \cdot 9 \cdot 10^9 \cdot 1 \cdot (1.6 \cdot 10^{-19})^2}{(1.054 \cdot 10^{-34})^2}} \tag{3.26}$$
$$= 13{,}738.$$

Should the hydrogen atom fall to the next lower stationary state, it will emit a photon of frequency

$$f_{ij} = \frac{k^2 Z^2 m_e e^4}{2\hbar^2 h}\left(\frac{1}{n_j^2} - \frac{1}{n_i^2}\right) \tag{3.27}$$

$$= 3.28 \cdot 10^{15}\left(\frac{1}{137372^2} - \frac{1}{137382^2}\right)$$

$$= 2530\,\text{Hz}.$$

The close correspondence between the classical value for the radiation's frequency and that obtained for the photon's frequency from Bohr's theory is evident.

Example 3.4: The Bohr Atom and De Broglie's Principle

In circular stationary orbits, the electron's motion is essentially one-dimensional; the only variable is its position in the orbit. Thus, the electron's circumstance is analogous to that of a particle trapped in a one-dimensional potential well, whose length is equal to that of the circumference of the electron's orbit, *viz.* $2\pi R_n$. Accordingly, its matter wave can be viewed as a one-dimensional standing wave whose permitted wavelengths are given by

$$n\lambda = 2\pi R_n \quad (n = 1, 2, 3\ldots). \tag{3.28}$$

The justification for this is that the orbiting electron's matter wave would interfere destructively with itself under any other condition (Fig. 3.12).

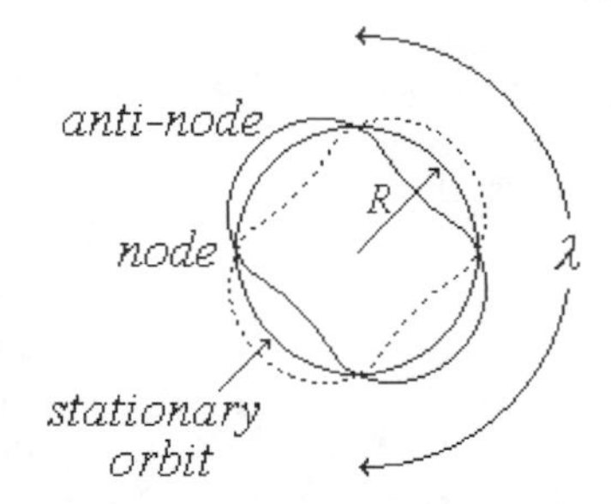

Fig. 3.12. Standing waves fitting a circular Bohr orbit of radius R in the case where $n = 2$. The circumference of the orbit exactly equals two wavelengths of the electron's matter wave.

Substituting the De Broglie equation, $\lambda = h/m_e v$, *in Eq.* (3.27) *gives*

$$m_e v R_n = n \cdot \frac{h}{2\pi}, \tag{3.29}$$

which is identical to Eq. (3.13) that Bohr derived from Planck's quantum hypothesis. The importance of this consistency cannot be over-emphasised.

3.2.2 The Zeeman Effect – Space Quantisation

Basing his argument on Maxwell's theory that electromagnetic waves are generated by accelerated charges, Lorentz had predicted that atoms must contain electrical charges. This being so, emission spectra should be affected by a magnetic field; the presence of the field should alter the motion of the charges. In 1896, a year before the discovery of the electron, Zeeman[31] confirmed this prediction when he discovered the splitting of the spectral lines of the light emitted from a source situated in a magnetic field; two or three adjacent lines are observed where, in the absence of the magnetic field, just one is seen. This phenomenon, called the *Zeeman effect*, was first explained classically by Lorentz. He showed that a magnetic field may increase the frequency of one component of an electric oscillator's vibrations by exactly the same amount as it decreases the opposite component. Consequently, in the presence of a magnetic field, an atomic oscillator may emit more than one frequency and so more than one spectral line may appear.

The Bohr model offers a different explanation. It attributes the Zeeman effect to a precessional top-like motion of the normal to the electron's orbital plane relative to the direction of the external magnetic field (Fig. 3.13). The angular momentum associated with this motion is quantised and the additional spectral lines arise from transitions between the energy sub-levels that accompany this.

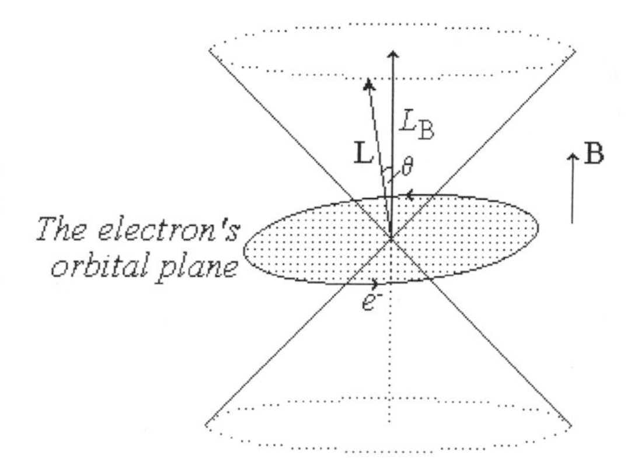

Fig. 3.13. The top-like precession of the normal to the plane of the electron's orbit and of its angular momentum vector, **L**, around an axis defined by the direction of the external magnetic field **B**. The normal describes the surface of a cone. L_B is the component of the electron's angular momentum parallel to the magnetic field such that $L_B = L\cos\theta$.

An orbiting atomic electron possesses a magnetic moment, μ, of magnitude

$$\mu = -\frac{eL}{2m_e} \tag{3.30}$$

[31] Pieter Zeeman 1865–1943, Dutch physicist who was awarded the Nobel prize in 1902 for his researches on the effect of magnetism on radiation.

which it confers on the atom.[32] Considered as a vector, this magnetic (dipole) moment, $\boldsymbol{\mu}$, is parallel to the angular momentum vector, **L**, i.e. it too is normal to the electron's orbital plane. In a magnetic field, **B**, a torque, $\boldsymbol{\tau} = \boldsymbol{\mu} \times \mathbf{B}$, acts on the magnetic moment which causes it to precess around the direction of the field, like the motion of a spinning top in a gravitational field. The frequency, f_L, of this precession, called the *Larmor*[33] *frequency*, is given by

$$f_L = \frac{\mu B}{2\pi L}. \tag{3.31}$$

Classically, the normal to the plane of the electron's orbit can precess at any angle, θ, relative to the direction of the magnetic field. However, on the Bohr model, the sub-levels responsible for the Zeeman splitting arise from the introduction of a further quantisation condition that has the effect of restricting the orbital plane's precession to certain specific angles relative to the direction of the magnetic field. This additional, apparently arbitrary, quantisation condition requires that the component, L_B, of the electron's angular momentum parallel to the direction of the external magnetic field be an integer multiple of $\hbar$:

$$L_B = m\hbar \quad (m = -k, -k+1, -k+2\ldots, +k). \tag{3.32}$$

It follows, from Fig. 3.13, that the normal to the electron's orbital plane and the total orbital angular momentum vector, **L**, can only take up specific inclinations, θ_m, relative to the direction of the external magnetic field, i.e. those that satisfy the condition

$$\cos\theta_m = \frac{L_B}{L} = \frac{m\hbar}{L}. \tag{3.33}$$

The phenomenon is called the *quantisation!of direction* or *space quantisation* and its existence was first explicitly demonstrated by Stern[34] and Gerlach[35] in an experiment carried out in 1921.[36]

[32] See Section 6.3.1.

[33] Sir Joseph Larmor 1857–1942, Irish physicist and mathematician.

[34] Otto Stern 1888–1969, German–American physicist who was awarded the Nobel prize in 1943 for the discovery of the magnetic moment of the proton.

[35] Walther Gerlach 1889–1979, German physicist.

[36] What they actually observed was the space quantisation associated with electron spin and not that which is a consequence of the quantisation of orbital angular momentum. Accordingly, the experiment is described in Chapter 3.4 which deals with electron spin.

The energy, E_μ, of a magnetic dipole moment, $\boldsymbol{\mu}$, situated in a magnetic field, $\mathbf{B}$, is given by

$$E_\mu = -\boldsymbol{\mu} \cdot \mathbf{B} = \mu B \cos\theta \tag{3.34}$$

where θ is the angle between the direction of the magnetic dipole and that of the field. Substituting the space quantisation provision, $\cos\theta_m = \frac{m\hbar}{L}$, in this equation gives the following simple expression for the energy sub-levels responsible for the Zeeman effect:

$$E_\mu = \frac{\mu B m\hbar}{L} \quad (m = -k, -k+1, -k+2\ldots, +k). \tag{3.35}$$

In terms of the dipole's Larmor frequency, $f_L = \frac{\mu B}{2\pi L}$, this reduces to

$$E_\mu = mhf_L \quad (m = -k, -k+1, -k+2\ldots, +k). \tag{3.36}$$

According to this model, the Zeeman splitting of the spectral lines occurs when transitions between the k states are accompanied by transitions between the sub-levels defined by the additional quantum number m. In each such transition a photon of energy hf_L will be absorbed or emitted. Although every k state is split into $(2k + 1)$ sub-levels, it transpires that only transitions that obey the selection rules, $\Delta m = \pm 1$ or $\Delta m = 0$, show up in observed spectra (Fig. 3.14).

An important corollary of the postulate of angular momentum quantisation is that there exists a natural unit of atomic magnetic moment. The absolute

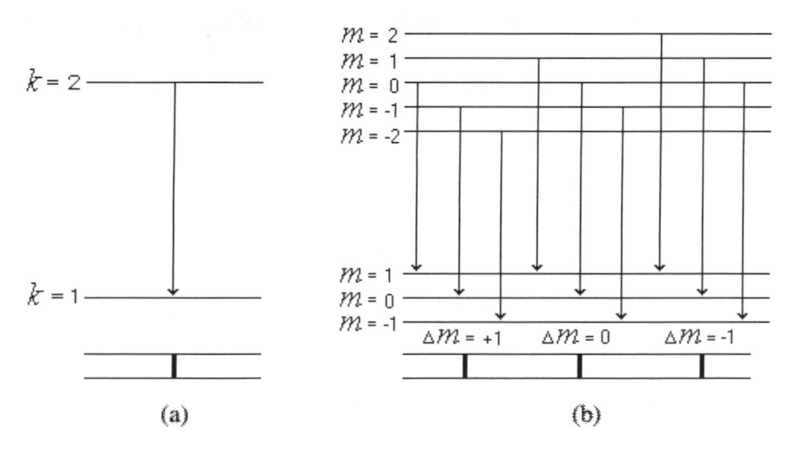

Fig. 3.14 The Zeeman effect. (a) The single spectral line observed in the absence of a magnetic field that corresponds to the transition from the $k = 2$ to the $k = 1$ state. (b) The three lines that appear in the presence of a magnetic field as a result of the splitting of the levels into sub-levels. Each line corresponds to a different value of Δm.

value of this unit, which is designated μ_B, is obtained by substituting $L = \hbar$ in Eq. (3.30):

$$\mu_B = \frac{e\hbar}{2m_e} = 9.27 \cdot 10^{-24}\mathrm{J} \cdot \mathrm{T}^{-1}. \tag{3.37}$$

This quantity is called the *Bohr magneton.*

3.2.3 Moseley's Experiment

The high intensity penetrating radiation emitted by X-ray tubes, characteristic of the metal from which the target anode is made, was first discovered by Barkla.[37] He found that when the tubes were operated at higher potentials, series of high intensity peaks, each of a specific wavelength, were superimposed on the spectrum of the continuous brehmstrahlung radiation (Fig. 3.15).

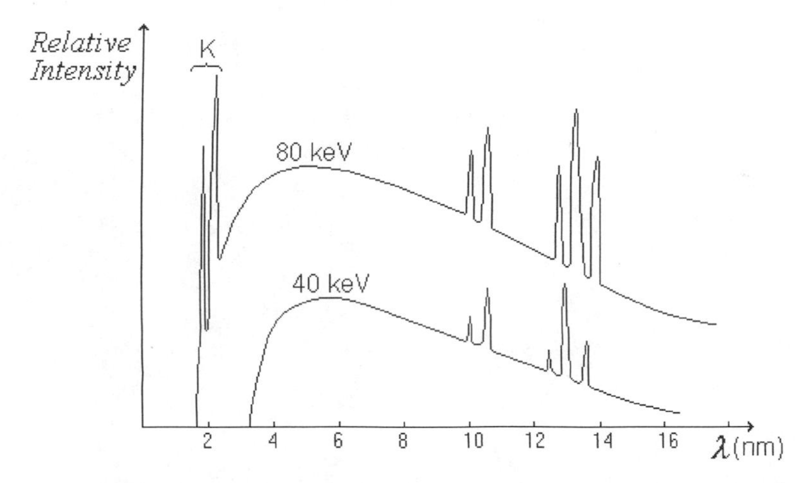

Fig. 3.15 The spectrum of the radiation emitted from a typical X-ray tube operated at potentials of 40keV and 80keV. The more penetrating K series only appears in the spectrum at the higher potential because the 40keV electrons have insufficient energy to dislodge electrons from the innermost K shell.

The phenomenon is analogous to the atomic line spectra seen in the visual region of the electromagnetic spectrum. Changing the metal or element from which the target anode in the X-ray tube is made alters the wavelengths at which the high intensity peaks occur. The most penetrating series in an element's

[37] Charles Barkla 1877–1944, English physicist who was awarded the Nobel prize in 1917 for the discovery of the characteristic X-ray radiation.

characteristic X-ray spectrum is called the *K series*; the second is the *L series*; the third the *M series* and so on.

At the time of their discovery, the available experimental techniques did not enable an accurate determination of the wavelengths or frequencies of the characteristic X-rays. However, after the Braggs had shown how such measurements could be made with the help of crystals (Section 2.2.5), Moseley[38] decided to use their technique in order to carry out a systematic examination of the characteristic radiation of as many elements as possible. He examined the X-ray spectra of the 38 elements from aluminium (Al) to gold (Au). As regards 15 of these elements, he studied just the K series; regarding another 17, just the L series; as to the remaining six elements, both series. He recorded the spectra on photographic plates.

Moseley discovered the following simple empirical relationship, illustrated in Fig. 3.16, between the frequencies, f, of the lines in each series and the ordinal number, N, of the element's position in the periodic table (starting

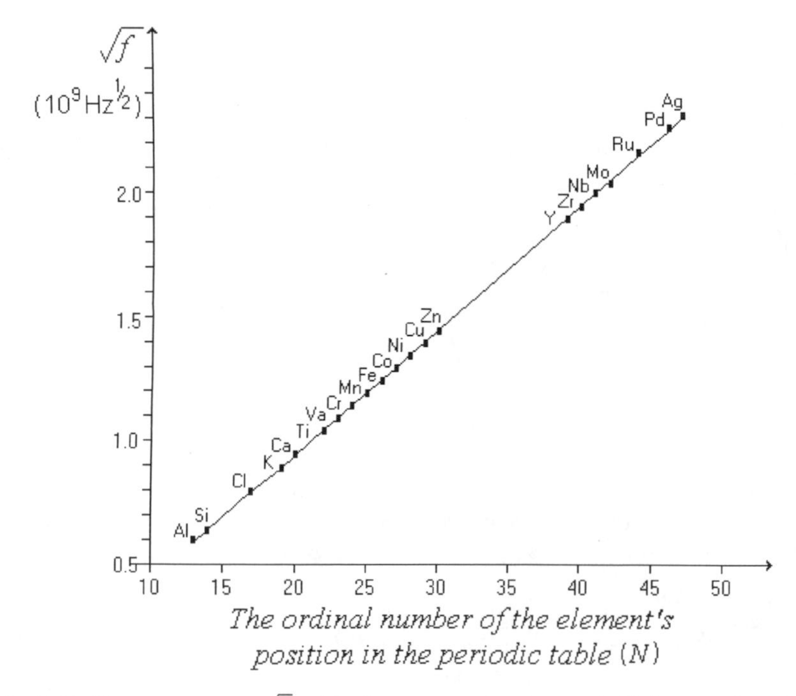

Fig. 3.16 The square root, $\sqrt{f}$, of the frequency of an element's K_α line as a function of the ordinal number, N, of its position in the periodic table.

[38]Henry Moseley 1887–1915, English physicist.

from hydrogen):

$$\sqrt{f} = a(N - \sigma). \tag{3.38}$$

σ is called the *screening constant.*

Moseley formed the opinion that some physical attribute of the atom must increase by a regular fixed amount, from one element to the next, rising through the periodic table. He postulated that this could only be the atom's nuclear charge. According to this hypothesis, the number N, that is the element's ordinal position in the periodic table, is equal to the number of natural units of positive electricity carried by the nuclei of the element, i.e. $N = Z$. The number Z is now called the *atomic number* of the element; it is equal to the number of protons in the element's nuclei.

Prior to Moseley's investigation, the elements were arrayed in the periodic table in the ascending order of their atomic weights and on the basis of their chemical properties. As a result of Moseley's research, which provided the first direct means of determining an element's atomic number, inaccuracies in the periodic table were discovered and corrected. For example, the positions of the transition metals cobalt ($Z = 27$) and nickel ($Z = 28$), that had been previously determined by the ascending order of their atomic weights, Ni $= 58.71$ and Co $= 58.93$, were switched. Similarly, empty positions were revealed in the table, corresponding to the yet undiscovered elements of atomic number 43, 61, 72 and 75.

The origin of the characteristic X-ray radiation is readily explained by the Bohr model of the atom. Let us assume that the electrons orbiting the nucleus in multi-electron atoms are arranged in shells, each electron having its specific slot in a shell. We will designate the innermost shell as the *K shell*; it corresponds to the principal quantum number $n = 1$. The second shell is designated the *L shell*; it corresponds to the principal quantum number $n = 2$. The third shell is designated the *M shell*; it corresponds to the principal quantum number $n = 3$, and so on. According to this model, all the electrons in a particular shell have the same energy and the closer a shell is to the nucleus the greater the energy binding its electrons to the atom. The electrons in the K shell are all in the energy level E_1, those in the L shell in the level E_2, and so on (Fig. 3.17).

When an electron, with sufficiently high energy, strikes an atom in the X-ray tube anode, it ejects an electron from one of the atom's inner shells, say the K shell. This leaves an empty slot or *hole* in this shell. One of the electrons from an outer shell, corresponding to an energy level E_n, can 'fall' into this hole, releasing an amount of energy, $E_n - E_1$, equal to the difference between the energy levels of the two shells; this energy is released as an X-ray photon of frequency

$$f = \frac{E_n - E_1}{h}. \tag{3.39}$$

On the basis of this explanation, the K series results from the filling of holes in the K shell ($n = 1$), with the K_α line corresponding to the hole being filled by an electron falling from the L shell; the K_β line to the electron falling from the M shell, and so on. Similarly, the L series results from the filling of holes in the L shell ($n = 2$).

The more electrons there are in an atom, the more series there will be in its X-ray spectrum and the more lines there will be in each series. An atom's electronic energy levels depend on the interaction between its electrons and its nuclear charge, $+Ze$, and so, since the atoms of each element carry a characteristic nuclear charge, each element exhibits a characteristic X-ray spectrum.

To a close approximation, the frequencies, f_{K_α}, of the K_α line – the most intense line in an element's K series – are given by the formula

$$f_{K_\alpha} = cR_0(Z-1)^2\left(\frac{1}{1^2} - \frac{1}{2^2}\right), \tag{3.40}$$

where R_0 is Rydberg's constant and c the speed of light. It is as though after the removal of an electron from the K shell, the electron in the L shell sees a nuclear charge of $(Z-1)$. The frequencies, f_{L_α}, of the L_α line – the most intense line in the element's L series – are given by the formula

$$f_{L_\alpha} = cR_0(Z-7.4)^2\left(\frac{1}{2^2} - \frac{1}{3^2}\right). \tag{3.41}$$

The similarity between these formulae and Eq. (3.20), that gives the lines in the hydrogen spectrum, is apparent.

On account of the relatively large nuclear charge of the anode metals ($Z > 30$), the motion of their innermost electrons is virtually unaffected by the outermost electrons. The innermost electrons interact almost exclusively with the nucleus, just like the electron in the hydrogen atom. Consequently, as in Bohr's elucidation of the hydrogen spectrum, the denominators of the fractions in Eqs (3.40) and (3.41) denote the principal quantum numbers of the energy levels between which the electronic transitions have taken place. The screening constant, σ, is greater in Eq. (3.41) because the electrons in the L shell are shielded from the nuclear charge by the innermost electrons in the K shell.

Accurate measurements show that each principal line in the characteristic X-ray spectrum is, in fact, composed of a fine structure of very close discrete

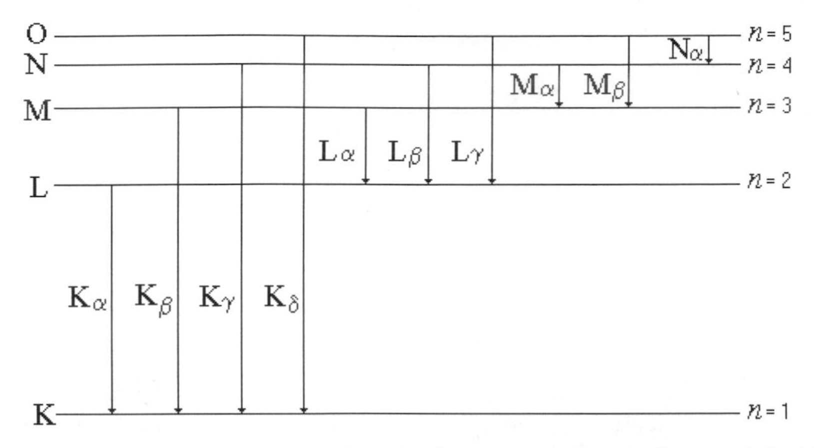

Fig. 3.17 The transitions that produce the characteristic X-ray radiations. A 'hole' in an inner shell of the atom's electronic structure is filled by an electron that 'falls' from a higher energy level; the difference in energy appears as a high energy photon.

lines. For example, the K_α line comprises two lines; it is a *doublet*. It follows that when an electron falls from the L shell to the K shell, there are two possible values for the difference, $E_2 - E_1$, between the energy levels of the shells. This indicates that there are sub-levels within the principle energy levels of the atom. A full explanation of the existence of these sub-levels requires the concept of electron spin (Chapter 3.4).

Example 3.5: The Characteristic X-ray Spectrum of Copper

Estimate:

1. the frequency of the K_β line in the spectrum of the X-rays emitted from an X-ray tube with a copper anode;
2. the minimum potential at which the tube must be operated for this line to appear.

Calculation: 1. The K_β line corresponds to a transition from the M shell to the K shell. It is apparent from Fig. 3.17 that

$$f_{K_\beta} = f_{K_\alpha} + f_{L_\alpha}.$$

The atomic number of copper is 29 and so from Eq. (3.40)

$$\begin{aligned} f_{K_\alpha} &= 3.291 \cdot 10^{15}(29 - 1)^2 \left(\frac{1}{1^2} - \frac{1}{2^2}\right) \\ &= 1.935 \cdot 10^{18}\,\text{Hz}, \end{aligned}$$

and from Eq. (3.40)

$$\begin{aligned} f_{\mathrm{L}\alpha} &= 3.291 \cdot 10^{15}(29 - 7.4)^2 \left(\frac{1}{2^2} - \frac{1}{3^2}\right) \\ &= 0.213 \cdot 10^{18}\,\mathrm{Hz}, \end{aligned}$$

which gives for the frequency of the K_β line:

$$\begin{aligned} f_{\mathrm{K}_\beta} &= (1.935 + 0.213) \cdot 10^{18} \\ &= 2.148 \cdot 10^{18}\,\mathrm{Hz}. \end{aligned}$$

2. For any of the K lines to appear in the X-ray spectrum, an electron must be dislodged from the K shell. The binding energy of these electrons can be estimated by assuming that the conditions under which they orbit the nucleus are comparable to those of the electron in a hydrogen-type atom, i.e. that they interact almost exclusively with the nucleus. Substituting $Z = 29$ in Eq. (3.17) gives for the energy level, E_1, of the innermost electrons

$$\begin{aligned} E_1 &= -13.6Z^2 \\ &= -13.6 \cdot 29^2 \\ &= -11{,}440\,\mathrm{eV}. \end{aligned}$$

Thus, in order to dislodge a K shell electron in copper, we might estimate that they must be struck by an electron that has been accelerated through a potential of 11,440 V at least.

To summarise Bohr's model:

1. Atoms comprise a very small and very dense nucleus that contains their entire positive charge, $+Ze$, and virtually all of their mass.
2. In a neutral atom, Z negative electrons revolve around the nucleus in stationary non-radiating orbits: these electrons, together with the nucleus, comprise the neutral atom.
3. The interaction between the atomic nucleus and the electrons is Coulombic.
4. An atom absorbs and emits energy in accordance with the quantum hypothesis.
5. The orbits of the electrons in multi-electron atoms are arranged around the nucleus in shells.

Though revolutionary and seminal, Bohr's theory was incomplete. Like Compton's explanation of the scattering of X-rays by free electrons, it was an amalgam, almost an alchemy, of the quantum hypothesis with classical and relativistic mechanics, that worked in a few simple instances but which had no clear and consistent conceptual basis of its own.

Quantitatively, the model was successful only in the case of the hydrogen atom, the simplest of atoms; its application to more complex systems was never better than semi-quantitative. Furthermore, the model was unable to account for the fact that some of the transitions it forecast didn't show up as spectral lines; nor could it account for the different intensities of the spectral lines that did appear. And perhaps most serious of all, experiments showed that in its ground state, $n = 1$, the angular momentum of the electron in the hydrogen atom is zero, and not $\hbar$ as predicted by the axiom of angular momentum quantisation.

Moreover, their stability was not the only problem presented by the nuclear atom's electrons. Parallel to the physicists' efforts to find an atomic model that would satisfy the requirements of spectroscopy and of the resonance of energy emission and absorption, the chemists were looking for a satisfactory explanation to valency and to the stability of the bonds between the component atoms of molecules and ionic crystals. Both the physicists and the chemists were dealing with the same atoms, however, the notion of electrons revolving at frequencies of billions of cycles per seconds appeared to preclude their participation 'in the process of binding atom to atom in the molecule'.

> '*The view based on the periodic law and the chemical behaviour of the elements leads to the picture of a relatively static atom . . . the electron as a whole is regarded as occupying a fixed position in the atom.*'[39]

The chemical stability of the non-radioactive elements, i.e. the fact that they do not undergo spontaneous chemical degradation, requires that the atom's electrons, together with its nucleus, constitute an unchanging self-sufficient unit. Furthermore, molecular architecture – the fixed shape and configuration of molecules – requires that an atom's valence electrons should be 'pointing' in specific directions in space and that they can be shared with another atom. The motion and energy of these bonding electrons will be determined not just by their parent atoms but also by:

1. their interactions with the positive nuclei of the atoms they bond;
2. their interactions with one another.

The significance of the tendency of electrons and of other sub-atomic particles to pair was first noted by the chemist Lewis[40] in 1916.

> ' . . . *whenever a radioactive atom emits one beta-particle [high-energy electron] it almost immediately emits another, again illustrating the instability of an unpaired*

[39] G.N. Lewis (1923) *Valence and the Structure of Atoms and Molecules*, New York: Dover Publications, p. 55.
[40] Gilbert Newton Lewis 1875–1946, American chemist.

electron within the nucleus. So also we find that in all the more stable states which atoms assume, the electrons occur in even numbers in the several inner shells ... the valence electrons almost invariably follow the same rule. The simplest explanation of these facts appears to lie in the assumption of a physical pairing of the electrons. There is nothing in the known laws of electric force, nor is there anything in the quantum theory of atomic structure, so far as it has yet been developed, to account for such a pairing ... there can be no question that the coupling of electrons is intimately connected with the magnetic properties of the electron orbits, and the explanation of this phenomenon must be regarded as one of the most important outstanding problems in quantum theory.'[41]

As we shall see, the explanation for this phenomenon, which is central to a fuller understanding of atomic and nuclear structure and of the nature of the chemical bond, was only found some 15 years later, when quantum mechanics was applied to atomic theory.

[41] G.N. Lewis (1923) *Valence and the Structure of Atoms and Molecules*, New York: Dover Publications, p. 55.

Chapter 3.3

The Quantum Mechanical Model of the Atom

Heisenberg had argued that in a consistent system of atomic mechanics, no entities should be introduced that are not physically observable. In his judgement, the Bohr model broke down because, in addition to dealing with the observable frequencies of the radiations emitted and absorbed by an atom, it also spoke of electron trajectories. However, as we have seen, minute particles such as electrons are inherently elusive. Fixed paths, orbits or trajectories cannot be assigned to them; their motion can only be construed in terms of probabilities.

A consistent atomic mechanics must, therefore, describe the motion of atomic electrons in terms of measurable probabilities and not in terms of trajectories. In principle, these probabilities can be computed from the wave functions (probability amplitudes) of the atomic electrons which, in turn, can be obtained from the solutions of the appropriate Schrödinger equations.

The electrons are held in an atom by the electric field generated by the positive nuclear charge; they lack the necessary energy to escape this attraction. Their situation is analogous to that of a particle confined to a small region of space – a three-dimensional potential well. As we have seen, a minute particle that is confined to a potential well can only occupy certain quantum states whose energies are sharply observable. These stationary quantum states are characterised by integers called *quantum numbers.*

A three-dimensional potential well allows for three integer quantum numbers. In the case of an atomic electron, the first of these integers – designated n – fixes the particle's energy eigenvalues. The other two quantum numbers, l and m_l, are associated with the electron's orbital angular momentum: l determines the magnitude of the orbital angular momentum, m_l its component parallel to an arbitrarily chosen axis. Thus, each atomic stationary state is characterised by a set of three parameters, the value of each being that enjoined by its particular quantum number. This is the essence of the quantum mechanical model of the atom.

Determining the quantum numbers that characterise the allowed states of the atomic electrons is the main task faced in developing the quantum mechanical model of the atom. We will illustrate how this is done by first considering its application to the hydrogen atom.

3.3.1 The Hydrogen Atom

In developing a quantum mechanical model of hydrogen and hydrogen-like atoms, we begin, as we did with the Bohr model, by assuming that the interaction between the atom's negative electron and its positive nucleus (proton) is Coulombic in nature, i.e. that the attractive force, F_E, acting between the electron and the nucleus is given by the equation

$$F_E = -\frac{kZe^2}{r^2}, \tag{3.42}$$

where r is the distance between them and k is the electrostatic constant of proportionality. The electron's potential energy, U, is given by

$$U = -\frac{kZe^2}{r}. \tag{3.43}$$

Substituting this value in the general time-independent Schrödinger equation – Eq. (2.104) – for a particle of mass m_e gives:

$$\frac{-h^2}{8\pi^2 m_e}\left(\frac{\partial^2\psi}{\partial x^2}+\frac{\partial^2\psi}{\partial y^2}+\frac{\partial^2\psi}{\partial z^2}\right) = \left(E+\frac{kZe^2}{r}\right)\psi. \tag{3.44}$$

In view of the spherical symmetry of the potential field, it is more convenient to denote the position of the electron in the hydrogen atom in terms of the polar coordinates, r, θ and ϕ, rather than the Cartesian coordinates, x, y and z (Fig. 3.18). The mathematical relationship between the two sets of coordinates is

$$\begin{aligned} x &= r\sin\theta\cos\phi \\ y &= r\sin\theta\sin\phi. \\ z &= r\cos\theta \end{aligned} \tag{3.45}$$

Placing the origin of the coordinate system at the hydrogen atom's nucleus makes the coordinate r equal to the distance between the atom's positive nucleus and its electron.

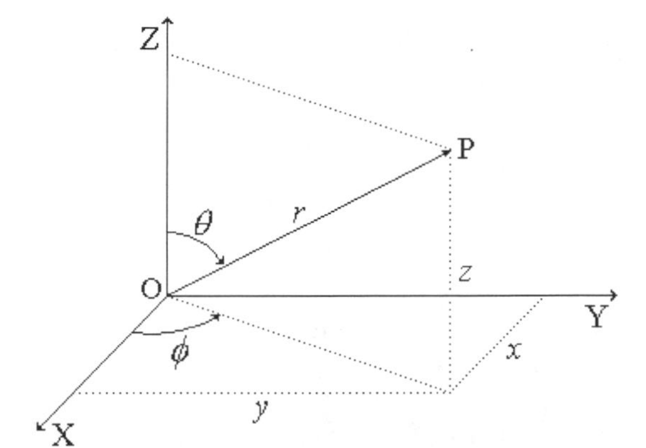

Fig. 3.18. The Cartesian coordinates of the point P are (x, y, z). The polar coordinates of the point are (r, θ, ϕ). θ is a 'latitude'; ϕ a 'longitude'.

The use of polar coordinates does not affect the total number of quantum numbers needed to describe the electron's allowed spatial states. Irrespective of the coordinates used to describe its motion, three quantum numbers will be required because the electron moves in three-dimensional space; there are three sets of boundary conditions that the electron's wave function must obey. Likewise, the requirement that the wave function must have only one value at each point in space must still be satisfied.

Converting this equation into the corresponding polar coordinates equation by substituting the appropriate partial derivatives obtained from Eqs (3.45) gives

$$\left[\frac{\partial^2\psi}{\partial r^2} + \frac{2}{r}\frac{\partial\psi}{\partial r} + \frac{1}{r^2\sin\theta}\frac{\partial}{\partial\theta}\left(\sin\theta\frac{\partial\psi}{\partial\theta}\right) + \frac{1}{r^2\sin\theta}\left(\frac{\partial^2\psi}{\partial\phi^2}\right)\right] + \frac{2m_e}{\hbar^2}\left(E + \frac{kZe^2}{r}\right)\psi = 0. \tag{3.46}$$

This awesome differential equation can be solved by assuming that the electron's wave function, $\psi(r, \theta, \phi)$, can be expressed as the product of three separate functions, $R(r)$, $\Theta(\theta)$ and $\Phi(\phi)$, each of which depends explicitly on just a single variable

$$\begin{aligned} \psi(r, \theta, \phi) &= R(r) \cdot Y(\theta, \phi) \\ &= R(r) \cdot \Theta(\theta) \cdot \Phi(\phi) \end{aligned} . \tag{3.47}$$

$R(r)$, called the *radial wave function*, depends just on r; the function $\Theta(\theta)$ depends only on the zenith angle, θ; the third function $\Phi(\phi)$ depends on the azimuth angle, ϕ, alone. However, instead of performing this laborious procedure, we will use our physical intuition to reveal some of the more important results of the quantum mechanical model of the hydrogen atom.[42]

The electric force that binds the electron to the nucleus of the hydrogen atom acts along the line joining them; it is a *central force*. In a classical two-body system governed by a central force, two dynamical quantities are conserved: the system's total energy, E, and its orbital angular momentum, $\mathbf{L}$, about the force centre. We might thus expect that in a hydrogen atom both of these quantities will be sharply observable and quantised.

As regards the electron's total energy, we have already seen that the expression given by the Bohr model for the energy of the stationary states in the hydrogen atom, $E_n = -(k^2 Z^2 m_e e^4/2n^2\hbar^2)$ $(n = 1, 2, 3, \ldots)$, correctly predicts the observed frequencies of its spectral lines. Moreover, it satisfies the correspondence principle; it makes the same predictions as classical physics when applied to those macroscopic systems for which classical physics is valid. Thus, we might well expect that the quantum mechanical model should give the same expression. And, indeed, the solution of the radial wave function, $R(r)$, requires that energy eigenvalues, E_n, be assigned to the bound electron that correspond exactly to those given by the Bohr model.

Thus, the first of the quantum numbers, n, has the same role in the quantum mechanical model of the hydrogen atom as the principal quantum number in the Bohr model, namely, determining the electron's total energy. The atomic energy levels described by the principal quantum number are often designated by the spectroscopic notation K, L, M, ... associated with X-ray spectra (Table 3.1).

Table 3.1. The spectroscopic notation of the principal states in the hydrogen atom.

Principal quantum number	Spectroscopic notation
$n = 1$	K
$n = 2$	L
$n = 3$	M
$n = 4$	N

Turning to the electron's orbital angular momentum, we find that obtaining eigenvalues for this classically conserved dynamical variable, without

[42]The reader is referred to Appendix 3.5.2 for a detailed elucidation of the solution of this differential equation.

contravening the uncertainty principle, restricts the way the other two quantum numbers can be used. For example, if the direction of the orbital angular momentum vector, **L**, was precisely known, it could be chosen to define the orientation of a coordinate axis, say the Z-axis[43] (Fig. 3.19). This would confine the electron to orbits in the XY plane and would unequivocally fix its z coordinate as equal to zero, implying a zero uncertainty, $\Delta z = 0$, in its z coordinate. Moreover, under these circumstances, the electron could have no linear momentum normal to the orbital plane. This, in turn, would unequivocally set the z-component, p_z, of the electron's linear momentum as equal to zero, implying an inherent zero uncertainty, $\Delta p_z = 0$, in its value. Clearly, a situation where both uncertainties, Δz and Δp_z, equal zero contravenes the uncertainty principle, $\Delta z \cdot \Delta p_z \geq \hbar$. This is precisely the situation in the Bohr model of the hydrogen atom and illustrates its fundamental dissonance from quantum mechanics.

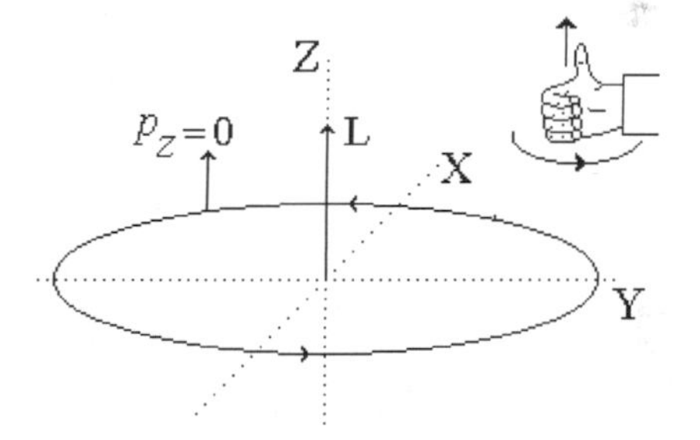

Fig. 3.19. The orientation of the Z-axis is defined by that of the axial vector, **L**. This confines the electron to the orbital plane XY, which in turn fixes the value of both its z coordinate and the z-component of its linear momentum at zero.

Evidently, precise knowledge of the direction of the orbital angular momentum vector, **L**, is forbidden by the uncertainty principle. What then are the sharp observables associated with the electron's angular momentum that can be characterised by the two remaining quantum numbers? What is the most information we can possess about the electron's angular momentum vector whilst still remaining ignorant as to its direction in space?

Clearly, we cannot know the value of all three components, L_x, L_y and L_z, of the electron's angular momentum, for that would be equivalent to knowing the direction in which the vector **L** is pointing. However, irrespective of its direction in space, or of the coordinate system used to define it, the magnitude of the electron's angular momentum will always satisfy the relationship

$$L^2 = L_x^2 + L_y^2 + L_z^2. \tag{3.48}$$

[43]The choice of the Z-axis as the reference direction is arbitrary but conventional.

Thus, the first of the sharp observables associated with the electron's angular momentum is the square of its magnitude, L^2.

The full solution of the Schrödinger equation for the hydrogen atom shows that the eigenvalues of the square, L^2, of the magnitude of the orbital angular momentum of an atomic electron are

$$L^2 = l(l+1)\hbar^2 \tag{3.49}$$
$$\text{where } l = 0, 1, 2, 3 \ldots, (n-1).$$

l is called the *orbital quantum number.* It follows that, in its ground state, where $n = 1$, the orbital quantum number of the electron in atomic hydrogen is zero: $l = 0$. Thus, its angular momentum in this state is zero. This value and not the magnitude $\hbar$ given by the Bohr model is the one borne out by experiment.[44]

The atomic sub-levels characterised by the l quantum numbers are also often designated by the spectroscopic notation s, p, d, f and g (Table 3.2).

Table 3.2. The spectroscopic notation of the quantum states in the hydrogen atom.

	$l=0$	$l=1$	$l=2$	$l=3$	$l=4$
$n=1$	$1s$				
$n=2$	$2s$	$2p$			
$n=3$	$3s$	$3p$	$3d$		
$n=4$	$4s$	$4p$	$4d$	$4f$	
$n=5$	$5s$	$5p$	$5d$	$5f$	$5g$
$n=6$	$6s$	$6p$	$6d$	$6f$	$6g$

Ignoring relativistic effects, all states in the hydrogen atom designated by the same principal number, n, possess the same energy, whatever their orbital quantum number, l. This congruence of energy levels is called *degeneracy.* We say that sub-levels sharing the same energy are *degenerate.* However, each degenerate state is described by a different wave function.

What information can the third quantum number give us? Accurate knowledge (consistent repeated measurements) regarding any one of the components of the angular momentum, say L_z, will still leave us ignorant as to the actual direction in which the vector $\mathbf{L}$ is pointing, for we still have no information regarding the components L_x and L_y. This is equivalent to saying that we have

[44]The relationship between the l quantum number and the k quantum number of the Bohr–Sommerfeld atom is $k = l + 1$.

no reliable information about the azimuth angle, ϕ, the angle in the XY plane. In terms of the uncertainty principle this can be expressed as

$$\Delta\phi \cdot \Delta L_z \approx \hbar. \tag{3.50}$$

Thus, the second of the sharp observables associated with the electron's angular momentum is its component parallel to an arbitrarily chosen Z-axis.

From the solution to the Schrödinger equation, we obtain that the eigenvalues of the L_z-component of the orbital angular momentum of an atomic electron are given by

$$L_z = m_l \hbar \tag{3.51}$$
$$\text{where } m_l = -l, (-l+1), (-l+2) \ldots, l.$$

m_l is called the *magnetic quantum number.* It follows that there are $(2l+1)$ sub-states corresponding to each l state. Thus, whereas the $l = 0$ state allows for only one sub-state, $m_l = 0$, the $l = 2$ state permits five sub-states: $m_l = -2, -1, 0, +1, +2$. The permitted sub-states in the hydrogen atom are summarized in Table 3.3.

Table 3.3. The sub-states in the hydrogen atom.

	m_l
$l = 0$	0
$l = 1$	$-1, 0, +1$
$l = 2$	$-2, -1, 0, +1, +2$
$l = 3$	$-3, -2, -1, 0, +1, +2, +3$
$l = 4$	$-4, -3, -2, -1, 0, +1, +2, +3, +4$

The picture that emerges from these two quantum numbers is of the angular momentum vector, **L**, precessing around the designated Z-axis at an angle θ given by

$$\cos\theta = \frac{L_z}{L} = \frac{m_l}{\sqrt{l(l+1)}}. \tag{3.52}$$

The angle θ is restricted to certain discrete values in accordance with the values of the quantum numbers l and m_l (Fig. 3.20). It must be emphasised that the space quantisation predicted by the quantum mechanical model of the atom is not a physical restriction; there is no actual precessional motion of the electron. Space quantisation is a fundamental property of space itself. It makes no difference how the direction of the Z-axis is defined; whether by reference to an external magnetic field or to any other effect, space quantisation will be observed.

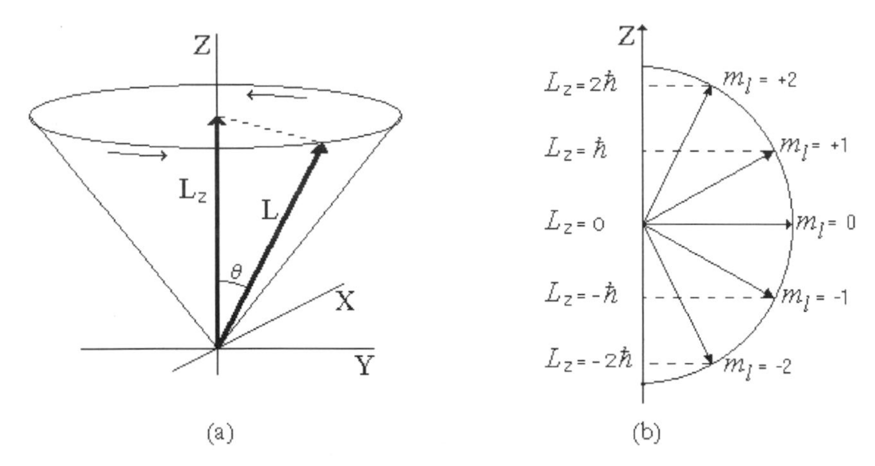

Fig. 3.20. The vector model of angular momentum quantisation. (a) The angular momentum vector, **L**, precesses around the Z-axis at certain discrete angles given by $\cos\theta = L_z/L$. In this visualisation, the components L_x and L_y are changing continuously. (b) The eigenvalues and orientations of the angular momentum vector in the case where $l = 2$, i.e. $L = \sqrt{6}\hbar$ and $L_z = m_l\hbar$.

We can now use the three quantum numbers, n, l and m_l, that we have devised in order to build a quantum mechanical model of the hydrogen atom. In terms of its quantum numbers, the ground state of the hydrogen atom is characterised by the set: $n = 1, l = 0, m_l = 0$. In order to indicate the quantum states it describes, an electron's wave function, $\psi(r, \theta, \phi)$, is usually designated as ψ_{n,l,m_l}. Thus, the ground state's wave function is designated by $\psi_{1,0,0}$.

The full solution of the Schrödinger equation for the ground state of the hydrogen atom gives the wave equation

$$\psi_{1,0,0} = \frac{1}{\sqrt{\pi a_0^3}} e^{-r/a_0}, \tag{3.53}$$

where a_0 is the Bohr radius. Note that this function is independent of both the zenith and azimuth angles, i.e. it is spherically symmetrical. Consequently, the probability of finding a ground state electron at a distance r from the nucleus will be the same in all directions.

In the ground state, the probability density at any point at a distance r from the hydrogen atom's nucleus is $|\psi_{1,0,0}|^2$. Because of the spherical symmetry of the wave function, the overall probability density, $P(r)$, at a distance r from the nucleus is equal to the probability density at any one point, $|\psi_{1,0,0}|^2$, multiplied by the surface area, $4\pi r^2$, of a sphere of radius r

$$P(r) = |\psi_{1,0,0}|^2 4\pi r^2. \tag{3.54}$$

$P(r)$ is called the *radial probability density*. Substituting the expression for the ground state wave function, $\psi_{1,0,0}$, gives

$$P(r) = \left(\frac{1}{\sqrt{\pi a_0^3}} e^{-r/a_0} \right)^2 4\pi r^2 = \frac{4r^2}{a_0^3} e^{-2r/a_0}. \tag{3.55}$$

This function has its maximum value when $r = a_0$. Thus, in the ground state the probability of finding the electron in the hydrogen atom is greatest at a distance a_0 from the nucleus. This is the significance of the Bohr radius in quantum mechanics (Fig. 3.21).

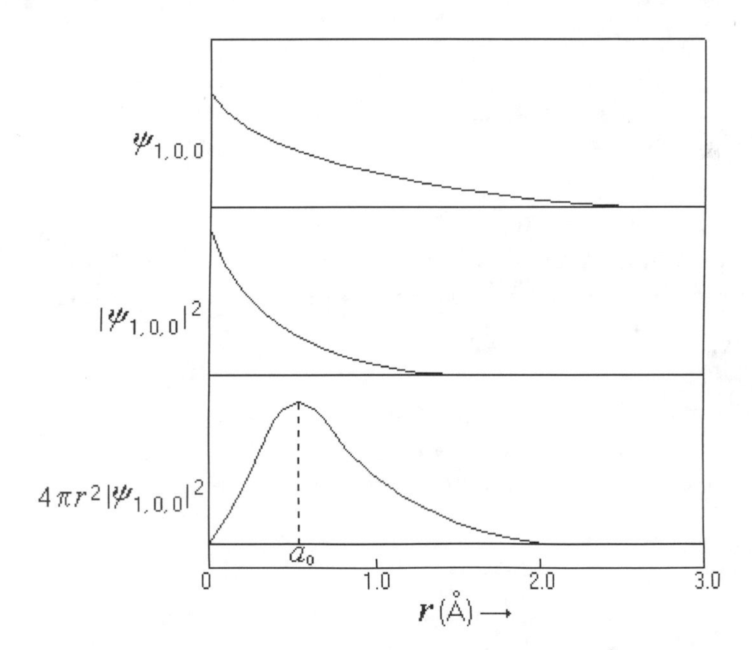

Fig. 3.21. The wave function, $\psi_{1,0,0}$, probability density, $|\psi_{1,0,0}|^2$, and radial probability density, $P(r) = |\psi_{1,0,0}|^2 \ 4\pi r^2$, of the electron in a hydrogen atom in the ground state.

The radial probability density, $P(r)$, in excited hydrogen atom states whose orbital quantum number $l = 0$, e.g. the $2s, 3s, \ldots, ns$ states, is also spherically symmetrical. However, the radial probability density in these states does show distinct maxima and minima at different distances from the nucleus (Fig. 3.22). In general, the higher the principal quantum number the further away from the nucleus is the largest probability density maximum.

In states having $l > 0$ there is no spherical symmetry (though there may be other symmetries). In these states, the probability density also depends on

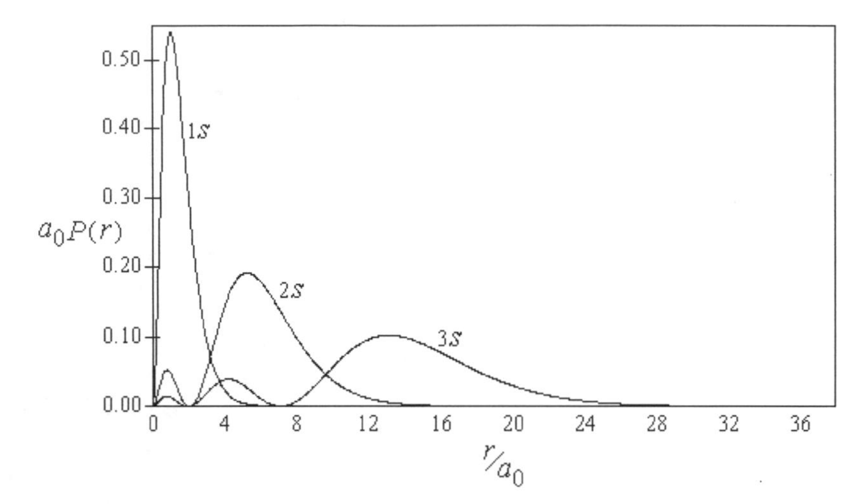

Fig. 3.22. The radial probability density functions of the $1s$, $2s$ and $3s$ states in the hydrogen atom.

the zenith and azimuth angles. The $2p$ state ($n = 2, l = 1$), in which the magnetic quantum number can take the values $m_l = -1, 0$ or $+1$, exhibits distinct directional characteristics. For example, in the $\psi_{2,1,0}$ state, the electron is found preferentially along the Z-axis; for this reason, this state is designated the $2p_z$ state.

The wave functions of the other two $2p$ states, $\psi_{2,1,+1}$ and $\psi_{2,1,-1}$, can combine[45] by addition or by subtraction to give two more wave functions, ψ_{2p_x} and ψ_{2p_y}, that possess distinct directional preferences:

$$\begin{aligned}\psi_{2p_x} &= \frac{1}{\sqrt{2}}(\psi_{2,1,+1} + \psi_{2,1,-1}) \\ \psi_{2p_y} &= \frac{1}{\sqrt{2}}(\psi_{2,1,+1} - \psi_{2,1,-1})\end{aligned} \tag{3.56}$$

Assigning the same amplitude to each of the component waves, $\psi_{2,1,+1}$ and $\psi_{2,1,-1}$, indicates that each makes an equal contribution to the new wave functions.[46] The preferred orientation of the electron in the ψ_{2p_x} function ($2p_x$ state) is along the X-axis, and that in the ψ_{2p_y} function ($2p_y$ state) is along the Y-axis (Fig. 3.23). This directional preference is of great import in the explanation of chemical bonding as it affords the possibility of directed bonds and

[45] Any superposition of wave functions is itself a possible wave function (Section 2.4.3).

[46] The functions ψ_{2p_x} and ψ_{2p_y} are normalised.

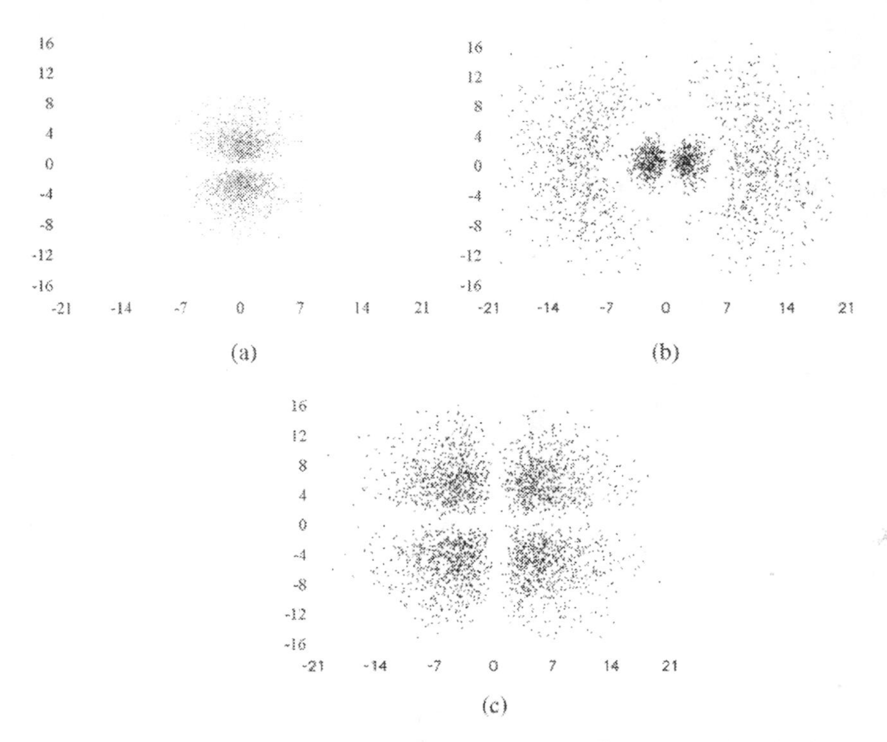

Fig. 3.23. Computer simulations of the image that would be revealed on looking through 2,000 slide photographs (transparencies) of the electron in a hydrogen atom in (a) the $2p_0$ state ($\psi_{2,1,0}$); (b) the $3p_{\pm1}$ state ($\psi_{3,1,\pm1}$); (c) the $3d_{\pm1}$ state ($\psi_{3,2,\pm1}$). The images were produced by plotting the probability density, $|\psi_{n,l,m_l}|^2$, of each state at 2,000 randomly selected points in the *zx* plane. The scaling shows the distance from the nucleus in Bohr radii.

of molecular architecture. As such, it marks one of the major improvements of the quantum mechanical model over the Bohr–Sommerfeld model.

Since, according to quantum mechanics, the electron's motion cannot be described in terms of trajectories, its distribution in the space around the nucleus is often pictured in terms of a 'cloud of negative charge' or an 'electron cloud', whose density is obtained by multiplying the electron's probability density, $|\psi_{n,l,m_l}|^2$, by the electronic charge, *e*. This *charge density* or 'thickness' of the electron cloud at a point expresses the probability of finding an electron there. However, it must be emphasised, that in any measurement that is made, just a single electron will be revealed and not a cloud. The electron clouds can be regarded as what would be seen on looking through a pile of thousands of slide photographs (transparencies) of the electron, each taken at a different moment (Fig. 3.23).

Example 3.6: The Average Distance of the Electron from the Hydrogen Nucleus

What is the average (expectation) value, $\langle r \rangle$, of the distance from the nucleus to the electron in the spherically symmetrical ground state in hydrogen?

Calculation:

The probability of finding the electron in an infinitesimal volume element, dV, is

$$P = |\psi_{1,0,0}|^2 dV.$$

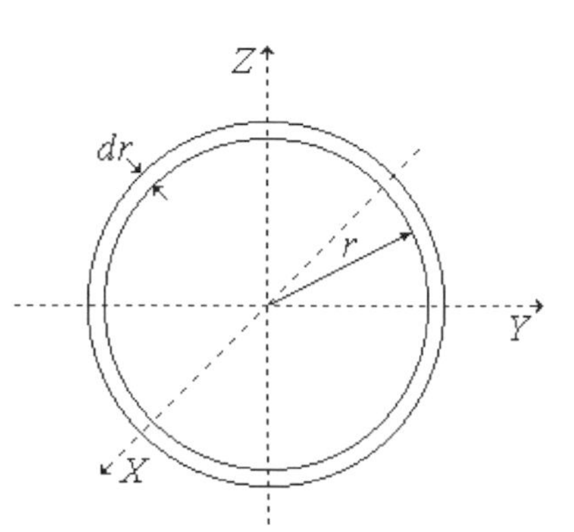

The volume, dV, of the thin spherical shell is the difference between that of the sphere of radius $r + dr$ and that of the sphere of radius r

$$\begin{aligned} dV &= \frac{4\pi}{3}[(r+dr)^3 - r^3] \\ &= 4\pi r^2 dr \end{aligned}$$

Hence, the probability, $P(r)dr$, of finding the electron at a distance r from the nucleus is

$$P(r)dr = |\psi_{1,0,0}|^2 4\pi r^2 dr. \tag{3.57}$$

By analogy from Eq. (2.80), the average value, $\langle r \rangle$, of the electron's distance from the nucleus is given by

$$\langle r \rangle = \int_0^\infty rP(r)dr = \int_0^\infty r|\psi_{1,0,0}|^2 4\pi r^2 dr. \tag{3.58}$$

Substituting $\psi_{1,0,0} = \frac{1}{\sqrt{\pi a_0^3}} e^{-r/a_0}$ from Eq. (3.52) gives

$$\langle r \rangle = \frac{4}{a_0^3} \int_0^\infty e^{-2r/a_0} \cdot r^3 dr. \tag{3.59}$$

The general solution of such integrals is

$$\int_0^\infty e^{-\beta r} \cdot r^n dr = \frac{n!}{\beta^{n+1}}. \tag{3.60}$$

Accordingly,

$$\langle r \rangle = \frac{4}{a_0^3} \cdot \frac{3!}{\left(2/a_0\right)^4} = 1.5a_0. \tag{3.61}$$

Thus, the average distance of the electron from the nucleus is greater than the Bohr radius.

Example 3.7: The Probability of Finding an Electron

What is the probability of finding the electron in a hydrogen atom in the ground state closer to the nucleus than the Bohr radius?

Calculation: The expression for the probability, P_{0a_0}, of finding the electron closer to the nucleus than the Bohr radius, a_0, in the ground state, $\psi_{1,0,0}$, is obtained from Eq. (2.78):

$$P_{0a_0} = \int_0^{a_0} P(r)\,dr = \int_0^{a_0} |\psi_{1,0,0}|^2 4\pi r^2\,dr. \tag{3.62}$$

Substituting $\psi_{1,0,0} = \frac{1}{\sqrt{\pi a_0^3}} e^{-r a_0}$ from Eq. (3.52) gives

$$P_{0a_0} = \frac{4}{a_0^3} \int_0^{a_0} e^{-2r/a_0} \cdot r^2\,dr. \tag{3.63}$$

Making the substitution $x = 2r/a_0$ gives the integral

$$P_{02} = 0.5 \int_0^2 e^{-x} \cdot x^2\,dx, \tag{3.64}$$

whose solution gives a probability $P_{02} = 0.32$ of finding the electron closer to the nucleus than the Bohr radius.

3.3.2 Atomic Spectra and Quantum Mechanics

Transitions between stationary states in which radiation is emitted or absorbed are called *radiative transitions*; they are the transitions that appear as spectral lines. The Bohr model relates the frequency of the spectral lines to the difference between the energies of the stationary states. However, it offers no real explanation for the actual occurrence of radiative transitions.

Why should atoms spontaneously shift from one stationary state to another? Apart from the intuitive feeling that systems ought to 'fall' from higher energy states to lower ones, why should excited atoms spontaneously 'drop' from higher stationary states into lower ones. On the contrary, it can be argued that since an atom's stationary states are all non-radiating, they are all equally 'desirable'. Moreover, other orbiting bodies such as planets don't spontaneously change their orbits, so why should orbiting electrons? And if electrons do in fact spontaneously 'drop' from higher stationary states into lower ones, how are the photons emitted in the transfers actually created?

The spontaneous emission of radiation has no classical analogue. It is a purely quantum phenomenon, a satisfactory explanation for which can only be given in terms of the quantum mechanical model of the atom. It is most readily understood as arising from the superposition of atomic wave functions.

The electron involved in a radiative transition can occupy either of the quantum states, i and j, between which it moves. There is no way we can tell which state it is occupying at any moment, the higher or the lower one.

Thus, the electron's time dependent wave function, Ψ_{ij}, is given by the linear combination (superposition) of the time dependent wave functions, Ψ_i and Ψ_j, corresponding to the two states

$$\Psi_{ij} = A_i\Psi_i + A_j\Psi_j, \tag{3.65}$$

where A_i^2 and A_j^2 are the probabilities of the electron being found in each state. Normalisation requires that $A_i^2 + A_j^2 = 1$ at all times.

When the electron is in the i state, $A_i = 1$ and $A_j = 0$; when it is in the j state, $A_i = 0$ and $A_j = 1$. In each of these instances, which correspond, respectively, to the i and j stationary states, the probability density, $|\Psi_{ij}|^2$, is constant. But what happens to the probability density when both A_i and A_j have non-vanishing values?

In general, the time dependent wave function, $\Psi_S(x, y, z, t)$, of a particle in a stationary state can be expressed in terms of its time independent wave function, $\psi(x, y, z)$, and its energy eigenvalue, E, in the form[47]

$$\begin{aligned}\Psi_S(x, y, z, t) &= \psi(x, y, z)e^{-iEt/\hbar}\\ &= \psi(x, y, z)e^{-i2\pi ft}\end{aligned} \tag{3.66}$$

where $f = E/h$ is the frequency of the particle's matter wave. This gives for the wave functions of the i and j stationary states:

$$\Psi_i = \psi_i e^{-i2\pi f_i t} \text{ and } \Psi_j = \psi_j e^{-i2\pi f_j t}, \tag{3.67}$$

where $f_i = E_i/h$ and $f_j = E_j/h$ are, respectively, the frequencies of the matter wave of the i state and the j state.[48] Substituting these expressions in Eq. (3.65) gives

$$\Psi_{ij} = A_i\psi_i e^{-i2\pi f_i t} + A_j\psi_j e^{-i2\pi f_j t}. \tag{3.68}$$

Compared to the frequencies of the matter waves, the difference between them, $f_i - f_j$, is very small.[49] Consequently, when both A_i and A_j have non-vanishing values, the superposition of the matter waves corresponds to the phenomenon of *beats*, the periodic amplitude modulation that arises from the superposition of two vibrations with neighbouring frequencies (Fig. 3.24).

[47] Equation (2.102) in Section 2.4.7.

[48] The letter i in the exponent is unrelated to the subscript i. The former indicates the imaginary number defined by $i^2 = -1$.

[49] f_i and f_j are both very large because E_i and E_j are the electron's total energy. Furthermore, the denominator h is very small.

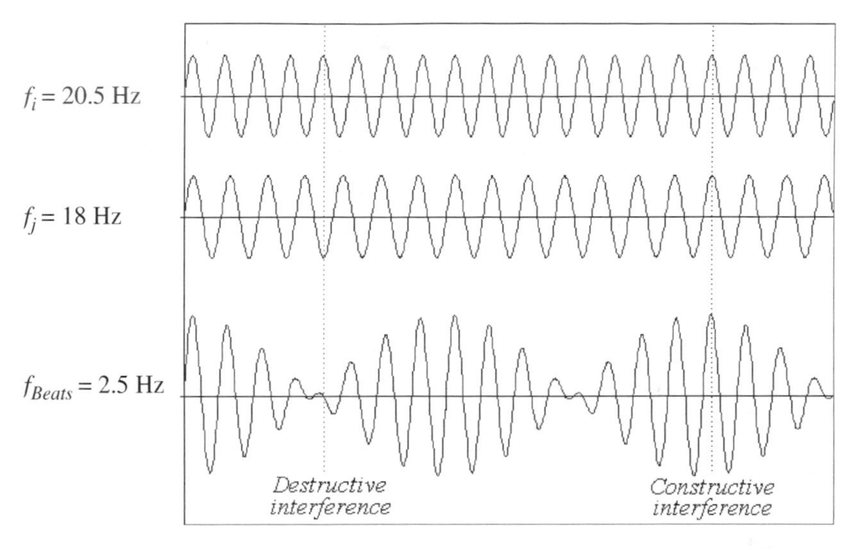

Fig. 3.24. The beats produced by the superposition, $\Psi_{ij} = \Psi_i + \Psi_j$, of two waves with similar frequencies: $f_i = 20.5\,\text{Hz}$; $f_j = 18\,\text{Hz}$. Hence, $f_{Beats} = 2.5\,\text{Hz}$.

The frequency, f, of the beats is equal to the absolute value of the difference in the frequencies of the two component waves.[50] In this case,

$$f = |f_i - f_j|. \tag{3.69}$$

Since, in general, the probability density of a state depends on the amplitude of its wave function,[51] the probability density of the ij state, $|\Psi_{ij}|^2$, which is usually called the *transition density*, will oscillate with the same frequency as does the amplitude of the resultant wave, i.e. with the beat frequency $f = |f_i - f_j|$.

In a stationary state, the charge density, $e|\Psi_{ij}|^2$, associated with the electron does not oscillate and so radiation is neither emitted nor absorbed. However, when the electron is in transit between the two states, i.e. when both A_i and A_j have non-vanishing values, the charge density oscillates with the beat frequency. These fluctuations in the *electronic cloud* around the nucleus, that arise from the superposition of the matter waves of the stationary states, are the quantum equivalent of the electric oscillations that, according to classical electromagnetic theory, must accompany the emission and absorption of radiation. Only when these fluctuations arise can radiation be emitted or absorbed. Intense spectral lines correspond to radiative transitions whose probability is

[50] See Supplementary Topic A – The Mathematical Description of Wave Motion for a full explanation of the phenomenon of beats.

[51] The probability density (the intensity of the matter wave) is given by the product of the wave function, $\Psi = Ae^{+i2\pi ft}$, and its conjugate, $\Psi^* = A^*e^{-i2\pi ft}$, *viz.* $|\Psi|^2 = \Psi\Psi^* = Ae^{+i2\pi ft}A^*e^{-i2\pi ft} = A^2$.

high; faint lines correspond to transitions of low probability; forbidden transitions are those whose transition density equals zero.

Thus, the frequency, f, of the radiation emitted or absorbed in a transition is determined by that of the periodic amplitude modulation (the beats) that arises from the superposition of the matter waves of the stationary states between which the transition occurs:

$$f = f_i - f_j = \frac{E_i - E_j}{h}. \tag{3.70}$$

In this way, Bohr's equation, $f_{ij} = \frac{E_i - E_j}{h}$, that was introduced as a postulate in Bohr's model of the atom, arises as a straightforward corollary of the wave nature of the electron in the quantum mechanical model of the atom.

However, the generation of a fluctuating charge density – an oscillating electric dipole – is not the only condition that must be satisfied for the emission or absorption of electromagnetic radiation, i.e. of photons. It transpires that radiative transitions never occur when the total angular momentum of both the initial and final states of the emitting or absorbing system equals zero. Assuming that angular momentum is conserved overall in these transitions, this indicates that the photons emitted or absorbed in radiative transitions must themselves possess angular momentum. The classical analogue of this intrinsic angular momentum of the photons is the circular polarisation of the electromagnetic wave. It follows that radiative transitions can only occur when the fluctuations in the electron cloud and the attendant emission or absorption of radiation obey the law of angular momentum conservation.

Intense spectral lines correspond to radiative transitions whose probability is high; faint lines correspond to transitions of low probability; forbidden transitions are those whose transition density equals zero. Allowed atomic transitions in hydrogen and hydrogen-like atoms obey the selection rules:

$$\Delta l = \pm 1 \quad \Delta m_l = 0, \pm 1.$$

Both rules ensure the conservation of angular momentum. In addition, the $\Delta l = \pm 1$ rule ensures that a non-vanishing electric dipole is generated by the fluctuations in the electron cloud (Fig. 3.25). Similar selection rules had been previously derived empirically by investigators of the Bohr–Sommerfeld model. They were explained at the time by applying the correspondence principle to classical harmonics; if the rules are valid for high quantum numbers, their validity can be postulated for all quantum numbers. However, it was the quantum mechanical model that first gave them an *a priori* theoretical basis.

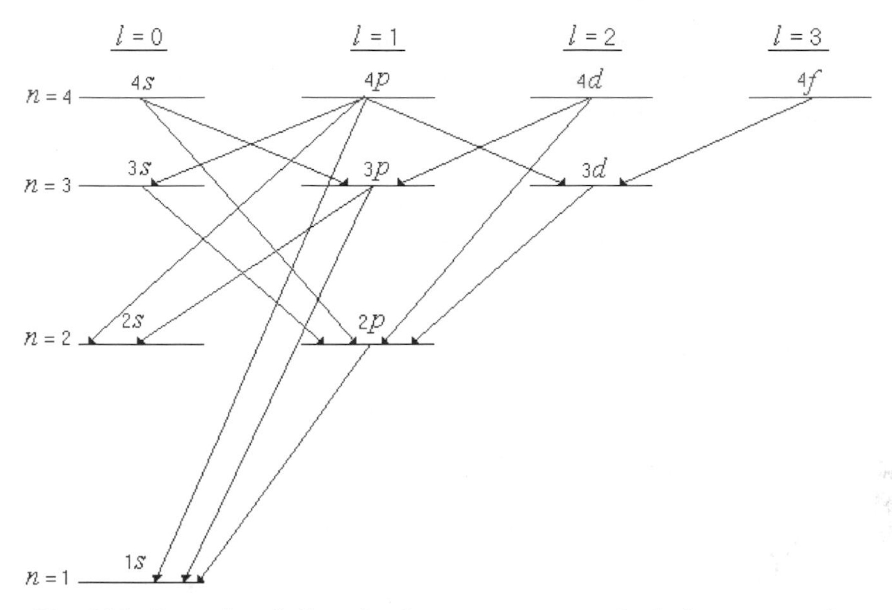

Fig. 3.25. Examples of allowed radiative transitions in the hydrogen atom that obey the rule $\Delta l = \pm 1$.

The average lifetime of an excited state is inversely proportional to the transition probability for a drop to a lower energy state. The average lifetime of a typical excited atomic state is $\sim 10^{-8}$s. The lifetime of the ground state is practically infinite for there is no energy level below it.

Of the excited states in the hydrogen atom, the $2s$ state is of particular interest. Being an excited state, its lifetime might be expected to be very short. However, there is no lower state to which the selection rule $\Delta l = \pm 1$ allows it to fall; the only state below it is the $1s$ state which shares the same $l = 0$ quantum number. Consequently, the $2s$ state has a longer than expected lifetime and is thus called a *metastable* state.

To summarise the quantum mechanical model of the hydrogen and hydrogen-like atoms:

1. Hydrogen and hydrogen-like atoms consist of a very small and very dense nucleus that contains the entire positive charge of the atom, $+Ze$, and virtually all of its mass.
2. A single negative electron, to which no specific trajectory can be assigned, occupies the space around the nucleus: this electron, together with the nucleus, comprises the atom.
3. The interaction between the nucleus and the electron is Coulombic.

4. The atom is described by a wave function which reflects the restrictions imposed by the Heisenberg uncertainty principle and from which the quantum numbers that characterise its stationary states can be derived.
5. The atom absorbs and emits energy in accordance with the probabilities prescribed by its wave function.

Extending the quantum mechanical model to the fine details of the spectra and to multi-electron atoms requires the introduction of a new concept – *electron spin.*

Chapter 3.4

Electron Spin

In 1925, two graduate students, Goudsmit[52] and Uhlenbeck,[53] suggested that apart from any orbital angular momentum they might have, electrons also possess an intrinsic (internal) angular momentum – *spin.* They were apparently unaware that this singular hypothesis, which provided a physical interpretation for the fourth quantum number[54] that Pauli[55] had already shown was needed to explain the observed splitting of fine structure spectral lines, had been suggested previously in 1921 by Richardson[56] in order to account for the magnitude of the Amperian currents in magnetic materials. Subsequently, the concept of electron spin became the basis for the elucidation of diverse phenomena such as the pattern of the periodic table of the elements and the pairing of electrons in chemical bonds.

3.4.1 Electron Spin

The intrinsic (internal) angular momentum of a spinning particle, its spin, is quantised in a way that is analogous to the quantisation of orbital angular momentum. The eigenvalues of the square, S^2, of the magnitude of a particle's spin are given by

$$S^2 = s(s+1)\hbar^2, \tag{3.71}$$

where the quantum number s is called the *spin quantum number.* The component, S_Z, of the particle's spin parallel to an arbitrarily chosen Z-axis is given by

$$S_Z = m_s\hbar, \tag{3.72}$$

where

$$m_s = -s, (-s+1), (-s+2), \ldots, s.$$

[52]Samuel Goudsmit 1902–1978, Dutch–American physicist.
[53]George Uhlenbeck 1900–1988, Dutch–American physicist.
[54]In addition to the quantum numbers n, l and m_l.
[55]Wolfgang Pauli 1900–1958, Austrian–American physicist who was awarded the Nobel prize in 1945 for the discovery of the rule known as the Pauli principle.
[56]Sir Owen Richardson 1879–1959, English physicist who was awarded the Nobel prize in 1928 for his work on the thermionic emission of electrons.

m_s is called the *spin magnetic quantum number*. It follows that the particle's intrinsic angular momentum vector can have $(2s+1)$ orientations in space.

In the case of the electron, the quantum number s can take only one value: $s=\frac{1}{2}$. This allows for just two orientations of the spin vector corresponding to $m_s=\pm\frac{1}{2}$: 'spin up' ($\uparrow$) and 'spin down' ($\downarrow$). It is customary to call the quantum number s the *spin of the electron* and to say that 'an electron has a spin of a half', this despite the fact that the magnitude of its spin is actually $S=\sqrt{3}\cdot(\hbar/2)$ (Fig. 3.26).

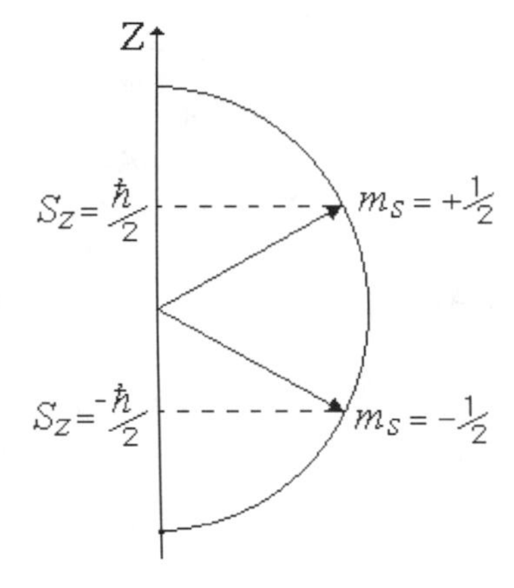

Fig. 3.26. The space quantisation of the electron's spin. In each orientation the magnitude of the spin is $S=\sqrt{3}\cdot(\hbar/2)$.

It is incorrect to picture the electron as a spinning negatively-charged solid sphere. The electron's intrinsic angular momentum is not the result of the spinning of an inertial mass. Electrons have no discernible dimensions; it is quite possible that they are point particles. Electron spin is a pure quantum phenomenon that is described by the *spin wave function*, χ.

In 1929, Dirac[57] showed that the existence of electron spin follows from the solution of the Schrödinger equation if the relativistic expression for the particle's energy, $E=\sqrt{m_e^2c^4+p^2c^2}+U$, is employed in its formulation instead of the classical expression, $E=p^2/2m_e+U$. The picture this solution gives is of a flow of energy and momentum around the axis along which the electron's matter wave propagates. It is this circulating flux that confers the intrinsic angular momentum on the electron.

In general, charged particles that possess an intrinsic angular momentum (spin), $\mathbf{S}$, have a magnetic moment, $\boldsymbol{\mu}_S$, associated with them that is given by

$$\boldsymbol{\mu}_S=\gamma_S\mathbf{S}\left(\frac{e}{2m}\right), \tag{3.73}$$

[57]Paul Dirac 1902–1984, English physicist who was awarded the Nobel prize in 1933 for innovations in atomic theory.

where m is the particle's mass. The factor γ_S is called the *spin gyromagnetic ratio*; its value varies from particle to particle. In the case of an electron, $\gamma_S = -2.0023$.

The component, $\mu_{S_Z(electron)}$, parallel to an arbitrarily chosen Z-axis of the magnetic moment associated with the electron's spin is given by

$$\mu_{S_Z(electron)} = \gamma_S S_Z \left(\frac{e}{2m_e}\right). \tag{3.74}$$

The space quantisation of the electron's spin, up or down, results in a corresponding restriction on the orientations of the magnetic moment associated with its spin; it too is restricted to just two possible orientations. This simple fact has profound repercussions.

3.4.2 The Stern–Gerlach Experiment

The phenomenon of space quantisation was first predicted by the Bohr–Sommerfeld model of the atom, in which it arises from the quantisation of the orbital angular momentum of the electron. Its actual existence was verified experimentally by Stern and Gerlach in 1921, several years before the development of quantum mechanics. However, as we shall see, what they actually observed was the space quantisation associated with the electron's intrinsic angular momentum (spin) and not that which is a consequence of the quantisation of its orbital angular momentum.

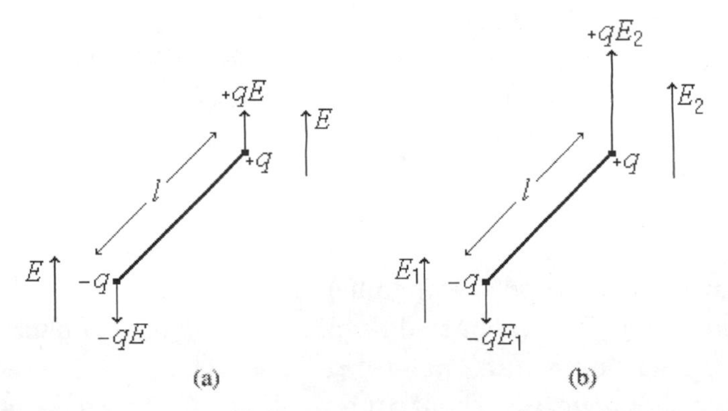

Fig. 3.27. The forces acting on an electric dipole in an electric field. (a) In a uniform field the resultant force is zero: $+qE + (-qE) = 0$. (b) In a non-uniform field the resultant force is not zero: $+qE_2 + (-qE_1) \neq 0$.

In a uniform field, the magnitude of the resultant force that acts on a dipole is zero. Thus, though the dipole may rotate as a result of the couple impressed upon it by the field, particles possessing a dipole moment that enter a uniform field will pass through in a straight line (Fig. 3.27a). This is equally true for an electric dipole in an electric field or a magnetic dipole in a magnetic field.

However, in a non-uniform field the magnitude of the resultant force acting on the dipole does not equal zero and so the dipole's centre of mass will be accelerated (Fig. 3.27b). Thus, particles possessing a dipole moment that enter a non-uniform field will be deflected from their paths. The extent of each particle's deflection will depend on (i) the degree of non-uniformity of the field; (ii) the strength of the particle's dipole moment; (iii) the orientation of the particle's dipole moment relative to the direction of the field.

Many atoms behave like magnetic dipoles, i.e. as though they possessed a magnetic moment. According to classical theory, in the gaseous state these magnetic moments point randomly in all directions. Consequently, passing a narrow collimated beam of gaseous neutral atoms through a non-uniform magnetic field should result in an overall broadening of the beam. The atoms should be deflected in random directions, such that the beam's cross-section should broaden uniformly.

However, space quantisation implies that the orientations of the atomic magnets are restricted to a discrete number of directions relative to any chosen axis. Depending on whether the Bohr–Sommerfeld or the quantum mechanical theory is applied, the magnetic moment of an atom possessing orbital quantum numbers k or l is limited, respectively, to $(2k+1)$ or $(2l+1)$ orientations relative to any defined direction. The most convenient reference axis is that defined by the direction of the magnetic field; this is usually designated the Z-axis. Consequently, when passing through a non-uniform magnetic field, a narrow collimated beam of atoms possessing magnetic moments should not broaden but split into an odd number of beams. If the orbital quantum number, k or l, equals one, the original beam should split into three beams; if it equals two, there should be five beams, and so on. If the orbital quantum number equals zero, there should be no beam splitting at all.

Stern and Gerlach produced a non-uniform magnetic field by means of an electromagnet whose poles were differently shaped; one pole was flat and the other knife-edged. A collimated beam of neutral silver atoms was passed through the field between the poles (Fig. 3.28). They found that the beam split into two distinct beams. Though this result demonstrated the quantisation of space, the even number of beams, two in this case, was inexplicable at the time.

The experiment was repeated some years later with hydrogen atoms; they too divided into two beams. In the Bohr–Sommerfeld model, the ground state in hydrogen has an orbital quantum number $k = 1$ and so the atoms should divide into three distinct beams on passing through the non-uniform magnetic field. At first sight, the quantum mechanical model appears to fare no better. It assigns an orbital quantum number $l = 0$ to the hydrogen atom's ground state. This implies that the original beam should not have split at all.

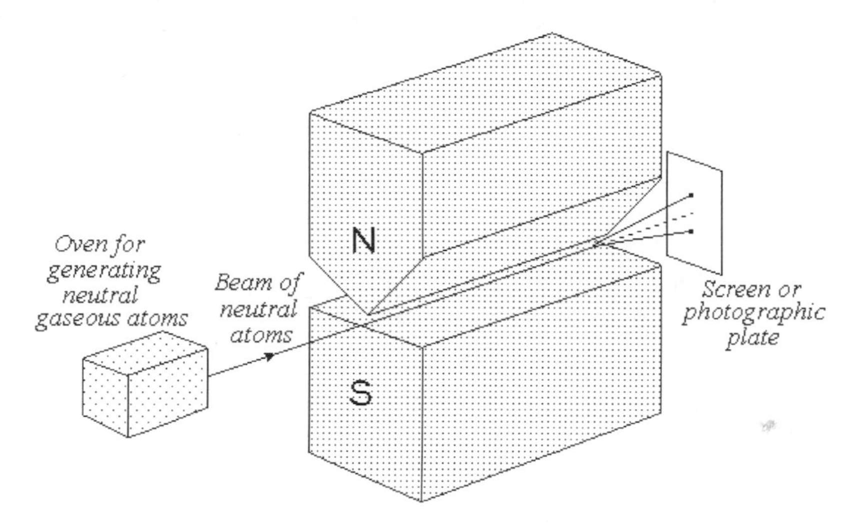

Fig. 3.28. The Stern–Gerlach experiment. The collimated beam of neutral silver atoms passes through the non-uniform magnetic field and splits into just two beams.

The splitting into two beams could be understood if, apart from its electron's orbital angular momentum, there was some other source of angular momentum in the hydrogen atom whose quantisation was characterised by a fractional quantum number, e.g. $\frac{1}{2}$. Were this so, the number of beams produced would be $2 \cdot \frac{1}{2} + 1 = 2$, as found in the experiment, even when $l = 0$.

Fractional quantum numbers were completely beyond the conception of the Bohr–Sommerfeld model. They are, however, anticipated by the relativistic solution of the electron's Schrödinger equation and by the matrix derivation of Heisenberg, both of which assign the fractional quantum number, $s = \frac{1}{2}$, to the electron's intrinsic angular momentum (spin). Thus, the splitting into two beams observed in the Stern–Gerlach experiment is actually due to the quantisation of the electron's spin and not to that of its orbital angular momentum.

Example 3.8: Electron Spin Resonance

Calculate the difference between the energy eigenvalues of the magnetic quantum states in the two beams produced by passing a beam of neutral silver atoms through a non-uniform magnetic field of magnitude 0.25 T. Radiation of what frequency could cause a transition between these two states?

Solution: The vector sum of the angular momentum of the electrons in the inner shells of the silver atom is zero. Hence, the net angular momentum possessed by the silver atom is just that attributable to the spin of its single valence *s* electron $(n = 5, l = 0)$. It follows that the component of the atom's magnetic moment along an arbitrary Z-axis defined by the direction of the magnetic field can only be that attributable to the electron's spin, $\mu_{S_Z(electron)}$. From Eq. (3.74), this component is given by $\mu_{S_Z(electron)} = \gamma_S S_Z(e/2m_e)$. Substituting $\gamma_S = -2.0024$ (≈ -2) and $S_Z = \pm\hbar/2$ gives

$$\mu_{S_Z(electron)} = \pm\frac{\hbar e}{2m_e}. \tag{3.75}$$

The quantity $\hbar e/2m_e$ is the Bohr magneton:

$$\mu_B = 9.27 \cdot 10^{-24}\,\mathrm{J} \cdot \mathrm{T}^{-1}. \tag{3.76}$$

The energy, E_μ, of a magnetic moment in a magnetic field of magnitude B is

$$E_\mu = -\mu B \cos\theta, \tag{3.77}$$

where θ is the angle between the magnetic moment and the field. Because of the space quantisation of the electron's spin the angle θ can only be 0° or 180°. Consequently, the difference, ΔE_μ, between the magnetic moment's energy in the two permitted orientations is

$$\begin{aligned}\Delta E_\mu &= -\mu B(\cos 0^\circ - \cos 180^\circ)\\ &= \pm 9.27 \cdot 10^{-24} \cdot 0.25 \cdot 2\\ &= \pm 4.635 \cdot 10^{-24}\,\mathrm{J}.\end{aligned} \tag{3.78}$$

Transitions between the two orientations can be induced by radiation of frequency

$$f = \Delta E_\mu/h = 7 \cdot 10^9\,\mathrm{Hz}, \tag{3.79}$$

i.e. by radiation in the microwave region of the spectrum. The phenomenon is called *electron spin resonance.*

3.4.3 Spin-Orbit Coupling

Physics permits of no preferred frame of reference; all reference frames are equivalent. Thus, although we have tended to describe the atom in terms of a stationary nucleus and a moving electron, i.e. from the typical reference frame of the nucleus, an observer on the electron would be just as entitled to describe it in terms of the apparent motion of the nucleus relative to him. Such an observer would note that, as a consequence of its motion around the electron, the charged nucleus induces a magnetic field in the region. The interaction of this magnetic field with the spin magnetic moment of the electron affects the electron's energy levels, causing some of them to split. The phenomenon is a kind of internal Zeeman effect, the magnetic field with which the atomic electron interacts being one that is created by the atom itself.[58]

Due to the coupling of the two magnetic moments, it is the electron's total angular momentum, $\mathbf{J} = \mathbf{L} + \mathbf{S}$, that is the physical quantity whose quantum numbers, j and m_j, determine the energy eigenvalues of the electron's quantum states. The permitted values of the square, J^2, of the magnitude of the electron's total angular momentum is characterised by the values

$$J^2 = j(j+1)\hbar^2 \tag{3.80}$$

where, for a single electron,

$$j = l \pm \frac{1}{2}.$$

The component, J_z, of the electron's total angular momentum is given by

$$J_z = m_j \hbar \tag{3.81}$$

where

$$m_j = -j, (-j+1), (-j+2) \ldots, j.$$

The total angular momentum quantum number, j, is always a fractional number. Consequently, the number, $2j + 1$, of spatial orientations of the electron's total angular momentum vector, $\mathbf{J}$, is always even. For example, when $l = 1$, the total angular momentum quantum number can have two values: $j = \frac{1}{2}$ or $j = \frac{3}{2}$ giving two and four orientations, respectively.[59]

The pre-eminent example of spectral line splitting associated with spin-orbit coupling is the *sodium doublet.* The yellow line in the sodium emission spectrum that corresponds to a transition from the $3p$ level to the $3s$ level is split into two. The twin adjacent lines are attributed to a splitting of the $3p$ level into two sub-levels. The upper $3p$ state corresponds to a total angular quantum number $j = \frac{3}{2}$; the lower state to $j = \frac{1}{2}$ (Fig. 3.29).

A similar splitting is observed in transitions from the $2p$ state to the $1s$ state in hydrogen. It follows that the quantum numbers n, l (magnitude of the orbital angular momentum), s (the spin) and j and m_j (magnitude and z-component of

[58] Atomic nuclei also possess spin and magnetic moments, though the latter are about 1,000 times smaller than the magnetic moments associated with atomic electrons. These nuclear moments lead to a further splitting of spectral lines called the *hyperfine structure* that is detectable only at very high resolutions.

[59] The quantum number j was the fourth quantum number that was postulated by Pauli to explain the splitting of the fine structure lines. But Pauli gave no explanation as to why the total angular momentum characterised by j differed from the orbital angular momentum characterised by l.

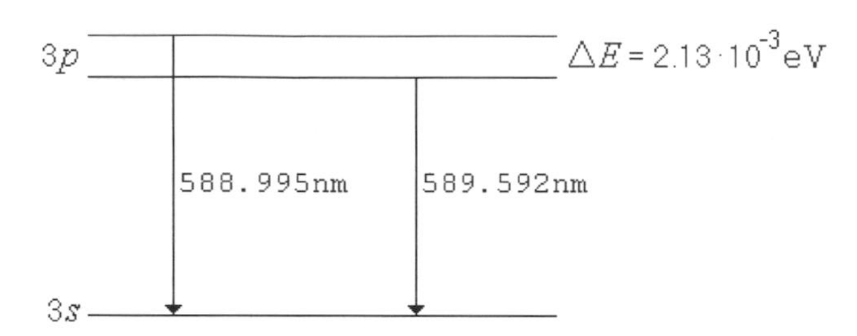

Fig. 3.29. The sodium doublet. Two adjacent yellow spectral lines are observed corresponding to transitions from the upper and lower sub-levels of the $3p$ state to the $3s$ state. There is no splitting of the s level since no internal magnetic field arises in a state with zero orbital angular momentum. When $l = 0$ the only possible value of j is $j = \frac{1}{2}$. Because the energy difference, ΔE, between the $3p$ sub-levels is about one thousandth that between the $3p$ and $3s$ levels the doublet is only revealed at high resolutions.

the total angular momentum) are 'good' quantum numbers for the hydrogen atom. Their values remain fixed so long as the electron remains in the same energy eigenstate. Thus, repeated measurements of the quantities they define, which correspond to the constants of motion in classical mechanics, yield the same result every time. On the other hand, the quantum numbers m_l and m_s, that define the z-components of the orbital and intrinsic angular momentum respectively, are not good quantum numbers; their values change. Repeated measurements of the z-components they define yield one or other of the values $m_l = 0, \pm 1, \pm 2, \ldots$ or $m_s = \pm\frac{1}{2}$, but which value emerges on each occasion is uncertain.

3.4.4 The Pauli Exclusion Principle and the Periodic Table

Solving the Schrödinger equation for an electron in a multi-electron atom presents great difficulties. The nucleus of a neutral multi-electron atom of atomic number Z carries a charge of $+Ze$ and is surrounded by a cloud of Z electrons. Each electron is influenced by the electric fields produced by the positive nuclear charge and the other $(Z - 1)$ negative electrons in the atom. The potential energy function, $U(r, \theta, \phi)$, that describes such a situation is so complex that a complete solution to the Schrödinger equation for such an atom is unattainable. The best that can be obtained is a series of estimates and approximations.

In making these estimates, it is reasonable to assume that in the ground state, each electron in the atom will occupy the lowest possible stationary state that it can. This begs the immediate question of why don't all the electrons in an atom pile into the lowest quantum state, the $1s$ state, where they would all share the set of quantum numbers $n = 1, l = 0, m_l = 0$ and $m_s = \pm 1/2$? That this is not the case is evident from the actual chemical and physical properties of the elements.

The chemistry of each of the three elements fluorine, neon and sodium is completely different even though they have almost the same number of electrons: 9, 10 and 11 electrons, respectively. Were all their electrons in the same $1s$ quantum state, the addition of one more or the subtraction of one would hardly explain the obvious differences in their chemical properties.

On the other hand, the elements helium, neon, argon, krypton, xenon and radon,[60] that have 2, 10, 18, 36, 54 and 86 electrons respectively, show striking similarities in all their properties, despite the considerable differences in the number of their electrons. Clearly, the deployment of electrons in the atom involves more than just apportioning them all to the lowest quantum state.

The solution to this problem was found in the framework of the following principle – the *exclusion principle* – formulated by Pauli in 1925 as a result of his researches in atomic spectroscopy:

> **In a system containing a number of electrons, no two electrons can occupy the same state; each electron must have a different set of quantum numbers.**

Thus, allotting all the electrons in a multi-electron atom to the $1s$ state or to any other single state is forbidden. The electrons must be arranged in the atom such that no two share the same set of quantum numbers; each electron in the atom must have its own unique set of the four quantum numbers, n, l, m_l and m_s. Table 3.4 lists all the different atomic states (sub-levels) that can be specified by these four quantum numbers for values of the principle quantum number from $n = 1$ to $n = 5$.

In the ground state of hydrogen, $H(Z = 1)$, the first element in the periodic table – Table 3.5 – the sole electron occupies the $1s$ state; $n = 1, l = 0, m_l = 0$. Its spin may be up or down.

The ordering of the electrons in the ground state of multi-electron atoms starts with the lowest energy state, the $1s$ state, and moves progressively from there up the energy scale until each of the atom's electrons has been assigned a unique set of quantum numbers. Thus, in the ground state, both electrons in the inert gas helium ($Z = 2$), He, occupy the $1s$ state, but as required by the exclusion principle, each has a different spin quantum number; their spins are 'paired'. The respective quantum numbers of the helium electrons are

[60] The inert or 'noble' gases.

	n	l	m_l	m_s
electron 1	1	0	0	$+\frac{1}{2}$
electron 2	1	0	0	$-\frac{1}{2}$

The electronic configuration of helium is designated by $1s^2$, i.e. two electrons in the $1s$ level.

Table 3.4. The electronic quantum states in an atom.

n	l	Spectroscopic symbol	m_l	Number of states[61]
1	0	$1s$	0	2
2	0	$2s$	0	2
	1	$2p$	$-1, 0, +1$	6
3	0	$3s$	0	2
	1	$3p$	$-1, 0, +1$	6
	2	$3d$	$-2, -1, 0, +1, +2$	10
4	0	$4s$	0	2
	1	$4p$	$-1, 0, +1$	6
	2	$4d$	$-2, -1, 0, +1, +2$	10
	3	$4f$	$-3, -2, -1, 0, +1, +2, +3$	14
5	0	$5s$	0	2
	1	$5p$	$-1, 0, +1$	6
	2	$5d$	$-2, -1, 0, +1, +2$	10
	3	$5f$	$-3, -2, -1, 0, +1, +2, +3$	14
	4	$5g$	$-4, -3, -2, -1, 0, +1, +2, +3, +4$	18

No more than two electrons can occupy the $1s$ state; a larger number is precluded by the exclusion principle. Thus, the third electron in lithium, Li ($Z = 3$), must be assigned to the level above, to the $2s$ state. The electronic configuration of lithium is $1s^2 2s^1$, i.e. two electrons in the $1s$ state and one in the $2s$ state. The two electrons in the $1s$ state constitute a complete or closed shell; this is the innermost K shell ($n = 1$). The single $2s$ electron is shielded from the nuclear charge by the closed K shell. Consequently, the effective nuclear charge binding it to the atom is much less than $+3e$ and its *first ionisation potential*[62] is correspondingly low, just 5.4 eV. In general, an atom's chemistry is determined by the configuration of its outermost electrons, i.e. by those outside closed shells. Lithium's chemical valency of $+1$ ensues from its single loosely bound $2s$ electron.

[61] Each of the m_l states can be occupied by an electron with spin up or down. Thus, the total number of possible states is twice the number of m_l states.

[62] The amount of energy that must be invested in the neutral atom to remove one electron from it.

Table 3.5. The periodic table of elements.

Group	I	II											III	IV	V	VI	VII	VIII
Period	1																	2
1	H																	He
2	3	4											5	6	7	8	9	10
	Li	Be											B	C	N	O	F	Ne
3	11	12											13	14	15	16	17	18
	Na	Mg											Al	Si	P	S	Cl	Ar
4	19	20	21	22	23	24	25	26	27	28	29	30	31	32	33	34	35	36
	K	Ca	Sc	Ti	V	Cr	Mn	Fe	Co	Ni	Cu	Zn	Ga	Ge	As	Se	Br	Kr
5	37	38	39	40	41	42	43	44	45	46	47	48	49	50	51	52	53	54
	Rb	Sr	Y	Zr	Nb	Mo	Tc	Ru	Rh	Pd	Ag	Cd	In	Sn	Sb	Te	I	Xe
6	55	56	57–	72	73	74	75	76	77	78	79	80	81	82	83	84	85	86
	Cs	Ba	71	Hf	Ta	W	Re	Os	Ir	Pt	Au	Hg	Tl	Pb	Bi	Po	At	Rn
7	87	88	89–	104	105	106	107	108	109									
	Fr	Ra	103	Rf	Ha	Sg	Bh	Hs	Mt									

Lanthanides	57	58	59	60	61	62	63	64	65	66	67	68	69	70	71
	La	Ce	Pr	Nd	Pm	Sm	Eu	Gd	Tb	Dy	Ho	Er	Tm	Yb	Lu
Actinides	89	90	91	92	93	94	95	96	97	98	99	100	101	102	103
	Ac	Th	Pa	U	Np	Pu	Am	Cm	Bk	Cf	Es	Fm	Md	No	Lr

The electronic configuration of the four electrons in beryllium, Be ($Z = 4$), is $1s^2 2s^2$. Because of the exclusion principle, the extra electron in boron, B ($Z = 5$), cannot be added to the $2s$ sub-shell; it must be placed in the $2p(l = 1, m_l = -1, 0, 1)$ sub-shell, which can accommodate a total of six electrons. As a rule, the three m_l states of a p sub-shell are progressively filled in such a way as leaves as many unpaired electrons as possible; this is known as *Hund's rule* (Table 3.6).

Table 3.6. Hund's rule.

Atom	Number of electrons	Electronic configuration	Spin of the p electrons
B	5	$1s^2 2s^2 2p^1$	↓
C	6	$1s^2 2s^2 2p^2$	↓ ↓
N	7	$1s^2 2s^2 2p^3$	↓ ↓ ↓
O	8	$1s^2 2s^2 2p^4$	↓↑↓ ↓
F	9	$1s^2 2s^2 2p^5$	↓↑↓↑↓
Ne	10	$1s^2 2s^2 2p^6$	↓↑↓↑↓↑

The inert gas neon, Ne ($Z = 10$), marks the end of the second period in the periodic table of elements. In neon the L shell ($n = 2$) is complete; all its states are occupied by paired electrons. Possessing no electrons outside a closed shell, the element neon is chemically inert. The stability and low energy of a closed electronic configuration is evinced by the neon atom's relatively high first ionisation potential of 21.5 eV. Fluorine, F, the element before neon in the periodic table, is short of just one electron to complete the L shell. The fluorine atom's *electron affinity*[63] of −3.4 eV is further evidence of the low energy associated with a closed shell.

Since there is no change in the shielding of the nuclear charge by the closed K shell as it increases from $+3e$ in lithium to $+10e$ in neon, the effective nuclear charge that acts on the outer L shell electrons increases. Thus, the outer electrons are less tightly bound in lithium than in neon and the first ionisation potential rises from 5.4 eV in lithium to 21.5 eV in neon.

The third and fourth periods commence with the elements sodium, Na, and potassium, K, respectively:

Na	11 electrons	$1s^2 2s^2 2p^6 3s^1$
K	19 electrons	$1s^2 2s^2 2p^6 3s^2 3p^6 4s^1$

Both sodium and potassium have low first ionisation potentials; 5.2 eV and 4.3 eV respectively. This is attributed to the screening of the nuclear charge by the closed inner shells, as in lithium. Since the magnetic quantum number, m_l, can take five discrete values when $l = 2$, the d sub-level that becomes available in the M shell ($n = 3$), can have a maximum population of ten electrons. However, in multi-electron atoms the energies of the electronic levels do not follow the same sequence as in hydrogen; they do not depend exclusively on the principal quantum number, n. As Fig. 3.30 shows, the energy of the $4s$ level in multi-electron atoms is, in fact, lower than that of the $3d$ level. Thus, electrons are introduced into the $4s$ level before they are assigned to the $3d$ level. As a result, only the $3s$ and $3p$ levels are filled across the third period and it comprises just eight elements, like the second period. It is only in the fourth period that the $3d$ levels are progressively filled in the group of elements from scandium, Sc, to zinc, Zn. These ten elements constitute the first row of the *transition metals* or *d-block elements* in the periodic table.

A similar situation arises in the N shell ($n = 4$). Since the magnetic quantum number, m_l, can take seven discrete values when $l = 3$, the f sub-level that

[63]The amount of energy emitted when a fluorine atom becomes a negative ion, F−.

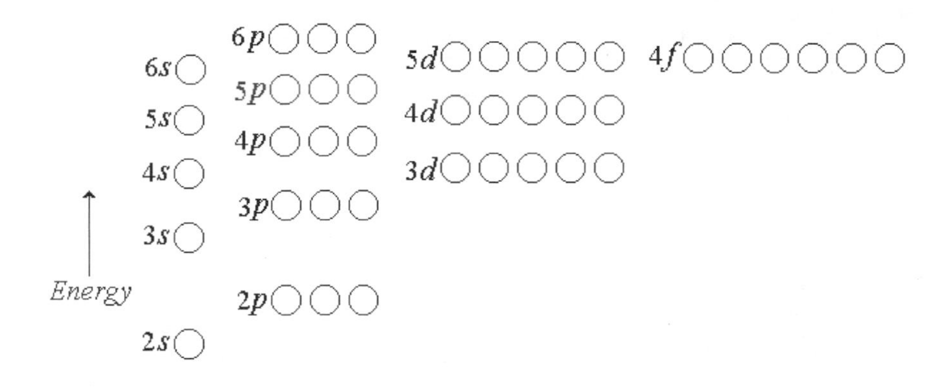

Fig. 3.30. The notional sequence of energy levels in multi-electron atoms. $1s < 2s < 2p < 3s < 3p < 4s < 3d < 4p < 5s < 4d < 5p < 6s < 4f < 5d < 6p < 7s$. Each such quantum state can be occupied by a pair of electrons possessing opposite spins. In highly ionised species the order may differ due to a change in the effective nuclear charge.

becomes available in this shell can have a maximum population of 14 electrons. However, the energy of the $4f$ level is greater even than that of the $6s$ level and so it is only in the sixth period that the $4f$ levels are progressively filled in the group of elements from cerium, Ce, to lutetium, Lu, called the *rare-earths* or *lanthanides.* As a result, the fifth period – the O ($n = 5$) shell – like the fourth period, contains 18 elements of which ten are the second row of d-block elements.

The sixth period – the P shell ($n = 6$) – contains 32 elements. These include the 14 rare earths and the third row of d-block elements, apart from the two $6s$ and six $6p$ elements. Since the chemical properties of an element are largely determined by its outermost electrons, the elements in each period of d-block elements are chemically similar, the main difference between them being in the number of electrons they possess in the shell below that of their valence electrons. The phenomenon is even more marked in the case of the rare-earth elements; they are almost chemically indistinguishable. The only difference between them is in the number of electrons they possess in the $4f$ level, i.e. in a level that is two principal quantum numbers lower that that of their valence electrons.

The seventh period – the Q shell ($n = 7$) – is incomplete; there are just not enough elements to fill it. The highest naturally occurring atomic number, Z, is 92, that of the element uranium, U. A further 17 elements – *transuranic*

Table 3.7. The electronic configuration of the elements.

				K	L		M			N				O				P			Q	
Period				*1s*	*2s*	*2p*	*3s*	*3p*	*3d*	*4s*	*4p*	*4d*	*4f*	*5s*	*5p*	*5d*	*5f*	*6s*	*6p*	*6d*	*7s*	
1	1	Hydrogen	H	1																		
	2	Helium	He	2																		
2	3	Lithium	Li	2	1																	
	4	Beryllium	Be	2	2																	
	5	Boron	B	2	2	1																
	6	Carbon	C	2	2	2																
	7	Nitrogen	N	2	2	3																
	8	Oxygen	O	2	2	4																
	9	Fluorine	F	2	2	5																
	10	Neon	Ne	2	2	6																
3	11	Sodium	Na	2	2	6	1															
	12	Magnesium	Mg	2	2	6	2															
	13	Aluminium	Al	2	2	6	2	1														
	14	Silicon	Si	2	2	6	2	2														
	15	Phosphorus	P	2	2	6	2	3														
	16	Sulphur	S	2	2	6	2	4														
	17	Chlorine	Cl	2	2	6	2	5														
	18	Argon	Ar	2	2	6	2	6														
4	19	Potassium	K	2	2	6	2	6		1												
	20	Calcium	Ca	2	2	6	2	6		2												
	21	Scandium	Sc	2	2	6	2	6	1	2												
	22	Titanium	Ti	2	2	6	2	6	2	2												
	23	Vanadium	V	2	2	6	2	6	3	2												
	24	Chromium	Cr	2	2	6	2	6	5	1												
	25	Magnesium	Mn	2	2	6	2	6	5	2												d-block
	26	Iron	Fe	2	2	6	2	6	6	2												elements
	27	Cobalt	Co	2	2	6	2	6	7	2												
	28	Nickel	Ni	2	2	6	2	6	8	2												
	29	Copper	Cu	2	2	6	2	6	10	1												
	30	Zinc	Zn	2	2	6	2	6	10	2												
	31	Gallium	Ga	2	2	6	2	6	10	2	1											
	32	Germanium	Ge	2	2	6	2	6	10	2	2											
	33	Arsenic	As	2	2	6	2	6	10	2	3											
	34	Selenium	Se	2	2	6	2	6	10	2	4											
	35	Bromine	Br	2	2	6	2	6	10	2	5											
	36	Krypton	Kr	2	2	6	2	6	10	2	6											
5	37	Rubidium	Rb	2	2	6	2	6	10	2	6			1								
	38	Strontium	Sr	2	2	6	2	6	10	2	6			2								
	39	Yttrium	Y	2	2	6	2	6	10	2	6	1		2								
	40	Zirconium	Zr	2	2	6	2	6	10	2	6	2		2								
	41	Niobium	Nb	2	2	6	2	6	10	2	6	4		1								
	42	Molybdenum	Mo	2	2	6	2	6	10	2	6	5		1								
	43	Technetium	Tc	2	2	6	2	6	10	2	6	5		2								d-block
	44	Ruthenium	Ru	2	2	6	2	6	10	2	6	7		1								elements
	45	Rhodium	Rh	2	2	6	2	6	10	2	6	8		1								
	46	Palladium	Pd	2	2	6	2	6	10	2	6	10										
	47	Silver	Ag	2	2	6	2	6	10	2	6	10		1								
	48	Cadmium	Cd	2	2	6	2	6	10	2	6	10		2								
	49	Indium	In	2	2	6	2	6	10	2	6	10		2	1							
	50	Tin	Sn	2	2	6	2	6	10	2	6	10		2	2							
	51	Antimony	Sb	2	2	6	2	6	10	2	6	10		2	3							
	52	Tellurium	Te	2	2	6	2	6	10	2	6	10		2	4							
	53	Iodine	I	2	2	6	2	6	10	2	6	10		2	5							
	54	Xenon	Xe	2	2	6	2	6	10	2	6	10		2	6							

(*Continued*)

Table 3.7. (*Continued*)

Period				K	L		M			N				O				P			Q	
				1s	2s	2p	3s	3p	3d	4s	4p	4d	4f	5s	5p	5d	5f	6s	6p	6d	7s	
6	55	Caesium	Cs	2	2	6	2	6	10	2	6	10		2	6			1				
	56	Barium	Ba	2	2	6	2	6	10	2	6	10		2	6			2				
	57	Lanthanum	La	2	2	6	2	6	10	2	6	10		2	6	1		2				
	58	Cerium	Ce	2	2	6	2	6	10	2	6	10	2	2	6			2				
	59	Praseodymium	Pr	2	2	6	2	6	10	2	6	10	3	2	6			2				
	60	Neodymium	Nd	2	2	6	2	6	10	2	6	10	4	2	6			2				
	61	Promethium	Pm	2	2	6	2	6	10	2	6	10	5	2	6			2				
	62	Samarium	Sm	2	2	6	2	6	10	2	6	10	6	2	6			2				
	63	Europium	Eu	2	2	6	2	6	10	2	6	10	7	2	6			2				
	64	Gadolinium	Gd	2	2	6	2	6	10	2	6	10	7	2	6	1		2				Lanthanides
	65	Terbium	Tb	2	2	6	2	6	10	2	6	10	9	2	6			2				(rare-
	66	Dysprosium	Dy	2	2	6	2	6	10	2	6	10	10	2	6			2				earths)
	67	Holmium	Ho	2	2	6	2	6	10	2	6	10	11	2	6			2				
	68	Erbium	Er	2	2	6	2	6	10	2	6	10	12	2	6			2				
	69	Thulium	Tm	2	2	6	2	6	10	2	6	10	13	2	6			2				
	70	Ytterbium	Yb	2	2	6	2	6	10	2	6	10	14	2	6			2				
	71	Lutetium	Lu	2	2	6	2	6	10	2	6	10	14	2	6	1		2				
	72	Hafnium	Hf	2	2	6	2	6	10	2	6	10	13	2	6	2		2				
	73	Tantalum	Ta	2	2	6	2	6	10	2	6	10	14	2	6	3		2				
	74	Tungsten	W	2	2	6	2	6	10	2	6	10	14	2	6	4		2				
	75	Rhenium	Re	2	2	6	2	6	10	2	6	10	13	2	6	5		2				d-block
	76	Osmium	Os	2	2	6	2	6	10	2	6	10	14	2	6	6		2				elements
	77	Iridium	Ir	2	2	6	2	6	10	2	6	10	14	2	6	7		2				
	78	Platinum	Pt	2	2	6	2	6	10	2	6	10	13	2	6	9		1				
	79	Gold	Au	2	2	6	2	6	10	2	6	10	14	2	6	10		1				
	80	Mercury	Hg	2	2	6	2	6	10	2	6	10	14	2	6	10		2				
	81	Thallium	Tl	2	2	6	2	6	10	2	6	10	14	2	6	10		2	1			
	82	Lead	Pb	2	2	6	2	6	10	2	6	10	14	2	6	10		2	2			
	83	Bismuth	Bi	2	2	6	2	6	10	2	6	10	14	2	6	10		2	3			
	84	Polonium	Po	2	2	6	2	6	10	2	6	10	14	2	6	10		2	4			
	85	Astatine	At	2	2	6	2	6	10	2	6	10	14	2	6	10		2	5			
	86	Radon	Rn	2	2	6	2	6	10	2	6	10	14	2	6	10		2	6			
7	87	Francium	Fr	2	2	6	2	6	10	2	6	10	14	2	6	10		2	6		1	
	88	Radium	Ra	2	2	6	2	6	10	2	6	10	14	2	6	10		2	6		2	
	89	Actinium	Ac	2	2	6	2	6	10	2	6	10	14	2	6	10		2	6	1	2	
	90	Thorium	Th	2	2	6	2	6	10	2	6	10	14	2	6	10		2	6	2	2	
	91	Protactinium	Pa	2	2	6	2	6	10	2	6	10	14	2	6	10	2	2	6	1	2	
	92	Uranium	U	2	2	6	2	6	10	2	6	10	14	2	6	10	3	2	6	1	2	
	93	Neptunium	Np	2	2	6	2	6	10	2	6	10	14	2	6	10	4	2	6	1	2	
	94	Plutonium	Pu	2	2	6	2	6	10	2	6	10	14	2	6	10	5	2	6	1	2	
	95	Americium	Am	2	2	6	2	6	10	2	6	10	14	2	6	10	6	2	6	1	2	
	96	Curium	Cm	2	2	6	2	6	10	2	6	10	14	2	6	10	7	2	6	1	2	
	97	Berkelium	Bk	2	2	6	2	6	10	2	6	10	14	2	6	10	8	2	6	1	2	
	98	Californium	Cf	2	2	6	2	6	10	2	6	10	14	2	6	10	10	2	6		2	
	99	Einsteinium	Es	2	2	6	2	6	10	2	6	10	14	2	6	10	11	2	6		2	
	100	Fermium	Fm	2	2	6	2	6	10	2	6	10	14	2	6	10	12	2	6		2	
	101	Mendelevium	Md	2	2	6	2	6	10	2	6	10	14	2	6	10	13	2	6		2	
	102	Nobelium	No	2	2	6	2	6	10	2	6	10	14	2	6	10	14	2	6		2	
	103	Lawrencium	Lw	2	2	6	2	6	10	2	6	10	14	2	6	10	14	2	6	1	2	
	104	Rutherfordium	Rf	2	2	6	2	6	10	2	6	10	14	2	6	10	14	2	6	2	2	
	105	Dubnium	Db	2	2	6	2	6	10	2	6	10	14	2	6	10	14	2	6	3	2	
	106	Seaborgium	Sg	2	2	6	2	6	10	2	6	10	14	2	6	10	14	2	6	4	2	
	107	Bohrium	Bh	2	2	6	2	6	10	2	6	10	14	2	6	10	14	2	6	5	2	
	108	Hassium	Hs	2	2	6	2	6	10	2	6	10	14	2	6	10	14	2	6	6	2	
	109	Meitnerium	Mt	2	2	6	2	6	10	2	6	10	14	2	6	10	14	2	6	7	2	

elements – have been manufactured by nuclear processes,[64] all except plutonium, Pu, in just microscopic quantities. The group of elements from thorium, Th, to lawrencium, Lw, constitute the *actinides*. They are analogous to the lanthanides, differing from one another only in the number of $5f$ electrons they possess.

The electronic configurations of the 109 named naturally occurring and artificially manufactured elements[65] are shown in Table 3.7 on pp. 292–293.

3.4.5 Spin, Identical Particles and Pauli's Principle

In classical mechanics, intrinsically identical particles, such as all the red balls on a snooker table, can be distinguished from one another by means of their individual trajectories. Should the particles collide, their trajectories after the collision can be predicted, in principle if not always in practice, from the laws of classical mechanics. However, in quantum mechanics, particles cannot be assigned trajectories; they cannot be tracked. Quantum mechanics pictures collisions in terms of the overlapping (superposition) of the particles' wave functions. On this view, we can never be sure, after the collision, which particle is which. There is no way of knowing exactly how the composite wave function formed at the collision 'unwinds' itself afterwards. How then, if at all, can intrinsically identical particles be distinguished from one another in quantum mechanics?

In order to investigate this question, let us suppose that following a collision between two identical minute particles, P_1 and P_2, one appears at A and the other at B (Fig. 3.31). There are two different ways in which this can occur:

1. The particle that arrives at point A after the collision is particle P_1 and that which arrives at point B is P_2.
2. The particle that arrives at point A after the collision is particle P_2 and that which arrives at point B is P_1.

In accordance with the rules for describing events in terms of wave functions, the wave function, Ψ, that describes the collision is

$$\Psi = \Psi_1 + \Psi_2, \tag{3.82}$$

[64]The theory and method of their manufacture is explained in Chapter 5.3.

[65]The creation of minute quantities of further elements, $Z = 110$ to 118, has been reported, but controversy surrounds the naming of these elements. The naming of a chemical element is influenced by national pride, professional rivalry and personal sensitivities; the process is comparable to that of selecting an international beauty queen. The names given in Table 3.7 are those approved by the International Union of Pure and Applied Chemistry (IUPAC).

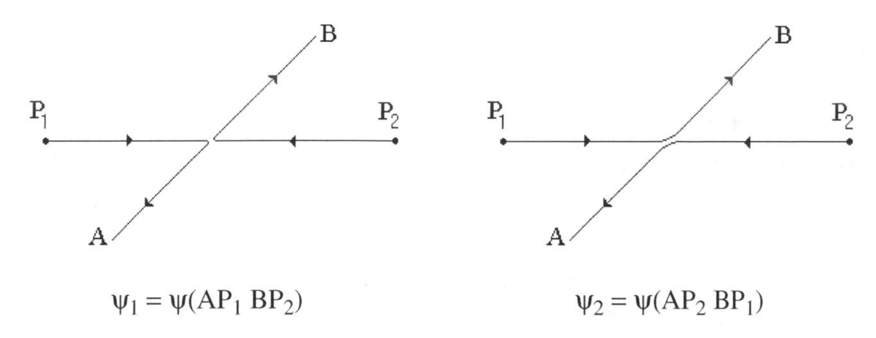

Fig. 3.31. A collision between two intrinsically identical particles after which one is found at point A and another at point B. One possibility is that particle 1 arrives at point A and particle 2 at point B; the other is that particle 1 arrives at point B and particle 2 at point A. Each of these two possibilities is characterised by its own wave function.

where $\Psi_1 = \Psi(\mathrm{AP}_1, \mathrm{BP}_2)$ and $\Psi_2 = \Psi(\mathrm{AP}_2, \mathrm{BP}_1)$ are the wave functions of the two alternative ways the event can occur. Both wave functions express the possibility of finding one particle at A and another at B after the collision.

Because the particles are identical (exchangeable), the two possible outcomes of the collision, $(\mathrm{AP}_1, \mathrm{BP}_2)$ and $(\mathrm{AP}_2, \mathrm{BP}_1)$, are indistinguishable and equally probable. Consequently, their probability densities must be equal:

$$|\Psi(\mathrm{AP}_1, \mathrm{BP}_2)|^2 = |\Psi(\mathrm{AP}_2, \mathrm{BP}_1)|^2. \tag{3.83}$$

There are only two ways that the wave functions, $\Psi_1 = \Psi(\mathrm{AP}_1, \mathrm{BP}_2)$ and $\Psi_2 = \Psi(\mathrm{AP}_2, \mathrm{BP}_1)$, can be related so as to satisfy Eq. (3.83).

1. The two wave functions are in phase with each other:

$$\Psi(\mathrm{AP}_1, \mathrm{BP}_2) = \Psi(\mathrm{AP}_2, \mathrm{BP}_1). \tag{3.84}$$

In this case the wave function, Ψ_S, that describes the collision is

$$\Psi_S = \frac{1}{\sqrt{2}}[\Psi(\mathrm{AP}_1, \mathrm{BP}_2) + \Psi(\mathrm{AP}_2, \mathrm{BP}_1)]. \tag{3.85}$$

This type of function is called a *symmetric wave function.*

2. The two wave functions are out of phase with each other, there being a phase difference of π between them:

$$\Psi(\mathrm{AP}_1, \mathrm{BP}_2) = -\Psi(\mathrm{AP}_2, \mathrm{BP}_1). \tag{3.86}$$

In this case the wave function, Ψ_A, describing the collision is

$$\Psi_A = \frac{1}{\sqrt{2}}[\Psi(\mathrm{AP}_1, \mathrm{BP}_2) - \Psi(\mathrm{AP}_2, \mathrm{BP}_1)]. \tag{3.87}$$

This type of function is called an *anti-symmetric wave function.*

In both cases the factor $1/\sqrt{2}$ is required to normalise the wave function.

Whether the wave function describing an event is symmetric or anti-symmetric depends on the nature of the particles. Those particles for which the appropriate wave functions are symmetric, like Eq. (3.85), are called *bosons*; those for which they are anti-symmetric, like Eq. (3.87), are called *fermions.*

Is there any chance that both particles might actually finish up at the same point, say at the point A, after the collision? Strange as this may appear to those accustomed to picturing all collisions between particles in terms of snooker or billiards, in quantum mechanics there is no *a priori* reason why this should not happen; the overlapping and subsequent unwinding of the wave functions does not of necessity preclude this outcome.

For example, the symmetric wave function, Ψ_S, does not collapse, i.e. it does not reduce to zero ($\Psi_S \neq 0$), if 'A' is substituted in place of 'B' in Eq. (3.85):

$$\Psi_A = \frac{1}{\sqrt{2}}\left[\Psi(\mathrm{AP}_1, \mathrm{AP}_2) + \Psi(\mathrm{AP}_2, \mathrm{AP}_1)\right] = 0. \qquad (3.88)$$

And so, strange as it may appear, there is a certain probability that after a collision between two bosons they will both turn up at the same place. Putting this another way, there is nothing to prevent two or even more bosons from simultaneously occupying the same state.

On the other hand, the anti-symmetric, Ψ_A, wave function does collapse if 'A' is substituted in place of 'B' in Eq. (3.83):

$$\Psi_A = \frac{1}{\sqrt{2}}\left[\Psi(\mathrm{AP}_1, \mathrm{AP}_2) - \Psi(\mathrm{AP}_2, \mathrm{AP}_1)\right] = 0. \qquad (3.89)$$

Thus, there is a zero probability of the event taking place in this way; two identical fermions can never simultaneously occupy the same state. However, this is exactly the import of Pauli's exclusion principle:

In a system containing a number of electrons, no two electrons can occupy the same state; each electron must have a different set of quantum numbers.

Let us suppose that two electrons collide. According to Pauli's principle, they cannot both finish up in the same state. This is equivalent to saying that the wave function which describes the collision must be an anti-symmetric wave function, Ψ_A. It follows that electrons are fermions and that Pauli's exclusion principle can be expressed alternatively as requiring that the wave function, which describes the electrons in a system, must be anti-symmetric.

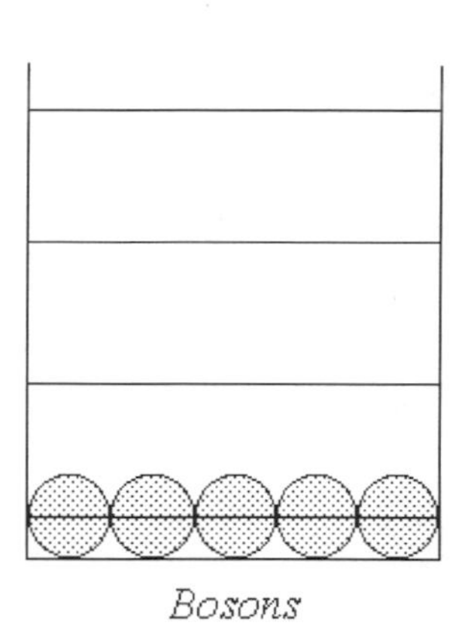

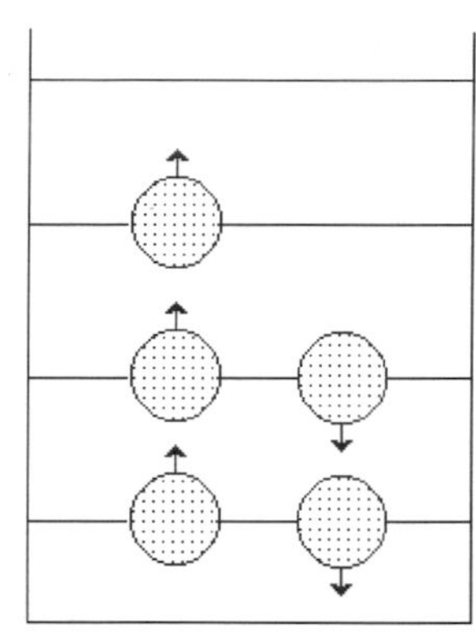

Fig. 3.32. Bosons and fermions in a quantum box. Each system tends to minimum energy. The system of bosons can achieve this by placing all the bosons in the ground state. However, the system of fermions can allocate a pair of particles at most to each of the energy levels.

In general, particles whose spin is an integer $(0, 1, 2, \dots)$ are bosons: photons, α particles and helium atoms are bosons. Pauli's exclusion principle does not apply to them, and so there is no restriction on the number of bosons that can occupy the ground state.

On the other hand, particles whose spin is an odd multiple of a half $(\frac{1}{2}, \frac{3}{2}, \frac{5}{2}, \dots)$ are fermions: electrons, protons and neutrons are fermions. Pauli's exclusion principle does apply to them, and the number of fermions that can occupy any state is restricted. Thus, in multi-electron atoms, higher electronic energy states must also be occupied (Fig. 3.32).

In order to determine the distribution of a system's particles between its energy levels, different statistical functions must be used depending on the type of particle they are: classical particles, bosons or fermions.

Since, under ordinary conditions, the average distance between the atoms and molecules of a gas is greater than the wavelength of the matter wave associated with them, they can be distinguished by means of their trajectories. It follows that they can be described by classical physics. It can be shown that the probability, $f(E_i)$, of a classical particle occupying the E_i energy level at an absolute temperature T is given by

$$f(E_i) = (Ae^{E_i/kT})^{-1}, \tag{3.90}$$

where the parameter A depends on the total number of particles in the system. This is the classic Maxwell–Boltzman distribution. In such systems there are no restrictions on the number of particles that can occupy a particular state, nor does the presence of a particle in a particular state have any effect on the other particles in the system.

The statistical function which describes a system of bosons is called *Bose–Einstein*[66] *statistics.* The probability, $f(E_i)_{Boson}$, of a boson occupying the E_i energy level at an absolute temperature T is given by

$$f(E_i)_{Boson} = (Be^{E_i/kT} - 1)^{-1}, \tag{3.91}$$

where the parameter B depends on the nature of the system. In a system of bosons there are no restrictions on the number of particles that can occupy a particular state. The presence of a particle in a particular state even increases the probability that other particles will be found in the same state. Consequently, there is a tendency for the lowest energy states to be occupied, and at very low temperatures most of the particles will be found in the ground state.

The photon gas that fills a constant-temperature cavity is an example of a system of bosons; the spectrum of black-body radiation can be explained in terms of the distribution of the photons in the cavity between the various energy levels as required by the Bose–Einstein function. Helium atoms can also behave as bosons; liquid helium, at temperatures below 2.18 K, exhibits strange properties that stem from this. At this very low temperature, $f(E_i)_{Boson} \to 0$ for all but the lowest energy level. The higher levels are, in effect, completely empty and all the helium atoms 'condense' into the lowest energy level where they cooperate with one another. This gives the liquid strange properties. It shows no viscosity and flows out of an open container by spontaneously climbing up its inside walls.

The statistical function which describes a system of fermions is called *Fermi–Dirac statistics.* In a system of fermions only one particle, and no more, can occupy each state, and so the presence of a particle in a particular state prevents other particles from being in that state. The probability, $f(E_i)_{Fermion}$, of a fermion occupying the E_i energy level at an absolute temperature T is given by

$$f(E_i)_{Fermion} = (Ce^{E_i/kT} + 1)^{-1}, \tag{3.92}$$

where the parameter C depends on the nature and temperature of the system. The electron gas in metal conductors is an example of a system of fermions.[67]

3.4.6 Total Spin and the Energy Levels in Atoms

Whereas in the elucidation of the structure of the periodic table we considered each atomic electron individually, it is often useful, in other instances, to regard

[66] Satyendranath Bose 1894–1974, Indian physicist.
[67] The electron gas in metals is discussed in some detail in Chapter 6.4.

the constituent electrons of an atom or molecule as a group or a whole. The key to this approach is the concept of *total spin.*

The total intrinsic angular momentum, i.e. the total spin, **S**, of an atom (or molecule) is given by the vector sum of the spins of all its electrons:

$$\mathbf{S} = \mathbf{S}_1 + \mathbf{S}_2 + \mathbf{S}_3 + \ldots, \tag{3.93}$$

where the eigenvalues of the square, S^2, of its magnitude are given by

$$S^2 = \mathsf{S}(\mathsf{S}+1)\hbar^2. \tag{3.94}$$

S is called the *total spin quantum number.*[68]

The component, S_Z, of the total spin parallel to an arbitrarily chosen Z-axis is given by

$$S_Z = M_{\mathsf{S}}\hbar, \tag{3.95}$$

where the quantum number M_{S} is equal to the sum of the spin magnetic quantum numbers, $m_{s_1}, m_{s_2} \ldots m_{s_N}$, of the N constituent electrons of the atom or molecule:

$$M_{\mathsf{S}} = m_{s_1} + m_{s_2} + \cdots + m_{s_N}. \tag{3.96}$$

The total spin quantum number, S, is subject to the following three rules:

1. a. If the number, N, of the constituent electrons is even, the total spin quantum number S is an integer: $\mathsf{S} = 0, 1, 2, \ldots$;
   b. if N is an odd number, the total spin quantum number S is an odd multiple of a half: $\mathsf{S} = \frac{1}{2}, \frac{3}{2}, \frac{5}{2} \ldots$
2. Because of Pauli's principle, the total spin quantum number of a closed shell or sub-shell, i.e. one that contains its full quota of electrons, must be zero: $\mathsf{S}_{Closed\ shell} = 0$.
3. For any number, N', of electrons outside a closed shell, the highest possible value for S is $N'/2$.

It follows that the only possible total spin quantum numbers for a pair of electrons outside a closed shell ($N' = 2$) are $\mathsf{S} = 1$ or $\mathsf{S} = 0$. In those cases where $\mathsf{S} = 0$, the spins of the two electrons are anti-parallel, i.e. one is directed upwards and the other downwards; their resultant is then zero. This is called a *singlet* state, since it allows for only one possible value for the quantum number M_{S}, namely, $M_{\mathsf{S}} = 0$ (Fig. 3.33a).

When $\mathsf{S} = 1$, the possible values of M_{S} are $-1, 0$, and $+1$; this is called a *triplet* state because it allows for three different values of M_{S}. In the $\mathsf{S} = 1$,

[68] It is customary to designate the quantum numbers of total spin by capital letters; a distinguishing font, S, is used to avoid confusion with the magnitude, S, of the total angular momentum.

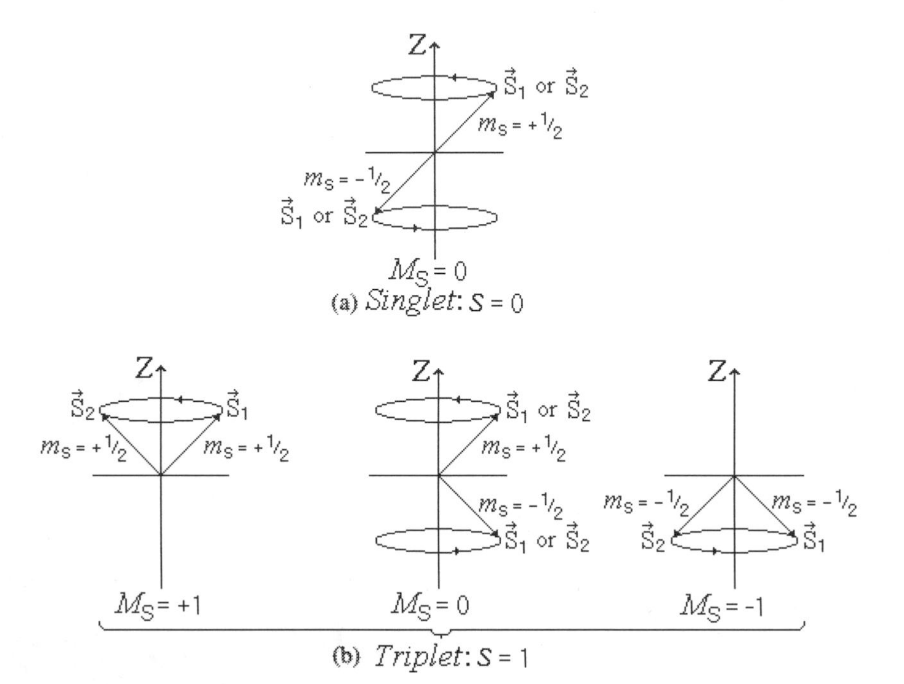

Fig. 3.33. The spin states of a pair of electrons. (a) $S = 0$; singlet state. The spins of the two electrons are anti-parallel. (b) $S = 1$; triplet state. In the $M_S = 0$ state the spins of the electrons will be anti-parallel; in the $M_S = -1$ and $M_S = 1$ states the electron spins will be parallel.

$M_S = 0$ state the electron spins are anti-parallel; in the $S = 1$, $M_S = -1$ and $S = 1$, $M_S = +1$ states they are parallel. Thus, in an external magnetic field, each $S = 1$ energy level splits into three states which show up as fine structure in the spectrum (Fig. 3.33b).

The *total spin function* of the pair of electrons comprising the singlet and triplet states can be expressed in terms of the one-electron spin wave functions, $\chi_{+\frac{1}{2}}$ or $\chi_{-\frac{1}{2}}$, that correspond, respectively, to the $m_s = +\frac{1}{2}$ and $m_s = -\frac{1}{2}$ states. Since the electrons are indistinguishable, these wave functions can be equally ascribed to both, such that those of the one can be designated $\chi_{+\frac{1}{2}}(e_1)$ or $\chi_{-\frac{1}{2}}(e_1)$ and those of the other $\chi_{+\frac{1}{2}}(e_2)$ or $\chi_{-\frac{1}{2}}(e_2)$. Whether the total spin function of the pair of electrons is symmetric, χ_S, or anti-symmetric, χ_A, depends on how their individual spin functions are combined.

In the singlet state the spins of the pair of electrons are always opposed. Hence, the total spin function of the singlet state must be anti-symmetric:

$$\chi_A = \frac{1}{\sqrt{2}}[\chi_{+\frac{1}{2}}(e_1)\chi_{-\frac{1}{2}}(e_2) - \chi_{+\frac{1}{2}}(e_2)\chi_{-\frac{1}{2}}(e_1)] \quad M_S = 0. \tag{3.97}$$

In the triplet state the three possible spin wave functions are all symmetric:

$$\chi_S = \begin{cases} \chi_{+\frac{1}{2}}(e_1)\chi_{+\frac{1}{2}}(e_2) & M_S = +1 \\ \frac{1}{\sqrt{2}}\left[\chi_{+\frac{1}{2}}(e_1)\chi_{-\frac{1}{2}}(e_2) + \chi_{+\frac{1}{2}}(e_2)\chi_{-\frac{1}{2}}(e_1)\right] & M_S = 0 \\ \chi_{-\frac{1}{2}}(e_1)\chi_{-\frac{1}{2}}(e_2) & M_S = -1 \end{cases} . \tag{3.98}$$

3.4.7 The Energy Levels in Multi-electron Atoms

Whereas the total energy of the single electron in a hydrogen atom depends just on the potential energy of its attractive interaction with the nucleus and its kinetic energy, that of the electrons in multi-electron atoms is also affected by the repulsive electrostatic interaction between the electrons themselves. This repulsive interaction raises the total energy of the atom.

The magnitude of the repulsive interaction between the electrons in multi-electron atoms depends on two factors:

1. the atom's total spin quantum number, S;
2. their spatial separation (the average distance between them).

Each of these factors can be represented by a wave function: the former by the total spin wave function, χ; the latter by the wave function, ϕ, usually referred to as the *spatial wave function* or *orbital*, for short. These wave functions can be symmetric, χ_S and ϕ_S, or anti-symmetric, χ_A and ϕ_A. In a triplet state, the spin wave function will be symmetric; in a singlet state, anti-symmetric. In like manner, if the electrons can share the same position in space, the spatial wave function – the orbital – will be symmetric; if they cannot, it will be anti-symmetric.

In accordance with the rules for describing quantum-mechanical systems, the overall wave function that describes the atomic electrons is given by the product, $\psi = \phi \times \chi$. In principle, this gives four possibilities for the overall wave function:

	Symmetric orbital, ϕ_S	Anti-symmetric orbital, ϕ_A
Singlet ($S = 0$)	$\psi_A = \phi_S \times \chi_A$	$\psi_S = \phi_A \times \chi_A$
Triplet ($S = 1$)	$\psi_S = \phi_S \times \chi_S$	$\psi_A = \phi_A \times \chi_S$

The overall wave function will be symmetric, ψ_S, if both factors are symmetric or both are anti-symmetric; it will be anti-symmetric, ψ_A, if one is symmetric and the other anti-symmetric.

Because they are fermions, the overall wave function that incorporates the repulsive interaction between the electrons in a multi-electron atom must be an anti-symmetric wave function. Clearly, in practice the only two possibilities are

$$\psi_{Triplet} = \phi_A \times \chi_S \tag{3.99}$$

and

$$\psi_{Singlet} = \phi_S \times \chi_A. \tag{3.100}$$

The simplest multi-electron atom is helium. In its ground state, the two electrons possessed by the atom share three identical quantum numbers, $n = 1$, $l = 0$ and $m_l = 0$. The electrons constitute a closed shell and in these circumstances (Rule 2 above), Pauli's principle does not permit a spin of $S = 1$. Their fourth quantum number – the spin magnetic quantum number – must distinguish between them: that of the one must be $m_s = +\frac{1}{2}$ and that of the other $m_s = -\frac{1}{2}$. As a result, the total spin magnetic quantum number must at all times be zero:

$$M_S = m_{s1} + m_{s2} = {}^1\!/_2 + \left(-{}^1\!/_2\right) = 0. \tag{3.101}$$

This restricts the ground state to the singlet state in which $S = 0$.

No such restriction applies to the excited states in the helium atom. Two distinct groups of excited states exist – singlet and triplet. In a sense, we may regard each group as comprising a different gas:

* *Parahelium* for which $S = 0$.
* *Orthohelium* for which $S = 1$.

In general, the energy levels of orthohelium (the triplet states of helium) are found to be lower than the corresponding parahelium (singlet state) energy levels (Fig. 3.34). This can be explained as follows.

The probability of the two helium electrons being found together at the same point in space is lower in the triplet state where the spatial function is anti-symmetric than in the singlet state where it is symmetric. Hence, on average, the electrons will be further apart in the triplet state and the repulsive electrostatic interaction between them will be correspondingly weaker. Consequently, the repulsive electrostatic interaction will raise the energy of the system less in the triplet state than in the singlet state.

Whereas transitions that conform to the selection rules $\Delta l = \pm 1$ and $\Delta m_l = 0, \pm 1$ are permitted and do occur within each helium state, singlet or triplet, spectral transitions between the triplet state and the singlet state do not occur under usual circumstances. Such transitions would require a spin rearrangement and so their probability is very low. On this basis we can formulate a further selection rule for spectral transitions:

Transitions obey the selection rule $\Delta S = 0$.

This rule always applies in systems where the potential energy does not explicitly include the spin. Because a fall from the lowest triplet states in helium violates this selection rule, these states are metastable. Unless a collision occurs, a helium atom can remain in these states for a relatively long time – a second or more – before radiating.

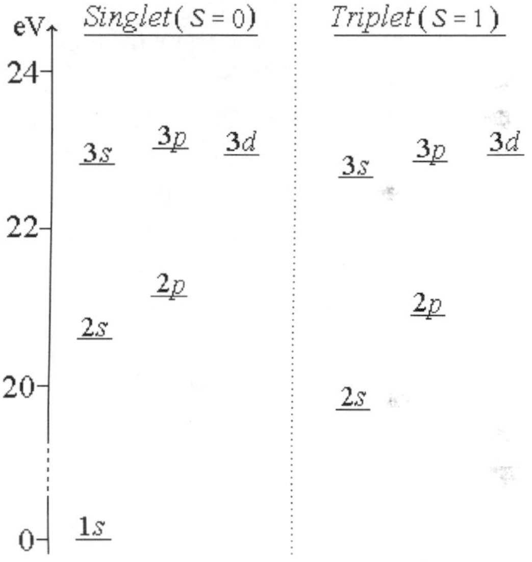

Fig. 3.34. The energy level diagram for helium for the principal quantum numbers $n = 1, 2$ and 3. The energies are given relative to the ground state. One of the atom's electrons is in the ground state and the other in the indicated *ns*, *np* or *nd* excited state. The energy levels of the singlet atoms (parahelium) are generally higher that those of the corresponding triplet (orthohelium) states.

In multi-electron atoms that possess two valence electrons, such as beryllium, magnesium and calcium, the situation is very similar to that in helium. Their energy levels also fall into two groups, singlet ($S = 0$) and triplet ($S = 1$); these correspond to the excitation of just one of the valence electrons. Transitions within the two groups that obey the selection rules $\Delta l = \pm 1$ and $\Delta m_l = 0, \pm 1$ readily occur; by contrast, those between the groups that would violate the rule $\Delta S = 0$ are very weak.

Electrostatic interactions are not the only ones that act upon the components of an atom; magnetic interactions also affect them. An example of this is the interaction of the spin magnetic moment of the electron with the magnetic field induced by the nucleus – spin-orbit coupling.[69] The spectral line splitting that results from this magnetic interaction is only a minor effect in

[69] See Section 3.4.3.

the atoms of elements at the beginning of the periodic table, especially of those which constitute organic material – hydrogen, carbon, nitrogen and oxygen. However, as the atomic number increases, so does the importance of this magnetic interaction. The total spin, S, of the electrons outside the closed shells then ceases to be a good quantum number with which to characterise the energy levels, and the selection rule $\Delta S = 0$ is no longer found to hold.

3.4.8 Total Spin and the Energy Levels in Molecules

Early attempts to explain the nature of the chemical bond, such as the *dualistic theory*, had supposed all bonds to be polar and to involve the transfer of charge from one of the linked atoms to the other. According to this theory, each bond between two atoms represents the transfer of a *valency electron* from one atom to the other. As a result, one of the atoms becomes positively charged, the other negatively charged. This creates an electrical dipole whose ends are attracted to one another. Although this approach succeeded in explaining the structure of ionic crystals, e.g. Na^+Cl^-, it proved incompetent to account for the behaviour of relatively non-polar compounds such as organic compounds.

Stark[70] first put forward the improved idea that the valency electrons released by the linked atoms attract their positively charged components concurrently. He regarded these electrons as being situated somewhere in the space between the atoms (Thomson-type atoms) and bonding them electrostatically. In an extension of this idea and of his own atomic model, Bohr imagined the chemical bond to be a ring of circulating electrons in an orbit whose plane was perpendicular to the line joining the atomic centres.

In 1907, Thomson had conjectured that atoms having the structure he had postulated might be bonded by electric forces without polarisation occurring. He suggested that in homonuclear molecules such H_2 or O_2 the atoms overlapped with their electrons situated symmetrically in the region of overlap. In this description, the spheres of positive charge were held together by the localisation and concentration of their electrons between them. This model did not require a molecular dipole to bond the atoms. On the contrary, it postulated the revolutionary notion that bonding arises from a localisation of the charges within each atom.

The modern conception of the chemical bond was first proposed by Lewis in 1916:

[70]Johannes Stark 1874–1957, German physicist who was awarded the Nobel prize in 1919 for discovering the Doppler effect in canal rays and the splitting of spectral lines by an electric field.

'The simple assumption [is] that the chemical bond is at all times and in all molecules merely a pair of electrons held jointly by two atoms.'[71]

However, an explanation of how two negatively-charged entities could cooperate to create a bond despite their mutual electrostatic repulsion, had to await the development of quantum mechanics. The key, in this case too, is the concept of total spin.

In contrast to the situation in atoms, where electrons in the triplet state generally have the lower energy levels, the study of molecules shows that the electrons that constitute covalent bonds have a lower energy in the singlet state than in the triplet state. It follows that the wave function of a covalent bond has to be of the form

$$\Psi_{CovalentBond} = \phi_S \times \chi_A, \tag{3.102}$$

i.e. the spin wave function of the electrons constituting the covalent bond is anti-symmetric, χ_A, and the spatial wave function (orbital) is symmetric, ϕ_S. In this state, the spatial wave functions (the orbitals) of the electrons overlap, a configuration that increases the electrostatic energy of repulsion between them. Notwithstanding, covalent bond formation is always accompanied by

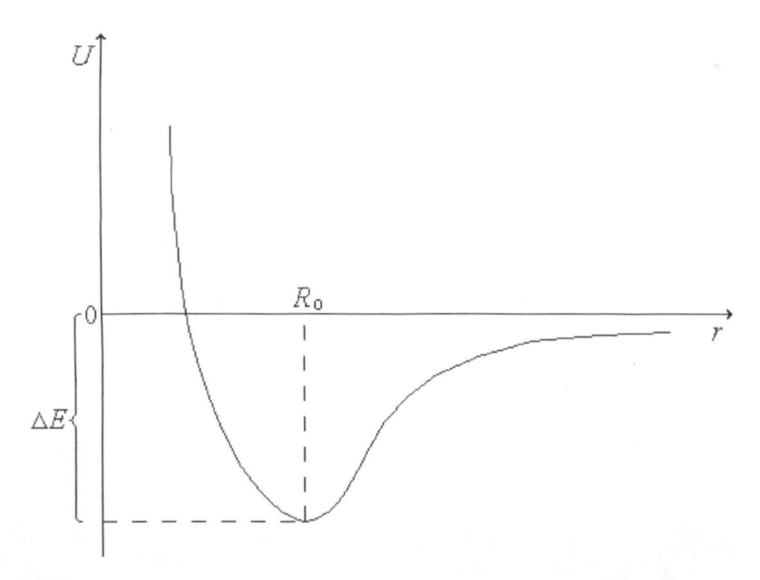

Fig. 3.35. The potential energy diagram of bond formation. As the atoms approach, the potential energy, U, falls. It reaches its minimum value at the inter-nuclear distance R_0 after which it rises once more as the repulsion between the nuclei becomes dominant. The energy released in the process, ΔE, is the bond energy.

[71] G.N. Lewis (Reprinted 1966) *Valence and the Structure of Atoms and Molecules*, New York: Dover Publications, p. 83.

the establishment of a minimum energy equilibrium (Fig. 3.35). How, then, does the system compensate for the increased energy of repulsion between the electrons and achieve the overall reduction in its energy?

The effects that serve to reduce the system's energy when a bond is formed arise from the fact that each electron comprising the bond is attracted by the two nuclei of the bonded atoms and not just by the nucleus of its 'parent' atom. These effects can be interpreted as having a Coulombic component and a purely quantum component.

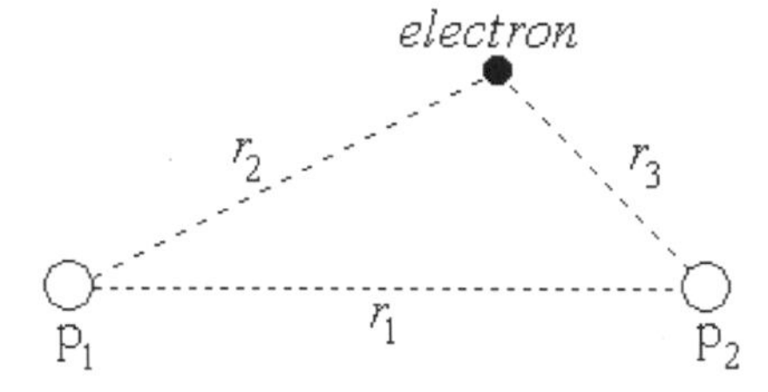

Fig. 3.36. The H_2^+ molecule formed when a hydrogen atom captures a proton. The two protons, p_1 and p_2, each carry a charge $+e$. The electron carries a charge $-e$.

The simplest such system in which an electron is shared by two nuclei is the molecule H_2^+ formed when a hydrogen atom captures a proton (Fig. 3.36). The formal procedure for establishing this molecule's stationary states is:

1. find the expression for its potential energy, U;
2. substitute this expression in the appropriate Schrödinger equation;
3. solve the resulting equation.

When the hydrogen atom and the proton are far apart, the system's energy will be just that of the hydrogen atom, i.e. $-13.6\,\text{eV}$. However, as they approach, the particles comprising the hydrogen atom and the second proton come under the influence of each other's electric fields and the following additional electrostatic interactions arise, each with its own potential energy:

* Between the nucleus of the hydrogen atom (its proton) and the second proton with a positive potential energy, $U_{p-p} = ke^2/r_1$.
* Between the electron and each proton with an overall negative potential energy, $U_{p-e} = -(ke^2/r_2 + ke^2/r_3)$.

The total potential energy of these interactions is

$$
\begin{aligned}
U &= U_{p-p} + U_{p-e} \\
&= ke^2\left(\frac{1}{r_1} - \frac{1}{r_2} - \frac{1}{r_3}\right).
\end{aligned}
\tag{3.103}
$$

This is the expression to be substituted in the Schrödinger equation. The solution of the resulting equation is complex and beyond the scope of this book.

We can, however, once again use our physical intuition in order to obtain an understanding of the solution and of the result it gives for the energy of the molecule's ground state.

Though it originally belonged to the hydrogen atom, the electron loses this identity in the formation of the H_2^+ molecule. Like the elusive particles that appear to pass simultaneously through the slits in experiments of the type carried out by Young, no definite trajectory can be assigned to the electron. It may even appear to be at two different places at the same time, orbiting both nuclei simultaneously. And, likewise, its spatial wave function can be viewed as arising from the superposition of a number of *base states*, $\phi = \phi_1 + \phi_2 + \phi_3 + \ldots$, with the corresponding probability density, $|\phi|^2$, involving interference terms in addition to the probability amplitudes of the base states. It transpires that most of the energy released when a covalent bond forms is associated with the interference terms that appear in the probability density. This energy, whose origin is a pure quantum effect, is termed *exchange energy*.

Let two base states of the electron in the H_2^+ molecule be described by the spatial wave functions ϕ_1 and ϕ_2 (Fig. 3.37). The effect of the interference term on the probability density of the state formed by the superposition of

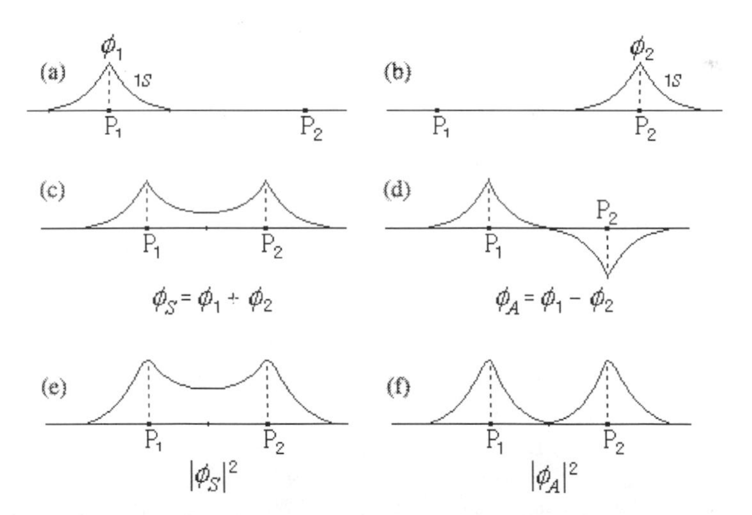

Fig. 3.37. A system of two protons and one electron. Two possible base states of the system are represented by (a) and (b), which show, respectively, the combination of the hydrogen atom P_1 with the proton P_2 and the combination of the proton P_1 with the hydrogen atom P_2. The spatial distribution of the electron in these states is described, respectively, by the wave functions, ϕ_1 and ϕ_2. In figures (c) to (f) the proximity of the protons corresponds to the situation in the H_2^+ molecule. Figure (c) shows the symmetric spatial wave function (molecular orbital) $\phi_S = \phi_1 + \phi_2$; Fig. (d) the anti-symmetric wave function $\phi_A = \phi_1 - \phi_2$. Figures (e) and (f) show the corresponding probability densities, $|\phi_S|^2$ and $|\phi_A|^2$.

ϕ_1 and ϕ_2 depends on whether the wave functions of these base states are in phase with one another or out of phase. If the wave functions are in phase, the resulting *molecular orbital* (spatial wave function), ϕ_S, is symmetric. If they are out of phase, the resulting orbital, ϕ_A, will be anti-symmetric. Ignoring the normalisation requirements, this gives

	Wave Function	Probability Density
In phase:	$\phi_S = \phi_1 + \phi_2$	$\lvert\phi_S\rvert^2 = \phi_1^2 + \phi_2^2 + 2\phi_1\phi_2$
Out of phase:	$\phi_A = \phi_1 - \phi_2$	$\lvert\phi_A\rvert^2 = \phi_1^2 + \phi_2^2 - 2\phi_1\phi_2$

In general, the average (expectation) value of the electron's potential energy depends on the probability density of the spatial wave function, $|\phi|^2$,

$$\begin{aligned} \langle U_{p-e}\rangle &\propto \int U_{p-e}|\phi|^2\, dr \\ &= \int U_{p-e}\phi_1^2\, dr + \int U_{p-e}\phi_2^2\, dr \pm \int U_{p-e}2\phi_1\phi_2\, dr. \end{aligned} \tag{3.104}$$

The contribution of the Coulombic interaction between the electron and the protons to the potential energy depends on the first two terms in this expression, those in which the probability densities of the two base states, ϕ_1^2 and ϕ_2^2, appear,

$$(U_{p-e})_{Coulomb} \propto \int U_{p-e}\phi_1^2\, dr + \int U_{p-e}\phi_2^2\, dr. \tag{3.105}$$

The contribution of the exchange energy depends on the third term in Eq. (3.104), that in which the interference factor, $2\phi_1\phi_2$, appears:

$$(U_{p-e})_{Exchange} \propto \pm \int U_{p-e}2\phi_1\phi_2\, dr. \tag{3.106}$$

If the molecular orbital is symmetric, this 'exchange' term has the same sign as that of the Coulombic interaction and adds to it: if it is anti-symmetric it has the opposite sign and opposes it.

$$\begin{aligned} (U_{p-e})_{Symmetric} &= (U_{p-e})_{Coulomb} + (U_{p-e})_{Exchange} \\ (U_{p-e})_{Anti-symmetric} &= (U_{p-e})_{Coulomb} - (U_{p-e})_{Exchange}. \end{aligned} \tag{3.107}$$

Thus, the total potential energy of the electron-proton interactions in the H_2^+ molecule depends on whether the molecular orbital is symmetric or anti-symmetric. This has the effect of splitting the molecular ground state into two energy levels.

Since the Coulombic interaction between the electron and the protons is attractive, its potential energy will be negative: $(U_{p-e})_{Coulomb} < 0$; likewise, the

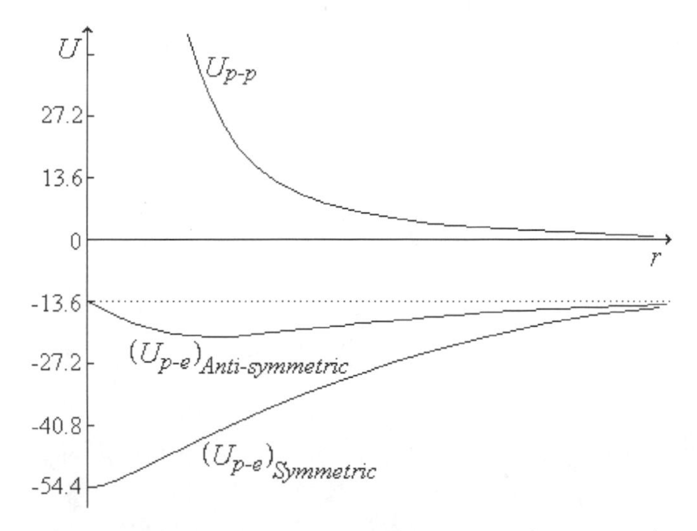

Fig. 3.38. The potential energy of the three basic interactions in the H_2^+ molecule as a function of the internuclear separation, r. U_{p-p} is the energy of the Coulombic interaction between the positive nuclei (the protons), $(U_{p-e})_{Symmetric}$ that of the symmetric proton–electron interaction and $(U_{p-e})_{Anti-symmetric}$ that of the anti-symmetric proton–electron interaction. When the internuclear separation is large, $r \to \infty$, the energy of the proton–electron interaction is essentially just that of the $1s$ ground sate of the hydrogen atom, i.e. $-13.6\,\text{eV}$.

exchange energy: $(U_{p-e})_{Exchange} < 0$. Hence, the effect of the minus sign in the expression for the potential energy of the anti-symmetric orbital – Eq. (3.107) – is to raise its value above that of the symmetric orbital:

$$(U_{p-e})_{Anti-symmetric} \geq (U_{p-e})_{Symmetric}. \tag{3.108}$$

Thus, the higher energy level corresponds to the anti-symmetric molecular orbital and the lower energy level to the symmetric orbital. Thus, the closer the nuclei are to one another, the greater is the contribution of the exchange energy and the wider the gap between these energy levels (Fig. 3.38).

The total potential energy of the anti-symmetric orbital in the H_2^+ molecule is

$$U_{Anti-symmetric} = (U_{p-e})_{Anti-symmetric} + (U_{p-p})_{Coulomb}. \tag{3.109}$$

This function does not exhibit a minimum at any point (Fig. 3.39). Hence, it cannot be the function that describes the formation of a bond between the nuclei. For this reason, the anti-symmetric orbital is termed an *anti-bonding orbital*.[72]

[72] In the literature it is often referred to as the 'σ^* state' and designated ψ_a.

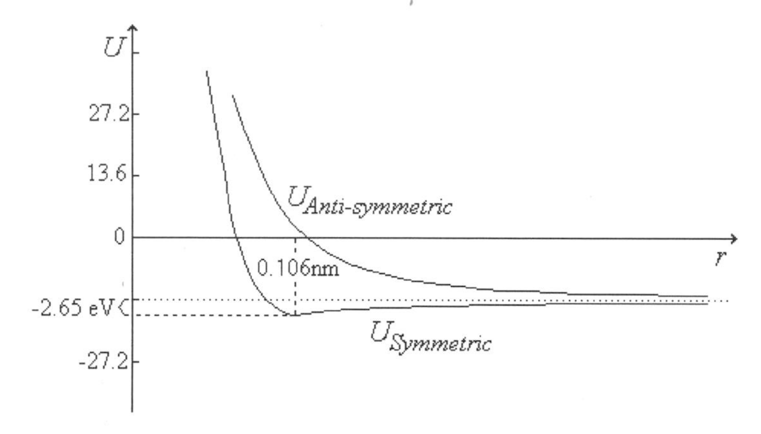

Fig. 3.39. The potential energy diagram of the symmetric (bonding) and anti-symmetric (anti-bonding) orbitals in the H_2^+ molecule.

The total potential energy of the symmetric molecular orbital is

$$U_{Symmetric} = (U_{p-e})_{Symmetric} + (U_{p-p})_{Coulomb}. \tag{3.110}$$

This function exhibits a distinct minimum at $U = -16.25\,\text{eV}$ and $r = 0.106\,\text{nm}$; the bond energy is 2.65 eV (Fig. 3.39). It is the function that describes the formation of the bond in the H_2^+ molecule. Hence, the symmetric orbital is termed a *bonding orbital*.[73]

The bond formed in the H_2^+ molecule is exceptional in that it is a *one-electron bond*. The simplest case of a typical covalent two-electron bond is that found in the hydrogen molecule, H_2. Because of the presence of the second electron in this molecule, three more factors must be taken into account when considering its nature:

1. A second Coulombic repulsive interaction between the electrons. This acts to raise the system's total energy.
2. More base states that must be taken into account when considering the molecular wave function. This increases the number of interference terms, which, in turn, enhances the contribution of the exchange energy to the system's total energy. In the hydrogen molecule this *resonance* between the base states provides some 90% of the of the bond's total energy of 4.75 eV; Coulombic effects contribute just 10%.
3. The overall wave function that describes the electrons comprising the bond must be anti-symmetric.

[73]It is also often referred to as the 'σ state' and designated ψ_b.

The repulsive interaction between the electrons adds a third term to the general expression for the system's potential energy,

$$U = U_{p-p} + U_{p-e} + U_{e-e}. \tag{3.111}$$

Substituting this expression in the appropriate Schrödinger equation produces an equation for which no exact solution has been found. However, by restricting the number of base states taken into account, a solution is obtained that is qualitatively similar to that found for the H_2^+ molecule. Above all, the inference that it is the symmetric orbital, ϕ_S, that leads to a bonded state remains valid.

Since the overall wave function that describes the pair of electrons must be anti-symmetric, the total spin function must be anti-symmetric; *ergo* the electrons must be in a singlet state. Thus, the quantum state that leads to bonding is that in which the electronic spins are opposed (Fig. 3.40).

Thus, we see that covalency is predominantly a quantum effect, a consequence of the fact that the electrons comprising the covalent bond are *exchangeable.* Though each originally belonged to one or other of the atoms, they lose their individual identities when the bond is formed. Similarly, the actual structure of a molecule can also be regarded as a hybrid, formed by the superposition of all the conceivable physically indistinguishable structures it can have. This

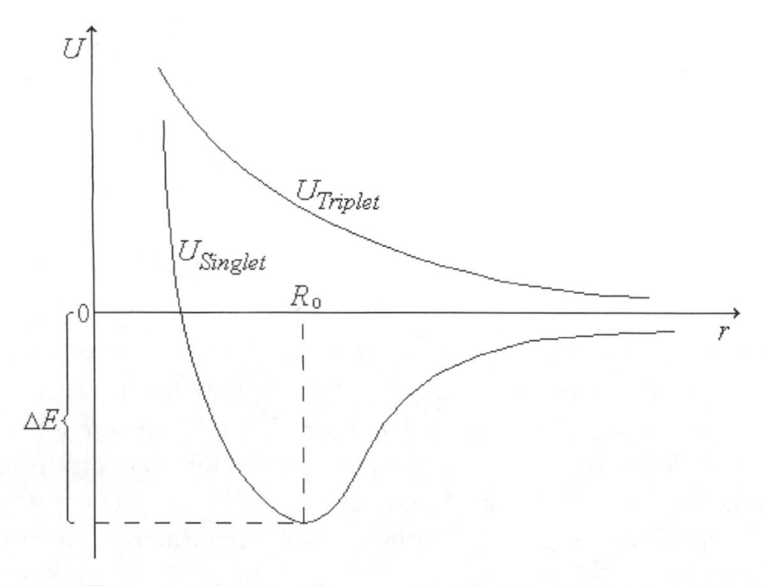

Fig. 3.40. The potential energy diagram of a pair of hydrogen atoms. Whereas the energy of the triplet state rises monotonically, that of the singlet state exhibits a minimum at an internuclear separation of $R_0 = 0.074\,\text{nm}$. The corresponding bond energy $\Delta E = 5.3\,\text{eV}$.

is the principle that underlies the concept of *resonance* postulated by Pauling[74] for elucidating molecular structures. In order that they be physically indistinguishable, the contributing structures can only differ in the positions of their electrons not in those of their atoms. Likewise, each must have the same number of paired and unpaired electrons as otherwise they would be magnetically distinguishable. The resultant spatial wave function is obtained by the weighted linear superposition of the orbitals of the contributing structures.

In determining which of the possible wave functions correctly describes a particular system, the principle adopted is:

> '*The actual structure of the normal state is that one, of all conceivable structures, that gives the system the lowest energy and the maximum stability.*'[75]

The classical example of this technique is given by its application to the structure of the benzene molecule.

The conventional Kekulé structural formula shows six quadrivalent carbon atom linked in a ring by alternating single and double bonds (Fig. 3.41). However, the chemical properties of benzene belie such distinct and localised bond-types. Moreover, the actual energy of a benzene molecule – its *resonance energy* – is about 1.6 eV less than it should be were this its structure.

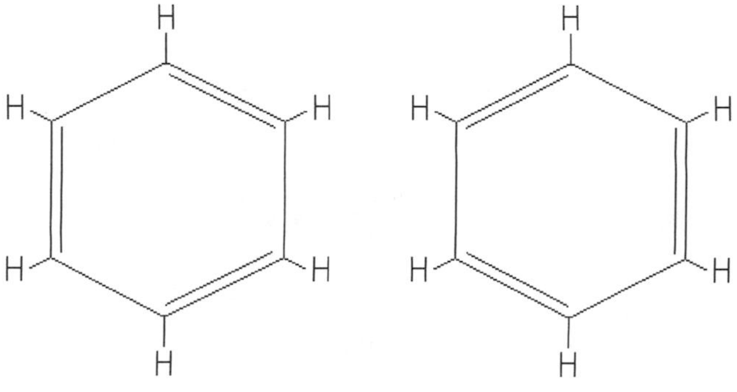

Fig. 3.41. The two possible forms of the Kekulé formula for benzene.

According to Pauling's concept of resonance, the actual structure of the benzene molecule is that represented by the weighted linear superposition of the wave functions of its base states. Two possible base states are the two forms of the Kekulé formula. These structures possess the same energy and so make equal contributions to the final state. Like the other *two-state systems* we have examined, such as the H_2^+ molecule, the linear superposition of their respective

[74] Linus Pauling 1901–1994, American chemist who was awarded the Nobel prize for chemistry in 1954 for his work on molecular structure and the Nobel prize for peace in 1963 for his efforts to promote world-wide nuclear disarmament.

[75] Linus Pauling (1960) *The Nature of the Chemical Bond*, Cornell University Press, p. 11.

wave functions gives rise to two distinct stationary states, one with an energy above the expected value and one with an energy below it. The lower state can be described by the orbital $\phi_{\mathrm{I}} + \phi_{\mathrm{II}}$; the higher state by $\phi_{\mathrm{I}} - \phi_{\mathrm{II}}$. In the case of benzene the difference between the energy levels is about 4.7 eV. At normal temperatures, the benzene molecule is in the lower energy stationary state. The existence of the higher energy level is evidenced by the fact that benzene absorbs ultra-violet radiation of frequency $1.15 \cdot 10^{15}$ Hz.

In the lower stationary state, the six electrons comprising the double-bonds – the π *electrons* – are free to circulate around the ring, above and below the plane of the ring. This delocalisation effectively enlarges the potential well in which these electrons can move and accounts for the reduction in the molecule's energy.

Appendices to Part Three

3.5.1 The Energy of an Orbiting Charged Particle[76]

In general, the electrical potential energy, U, of a particle carrying a charge q situated in an electric field at a point of potential V is

$$U = qV. \tag{3.112}$$

In the case where the electric field is that produced by a point charge q', the potential at a distance r from the charge is

$$V = \frac{kq'}{r}, \tag{3.113}$$

where $k = 1/4\pi\varepsilon$ is the electrostatic constant of proportionality. Thus, the electric potential energy, $U(r)$, associated with the charge q at a distance r from the charge q' is given by the formula

$$U(r) = \frac{kqq'}{r}. \tag{3.114}$$

Suppose that a particle of mass m carrying a charge $-q$ circles a fixed charge q' at a distance r with velocity v, and that the only force acting in the system is the Coulombic force of attraction, $F_E = -(kqq'/r^2)$, between the two charged particles. The Coulombic force will constitute the centripetal force, F_C, required for the particle's circular motion

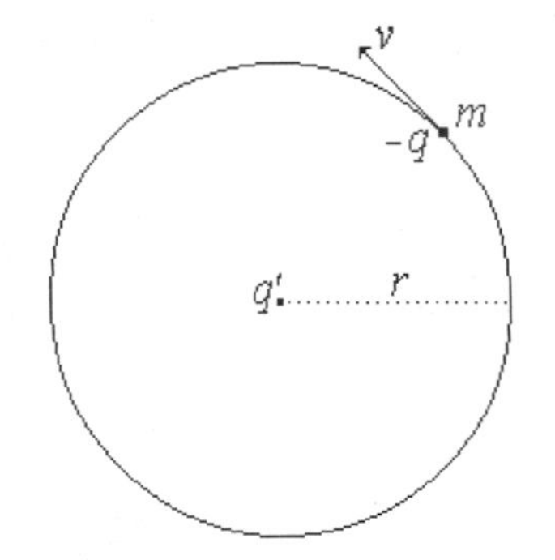

$$F_C = |F_E| = \left| -\frac{kqq'}{r^2} \right|. \tag{3.115}$$

The magnitude of the circling particle's centripetal acceleration is v^2/r. Substituting in Newton's second law of motion, $F = ma$, gives

$$F_C = \frac{kqq'}{r^2} = \frac{mv^2}{r}. \tag{3.116}$$

[76]The calculation is based on classical mechanics.

It follows that the kinetic energy, K, of the circling particle will be given by

$$K = \frac{mv^2}{2} = \frac{kqq'}{2r}. \tag{3.117}$$

The electric potential energy, U, of the system is

$$U = \frac{-kqq'}{r}. \tag{3.118}$$

Thus, its total energy, E, will be

$$\begin{aligned} E &= U + K \\ &= \frac{-kqq'}{r} + \frac{kqq'}{2r} \\ &= \frac{-kqq'}{2r} \end{aligned} \tag{3.119}$$

or

$$E = \frac{U}{2} = -K. \tag{3.120}$$

In terms of its total energy, the velocity of revolution of the circling particle is

$$v = \sqrt{\frac{kqq'}{mr}} = \sqrt{\frac{2|E|}{m}}. \tag{3.121}$$

3.5.2 The Schrödinger Equation for the Hydrogen Atom

As we showed in Section 2.4.6, the Schrödinger equation can be regarded as a synthesis of the law of energy conservation, De Broglie's equation and the principle of linear superposition. By analogy with the classical expression for energy conservation, $E = U + K$, the spherical polar coordinate form of the Schrödinger equation for the hydrogen and hydrogen-like atoms – Eq. (3.46) – can be written in the form

$$E\psi = \left(\frac{-Zke^2}{r}\right)\psi + \left(\frac{-\hbar^2}{2m_e}\Delta\psi\right), \tag{3.122}$$

where the operator

$$\Delta = \frac{\partial^2}{\partial r^2} + \frac{2}{r}\frac{\partial}{\partial r} + \frac{\Lambda(\theta,\phi)}{r^2}. \tag{3.123}$$

The term $-\hbar^2/2m_e\Delta\psi$ describes the kinetic energy of the atomic electron and is the quantum mechanical analogue of the familiar classical expression $K = p^2/2m_e$.

The hydrogen atom's quantum numbers, n, l and m_l, and its energy levels, E_n, can be obtained from the following solution to Eq. (3.122).

1. Separating the variables

Since the operator Δ separates into a radial part, $\dfrac{\partial^2}{\partial r^2} + \dfrac{2}{r}\dfrac{\partial}{\partial r}$, and an angular part, $\dfrac{\Lambda(\theta\phi)}{r^2}$, and since the potential energy, $U = -kZe^2/r$, is a function of r alone, the solution, $\psi(r, \theta, \phi)$, to the hydrogen atom's Schrödinger equation can be separated into a product of two functions, one depending on r alone the other on θ and ϕ alone:

$$\psi(r, \theta, \phi) = R(r) \cdot Y(\theta, \phi). \tag{3.124}$$

That this is so can be shown by substituting this product in the Schrödinger equation – Eq. (3.122) – employing, for convenience, the shorthand symbols $\psi = RY$. This gives

$$ERY = \left(-\frac{kZe^2}{r}\right) RY - \frac{\hbar^2}{2m_e}\left(R''Y + \frac{2}{r}R'Y + \frac{R\Lambda Y}{r^2}\right), \tag{3.125}$$

where R' and R'' are, respectively, the first and second partial derivatives of the function R. Dividing by RY, multiplying by r^2 and collecting terms gives

$$r^2\left[\frac{R''}{R} + \frac{2m_e}{\hbar^2}\left(E + \frac{kZe^2}{r}\right)\right] + 2r\frac{R'}{R} + \frac{\Lambda Y}{Y} = 0, \tag{3.126}$$

The first two terms in this expression depend on r alone; the third term, $\dfrac{\Lambda Y}{Y}$, depends on (θ, ϕ) alone. Thus, we have obtained an equation whose form is $F(r) + G(\theta, \phi) = 0$. Since this equation must hold for all values of r and θ, ϕ the implication is that each separate part, $F(r)$ or $G(\theta, \phi)$, is a constant.

2. The quantum number l

The function $Y(\theta, \phi)$ is called a *spherical harmonic.* Such functions appear as solutions of $\Delta\psi(r, \theta, \phi) = 0$. Let us look for solutions of the form

$$\psi_l(r, \theta, \phi) = r^l Y_l(\theta, \phi), \tag{3.127}$$

where $l \geq 0$ is an integer. Performing the operation Δ, as defined by Eq. (3.123), on this function gives

$$l(l-1)r^{l-2}Y_l + 2lr^{l-2}Y_l + r^{l-2}\Lambda Y_l = 0, \tag{3.128}$$

from which

$$\frac{\Lambda Y_l}{Y_l} = -l(l+1). \tag{3.129}$$

Hence, functions of the suggested form do indeed give values of $\Lambda Y/Y$ that are constants for any particular value of l. Such functions are called *eigenfunctions.*

3. The quantum number m_l

The operator, Λ, is given by

$$\Lambda = \frac{\partial}{\partial\theta^2} + \cot\theta\frac{\partial}{\partial\theta} + \frac{1}{\sin^2\theta}\frac{\partial}{\partial\phi^2}. \quad (3.130)$$

The similarity between the form of this operator and that of the operator Δ is apparent; it too separates into two parts, $\frac{\partial}{\partial\theta^2} + \cot\theta\frac{\partial}{\partial\theta}$ and $\frac{1}{\sin^2\theta}\frac{\partial}{\partial\phi^2}$, each depending on a different variable. Hence, just as we could separate the radial part of the Schrödinger equation from the angular part, we can likewise separate the spherical harmonic, $Y(\theta, \phi)$, into two parts, one that depends on θ and the other on ϕ:

$$Y(\theta, \phi) = \Theta(\theta) \cdot \Phi(\phi). \quad (3.131)$$

That this is so can be shown by the same rationale as we used above, namely:

a. substituting the product $\psi = R\Theta\Phi$ (employing the shorthand symbols) in the Schrödinger equation – Eq. (3.122);
b. performing the operation Λ on the resulting function;
c. dividing the result by $R\Theta\Phi$ and multiplying it by $\sin^2\theta$.

After collecting terms, this gives

$$\sin^2\theta\left[\frac{\Theta''}{\Theta} + \frac{2m_e}{\hbar^2}\left(E + \frac{kZe^2}{r}\right)\right] + \frac{\sin 2\theta}{2}\cdot\frac{\Theta'}{\Theta} + \frac{\Phi''}{\Phi} = 0. \quad (3.132)$$

The first two terms depend on θ alone; the third term, $\frac{\Phi''}{\Phi}$, on ϕ alone. Since this equation must hold for all values of θ and ϕ, the implication is that

$$\frac{\Phi''}{\Phi} = \text{a constant}. \quad (3.133)$$

It is convenient to designate this constant $-m_l^2$, from which we obtain the differential equation

$$\Phi'' + m_l^2\Phi = 0, \quad (3.134)$$

whose solution is

$$\Phi(\phi) = Ae^{im_l\phi}, \quad (3.135)$$

where A is the constant of integration or normalisation. From Fig. 3.18 it is evident that ϕ and $\phi + 2\pi$ must both identify the same 'longitude'. Hence, it must be true that $\Phi(\phi) = \Phi(\phi + 2\pi)$ or

$$Ae^{im_l\phi} = Ae^{im_l(\phi+2\pi)}. \quad (3.136)$$

This can only happen when

$$m_l = 0, \pm 1, \pm 2, \pm 3 \ldots. \tag{3.137}$$

4. The relation between the two quantum numbers, l and m_l

In terms of the Cartesian coordinates, x, y and z, the solutions to the equation $\Delta\psi(r,\theta,\phi) = 0$ of the form $\psi_l(r,\theta,\phi) = r^l Y_l(\theta,\phi)$ are homogeneous polynomials of degree l, i.e. polynomials each of whose terms is of the form $a_{jk}x^j y^k z^{l-j-k}$, where a_{jk} is a constant. From the relationship – Eq. (3.45) – between the Cartesian coordinates, x, y and z, and the polar coordinates, r, θ and ϕ,

$$\begin{aligned} x &= r\sin\theta\cos\phi \\ y &= r\sin\theta\sin\phi \\ z &= r\cos\theta \end{aligned}$$

we see that z is independent of ϕ. Thus, in Cartesian terms, the spherical harmonics which appear in the solutions of the equation $\Delta\psi(r,\theta,\phi) = 0$ and which are functions of θ and ϕ alone (for $r = 1$), will be homogeneous polynomials of degree l. The dependence on ϕ arises from products of x and y. Choosing a linear combination of the form $x \pm iy$ gives (for $r = 1$)

$$(x \pm iy)^{m_l} = (\sin\theta)^{m_l}(\cos\phi \pm i\sin\phi)^{m_l} = (\sin\theta)^{m_l} e^{\pm m_l i\phi}. \tag{3.138}$$

Clearly, the highest degree this polynomial can possess is l. Hence, m_l can have all the integer values between $-l$ and $+l$,

$$m_l = -l, -l+1, -l+2, \ldots, 0, 1, 2, \ldots, l, \tag{3.139}$$

i.e. $(2l+1)$ values.

5. The solution of the radial wave function: the quantum number n

Substituting $\frac{\Lambda Y_l}{Y_l} = -l(l+1)$ in Eq. (3.125) gives

$$r^2\left[\frac{R''}{R} + \frac{2m_e}{\hbar^2}\left(E + \frac{kZe^2}{r}\right)\right] + 2r\frac{R'}{R} - l(l+1) = 0. \tag{3.140}$$

When solving this equation, it is convenient to use the following substitutions, once again employing shorthand symbols:

$$\begin{aligned} Q &= rR \\ Q' &= rR' + R \\ Q'' &= rR'' + 2R'. \end{aligned} \tag{3.141}$$

Making these substitutions in Eq. (3.140), multiplying by r and collecting terms gives

$$Q'' + \frac{2m_e}{\hbar^2}\left(E + \frac{kZe^2}{r}\right)Q - \frac{l(l+1)}{r^2}Q = 0. \tag{3.142}$$

Let us first consider how this equation appears at very large values of r, namely, as $r \to \infty$. Under these conditions terms involving $1/r$ and $1/r^2$ can be disregarded. The remaining terms give

$$Q'' + \frac{2m_e E}{\hbar^2}Q = 0. \tag{3.143}$$

At these large values of r, the atomic system can be regarded as classical. As such, the total energy of the orbiting electron will be negative, i.e. $E < 0$. Under these circumstances we can write the equation in the form

$$Q'' - \alpha^2 Q = 0, \tag{3.144}$$

where $\alpha^2 = 2m_e|E|/\hbar^2$. The solution of this equation is

$$Q = e^{-\alpha r}. \tag{3.145}$$

Clearly, α has the dimensions of inverse length. Accordingly, we set

$$\alpha^2 = \frac{1}{4r_0^2} = \frac{2m_e|E|}{\hbar^2}, \tag{3.146}$$

and make the definitions

$$\begin{aligned} r &= r_0\rho \\ \frac{d}{dr} &= \frac{1}{r_0}\frac{d}{d\rho} \\ \frac{d^2}{dr^2} &= \frac{1}{r_0^2}\frac{d}{d\rho^2}. \end{aligned} \tag{3.147}$$

Substituting these definitions back into Eq. (3.142) together with the condition that $E < 0$ gives

$$\frac{1}{r_0^2}Q'' + \frac{2m_e}{\hbar^2}\left[\frac{-\hbar^2}{8m_e r_0^2} + \frac{kZe^2}{r_0\rho}\right]Q - \frac{l(l+1)}{r_0^2\rho^2}Q = 0, \tag{3.148}$$

or

$$Q'' + \left[-\frac{1}{4} + \frac{B}{\rho} - \frac{l(l+1)}{\rho^2}\right]Q = 0, \tag{3.149}$$

where

$$B = \frac{2km_e Ze^2 r_0}{\hbar^2}. \tag{3.150}$$

The next step we take in our search for a complete solution to the radial wave function is to examine a function that not only incorporates the circumstances at $r \to \infty$ but also those found at smaller values of r. Given that the atomic system is well described at large values of r by the function $e^{-\alpha r} = e^{-\rho/2}$, we will search for a function of the form

$$\begin{aligned} Q(\rho) &= e^{-\rho/2} W(\rho) \\ Q'(\rho) &= e^{-\rho/2}\left[-\frac{1}{2}W(\rho) + W'(\rho)\right] \\ Q''(\rho) &= e^{-\rho/2}\left[\frac{1}{4}W(\rho) - W'(\rho) + W''(\rho)\right]. \end{aligned} \tag{3.151}$$

Substitution in Eq. (3.149), employing shorthand symbols and dividing by $e^{-\rho/2}$, gives

$$W'' - W' + \left(\frac{B}{\rho} - \frac{l(l+1)}{\rho^2}\right) W = 0. \tag{3.152}$$

In order to avoid a singularity of this equation as $\rho \to 0$, the function W must vanish at $\rho \to 0$. This condition would be fulfilled if W is a function of the form

$$\begin{aligned} W &= \rho^{\nu} \sum a_i \rho^i = \sum a_i \rho^{(\nu+i)} \\ W' &= \sum (\nu + i) a^i \rho^{(\nu+i-1)} \\ W'' &= \sum (\nu + i)(\nu + i - 1) a_i \rho^{(\nu+i-2)}. \end{aligned} \tag{3.153}$$

Substitution in Eq. (3.152) gives the power series

$$\begin{aligned} &\sum (\nu + i)(\nu + i - 1) a_i \rho^{(\nu+i-2)} - \sum (\nu + i) a_i \rho^{(\nu+i-1)} \\ &\quad + B \sum a_i \rho^{(\nu+i-1)} - l(l+1) \sum a_i \rho^{(\nu+i-2)} = 0. \end{aligned} \tag{3.154}$$

This relationship holds for any value of ρ if the sum of the coefficients of each power of ρ equals zero. Note, that when summing the coefficients of $\rho^{(\nu+i-1)}$, the appropriate coefficients for the terms that include $\rho^{(\nu+i-1)}$ are those which include a_i, and for the terms that include $\rho^{(\nu+i-2)}$ those which include a_{i+1}. This gives

$$(\nu + i + 1)(\nu + i)a_{i+1} - (\nu - i)a_i + Ba_i - l(l+1)a_{i+1} = 0. \tag{3.155}$$

The lowest possible power of ρ is $(\nu-2)$, i.e. that of the factor $\rho^{(\nu+i-2)}$ when $i = 0$. Putting the sum of the coefficients of this factor equal to zero gives

$$\nu(\nu - 1)a_0 - l(l+1)a_0 = 0. \tag{3.156}$$

The two possible solutions of this equation are $\nu = -l$ and $\nu = l+1$. The former is unacceptable because it would make ψ rise to infinity at $\rho \to 0$. Recall, that ψ is proportional to W/r, and the function $W/r = \sum a_i\rho^{(-l+i-1)} \to \infty$ at $\rho \to 0$. However, ψ should remain finite at $\rho \to 0$.

To avoid an infinite value $Q(\rho)$ at $\rho - \infty$ the contribution $W(\rho)$ makes must cease at some point; the power series must terminate at some value of i. At this point, all the terms that include a_{i+1} can be ignored when summing the coefficients of $\rho^{(\nu+i-1)}$. Substituting this condition in Eq. (3.155) gives

$$-(\nu + i) + B = 0. \tag{3.157}$$

Combining this with the solution $\nu = l + 1$ already found for the case when $i = 0$ gives

$$B = l + 1 + i. \tag{3.158}$$

We denote

$$n = l + 1 + i. \tag{3.159}$$

Clearly, n is a positive integer. The equality $B = n$ gives

$$n = \frac{2km_eZe^2}{\hbar^2} \cdot r_0 = \frac{2km_eZe^2}{\hbar^2} \cdot \sqrt{\frac{\hbar^2}{8m_e|E|}}, \tag{3.160}$$

and finally

$$E = -\frac{k^2m_eZ^2e^4}{2\hbar^2n^2}. \tag{3.161}$$

This is the same result as Bohr obtained for the energy levels of hydrogen and hydrogen-like atoms – Eq. (3.16). The lowest value of n, the principal quantum number, is $n = 1$; this is obtained when $l = 0$ and $i = 0$.

This derivation also establishes the rule that $l \leq n - 1$, namely, that for a given n the possible values of l are

$$l = 0, 1, \ldots (n - 1). \tag{3.162}$$

3.5.3 The Angular Momentum of an Orbiting Particle

Relative to the body it is orbiting, the angular momentum, **L**, of a Keplerian satellite is given by the vector (cross) multiplication of the radius vector **r** and the linear momentum vector **p**:

$$\mathbf{L} = \mathbf{r} \times \mathbf{p} \tag{3.163}$$

from which

$$L = rp \sin\theta \tag{3.164}$$

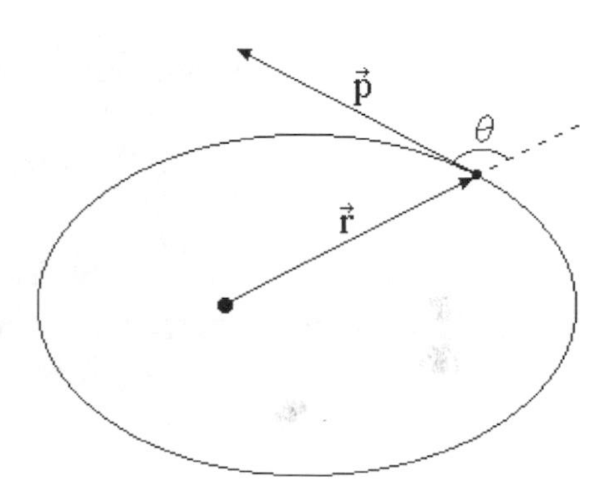

The scalar (dot) multiplication of the two vectors gives

$$\begin{aligned}\mathbf{r} \cdot \mathbf{p} &= rp \cos\theta \\ &= rp_r,\end{aligned} \tag{3.165}$$

where p_r is the component of the satellite's linear momentum parallel to the radius vector.

From the fact that the square of the vector product is $r^2p^2 \sin^2\theta$ and that of the scalar product is $r^2p^2 \cos^2\theta$, we can write the identity

$$\begin{aligned}r^2p^2 &= (\mathbf{r} \cdot \mathbf{p})^2 + (\mathbf{r} \times \mathbf{p})^2 \\ &= r^2p_r^2 + L^2.\end{aligned} \tag{3.166}$$

In terms of its kinetic energy, the linear momentum of a particle of mass m is given by[77]

$$p = \sqrt{2mK}. \tag{3.167}$$

Substituting this expression for p in Eq. (3.164) and rearranging gives

$$K = \frac{1}{2m}\left(p_r^2 + \frac{L^2}{r^2}\right). \tag{3.168}$$

[77]See Appendix 2.5.1.

To appreciate the significance of this result, let us return to the analogy we noted in Appendix 3.5.2 between the classical expression for energy conservation, $E = U + K$, and the quantum mechanical expression given by the Schrödinger equation. Comparing the quantum mechanical expression that describes the kinetic energy of the electron in a hydrogen atom, $\frac{-\hbar^2}{2m_e}\left(\frac{\partial^2}{\partial r^2} + \frac{2}{r}\frac{\partial}{\partial r} + \frac{\Lambda(\theta,\phi)}{r^2}\right)$, with that given by classical mechanics, $\frac{1}{2m_e}\left(p_r^2 + \frac{L^2}{r^2}\right)$, and making the correspondence

$$\frac{\Lambda(\theta,\phi)}{r^2} \rightarrow \frac{L^2}{r^2},$$

indicates that the operator $\Lambda(\theta,\phi)$ is proportional to L^2, the square of the electron's orbital angular momentum, the constant of proportion being $-\hbar^2$.

In the expression for the radial wave function – Eq. (3.130) – the operator $\Lambda(\theta,\phi)$ is replaced by $-l(l+1)$. By inference, we may therefore conclude that

$$L^2 = l(l+1)\hbar^2, \tag{3.169}$$

which is the rule for the quantisation of the square of the electron's orbital angular momentum.

Part Three: Questions, Exercises and Problems

Questions

1. In the Bohr model, when an electron 'jumps' from the state n_1 to the state n_2, there is a change in angular momentum. According to classical mechanics, how can an isolated system spontaneously change its angular momentum? What happens if the system splits in two? Does this suggest another flaw in the Bohr model? How does the quantum mechanical model deal with the changes in angular momentum that accompany atomic transitions?
2. In which Bohr orbit does the electron in the hydrogen atom have the largest energy? Are we justified in ignoring relativistic effects in this case?
3. Is the correspondence principle a necessary axiom of physics or just the fortunate agreement of different mathematical expressions?
4. Why can the Bohr–Sommerfeld model explain the characteristic X-ray spectrum of multi-electron atoms, although it is incompetent to explain their spectra in the visible and UV regions?
5. Compare the dynamical quantities that are constant for a particular value of the principal quantum number and the orbital quantum number in the Bohr model of the atom and the quantum mechanical model of the atom.
6. The Bohr model computes the atom's total energy from a knowledge of its potential and kinetic energies. Which of the following is constant in quantum mechanics for a given set of the quantum numbers n, l: (i) total energy; (ii) kinetic energy; (iii) potential energy?
7. To which states can a hydrogen atom in the $4p$ state pass by the emission of a photon?
8. Into how many beams will the original beam split in a Stern–Gerlach experiment carried out with carbon atoms; nitrogen atoms; oxygen atoms?
9. Spin-orbit coupling splits all states into doublets except s states. Why is this?
10. What is the maximum and minimum value of the orbital angular momentum of an electron in the $n = 1$; $n = 2$; and $n = 3$ state?
11. What is the magnitude, L, of the orbital angular momentum vector and what is the maximum value of its component L_Z in the s, p, and d states?
12. By considering the way an electron's orbital and intrinsic angular momentum couple explain why filled (closed) subshells make no contribution to an atom's magnetic properties.
13. Why is the first ionisation potential of neon, Ne, much greater than that of sodium, Na?
14. The first ionisation potentials of the elements with atomic numbers 20 to 29 are (in eV):

Ca	Sc	Ti	V	Cr	Mn	Fe	Co	Ni	Cu
6.1	6.6	6.7	6.7	6.7	7.3	7.7	7.7	7.5	7.6

Explain the similarity between the values and the rising trend.

15. Suppose that electrons had no spin but were still subject to the Pauli exclusion principle. How would this affect the shape of the periodic table of elements and the properties of the elements? Give examples.
16. Which neutral atom has the greater volume: Li or F; Li or Na; F or Cl?
17. The first ionisation potentials of the Group I elements are (in eV):

Li	Na	K	Rb	Cs
5.4	5.1	4.3	4.2	3.9

Explain the falling trend in the values.
18. The Van der Waals radii of the first two rows of *p*-elements are (in nm):

C	N	O	F	Ne
0.185	0.150	0.140	0.135	0.160
Si	P	S	Cl	Ar
0.224	0.190	0.185	0.180	0.192

Suggest an explanation for these figures and for the trends they show.
19. Which of the following will exert the greatest and the least pressure at a given temperature:

a. a gas composed of classical particles (molecules);
b. a photon (boson) gas;
c. an electron (fermion) gas?

Exercises

1. Calculate the first ionisation potential of Be^{3+}.
2. What is the longest wavelength that can be associated with a transition from an excited state to the $n = 3$ state in hydrogen?
3. What is the wavelength associated with the transition from $n = 6$ to $n = 4$ in He^+?
4. How much energy must be invested to remove an electron from the $n = 2$ level in (i) hydrogen; (ii) He^+?
5. An electron having no kinetic energy is captured by a He^{2+} ion. What is the wavelength of the photon emitted if the electron falls to the ground state?
6. A sample of atomic hydrogen also contains some deuterium (^2H) and tritium (^3H) atoms. Calculate the shift of the spectral line corresponding to the transition from $n = 3$ to $n = 2$ in deuterium and tritium?
7. Calculate the energies and the wavelengths of the photons that will be emitted from an excited hydrogen atom with an energy of -1.51 eV.
8. Calculate:

a. the magnitude of the orbital angular momentum;
b. the magnitude of the possible z-components of the orbital angular momentum;
c. the magnitude of the possible angles between the vector of the orbital angular momentum and the Z-axis of the hydrogen atom electron in the $4f$ state.

9. What values can the quantum numbers n, m_l and m_s take in a hydrogen atom in the $l = 3$ state?

10. What is the probability of finding the hydrogen atom's electron in the ground state between $r = a_0$ and $r = 1.02a_0$?
11. Two of the electrons in lithium have the quantum numbers $n = 1$, $l = 0$, $m_l = 0$, $m_s = \pm\frac{1}{2}$. What can the quantum numbers of the third electron be

    a. in the ground state;
    b. in the first excited state?

12. The wavelength of the K_α line in iron is $1.94 \cdot 10^{-10}$ m

    a. List the electronic configuration of the iron atom before and after the emission of the X-rays.
    b. Calculate the difference between the atom's energy before and after the emission.

13. Which element has a K_α line of wavelength 0.144 nm?
14. Calculate the wavelength of the radiation that will cause the direction of the spin of an $l = 0$ electron to flip in a magnetic field of magnitude 0.2T.
15. Draw a diagram showing the transitions by which a hydrogen atom in the $5d$ state can fall to its ground state.
16. What is the ground state total spin quantum number of chromium, Cr?

Problems

1. Show that in a Thomson hydrogen atom, the net force exerted on the electron is directly proportional to its displacement from the centre of the atom. Calculate the frequency of the radiation that would be emitted by the vibrations of the electron. Assume the atom's radius is 10^{-10} m.
2. The muon particle, μ^-, whose mass is 207 times greater than that of the electron, can combine with a proton to form an atom.

    a. Calculate the muonic atom's Bohr radius.
    b. How much energy is emitted when the muonic atom drops to the ground state from the $n = 2$ state?
    c. What difference, if any, would you expect between the Zeeman effect in a muonic atom and a hydrogen atom?

3. The positronium atom comprises a positron (instead of a proton) and an electron.

    a. Compare the ionisation energy of the positronium atom to that of the hydrogen atom.
    b. Compare the wavelength of the radiation emitted in the transition from $n = 3$ to $n = 2$ by a positronium atom to that emitted by hydrogen.

4. By considering the visible spectrum of hydrogen and He^+ show how you could determine spectroscopically if a sample of hydrogen was contaminated with helium.
5. What spectral lines might be observed in a Franck–Hertz experiment in which hydrogen was used instead of mercury vapour and the electrons were accelerated through a maximum potential of 12.5 V?
6. Show that the greatest number of electrons having the same principal quantum number, n, that an atom can possess is $2n^2$.

7. Show that the recoil energy a hydrogen atom acquires when it falls from the $n = 2$ state to its ground state is negligible compared to the energy of the photon it emits in the process.
8. Show that in the presence of a magnetic field, the 4f $\rightarrow$ 3d transition in hydrogen appears as three spectral lines.
9. Assuming that the life-time of the exited state is 10^{-8}s, how many revolutions would the electron in a Bohr hydrogen atom make before falling from the $n = 2$ state to the ground state?
10. At which temperature is the average kinetic energy of a particle equal to the ionisation energy of hydrogen?
11. The threshold of the K series in an atom is $0.178 \cdot 10^{-10}$ m. The average wavelengths of the series are:

$$K_\alpha = 0.210 \cdot 10^{-10}\ \text{m}$$
$$K_\beta = 0.184 \cdot 10^{-10}\ \text{m}$$
$$K\gamma = 0.179 \cdot 10^{-10}\ \text{m}$$

Draw an energy level diagram for the atom and calculate the wavelength of the L_α line.

12. By calculating the tangential velocity at its equator, show that the notion that the electron is a sphere of radius $3 \cdot 10^{-15}$ m (as was once thought) is incompatible with the quantisation of its spin.
13. Clouds of cold interstellar hydrogen are detected by means of the 21 cm line they emit in the radio region of the spectrum. This line is attributed to the flipping of the electron in the hydrogen atom between spin states. On this basis calculate the magnitude of the magnetic field 'felt' by the electron in a hydrogen atom.
14. Calculate the frequency of the radiation that could cause the spin of the protons in a water molecule to change its direction in a magnetic field of magnitude 1T. The magnetic moment of the proton is $1.41 \cdot 10^{-26}$J/T.

Part Four Interactions of Electromagnetic Radiation and Matter

In nature, energy is being continuously exchanged; between the atoms and molecules that comprise material objects and the electromagnetic radiation that fills the space around them. The interaction of sunlight with the objects on the earth's surface accounts for most of the phenomena we see around us, including life itself.

It was in order to account for the interaction between heat radiation and matter that the quantum hypothesis was first put forward by Planck. Subsequently, Einstein extended Planck's idea to account for the photoelectric effect. But it was only when Bohr combined the quantum hypothesis with the nuclear model of the atom in order to explain atomic spectra and to provide the first consistent atomic theory, that its full power as an investigative and interpretive tool became apparent.

In the following pages we will show how this synthesis has been applied to the elucidation of a number of physical phenomena, all of which involve an interaction between radiation and matter. We will restrict ourselves, at this stage, to phenomena involving radiation in the microwave, infra-red, visible and ultra-violet regions of the electromagnetic spectrum. Interactions between matter and high-energy radiation such as X-rays and γ-rays are discussed in Part Five of this book.

Many of the research techniques employed in the earth and life sciences, such as spectroscopy and fluorescence labelling, are based on these phenomena. These too are dealt with in the following pages. A special section deals with fluorescence in biological systems and its research applications.

In the second half of this century, the synthesis of quantum theory and the nuclear atom was widened to encompass the theory of relativity. The new quantum theory of light and matter that this spawned is known as *quantum electrodynamics* or *QED* for short. This 'strange theory of light and matter', as one of its creators dubbed it, 'describes *all* the phenomena of the physical world except the gravitational effect... and radioactivity.'[1] In homage to this remarkable union, a new chapter is devoted to a brief exposition of its principles.

[1]Richard P. Feynman (1988) *QED – The Strange Theory of Light and Matter*, New Jersey: Princeton University Press, p. 8.

Chapter 4.1

The Passage of Radiation through Matter

In general, when electromagnetic radiation is incident on a material object, some is reflected and some transmitted. The passage of the transmitted beam through the material medium attenuates its intensity (Fig. 4.1).[2] This reduction in the beam's intensity depends on the following factors:

1. the extent of the beam's passage through the material medium;
2. the nature of the material medium;
3. the frequency of the radiation.

Absorption is the general term used to describe this reduction in intensity. The mechanism by which the electromagnetic radiation interacts with the material medium depends on its energy and frequency and on the nature of the material (Section 4.1.2).

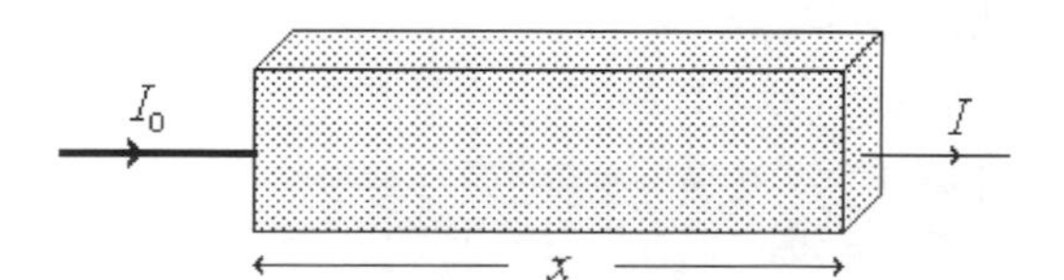

Fig. 4.1 The passage of a beam of radiation through matter. After passing through a thickness x, the intensity, I_0, of the incident beam is reduced to I.

Selective absorption is the preferred absorption of radiation of a particular frequency; it is an example of the capture of energy through resonance.[3] Atoms and molecules absorb radiation preferentially when its frequency coincides with one of their natural frequencies. Glass is transparent to visible light but not to infra-red radiation; it does not absorb visible light but it does absorb infra-red radiation. This accounts for the warming of glass-houses and of cars left out in the sun – the *glasshouse effect*. Sunlight enters through their windows and is absorbed by their contents. These warm up and, in turn, emit infra-red

[2]The laser is an exception to this rule – see Chapter 6.1.

[3]Not to be confused with the resonance between the base states of a molecule postulated by the quantum mechanical theory of chemical structure.

radiation, but this radiation cannot escape back through the glass. It is trapped and, in the absence of convection, raises the inside temperature still more.

White light, such as sunlight, is a mixture of coloured lights – the colours of the rainbow. It comprises a continuous visible spectrum that extends from the colour red through orange, yellow, green, blue and violet. Objects that are not themselves light sources are only visible by virtue of the light they reflect or scatter. An object that returns all the light incident upon it is a white object; one that returns none of the light is a black object. Coloured objects absorb light of all colours except their own, i.e. they selectively absorb all the colours in sunlight except those they return to the eye. For example, a green leaf selectively absorbs the red and blue constituents of white sunlight; the green constituent is left as the dominant colour in the light it returns and so the leaf appears green to our eyes.

Substances which give a colour to objects are called *pigments*. Pigment particles selectively absorb certain of the constituents of white light and transmit the rest; the pigment's colour is that of the transmitted constituents. Some of the white light incident on a coloured object or particle is reflected at its surface; the remainder enters the surface and its unabsorbed constituents are subsequently returned. Thus, when a blue copper sulphate crystal is ground up a white powder is obtained. Incident light is scattered (reflected) in all directions at their surfaces. Consequently, little or no light enters the crystals or is absorbed by them.

The colour of a green object such as a leaf is called *body colour*. A thin gold foil appears yellow when looked at by the light it reflects; this is its *surface colour*. However, the foil appears green when it is used as a light filter and a source of white light is viewed through it; this is its body colour. Gold foil reflects yellow light but transmits green light.

The absorption of radiation by matter is typically studied by means of a *spectrophotometer*. The arrangement of a simple spectrophotometer that operates in the visible and near ultra-violet range is shown schematically in Fig. 4.2.

4.1.1 The Attenuation of Radiation by Matter

The more absorbers there are in the radiation's path the greater is the probability it will be absorbed. Thus, irrespective of the mechanism of its absorption, the reduction, $-dI$, in the intensity of a beam of radiation as it passes through an infinitesimal element of matter of thickness dx is given by

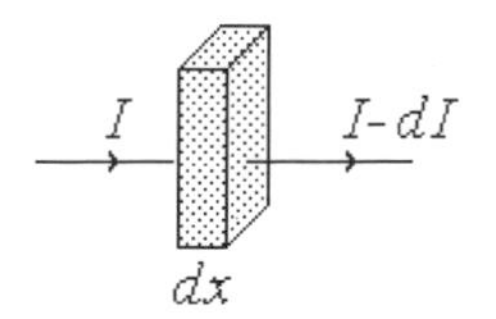

$$-dI = \mu I dx, \tag{4.1}$$

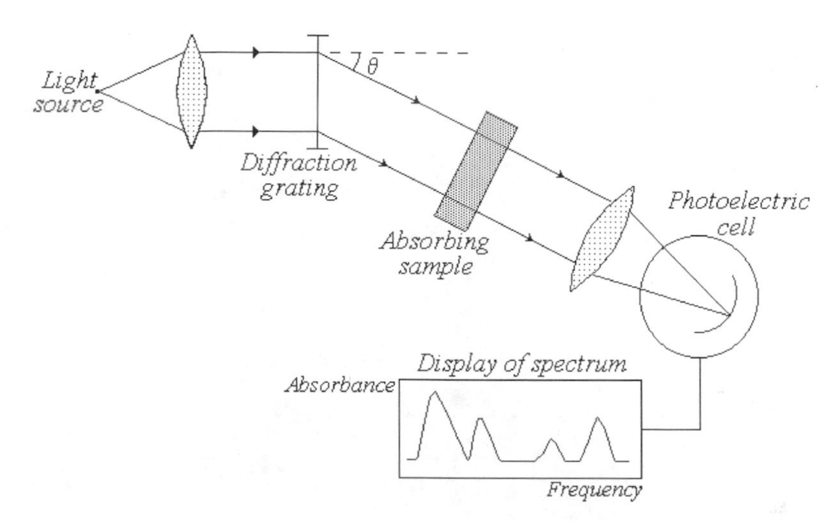

Fig. 4.2 A diagram of a spectrophotometer that operates in the visible region. The assembly of the diffraction grating and/or the absorbing sample, together with the photoelectric cell, is rotated so that the sample is successively exposed to different parts of the spectrum. The intensity of the light emerging from the sample is measured by the photoelectric cell. The absorption spectrum is usually displayed in terms of the % absorbance as a function of the radiation's frequency.

where I is the intensity of the incident beam. The constant of proportionality μ is called the *absorption* or *attenuation coefficient*; it has the SI units m^{-1} but is usually given in cm^{-1}.

The solution of Eq. (4.1) is obtained from the integration

$$\int \frac{dI}{I} = -\mu \int_0^x dx, \tag{4.2}$$

which gives

$$I(x) = I_0 e^{-\mu x}, \tag{4.3}$$

where I_0 is the beam's initial intensity and $I(x)$ is its intensity after passing a distance x through the matter (Fig. 4.3).

The quotient

$$\log_{10}\left(\frac{I_0}{I(x)}\right) = \frac{\mu x}{2.303} \tag{4.4}$$

is called the *absorbance* or *optical density* (OD).

In dilute solutions, where the solute particles are the principal absorbers, the absorption coefficient, μ, depends on their molar concentration, C,

$$\mu = 2.303\varepsilon C. \tag{4.5}$$

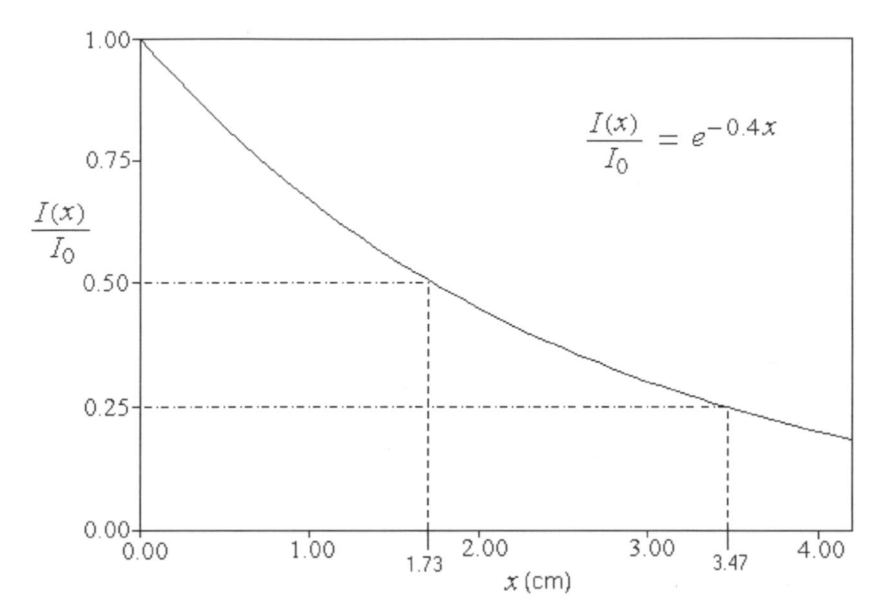

Fig. 4.3 The attenuation of the intensity, I, of a beam of radiation as a function of the thickness, x, of the material medium in the case where the attenuation coefficient $\mu = 0.4\,\text{cm}^{-1}$.

The quantity ε is called the *molar extinction coefficient.* Combining Eqs (4.4) and (4.5) gives

$$I(x) = I_0 e^{-2.303\varepsilon Cx}. \tag{4.6}$$

In this form, the relationship is called *Beer–Lambert's law.*

Example 4.1: The Attenuation of Ultra-Violet Radiation by a Glass Sheet

The intensity of a beam of ultra-violet radiation is halved when it passes through a sheet of glass of thickness 0.9 cm. What thickness of glass would be needed to reduce the beam's intensity to just 1% of its original value?

Calculation: The attenuation coefficient, μ, for the process can be calculated by substituting $x = 0.9\,\text{cm}$ and $I(x) = I_0/2$ in Eq. (4.3):

$$\frac{I_0}{2} = I_0 e^{-0.9\mu},$$

from which we obtain

$$\ln 2 = 0.9\mu$$
$$\mu = \frac{\ln 2}{0.9} = 0.77\,\text{cm}^{-1}.$$

Substituting $I(x) = 0.01 I_0$ in Eq. (4.3) gives

$$0.01 I_0 = I_0 e^{-0.77x},$$

from which

$$x = \frac{\ln 10^2}{0.77} = 5.98\,\text{cm}.$$

An alternative method is first to calculate the number of times, n, that the intensity must be halved until it is just 1% of its original value,

$$\frac{0.01 I_0}{I_0} = \left(\frac{1}{2}\right)^n$$

$$n = 6.64.$$

It follows that 6.64 glass sheets each of thickness 0.9 cm are required, i.e. a total thickness of $6.64 \cdot 0.9 = 5.98$ cm.

4.1.2 Mechanisms of the Absorption of Radiation

Radiation in the microwave, infra-red, visible and ultra-violet regions of the electromagnetic spectrum is absorbed/attenuated by one or more of the following three mechanisms: ionisation, excitation and scattering. Each of these mechanisms is characterised by its own particular absorption/attenuation coefficient. An absorbing medium's overall absorption coefficient, μ, is the sum of these coefficients:

$$\mu = \mu_{Scattering} + \mu_{Excitation} + \mu_{Ionisation}. \tag{4.7}$$

1. Ionisation

This can occur if the energy of the photons in the incident radiation is equal to or greater than the ionisation energy of the atoms or molecules in the absorbing medium. Electrons are then released from the atoms and molecules and photons are annihilated, resulting in a corresponding reduction in the intensity of the beam of radiation.

2. Excitation

At room temperature, the atoms and molecules that comprise matter are usually in their ground state. Photons possessing the right amounts of energy can excite the atoms and molecules to higher energy levels. When this occurs, radiation is absorbed and the intensity of the transmitted beam is reduced accordingly.

This mechanism is very selective; a given material only absorbs certain frequencies. Consequently, this phenomenon can be made the basis for the identification of materials and for determining the concentration of the absorbing solutes in solutions.

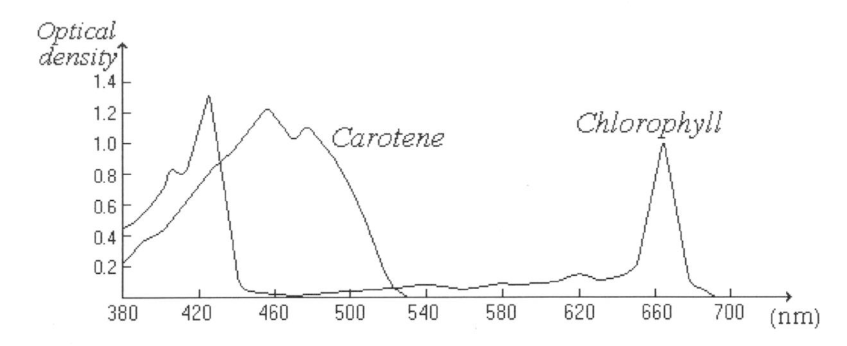

Fig. 4.4. The absorption spectrum of chlorophyll and carotene in the visible range. Chlorophyll has absorbance maxima at around 435 nm and 665 nm. Carotene has absorbance maxima at around 450 nm and 490 nm.

Excitation occurs in gases when the energy of the photons in the incident radiation exactly equals the difference between the energy of an excited state and that of the ground state. Only such photons are absorbed by the gas and so the absorption spectra of gases are line spectra that comprise discrete frequencies.

Because of the much more complicated patterns of their energy levels, liquids and solids absorb whole ranges of frequencies. Thus, their absorption spectra are typically *band spectra* (Fig. 4.4). Nevertheless, the absorption is still sufficiently selective in many cases for these spectra to still be used for identifying the materials.

Having risen to an excited state, the atom or molecule can lose its excess energy and return to the ground state in three different ways:

a. By non-radiative transitions such as those that result from collisions between the excited particles and other atoms or molecules in the material. In the process, the excess energy is transferred and shared out between other energy levels in the system. The overall result is an increase in the average kinetic energy of the particles comprising the matter, i.e. the conversion of the incident radiation's energy to heat energy. This is the most common method by which excited particles lose their excess energy.
b. The excess energy is converted to chemical energy. For example, the excess energy possessed by the excited particle can initiate a chemical reaction such as a photochemical reaction.
c. The excited particles may return to the ground state by spontaneous or stimulated emission of a photon. Generally, the photon will be emitted in a different direction from that of the incident beam.[4]

[4]The laser is an exception to this rule – see Chapter 6.1.

The average lifetime of an excited state is from 10^{-9} to 10^{-6} seconds. If nothing disturbs the particle during this time, it ultimately returns to the ground state by emitting a photon, namely, by the third of the above methods. However, this does not usually happen. During its lifetime, short though it may be, thousands of collisions will have taken place between the excited particle and the other particles in the system. Consequently, it is by the first method – non-radiative transitions – that excited particles most commonly lose their excess energy.

3. Scattering

We discussed the scattering of electromagnetic radiation by free electrons in Chapter 2.3 under the heading of the Compton Effect. Here, we will be concerned with the scattering of radiation by bound electrons. This phenomenon is most easily interpreted in terms of the wave theory of radiation.

When a periodic electromagnetic wave impinges on an object, the electric and magnetic fields associated with it interact with the electrons in the atoms it comprises. To a first approximation, the effect of the magnetic field can be ignored, its magnitude in the case of a plane electromagnetic wave being of the order $1/c$ that of the electric field. The electric field induces oscillating electric dipoles in the object's atoms and molecules. Acting like antennas, these dipoles, in turn, radiate secondary or *scattered* waves in all directions. As a result, the intensity of the radiation in the direction of the original beam is attenuated. The frequency of the scattered waves is generally the same as that of the original wave.

The dipole oscillations are induced parallel to the electric field of the incident wave, i.e. in the plane normal to its direction of propagation. Consequently, the secondary radiation is emitted at maximum intensity in the direction perpendicular to the dipole's axis and at minimum intensity in the direction parallel to the axis. Because of this, the scattered radiation is always polarised to some degree (Fig. 4.5).

The intensity of the scattered radiation is proportional to the square of the amplitude of the vibrations of the induced electric dipole. The closer the frequency of the incident radiation to the natural frequency of the dipole, the greater the amplitude of the induced vibrations; this is a straightforward example of *amplitude resonance.*

The natural frequencies of atomic electric dipoles are in the ultra-violet region of the spectrum. Hence, the blue and violet constituents in white light are scattered more than the shorter frequency red and yellow constituents. This explains why when a beam of white light is scattered by smoke particles, the light scattered at right angles to the direction of the incident beam has a

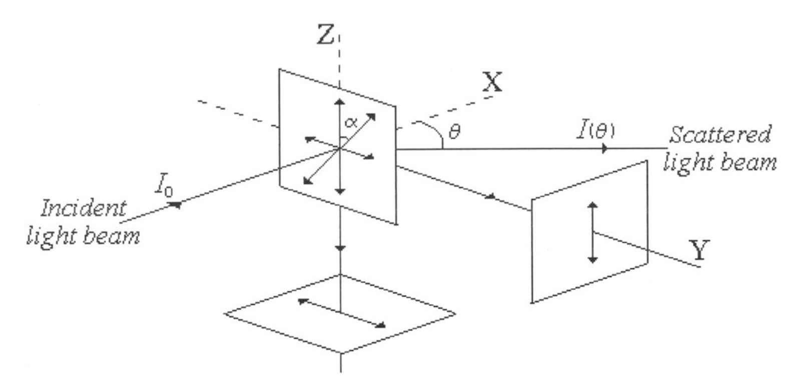

Fig. 4.5. The scattering of a beam of unpolarised electromagnetic radiation. The incident beam of intensity I_0 propagates along the X-axis. The figure shows an oscillating electric dipole that is induced in the ZY plane at an angle α to the Z-axis. The linearly polarised radiation scattered along the Y-axis is that produced by the z-component of the dipole. Similarly, the linearly polarised radiation scattered along the Z-axis is that produced by the y-component of the dipole. The radiation that continues along the X-axis is unpolarised. The radiation of intensity $I(\theta)$ scattered at an angle θ is partially polarised.

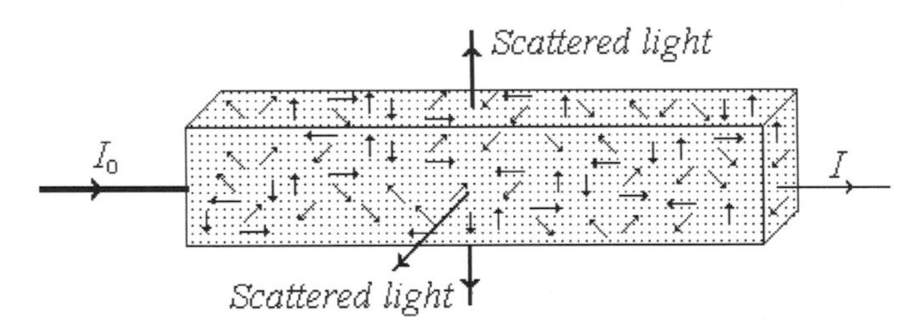

Fig. 4.6. The scattering of an incident beam of white light of intensity I_0 by smoke particles. The scattered light of overall intensity I_S is polarised and has a blue hue. The transmitted beam of intensity $I = I_0 - I_S$ is unpolarised and has a red/yellow tint. The intensity of the transmitted beam decreases exponentially with the length of its path through the smoke.

blue hue and is linearly polarised, whereas that which is transmitted along the original direction has a red/yellow tint and is unpolarised (Fig. 4.6).

When viewed from an orbiting spacecraft or from the moon's surface, the sky – the dome overhead – is black at day as well as at night. The blue of the sky seen from the earth's surface is a result of the scattering of sunlight by the atmosphere. The blue and violet constituents of sunlight, i.e. the shorter wavelengths, are scattered more by the particles comprising the earth's atmosphere than are the other constituents. Thus, an observer on the earth's surface sees a blue sky overhead in all directions, except in that directly towards the sun. At sunrise and sunset, when the sunlight's path through the atmosphere to the observer's eye is longest, even more of the blue and violet constituents are scattered, leaving the transmitted beam with little of these colours. Thus, at these times, the sun's disc appears red.

The blue of the sky is also polarised, the direction of polarisation at each point changing during the day as the sun moves across the sky. It has been suggested that this changing pattern in the sky's polarisation is used as a navigational reference grid by those creatures, such as bees, whose eyes, unlike the human eye, can distinguish between different modes of light polarisation.

With an absolutely homogeneous medium, such as perfect crystal, the secondary waves emitted by the oscillations induced in the electrons all interfere with one another destructively. Hence, scattering is not observed. The incident beam is just reflected and refracted according to the laws of geometrical optics. If the earth's atmosphere was absolutely homogeneous, the sky would also appear black to observers looking upwards from its surface during daytime.

The quantitative treatment of the scattering of light by particles, whose dimensions are small relative to the wavelength of the radiation incident upon them, was accomplished by Rayleigh.[5] He showed that the intensity of the scattered light depends on $1/\lambda^4$ where λ is the wavelength of the incident light.[6] If the particles are larger, such as dust particles, Rayleigh's treatment is no longer valid. Under these circumstances, scattered white light is untinted; it is white. Thus, though water vapour in the atmosphere contributes to the light scattering that produces the blue of the sky, when the vapour condenses into fine drops whose dimensions are relatively large, white clouds are seen.

Experiments show that though all electromagnetic waves propagate through empty space with velocity c, their velocity, v, through material mediums depends on the nature of the material and on the frequency, f, of the wave; the phenomenon is called *dispersion*. The ratio, n, between the two velocities is called the *refractive index*:

$$n = \frac{c}{v}. \tag{4.8}$$

Refractive indices are always equal to or greater than unity.

The colour of light, i.e. the energy of its photons and, hence, its frequency, is unaffected by the changes in velocity that occur when light passes from one medium to another; a blue swimming costume remains blue in the water. However, if its wavelength is measured it is found to be different. The light's wavelength, λ', in the material medium is always less than that its wavelength,

[5]Lord John Rayleigh 1842–1919, English physicist who was awarded the Nobel Prize in 1904 for the discovery of the element argon.

[6]See Appendix 4.3.1.

λ, in empty space, such that the fundamental relationships

$$f = \frac{c}{\lambda}, \ f = \frac{v}{\lambda'} \ \text{and} \ \lambda = n\lambda' \tag{4.9}$$

are respectively satisfied.

Dispersion arises from the fact that the secondary waves radiated by the vibrations of the induced dipoles in the material medium are out of phase with the incident wave; a forced oscillator is not generally in phase with the driving force. As a result, the actual wave produced in the material medium from the superposition of the incident wave and the secondary waves exhibits a phase velocity, v, different from the phase velocity, c, of electromagnetic waves *in vacuo.*

Although no particle or signal can move at a velocity greater than that of electromagnetic waves *in vacuo*, c, the velocity of high energy charged particles in a material medium can exceed the phase velocity, $v = c/n$, at which electromagnetic waves propagate in the medium. Consequently, if the particle itself emits electromagnetic waves, these will form a shock-wave analogous to that formed when an object moves at supersonic speeds through the air. The electromagnetic shock-waves, called *Çerenkov*[7] *radiation*, propagate along conical surfaces that make an angle α relative to the direction of the particle such that

$$\sin\alpha = \frac{v_{radiation}}{v_{particle}}. \tag{4.10}$$

This phenomenon has been employed in techniques for measuring the velocities of high-energy elementary particles.

[7]Pavel Alekseyevich Çerenkov 1904–1990, Russian physicist who was awarded the Nobel prize in 1958 for the discovery of the radiation that bears his name.

Chapter 4.2

Molecular Spectra

4.2.1 Molecular Energies

We can picture a molecule as an assembly of small, almost point, masses – the nuclei of its component atoms, linked together by almost massless springs – the bonding electrons (Fig. 4.7). In this model, the total energy of a molecule, $E_{molecule}$, can be viewed as the sum of its electronic energy, $E_{electronic}$, its vibrational energy, $E_{vibration}$ and its rotational energy, $E_{rotation}$:

$$E_{molecule} = E_{electronic} + E_{vibration} + E_{\text{rotation}}. \tag{4.11}$$

The vibrational and rotational energies are both related to the motion of the nuclei of the molecule's constituent atoms.

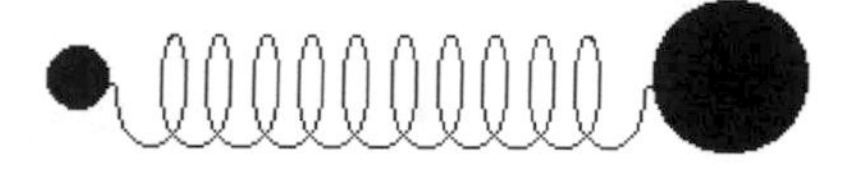

Fig. 4.7. A diatomic molecule pictured as two spherical masses linked by a massless spring.

Like its electronic energy, a molecule's vibrational and rotational energies are also quantised, i.e. they can only take certain discrete values. Thus, every molecule has a characteristic set of vibrational energy levels and rotational energy levels. Typical values for the differences, ΔE_{elec}, ΔE_{vib} and ΔE_{rot}, between adjacent electronic, vibrational and rotational energy levels are, respectively,

$$\Delta E_{elec} = 1 - 10\,\text{eV}$$
$$\Delta E_{vib} \approx 10^{-1}\,\text{eV}$$
$$\Delta E_{rot} \approx 10^{-4}\,\text{eV},$$

i.e. $\Delta E_{elec} \gg \Delta E_{vib} \gg \Delta E_{rot}$. Consequently, each electronic level can accommodate an array of vibrational sub-levels and, similarly, each vibrational level can accommodate an array of rotational sub-levels (Fig. 4.8).

In common with atoms, molecules too can absorb and emit radiation by means of transitions between the energy levels associated with their electrons, i.e. through changes in their electronic energy. However, there are two further modes by which molecules can absorb or emit radiation:

1. By means of changes in the internal vibrational energy of the molecules.
2. By means of changes in the rotational energy of the molecules.

The photons associated with transitions between electronic levels are in the visible and near ultra-violet range; those associated with transitions between vibrational levels in the same electronic level are in the infra-red; those associated with transitions between rotational levels in the same vibrational level are in the microwave region. The probabilities of the various transitions are expressed by appropriate selection rules.

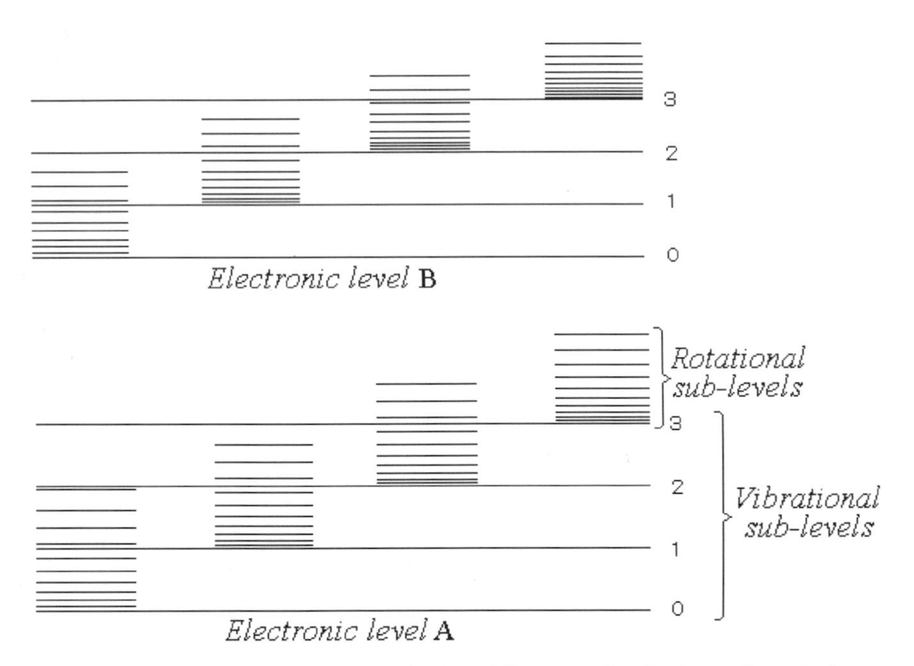

Fig. 4.8 Two electronic energy levels, A and B, in a molecule. A number of vibrational energy levels, $1, 2, 3, \ldots$, are associated with each electronic energy level. Similarly, a number of rotational sub-levels are associated with each vibrational energy level.

Transitions between a molecule's electronic energy levels are usually accompanied by changes in its vibrational and rotational energy. Consequently, when a molecule 'falls' from an excited electronic state to a lower one, its vibrational and rotational sub-levels can also change. Similarly, transitions between vibrational energy levels are usually accompanied by changes in the molecule's rotational energy. Hence, transitions between vibrational levels often occur with changes in the rotational sub-level.

To a first approximation, only molecules possessing an electric dipole moment can absorb or emit electromagnetic radiation. Consequently, non-polar diatomic molecules such as H_2 and Cl_2, and symmetric polyatomic molecules such as CO_2 and CH_4 do not exhibit rotational or vibrational spectra; radiative transitions between rotational and vibrational states in such molecules do not occur. Non-radiative transitions can occur as a result of collisions. Information about vibrational transfers in such molecules can be obtained from Raman scattering (Section 4.2.5 below).

4.2.2 Rotational Spectra

Figure 4.9 shows a typical rotational absorption spectrum, that of gaseous hydrogen chloride, $HCl_{(g)}$. The spectrum exhibits two principal characteristics:

1. The spectrum comprises many lines, all of comparable intensity.
2. The absorption lines are equally spaced along the frequency axis.

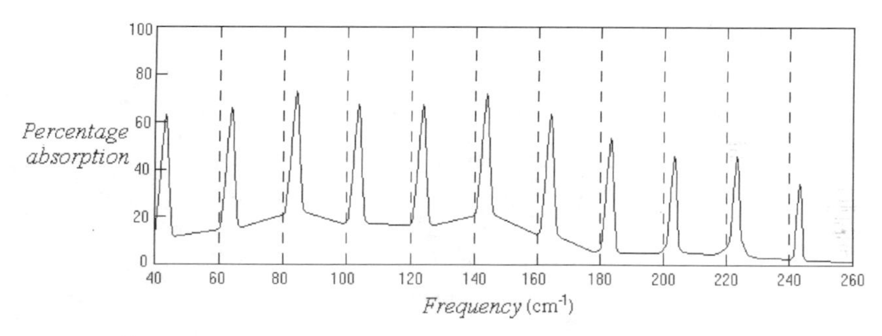

Fig. 4.9 Rotational absorption spectrum of $HCl_{(g)}$. The horizontal frequency axis is scaled in units of the wave number, $1/\lambda$, where λ is the radiation's wavelength in centimetres ($1\,\text{cm}^{-1} \equiv 3 \cdot 10^{10}\,\text{Hz}$).

By analogy with the expression $K = p^2/2m$, that relates an object's kinetic energy, K, to its linear momentum, p, the rotational energy, E_{rot}, of a diatomic molecule ('dumb-bell'), such as that pictured schematically in Fig. 4.10, revolving around an axis through its centre of mass is given by

$$E_{rot} = L^2/2I, \tag{4.12}$$

where L is the molecule's angular momentum and I is its moment of inertia about this axis.

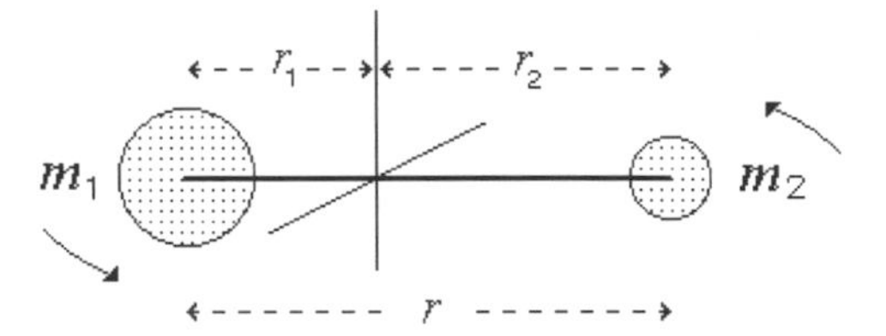

Fig. 4.10. Schematic representation of a diatomic molecule such as HCl rotating around an axis through its centre of mass.

The rotational energy eigenvalues of the molecule emerge from the quantisation of its angular momentum. The square, L^2, of the magnitude of the molecule's angular momentum obeys the quantisation condition

$$L^2 = J(J+1)\hbar^2, \tag{4.13}$$

where the *rotational quantum number* J can take the values $J = 0, 1, 2, 3, \ldots$ Substituting the allowed values of L^2 in Eq. (4.12) gives for the rotational energy eigenvalues (the rotational energy levels):

$$E_{rot} = \frac{J(J+1)\hbar^2}{2I} = J(J+1)Bh, \tag{4.14}$$

where the constant B is defined by

$$B = \frac{h}{8\pi^2 I}. \tag{4.15}$$

The difference, ΔE_{rot}, between two adjacent rotational energy levels, the J level and the $J+1$ level, is

$$\begin{aligned} \Delta E_{rot} &= [(J+1)(J+2) - J(J+1)]Bh \\ &= 2Bh(J+1). \end{aligned} \tag{4.16}$$

It follows that the difference, ΔE_{rot}, increases by an amount $2Bh$ with each unit increase in the quantum number J, i.e. the differences between adjacent rotational energy levels increase linearly with the quantum number J.

The selection rule for radiative rotational transitions is $\Delta J = \pm 1$. Hence, the frequencies, f, of the photons associated with such transitions will be

$$f = \frac{\Delta E_{rot}}{h} = 2B(J+1), \tag{4.17}$$

and the difference, Δf, between the frequencies of successive spectral lines a constant

$$\Delta f = 2B. \tag{4.18}$$

This result is consistent with the second of the principal characteristics of the rotational spectra noted above, the equal spacing of the absorption lines along the frequency axis.

Bearing in mind that a system cannot absorb the photon corresponding to a 'jump' from a given state unless that state is already occupied, the abundance of lines in the rotational absorption spectrum indicates that a substantial number of rotational states, apart from the ground state, are extensively occupied under normal conditions.

Furthermore, the intensities of these lines indicate that these states are occupied to a comparable extent. In a sample of gaseous molecules, the ratio between the number, N_J, of molecules occupying the J rotational state to the number, N_{J+1}, occupying the adjacent state $J+1$ is given by the Maxwell–Boltzmann distribution

$$\frac{N_{J+1}}{N_J} = e^{-\Delta E_{rot}/kT}, \tag{4.19}$$

where k is the Boltzmann constant and T is the sample's absolute temperature.[8]

In the case of HCl(g), the constant $B = 3.0 \cdot 10^{11}$ Hz.[9] Substitution in Eq. (4.16) gives, for the difference, ΔE_{rot}, between the energy of the $J = 0$ and $J = 1$ states,

$$\begin{aligned} \Delta E_{rot} &= 2Bh(J+1) \\ &= 2 \cdot 3.0 \cdot 10^{11} \cdot 4.14 \cdot 10^{-15} \cdot 1 \\ &= 2.5 \cdot 10^{-3}\,\text{eV}. \end{aligned} \tag{4.20}$$

At room temperature (25°C), $kT = 2.6 \cdot 10^{-2}$ eV. Substituting these values in Eq. (4.19) gives

$$\begin{aligned} \frac{N_{J+1}}{N_J} &= e^{-2.5 \cdot 10^{-3}}/2.6 \cdot 10^{-2} \\ &\approx 0.9, \end{aligned} \tag{4.21}$$

i.e. the $J = 0$ and $J = 1$ states are occupied almost to the same extent. A similar calculation shows that even the $J = 5$ state has a population about 20% that of the ground state. This explains the abundance of absorption lines in the rotational spectra, the first of the principal characteristics noted above.

We have examined only the rotational spectrum of diatomic molecules. The spectra of polyatomic molecules are more complex but involve similar characteristics to those involved in the diatomic case.

---

[8] See Section 2.1.6 – Eq. (2.20).

[9] See Example 4.2.

Example 4.2: The Interatomic Distance in the HCl Molecule

The spacing, $\Delta(1/\lambda)$, between the lines in the rotational spectrum of HCl(g) (Fig. 4.9) is $20\,\text{cm}^{-1}$. Calculate the distance between the nuclei (the length of the bond) in the HCl molecule.

Calculation: The moment of inertia, I, relative to its centre of mass, of a diatomic (dumb-bell) molecule can be calculated (i) from its rotational absorption spectrum and (ii) from mechanics. The interatomic distance can be computed by equating the mathematical expressions for the two methods.

1. Combining Eqs (4.15) and (4.18) gives the following expression for the molecule's moment of inertia, I, relative to its centre of mass:

$$I = \frac{h}{4\pi^2 \Delta f}. \tag{4.22}$$

In terms of the frequency, the spacing, Δf, between the spectral lines is

$$\begin{aligned} \Delta f &= \Delta\left(\frac{1}{\lambda}\right) \cdot c \\ &= 20 \cdot 10^2 \cdot 3 \cdot 10^8 \\ &= 6.0 \cdot 10^{11}\,\text{Hz}. \end{aligned} \tag{4.23}$$

Substituting this value in Eq. (4.22) gives

$$\begin{aligned} I &= \frac{6.62 \cdot 10^{-34}}{4\pi^2 \cdot 6.0 \cdot 10^{11}} \\ &= 2.8 \cdot 10^{-47}\,\text{kg} \cdot \text{m}^2. \end{aligned} \tag{4.24}$$

2. The moment of inertia, I, relative to its centre of mass, of a dumb-bell, comprising two masses, m_1 and m_2, at a distance r from one another is[10]

$$I = m' r^2, \tag{4.25}$$

where m', the molecule's reduced mass, is given by

$$m' = \frac{m_1 m_2}{m_1 + m_2}. \tag{4.26}$$

In the case of the HCl molecule,

$$\begin{aligned} m_1 &= m_{\text{H}} = 1.66 \cdot 10^{-27}\text{kg} \\ m_2 &= m_{\text{Cl}} = 35.0 \cdot 1.66 \cdot 10^{-27}\text{kg}. \end{aligned}$$

[10] See Appendix 4.3.2.

Hence,

$$m' = \frac{1 \cdot 35.0}{1 + 35.0} \cdot 1.66 \cdot 10^{-27} \\ = 1.61 \cdot 10^{-27}\,\text{kg} \quad . \tag{4.27}$$

Substituting the value of the reduced mass, m', and that of the moment of inertia, I, obtained from the rotational spectrum in Eq. (4.25) gives for the bond length, r,

$$r = \sqrt{\frac{I}{m'}} \\ = \sqrt{\frac{2.8 \cdot 10^{-47}}{1.61 \cdot 10^{-27}}} \\ = 1.3 \cdot 10^{-10}\text{m} \quad . \tag{4.28}$$

4.2.3 Vibrational Spectra

Whereas diatomic molecules can vibrate in just one mode, to and fro along the axis joining their atomic centres, polyatomic molecules comprising n atoms have $(3n - 6)$ vibrational modes. If the molecule is also linear it has one more mode, i.e. $(3n - 5)$ modes altogether (Fig. 4.11).

At low energies, the vibrations of a diatomic molecule can be described in terms of a harmonic oscillator. According to Planck's quantum hypothesis, the energy, E, of a harmonic oscillator of natural frequency f_0 can only take the values $E = nhf_0$ where $n = 0, 1, 2, 3, \ldots$. This implies that in its ground state, $n = 0$, the energy of the oscillator should equal zero and that the oscillator should be completely at rest. However, this would contravene the Heisenberg uncertainty principle; complete rest implies no uncertainty whatsoever as to the system's position and linear momentum. This lacuna in the original quantum hypothesis is corrected by quantum mechanics in which the correct expression for the energy, E_{vib}, of the oscillator is obtained from the solution of the appropriate Schrödinger equation.

The potential energy, U, of a simple harmonic oscillator is given by

$$U = \frac{kx^2}{2}, \tag{4.29}$$

where k is the oscillator (spring) constant. Substituting this expression for the potential energy in the one-dimensional time independent Schrödinger

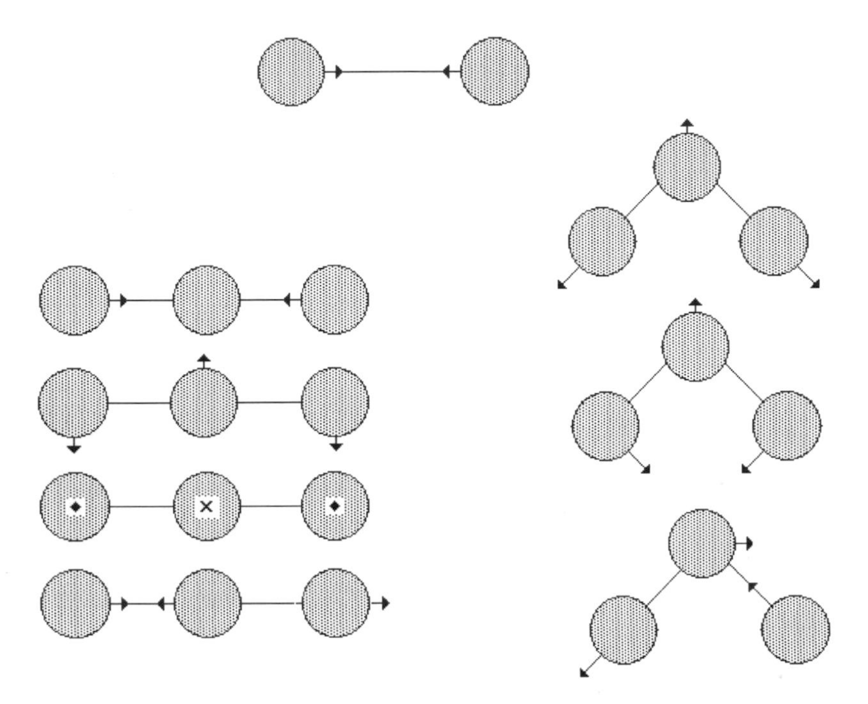

Fig. 4.11 The vibrational modes of a diatomic and triatomic molecule.

equation gives

$$\frac{d^2\psi}{dx^2} + \frac{8\pi^2 m'}{h^2}\left(E_{vib} - \frac{kx^2}{2}\right)\psi = 0. \tag{4.30}$$

The energy eigenvalues given by the solution to this equation are

$$E_{vib} = \left(v + \frac{1}{2}\right)hf_0, \tag{4.31}$$

where the oscillator's natural frequency, f_0, is given by

$$f_0 = \frac{1}{2\pi}\sqrt{\frac{k}{m'}}. \tag{4.32}$$

The integer v, is called the *vibrational quantum number* and can take the values $v = 0, 1, 2, 3, \ldots$. Thus, in its ground state, $v = 0$, the oscillator has a 'zero-point' energy of $hf_0/2$ and so is never completely at rest.

At low energies, where the diatomic molecule can be regarded as a harmonic oscillator, the vibrational energy levels are equally spaced. However, at higher energies, i.e. higher vibrational quantum numbers, the oscillator becomes anharmonic, as is shown in Fig. 4.12, and the energy levels close up. As the energy increases, so does the amplitude of the vibrations. At the

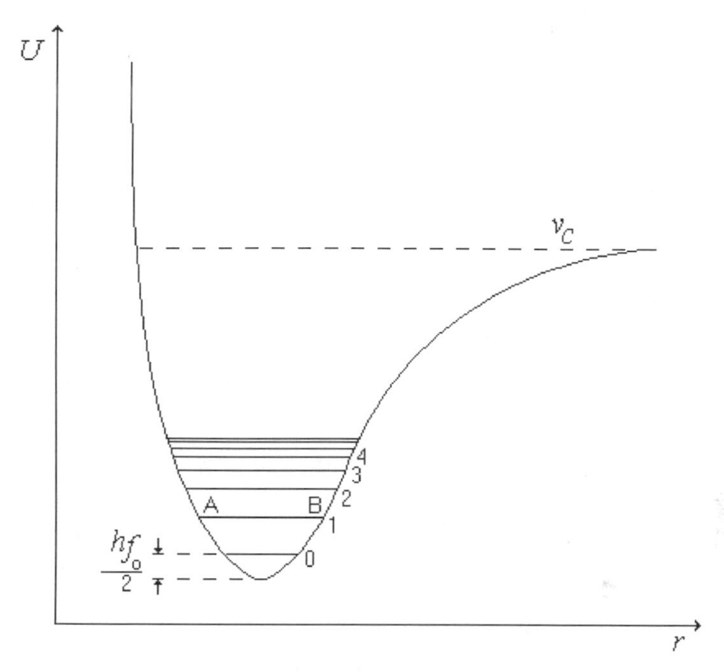

Fig. 4.12 The potential energy, U, of a diatomic molecule as a function of the interatomic distance, r. The horizontal lines represent the vibrational energy levels, E_{vib}, of the molecule. At low energies the curve is parabolic, consistent with harmonic oscillations.

energy corresponding to the quantum number v_c, the bond linking the atoms breaks.

In practice, the vibrational spectra of diatomic molecules are very simple, usually comprising just a single strong line called the *fundamental line.* There are two reasons for this:

1. The selection rule for radiative vibrational transitions is $\Delta v = \pm 1$. It follows, from Eq. (4.31), that only photons of frequency f_0 will be absorbed or emitted in transitions between low-energy vibrational states.
2. A typical value for the difference between two adjacent vibrational energy levels is $\Delta E_{vib} = 0.5\,\text{eV}$. Consequently, at room temperature (25°C), the numerical value of the factor $e^{-\Delta E_{vib}/kT}$ in the Maxwell–Boltzmann distribution function is about 10^{-8}, i.e. the population of the excited states is negligible. All the molecules are in the ground state and so there are no anharmonic vibrations or transitions between higher states; only transitions between the $v = 0$ and $v = 1$ states take place.

At low resolution, the fundamental line appears as a broad band. This is because the transitions between the vibrational states are accompanied by transitions between rotational states (Fig. 4.13). In these transitions, photons with energies in the range $E_{Photon} = \Delta E_{vib} \pm \Delta E_{rot}$ where ΔE_{rot} is the energy involved in these rotational transitions, are absorbed or emitted by the molecule. This range of photon energies shows up as a broadening of the fundamental line.

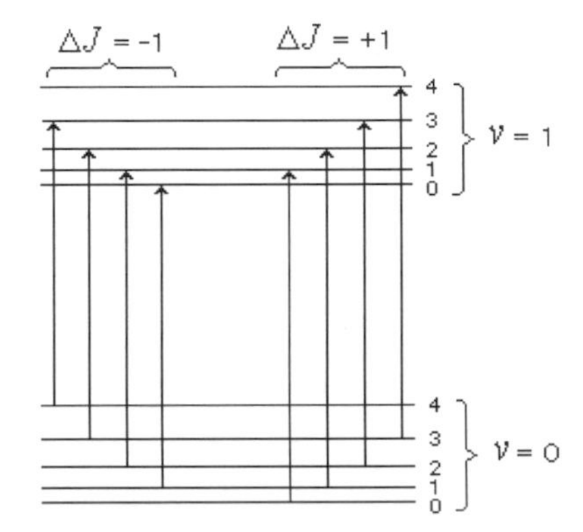

Fig. 4.13. Vibrational–rotational radiative transitions in a diatomic molecule. The transitions obey the selection rules $\Delta v = \pm 1$ and $\Delta J = \pm 1$.

At higher resolutions, the band of the fundamental line is seen to comprise a series of fine lines, each corresponding to a rotational transition between different rotational sub-levels (Fig. 4.14). The frequencies, f, of these fine lines are given by:

$$f = f_0 \pm 2B(J + 1). \tag{4.33}$$

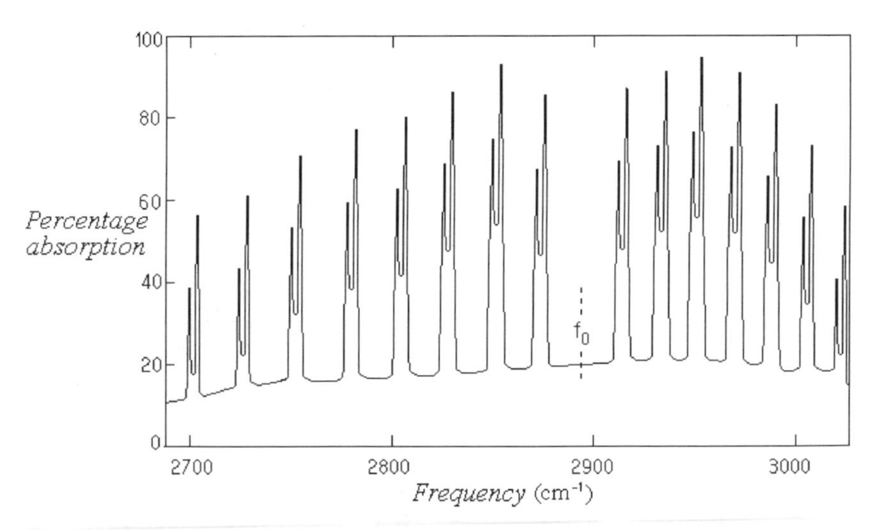

Fig. 4.14 The infra-red absorption spectrum of $HCl_{(g)}$ at high resolution. Each line that accords to a different rotational transition is itself split into two peaks, corresponding to the two isotopes Cl-35 and Cl-37 found naturally.

Note that the $f = f_0$ line does not appear since it requires that $\Delta J = 0$ which contravenes the selection rule $\Delta J = \pm 1$.

A complete elucidation of the infra-red spectrum can only be obtained in the case of simple molecules. Nevertheless, it is often possible to identify the presence of certain functional groups in polyatomic molecules by the approximate constancy of the frequency of the absorption band associated with them, irrespective of the compound in which the group appears. For example, the $-C=O$ carbonyl group vibrates with a frequency of $1{,}706\,\text{cm}^{-1}$ in acetone, $1{,}715\,\text{cm}^{-1}$ in acetaldehyde, $1{,}663\,\text{cm}^{-1}$ in acetic acid and $1{,}736\,\text{cm}^{-1}$ in methyl acetate ($1\,\text{cm}^{-1} \equiv 3 \cdot 10^{10}\,\text{Hz}$). This approximate constancy is the basis of the widespread use of infra-red spectroscopy in the identification of organic compounds and in the elucidation of their structure.

Example 4.3: Vibrational Spectrum of CO

An intense line of wave number $2{,}144\,\text{cm}^{-1}$ appears in the infra-red spectrum of carbon monoxide gas, $CO_{(g)}$.

Calculate:

1. the period of the molecule's vibrations;
2. the force constant of the molecule's bond;
3. the zero-point energy per mole of carbon monoxide.

1. The period, T, of the molecule's vibrations is inversely proportional to the frequency, f_0, of its vibrations, i.e.

$$T = \frac{1}{f_0} = \frac{1}{2144 \cdot 10^2 \cdot 3 \cdot 10^8} = 1.55 \cdot 10^{-14}\,\text{s}.$$

2. From Eq. (4.38), the force constant, k, of the bond in a diatomic molecule is given by

$$k = 4\pi^2 m' f_0^2.$$

Substituting in Eq. (4.32) gives for the reduced mass, m', of the carbon monoxide molecule

$$m' = \frac{12 \cdot 16}{12 + 16} \cdot 1.66 \cdot 10^{-27} = 1.14 \cdot 10^{-26}\,\text{kg}.$$

Thus, the molecule's force constant is

$$k = 4\pi^2 \cdot 1.14 \cdot 10^{-26} \cdot \left(\frac{1}{1.55 \cdot 10^{-14}}\right)^2 = 1862\text{N} \cdot \text{m}^{-1}.$$

3. In its ground state, $v = 0$, an oscillator has a zero-point energy of $E_0 = hf_0/2$. The zero-point energy of a mole of carbon monoxide molecules is, therefore,

$$E_0 = \frac{6.63 \cdot 10^{-34}(10^{14}/1.55)}{2} \cdot 6.02 \cdot 10^{23} = 1.28 \cdot 10^4\,\text{J/mol}.$$

4.2.4 Electronic Spectra

Transitions between molecular electronic levels ($\Delta E_{elec} = 1 - 10\,\text{eV}$) are usually accompanied by transitions between the molecule's vibrational and rotational levels. Accordingly, the frequencies of the photons emitted and absorbed in such transitions are given by

$$f = \frac{\Delta E_{elec} + \Delta E_{vib} + \Delta E_{rot}}{h}. \tag{4.34}$$

These frequencies are in the visible and the near ultra-violet. Transitions between electronic states appear as a spectral band. At higher resolutions these bands are seen to comprise closely packed lines corresponding to the various upper and lower vibrational and rotational levels.

All molecules, including homonuclear ones such as H_2 and N_2, exhibit electronic spectra since any change in the molecule's electronic configuration is always accompanied by a change in dipole moment.

An analysis of the fine structure of electronic spectra shows that the selection rule that applies to pure vibrational transitions, $\Delta v = \pm 1$, does not hold when the vibrational transition occurs together with an electronic transition. This anomaly can be explained in terms of the following two general rules known as the *Franck–Condon principle*.

1. The absorption of electronic energy occurs rapidly; a typical electronic 'jump' lasts just 10^{-16} s. On the other hand, bond vibrations are relatively sluggish, $\sim 10^{-13}$ s. Thus, it can be assumed, the position and velocity of the atomic nuclei in the molecule do not change during an electronic transition.
2. The probability density in the ground state is greatest at the mid-point of the atomic nuclei's vibrations. Thus, it is from this point that transitions from the vibrational ground state, $v = 0$, will most probably occur. In other vibrational states, the probability density is greatest at the extremes of the vibration and it is from these points that the vertical lines representing the transitions from these states should be drawn – points A and B in Fig. 4.12.

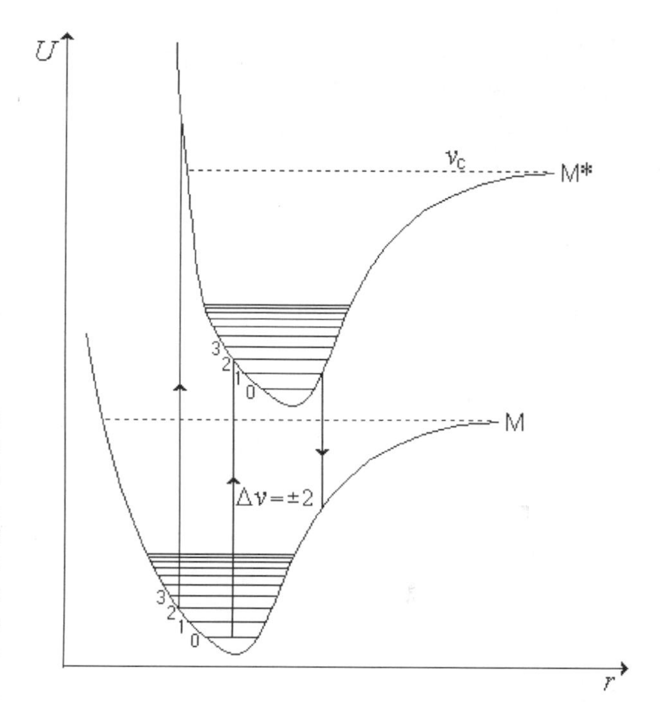

Fig. 4.15. M and M* are the potential energy curves of the ground state and of an excited electronic state, respectively, as a function of the interatomic distance, r. It is apparent that a transition from one electronic state to another may well be accompanied by a transition between vibrational sub-states.

Figure 4.15 shows the potential energy curves of the ground state, M, and an excited electronic state, M*, of a hypothetical molecule. Transitions between electronic states can be represented by vertical lines on the potential energy diagrams. Note that the dimensions of the excited state are different from those of the ground state; electronic excitation often affects molecular geometry. Because the geometry of the two states changes only slowly, a molecule at the mid-point of the vibrational sub-state $v = 0$ in the electronic ground state can rise to the vibrational sub-state $v = 2$ in the excited electronic state, or can fall back from there into the ground state, even though both of these transitions are contrary to the selection rule $\Delta v = \pm 1$. Similarly, the excitation of a molecule occupying an elevated vibrational sub-state in the electronic ground state can lead to the molecule's dissociation. This occurs when the resulting vibrational sub-state in the excited electronic state has a quantum number greater than v_c, which can be ascribed to the level asymptotic to the potential energy curve.

The electronic spectra of molecules composed of many atoms are very complex and are exceedingly difficult to analyse; their spectra are generally made up of a number of broad bands each corresponding to a particular electronic transition and each resolvable into numerous vibrational lines. Almost all such molecules absorb ultra-violet radiation at wavelengths less than 150 nm.

Coloured materials, such as pigments and dyes, absorb and emit components of visible light, i.e. radiation of wavelengths in the range 400 nm to 700 nm. This radiation corresponds to photon energies from 3.0 eV down to 1.7 eV. These materials must, therefore, possess at least one electronic energy level that is between 1.7 eV and 3.0 eV above their electronic ground state. This amount of energy is substantially less than the bond energy of a typical covalent bond. Consequently, a material's colouration indicates that at least one of its electrons is relatively loosely bound and can be elevated to a higher level with comparative ease. This electron is typically an unpaired electron or an electron in the π orbital associated with multiple bonds.

Example 4.4: The Excitation of π Electrons

Estimate the first excitation energy of a π electron in butadiene,

$$CH_2{=}CH{-}\ CH{=}CH_2.$$

Butadiene is an example of a conjugated molecule; see Fig. 4.16. In such molecules, the carbon atoms comprising the molecule's primary structure are linked by covalent σ bonds, all of roughly the same length, $\sim$1.45 Å. The four electrons associated with the two double bonds constitute π bonds that are delocalised over the whole length of the molecule. Allowing for the extension of the π bonds by half a σ bond length beyond the molecule's ends, these π electrons can be considered to be confined to a one-dimensional potential well of length $4 \cdot 1.45 \approx 5.8$ Å.

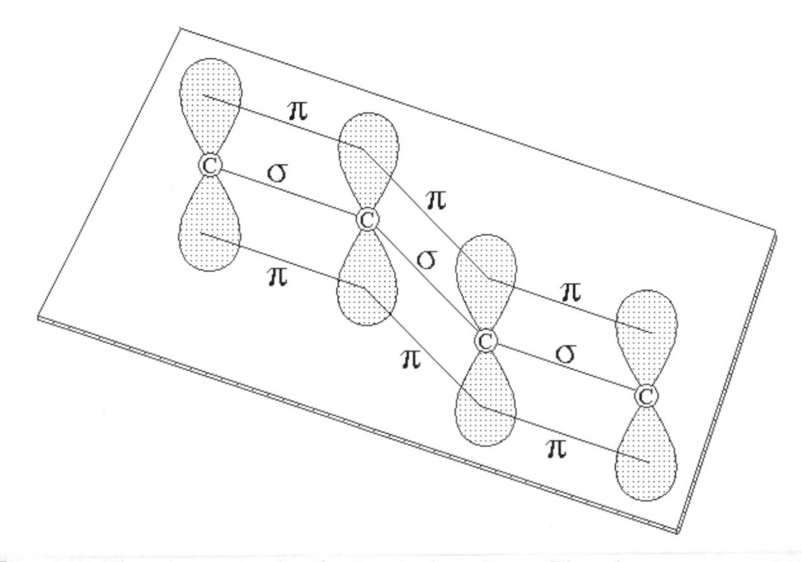

Fig. 4.16 The electronic distribution in butadiene. The electrons comprising the σ bonds are localised between the carbon atoms. The electrons comprising the π bonds are delocalised along the whole length of the molecule.

In the molecule's ground state, two of the four π electrons will fill the $n = 1$ level and the other two the $n = 2$ level, as required by the exclusion principle. Consequently, first excitation will involve raising an electron from the $n = 2$ to the $n = 3$ level.

The energy levels of a particle confined to a one-dimensional potential well are given by Eq. (2.95)

$$E_n = \frac{n^2 h^2}{8mL^2} \quad (n = 1, 2, 3, \ldots).$$

The first excitation energy, ΔE, of the π electrons in butadiene is, therefore,

$$\begin{aligned}\Delta E &= \frac{h^2}{8mL^2}(3^2 - 2^2)\\ &= \frac{(6.6 \cdot 10^{-34})^2}{8 \cdot 9.1 \cdot 10^{-31} \cdot (5.8 \cdot 10^{-10})^2 \cdot 1.6 \cdot 10^{-19}} \cdot (9 - 4)^{\cdot}\\ &= 5.5\,\text{eV}\end{aligned}$$

This excitation energy corresponds to the emission or absorption of a photon of wavelength 225 nm. Butadiene strongly absorbs ultra-violet radiation of wavelength 217 nm which indicates that our crude model is in fair agreement with experiment.

4.2.5 Raman Spectra

In 1928, Raman[11] discovered that the light scattered by molecules contains frequencies other than those of the incident radiation; the phenomenon is called *Raman scattering.* The source of the additional frequencies is in the internal structure of the scattering particles. Molecules both vibrate and rotate, and these motions can add to the vibrations induced by the electric field in the incident radiation. The phenomenon is best understood in terms of photons.

As a first possibility, a photon of energy hf striking a molecule will be scattered without any change in its energy (frequency). The molecule rises to a higher energy state from which it returns at once to its ground state by emitting an identical photon. This is the mechanism of the normal *Rayleigh scattering.* A second possibility is that the excited molecule emits a photon of lower energy (frequency) than that which struck it, using the surplus to enhance the energy of its vibrations and rotations. In this case, the energy of the emitted photon is $h(f - f_v)$ and the increase in the molecule's vibrational and/or rotational energy is hf_v (Fig. 4.17).

In rare instances, the incident photon will strike a molecule whose vibrational and/or rotational energy has already been enhanced. In this case, the

[11] Sir Chandrasekhara Raman 1888–1970, Indian physicist who was awarded the Nobel prize in 1930 for his researches on light scattering.

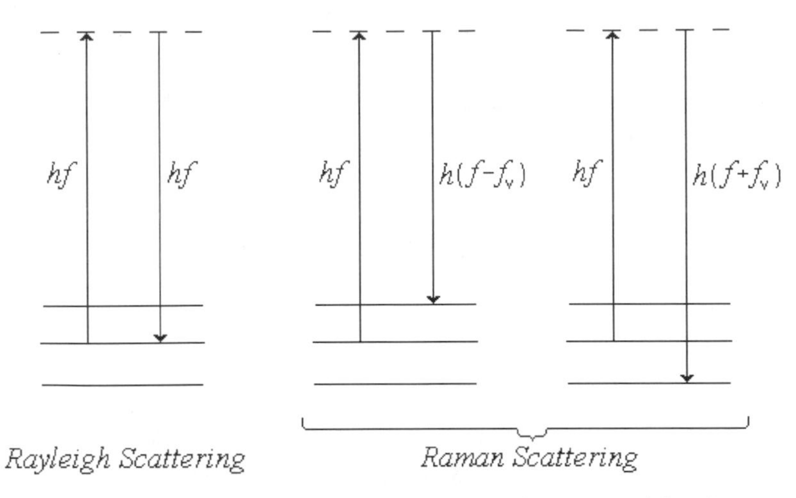

Fig. 4.17 Raman and Rayleigh scattering interpreted in terms of the absorption and emission of photons. In Rayleigh scattering the photon emitted is identical to that absorbed. In Raman scattering the emitted photon can have a slightly higher or lower energy than that of the incident photon.

molecule can emit a photon of energy $h(f + f_v)$. Thus, the scattered radiation can contain frequencies that are both slightly higher, $f + f_v$, and slightly lower, $f - f_v$, than that of the incident radiation.

Raman scattering is employed in the elucidation of molecular structures. For example, in theory, the molecule N_2O could have either of the two possible linear configurations: NON and NNO. The first configuration is symmetrical; the second asymmetrical. This difference in their structures expresses itself as a difference in their vibrational modes. The Raman scattering of N_2O corresponds to vibrations of the asymmetrical configuration and thus could the molecule's actual structure be deduced.

The study of Raman spectra has many advantages. Firstly, they are comparatively simple to observe; the incident radiation can be chosen such that the scattered radiation is in a region of the spectrum where it can be conveniently viewed. Secondly, homopolar diatomic molecules exhibit Raman spectra even though they do not interact directly with infra-red radiation as do other more complicated molecules whose vibrations are absent in normal spectra.

Chapter 4.3
Fluorescence and Phosphorescence

When white light is passed through an aqueous solution of fluorescein, the transmitted beam is seen to have a red hue. Evidently, the blue component of the incident light is absorbed by the fluorescein. However, when viewed from the side, the same solution appears green. Thus, though it absorbs blue light, it re-emits green light (Fig. 4.18).

The phenomenon is called *fluorescence.* In general, the light emitted by a fluorescing material is of a longer wavelength than that it absorbs. For example, in the case of fluorescein, $(\lambda_{\text{green}})_{\text{Emitted}} > (\lambda_{\text{blue}})_{\text{Absorbed}}$.

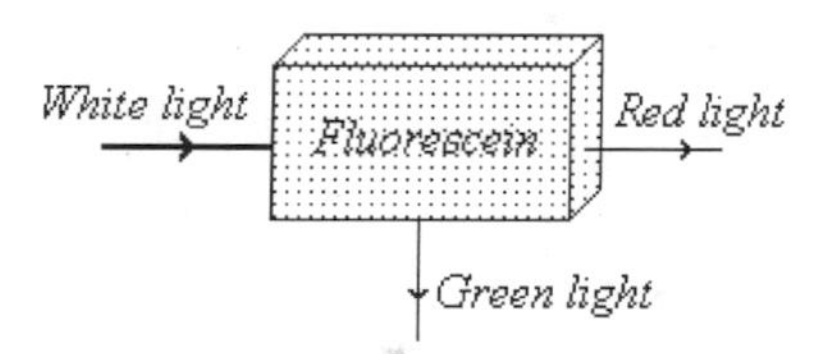

Fig. 4.18. The passage of a beam of white light through a fluorescein solution.

When illuminated with ultra-violet radiation, certain naturally occurring minerals, such as fluorite and calcite, fluoresce and emit visible light. Some washing powders contain fluorescent compounds that enhance the 'whiteness' of the laundered items in sunlight. The familiar fluorescent lamp is a low-pressure mercury vapour discharge tube whose inside walls are coated with a fluorescent material. The bombardment of the mercury atoms in the tube by the accelerated electrons elevates them to an excited state. As the atoms return to their ground state, they emit ultra-violet radiation, which impinges on the fluorescent coating on the inside of the tube. This, in turn, emits white light.

Because fluorescent materials can be easily detected in the dark by the light they emit when irradiated with invisible ultra-violet radiation, fluorescence is widely used in research and industry, particularly for identification and labelling.

Fluorescence is readily explained in terms of molecular energy levels. Whereas atomic electronic energy levels are fine and distinct, molecular energy levels are broadened and diffused as a result of the many vibrational and rotational sub-levels they comprise. Typical atomic and molecular energy level diagrams are compared in Fig. 4.19.

A molecule in an excited state can, just like an atom, return directly to its ground state by emitting a photon of the same frequency as that it absorbed. However, a molecule has another alternative. Before emitting the photon and returning to its ground state, it can lose some of its energy in collisions with other molecules in its environment, thereby dropping from a higher vibrational sub-level to a lower one within the same electronic level.

For example, consider a molecule in a liquid or a solid that absorbs a photon and rises from its ground state to the excited E_3 electronic state. Whereas the average lifetime of the excited electronic E_3 state of the molecule is about 10^{-8} seconds, that of the vibrational sub-states is just 10^{-12} seconds. Because of the relatively long lifetime of the E_3 state, the probability that the molecule will return directly to the E_1 ground state is small. During the time the molecule remains in this state it can lose vibrational energy as a result of collisions with the adjoining molecules. In these collisions it falls from one vibrational state to another within the same electronic level. It may even fall to the lower electronic level such as E_2 and drop from there to the ground state. However, whatever happens, the energy of the photon the molecule emits will always be less than that of the photon it originally absorbed, i.e. the wavelength of the radiation the material fluoresces will be greater than that of the radiation it absorbs.

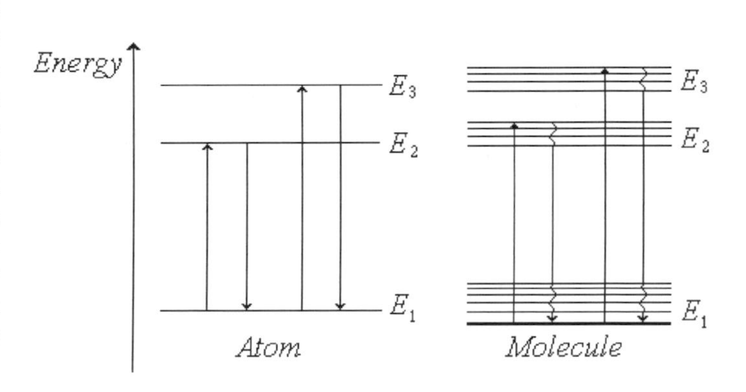

Fig. 4.19. Hypothetical atomic and molecular energy level diagrams. A molecule in a fluorescent material that is excited to the E_2 or E_3 electronic state can lose energy in non-radiative transitions within the state (represented by the squiggly line) before emitting a photon and returning to the ground state, E_1.

Fluorescence ceases when the radiation incident on the material is turned off. However, some materials continue to glow even after the incident radiation has been extinguished; this phenomenon is called *phosphorescence.* The essential difference between the phenomena is that whereas the total spin of the molecule is conserved in fluorescence transitions, in phosphorescence the excited states generally have a total spin different from that of the ground state.

In most instances, the ground state of a molecule is a singlet state, $S = 0$. Absorbing a photon does not change the molecule's spin state. However, in phosphorescent materials (generally solids) the non-radiative transitions that occur within the excited electronic state to which the system is elevated may result in its transition to a triplet state, $S = 1$. For instance, whereas the E_3 level in Fig. 4.20 is a singlet state, the E_2 to which the system may descend as a result of collisions is a triplet state. The probability of a transition from this triplet state back to the singlet state of the ground state is low, and so the lifetime of this intermediate triplet state is relatively long, from a thousandth of a second to even a few minutes. Thus, the emission of the photon is delayed and the material fluoresces even after the incident radiation has been extinguished.

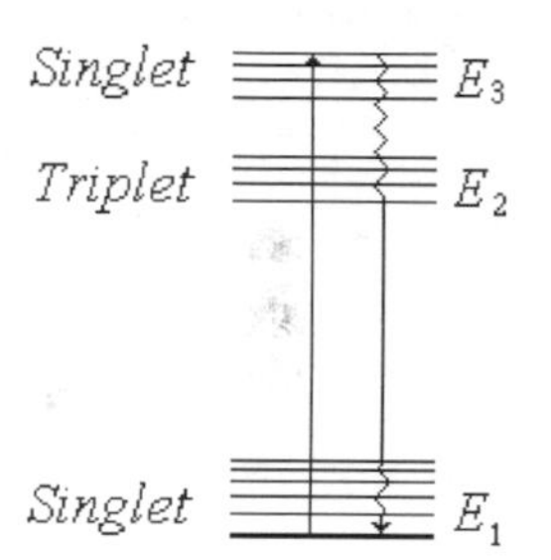

Fig. 4.20. Phosphorescence. A system, excited from the singlet E_1 ground state to the E_3 state, falls by non-radiative transitions to the triplet E_2 state where it remains for a relatively long time because of the low probability of the 'forbidden' transition from a triplet to singlet state.

4.3.1 Fluorescence in Biological Systems

1. Photosynthesis

Chlorophyll A strongly absorbs both blue (430nm) and red (660nm) light. About 10% of this light energy is re-emitted as crimson fluorescent radiation of wavelength 680–700nm and cannot be converted into chemical energy. The remaining light energy absorbed by chlorophyll molecules is either lost as heat or, in the presence of suitable acceptor materials, is converted into chemical energy.

Although blue and red light of the wavelengths strongly absorbed by chlorophyll are the most efficient promoters of photosynthesis, light of other wavelengths such as green and yellow (~500nm) also promotes the reaction in plants (Fig. 4.21). It has been suggested that this activity is the result of a transfer of energy to the chlorophyll molecules from the carotene molecules present in the leaves. Carotene strongly absorbs green and yellow light. The energy absorbed by the carotene is transferred to the chlorophyll which can re-emit it as fluorescence or convert it into chemical energy. Thus, for example, when plants are illuminated with light of wavelength 500nm the typical crimson fluorescence of chlorophyll is observed.

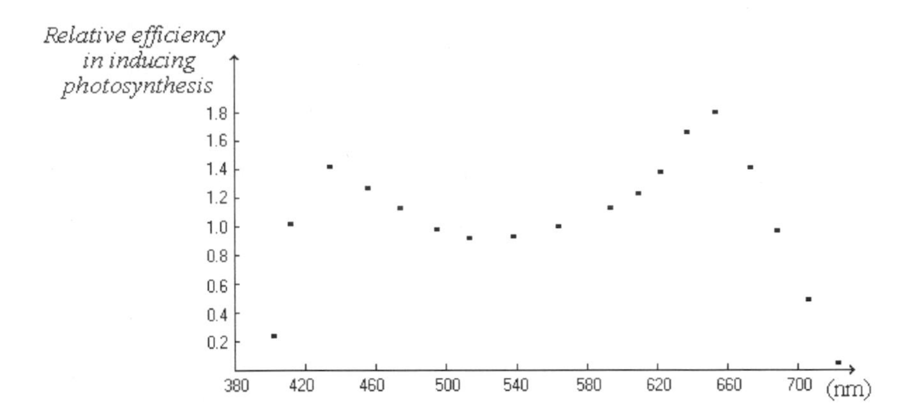

Fig. 4.21 Relative efficiency of light of different wavelengths in inducing photosynthesis.

2. Fluorescence spectroscopy in the study of membrane structure and function

Much information about membrane structure and about the processes occurring inside membranes has been revealed by fluorescence spectroscopy. Among the advantages of this procedure are:

∗ its sensitivity;
∗ that the membrane is virtually undisturbed by the fluorescent probes;
∗ that the kinetics of the processes can be monitored continuously.

The versatility of fluorescent techniques in membrane research is illustrated by the following examples.

a. It has provided details of the phase transitions from more ordered and viscous states (gels) to more fluid and less ordered (liquid crystalline) states that accompany temperature changes in membranes.
b. Molecular motions within biological and artificial membranes have been studied by introducing fluorescent probes at surface concentrations of about one mole percent into the membranes.
c. The binding of peptides and proteins to membranes can be determined without the need to separate the free from the bound molecules. This procedure utilises the increased fluorescence of probes that occurs when they are transferred from a hydrophilic to hydrophobic environment. By labelling peptides and proteins with these probes, the amount of peptide binding to the membranes can be directly measured from the increase in fluorescence intensity that accompanies the change in environment at the membrane. The fluorescence of tryptophan depends on its environment and so peptides containing tryptophan are intrinsically labelled.

3. Membrane fusion

Membrane fusion, i.e. the merging of two membranes, occurs in many cellular processes. For example, endocytosis, exocytosis, neurotransmitter release, fertilisation, the formation of intercellular bridges and the production of multinucleated fibres in skeletal muscles.

The first stage in the infection of cells by viruses such as the influenza or human immunodeficiency virus (HIV, the cause of AIDS) involves the fusion of the viral membrane with a cellular membrane. Mechanisms of membrane fusion have been usefully studied by exploring the fusion of artificial membranes, *viz.* vesicles, whose composition and size can be manipulated, and by studying the fusion of viruses with such vesicles (Fig. 4.22).

The mixing of the aqueous contents of fusing vesicles is commonly assayed by means of the relatively intense fluorescence of the complex $[Tb(DPA)_3]^{3-}$ formed from the union of a terbium cation, Tb^{3+}, with three anions, $(DPA)^{2-}$, of dipicolinic acid. As Fig. 4.23 shows, the system consists initially of two distinct populations of vesicles. A terbium salt is pre-encapsulated in one of the populations and dipicolinic acid in the other. Although the bulk concentration of these reactants is typically of the order of $1\,\mu$M, inside the vesicles their concentration can be as high as several mM.

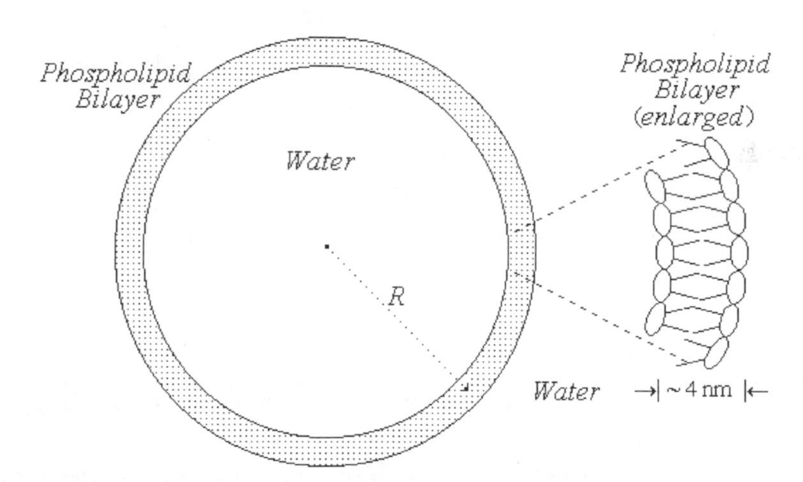

Fig. 4.22 Representation of a unilamellar phospholipid vesicle. For small vesicles, the radius, R, varies from 12.5 nm to 50 nm; for the larger ones $R > 50$ nm. The thickness of the bilayer ranges from 4–5 nm depending on the temperature and other factors.

When vesicle fusion commences, for instance as a result of the addition of calcium ions or a protein to the medium, some of the vesicles containing terbium ions will fuse with others containing the dipicolinic acid. As the contents of the two vesicles merge, the fluorescent complex, $[Tb(DPA)_3]^{3-}$, is produced. The vesicles are usually impermeable. Nevertheless, in order to mop up any fluorescent complex that might somehow be formed outside the vesicles or that might have leaked from them, a small concentration (~1 mM) of the chelator EDTA, which decomposes the fluorescent complex, is added to the medium. Thus, the rate at which the intensity of the fluorescence increases reflects the mixing of the aqueous contents of the vesicles that results from their fusion.

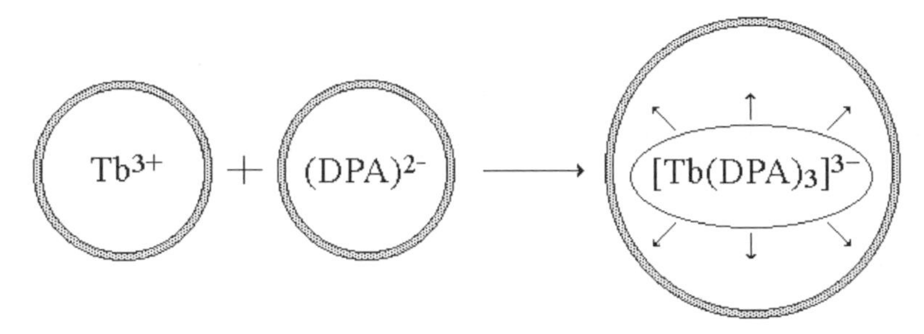

Fig. 4.23 Schematic representation of the Tb/DPA assay for the mixing of the aqueous contents of phospholipid vesicles.

4. Energy transfer in the study of membrane fusion and particle aggregation

Particle aggregation and/or dispersion occurs in practically every biological and physico-chemical system. Examples are:

* aggregation or coagulation of clay particles in the soil;
* aggregation of cells and bacteria in suspension;
* aggregation of particles or their close approach that is the first stage in membrane fusion;
* aggregation of glycoproteins in membranes;
* aggregation of peptides that form pores in membranes.

These processes can be conveniently monitored by fluorescence assays based on the phenomenon known as *energy transfer.*

Consider two types of fluorescent molecules, designated 'A' (acceptor) molecules and 'D' (donor) molecules, such that the emission peak of A occurs at a longer wavelength than that of D, and the excitation peak of A overlaps the emission peak of D to some extent (Fig. 4.24). These molecules can be in solution or can constitute a small fraction of the components of a membrane.

It is observed that when type D molecules are excited by irradiation in the presence of type A molecules, the intensity at which they fluoresce is significantly reduced or 'quenched'; the quenching increases with the

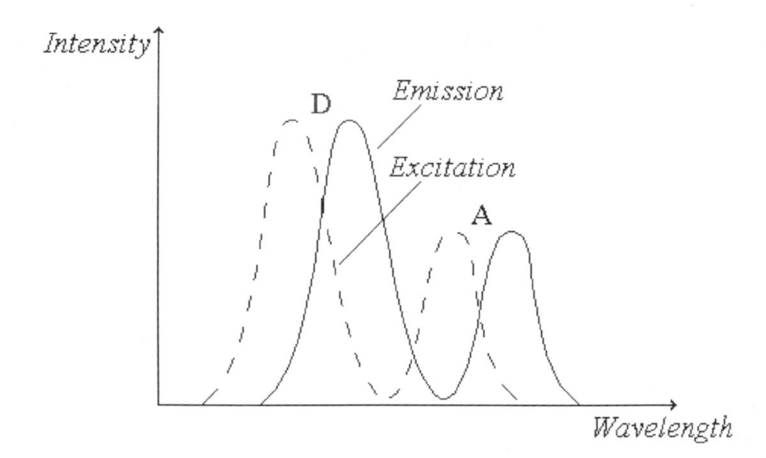

Fig. 4.24 The absorption (excitation) and emission spectra of the two fluorescent molecules D (donor) and A (acceptor). The continuous line is the emission spectrum; the dashed line is the absorption spectrum.

concentration of A molecules in the system. On the other hand, despite the fact that the incident radiation does not contain wavelengths that excite A molecules, wavelengths typical of the emission spectrum of A molecules are found in the fluorescent light emitted by the system. It appears that energy is being transferred to the acceptor molecules of type A from the donor molecules of type D.

The energy transfer between the molecules can be described schematically as

$$D^* + A_0 \rightarrow A^* + D_0,$$

in which D* denotes the excited state of the molecule D. After its excitation by the incident radiation, the molecule arrives at the D* state following a rapid ($\sim 10^{-12}$ s) succession of radiationless transitions. The energy transfer then takes place as the D* molecules drop to the D_0 state.

After being elevated to an excited state by the transfer of energy from the type D molecules, the type A molecules also undergo a series of rapid radiationless transitions that bring them down to the A* state. Their subsequent descent from the A* state to the A_0 state is accompanied by the emission of photons that are detected as the typical fluorescence of A type molecules (Fig. 4.25).

Although the process does not involve the absorption of a photon by the A type molecules, efficient energy transfer requires a degree of overlap between the excitation (absorption) spectrum of A and the emission spectrum of D. This is required for the interaction or coupling state to exist between the donor and acceptor molecules.

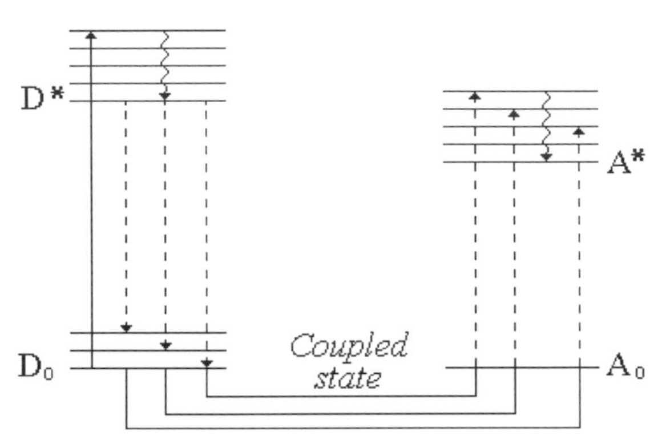

Fig. 4.25. Energy transfer from type D donor molecules to type A acceptor molecules. The wavy arrow represents radiationless transitions and the broken lines denote the energy transfers that arise from the coupling of the molecules.

Under certain conditions, the energy transfer requires physical contact between the donor and acceptor molecules. However, in other cases, energy transfer can occur even at separations as great as 10 nm. The mechanism involved in these distanced energy transfers is called the *Förster mechanism.* The probability of an energy transfer occurring by the Förster mechanism is proportional to $1/R^6$, where R is the distance between the centres of the pair of A and D molecules. The efficiency of the energy transfer may also depend on the relative orientation of the molecules.

The kinetics and final extents of vesicle aggregation can be studied by employing two populations of vesicles, one having type D molecules in its membranes, the other having type A. At close encounters between vesicles of the two types, energy transfer will occur between the molecules. The mechanism and rate of aggregation can be studied either by monitoring the reduction in the intensity of the fluorescence of the D type molecules or the increase in the overall fluorescence of the A type molecules. For example, in the case of a reversible aggregation, the fluorescence intensity should return to its original value when the aggregates disperse. However, should some A or D molecules have passed from their parent vesicles to vesicles of the other type during the aggregation, the fluorescence intensity will not return to its original value. This might indicate that membrane fusion had also occurred or that merely diffusion had taken place; both have been observed in different systems.

In fusion assays that monitor membrane mixing, both type A and type D are initially present in the membranes of certain vesicles (Fig. 4.26). The concentration of the fluorescent probes in the new membrane formed when these labelled vesicles fuse with unlabeled vesicles or virions and their membranes merge is lower than it was in the original labelled vesicle. As a result, the fluorescence of the donor type D molecules increases; it is *dequenched.* The kinetics of membrane fusion can be monitored by following the rate of fluorescence increase.

At high concentrations, certain fluorescent materials such as fluorescein exhibit the phenomenon known as *self-quenching.* Due to their close proximity to each other, excited fluorescent molecules can lose their energy in radiationless transitions instead of by emitting a photon. As a result, the intensity of the fluorescence emitted is reduced. This self-quenching can also be utilised to monitor membrane fusion. The surface concentration of the fluorescent probe molecules in labelled virus particles can be so great that self-quenching occurs and only a low level of fluorescence is emitted. However, the surface concentration of the fluorescent probes is diluted as the virus particles fuse with target membranes. This results in an increase in the intensity of the fluorescence which can be used to follow the rate of the fusion process.

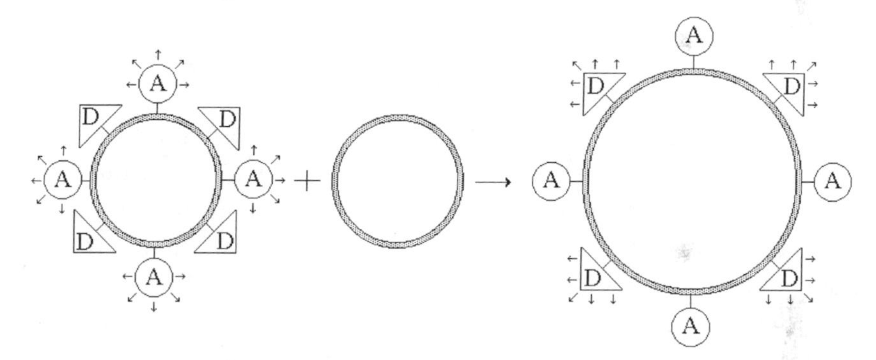

Fig. 4.26 The mixing of bilayer lipids during the fusion of phospholipid vesicles. The triangles represent type D donor molecules and the circles type A acceptor molecules. Membrane fusion amounts to surface dilution of the probes which results in an enhancement of the fluorescence intensity of D molecules and a reduction in that of A molecules.

5. Energy transfer in the reduction of the photodegradation of pesticides

Many potent pesticides, such as bioresmethrin, rapidly deactivate in sunlight. As a result, adequate pest control requires the use of large amounts of the pesticide or, alternatively, its repeated application. The detrimental effects of these practices on the environment, such as the contamination of soil and water by the pesticides and their degradation products, has already been well documented.

A possible solution to this problem, that has recently been investigated, involves the production of pesticide formulations containing the photodegradable biologically active molecules in close proximity with appropriate type A acceptor molecules. The biologically active pesticide molecules act as type D donor molecules for the acceptor molecules. When the formulation is exposed to sunlight, the light energy initially absorbed by the type D pesticide molecules is transferred to the type A acceptor molecules before the former can photodegrade. As a result, the overall photodegradation of the pesticide is significantly reduced.

In one particularly effective formulation, the pesticide is absorbed onto a fine clay that has been treated with a cationic chromophore that acts as the energy acceptor. The clay substrate ensures that the separations and orientations of the type D pesticide molecules and the type A chromophore molecules are the optimal conditions for efficient energy transfer between them.

6. Pore formation by peptides: a fluorescence assay and model

Cells and many organelles are enclosed by a semipermeable membrane, which enables them to retain particular molecules. The release of cellular or vesicular contents can be complete or may be limited to certain sizes and shapes of the molecules encapsulated. The release into the medium can be monitored by an ANTS/DPX fluorescence assay based on pre-encapsulation of molecules of ANTS (8-aminonaphtalene-1,2,3-trisulfonic acid) which are quenched by the presence of molecules of DPX (p-xylylenebis pyridinium bromide). When leakage occurs both ANTS and DPX molecules are diluted in a large volume and the quenching of ANTS becomes negligible, thus yielding an increase in its fluorescence intensity. Zero leakage corresponds to the initial fluorescence of the vesicles; total leakage corresponds to the fluorescence value obtained after the addition of a detergent which causes a complete disintegration of the engulfing membranes.

A particular model for pore formation by peptides assumes that certain peptides added into a vesicle suspension first bind, then become incorporated within the bilayer, and finally aggregate. When an aggregate within a membrane has reached a critical size, i.e. it consists of a sufficient number, M, of peptides, a pore can be created within the membrane and the leakage of encapsulated molecules can ensue. It is assumed that the process of peptide binding is rapid and once a pore has been formed in a vesicle, all its contents will quickly leak. Since, in most cases, bound peptides do not exchange between vesicles, the leakage is characterised by an all or none mechanism, i.e. the population of vesicles consists of those that did not leak at all, and those who have leaked all of their contents. Furthermore, the leakage ceases after a certain period of time. The final extent of the leakage depends on the particular peptide-to-lipid ratio. The rate and extent of the leakage is assumed to be limited by the rate and extent of the formation of surface aggregates of M or more peptides. The number M and geometrical considerations dictate the upper size of leaking molecules.

In most cases, the surface aggregation of the peptides is not irreversible and depends on the ratio between the on and off rate constants of surface aggregation. Simulations of the final extents and kinetics of leakage as a function of peptide/lipid ratios for several peptides and liposomes have been computed. The model could also explain the dependence of this leakage on the sizes of the pre-encapsulated molecules. One confirmed prediction of the model was that large vesicles yield a larger extent of leakage for the same peptide per lipid ratio, because they incorporate a larger number of peptides per vesicle. This pattern is different from other types of natural or induced leakage, where small vesicles, which are considered less stable, exhibit greater leakage.

7. Environmental Applications

Research on clay–liposome interactions has potential applications in several areas:

* Synthesis of mesoporous materials which can be used as catalysts.
* Hosts for inclusion of compounds and molecular sieves.
* Environmental technologies for water purification and reduction of pollutants in soils.

Liposome-clay formulations of anionic herbicides were designed for slow release by incorporation of the herbicides in positively charged vesicles, or neutral ones which were adsorbed on a negatively charged clay. Clay–liposome interactions were addressed by combining adsorption and X-ray diffraction (XRD) measurements with fluorescence studies for vesicles interacting with the clay-mineral montmorillonite. This clay-mineral consists of platelets of lateral dimension of the order of a micron and a width of about one nanometre. Freeze fracture electron microscopy demonstrated the existence of vesicles on external clay surfaces and aggregated structures. XRD results for vesicles interacting with montmorillonite implied transformations of the vesicular structure to lipid bilayers after long incubation times of three days.

Additional structural details on the liposome-clay system were elucidated by several fluorescence methods, which provided an idea about the kinetics of processes, such as binding, aggregation, sedimentation and transformations of the vesicular structure. Using vesicles labelled by a fluorescent probe revealed that adsorption was almost complete within five minutes for a partial loading of the surface. Aggregation and sedimentation of clay-lipid particles occurred within several minutes. Fluorescent measurements of supernatants indicated decomposition of vesicles at a high clay/lipid ratio due to rapidly adsorbing lipid monomers. This phenomenon is due to the fact that in a system of vesicles, or micelles, monomers of the constituents are also present in a certain equilibrium distribution. Adsorption of the monomers on excess clay results in their depletion from solution, and hence in the decomposition of the liposomes or micelles according to the equilibrium distribution.

Fluorescence studies on the kinetics of energy transfer between vesicles labelled by a donor and montmorillonite labelled by an acceptor yielded that aggregation of lipid-clay particles and structural changes of the vesicles occurred at times of minutes to hours. Experiments also monitored the reduction of the donor (NBD) fluorescence by addition of dithionite, which quenches its fluorescenece. Dithionite can only interact with the external surfaces of vesicles in suspension, since it cannot penetrate into their interiors. However, for vesicles interacting with the clay the additional reduction of fluorescence of the liposomes within several hours indicated penetration of dithionite into the interior monolayers, due to permeabilisation of the vesicles. This outcome was also confirmed by leakage from the vesicles of fluoresceinated dextran, which is a relatively large molecule.

Chapter 4.4
Quantum Electrodynamics

The need that was felt during the early years of the 20th century for a unifying theory that would reconcile the differing notions of quantum theory, Maxwellian electromagnetism and special relativity regarding the entities that underlie electromagnetic phenomena – photons or fields – brought about the development of what are now known as *quantum field theories.*

The first of these relativistic quantum-mechanical hypotheses, and to date the most successful – certainly in terms of the precision of its predictions – was *quantum electrodynamics* (*QED*). Its principles and techniques were subsequently employed in the development of other quantum field theories such as *quantum chromodynamics*, which is concerned with the interactions associated with the strong (nuclear) force. The *standard model* of particle physics is also formulated as a quantum field theory.[12]

In his celebrated popular exposition of *The Strange Theory of Light and Matter*, as he termed quantum electrodynamics, the eminent physicist Richard Feynman, boldly asserted that 'the theory describes all the phenomena of the physical world except the gravitational effect... and radioactive phenomena.'[13]

4.4.1 The Fine Structure Constant

The first application of a combination of quantum theory and special relativity to the elucidation of a physical system was Sommerfeld's analysis in 1916 of the elliptical electron orbits in the Bohr model of the hydrogen atom.[14] From the results he obtained, Sommerfeld was able to provide the first explanation for the splitting of the spectral lines in the Balmer series, their *fine structure*, that had been first observed by Michelson in 1887.

In Sommerfeld's calculation of the energy levels associated with these elliptical orbits, there emerged a dimensionless quantity, $\alpha = v_1/c$, where v_1 is the velocity of the electron in the first circular Bohr orbit ($n = 1$) and c is the

[12] See Sections 5.1.6 and 5.3.7 below.

[13] Richard P. Feynman (1985) *The Strange Theory of Light and Matter*, New Jersey: Princeton University Press, p. 8.

[14] See Section 3.2.1.

speed of light, whose value determined the magnitude of the line splitting. This quantity became known as Sommerfeld's *fine structure constant.*

Sommerfeld's theory was short-lived; the fixed Bohr orbits it incorporated had no place in the new quantum mechanics of Schrödinger and Heisenberg. It was superseded in 1928 by Dirac's innovative synthesis of quantum theory and special relativity which not only incorporated the quantum elusiveness of the electron but also predicted that it would possess an intrinsic angular momentum (spin).

Like Sommerfeld's earlier theory, Dirac's also provided an explanation for the fine structure observed in hydrogen atom spectra but it did so on the basis of a completely different phenomenon. Whereas Sommerfeld's calculations were derived from the periodic variations in the electron's relativistic mass as its velocity varied in its elliptical orbit, those given by Dirac's theory came from considerations of the coupling of the electron's orbital and spin magnetic moments (spin-orbit coupling).[15]

Considering their totally different physical bases, it was remarkable that both theories produced essentially identical expressions for the energy level differences that accounted for the spectral line splitting and, furthermore, that both of these expressions incorporated the *fine structure constant*, α. A further aura of mystery was added to this 'mathematical coincidence' when it was noted that the fine structure constant is the only dimensionless quantity that can be formed, without reference to any specific physical system, from a combination of the unit of electric charge, e, the permittivity of empty space ε_0, Planck's constant, h, and the speed of light, c:

$$\alpha = \frac{e^2}{2\varepsilon_0 hc} \approx \frac{1}{137}. \tag{4.35}$$

Writing in 1933, the physicist Max Born noted:

> '*There seems to be little doubt that the existence of this dimensionless number, the only one which can be formed from e, c and h, indicates a deeper relation between electrodynamics and quantum theory than the current theories provide, and the theoretical determination of its numerical value is a challenge to physics. The solution to this problem seems to be closely connected with a future theory of elementary particles in general. [However] all attempts have so far been in vain.*'[16]

Repeated attempts have been made during the past 100 years, some more hare-brained and some less so, to derive the value of the fine structure constant from first principles, many of them by some of the leading physicists, mathematicians

[15] See Section 3.4.3.
[16] M. Born (1959) *Atomic Physics*, New York: Dover Publications, p. 169.

and philosophers of their time; however, without exception, all have proved unsuccessful.

By contrast, Born's conjecture that the fine structure constant, α, would be closely connected with a future theory of elementary particles has been fully realised. This came about in the next generation of theories, the quantum field theories, in which it appears as the *coupling constant* that determines the observed strength of the interaction between electrons and photons at low energies. However, whilst its value can be *estimated* from the values of the physical constants e, c and $\hbar$, the theories do not predict a value.

4.4.2 The Strange Theory of Light and Matter

The physical nature of the electron, the lightest of all the elementary particles, has been a conundrum ever since its existence was first postulated. Is it a point-sized particle or does it have a finite size and structure? Either way there appear to be insurmountable problems.

If it is point-sized, the electron's electrostatic energy should be infinite[17] and since energy and mass are equivalent, it should, correspondingly, exhibit a infinite mass. If, on the other hand, it has a finite size and structure, why does it not blow itself apart due to internal electrostatic repulsion? And, irrespective of its dimensions, what is the fundamental nature of its mass and charge; and what is the mechanism that underlies the electron's interactions with an electromagnetic field, including its own? These questions were only partly answered, if at all, by the classical theory of the electron.

The advent of quantum mechanics presented new possibilities, chief among them the notion of the quantisation of the electromagnetic field. According to this concept, the classical continuum of magnetic and electrical fields is replaced by a 'sea' of transient photons. It was conjectured that photons of indeterminate energies, whose fleeting existence is allowed for by the Heisenberg uncertainty principle, are emitted and absorbed by charged particles ceaselessly and it is they that mediate the interactions between such particles. Photons emitted by one particle may be absorbed by another or even reabsorbed by the original emitting particle. These *exchanged photons* do not appear in either the initial or final states of the system and so are undetectable; they are usually referred to as *virtual photons*. The exchange of these virtual photons is manifested as the 'force' in the electromagnetic interaction.[18]

---

[17] The electrostatic energy of a charged spherical particle is proportional to q^2/r where q is its charge and r its radius.

[18] See Section 2.4.2.

It was further conjectured that virtual photons can, in turn, morph into fleeting virtual *electron–positron pairs* ($e^- e^+$). The spontaneous creation and subsequent mutual annihilation of such short-lived particle–antiparticle pairs, even in a vacuum, is also allowed for by the uncertainty principle; the phenomenon is called 'vacuum polarisation.' The virtual positrons so produced are attracted back towards the original electron, whilst the virtual electrons are repelled from it. This results in a screening of the 'bare' electron. The energies involved in typical experimental determinations of the electron's parameters are much too low to penetrate this polarisation cloud. As a result, the values obtained for these parameters in typical experiments are those of the screened 'quasi-particle'. However, in high-energy experiments the screening may be penetrated and higher values obtained for the strength of the electromagnetic interaction since the effective electron charge is greater. This effect has been confirmed in experiments conducted in particle accelerators at high energies; the fine structure constant was found to exhibit an increased value at these high energies indicative of stronger electron–photon interactions.

The novel effects allowed for by the quantisation of the electromagnetic field were incorporated into a new theory of the electron and its interactions known as 'quantum electrodynamics.' In its simplest form, this theory asserts that all the interactions that occur between electromagnetic radiation and matter can be reduced to just three basic processes:

1. A photon moves from one place to another.
2. A electron moves from one place to another.
3. An electron emits or absorbs a photon.

These three processes are most conveniently depicted by the space–time representations first employed by Feynman from which the stylised depictions of particle interactions known as *Feynman diagrams* were later developed (Fig. 4.27).

It is further postulated that a probability amplitude (wave function) can be assigned to each of these basic processes. We will designate the probability amplitude for a photon to go from one point to another by Π and that for an electron by Φ. The probability amplitude for an electron to emit or absorb a photon, the *coupling constant*, is a number which we will designate by the letter j.[19] It is postulated that the overall probability amplitude, Ψ, of a

[19]Like all probability amplitudes, Π, Φ and j have no direct physical meaning. The value of j is about -0.1 and so higher powers make a diminishing contribution: $j^4 \approx 0.0001$.

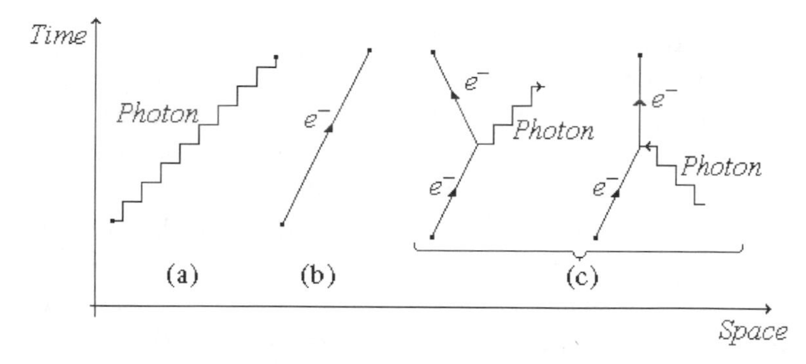

Fig. 4.27. In space–time diagram depictions, photon displacements are represented by wavy lines, those of electrons by straight lines and particle–photon interactions by a vertex. (a) A photon going from one point in space–time to another. (b) An electron going from one point in space–time to another. (c) The emission or absorption of a photon by an electron. It is important not to over-interpret these diagrams. Nothing is implied about *how* a particle gets from one point to another. We do not know how these things happen; the theory only tells us about the probabilities of them happening.

phenomenon or event and, hence, its probability, $|\Psi|^2$, can be computed from the appropriate combination of these amplitudes.

The first step in this procedure is to delineate all the combinations and permutations of the three basic processes that could possibly give rise to the phenomenon or event. These are most conveniently depicted by Feynman (space–time) diagrams; one for each possible way. In marking out these diagrams there are, in principle, no constraints on where a particle can go; even places that could apparently only be reached at speeds greater than the speed of light can or even should be considered when delineating the possibilities.

The next step is to calculate the probability amplitude of each of the possible options. Finally, the overall probability amplitude of the phenomenon or event, Ψ, is obtained from the appropriate combination of the computed amplitudes.[20] This may all sound a very laborious technique, as indeed it is, but it has produced the most accurate numerical predictions yet made by any physical theory. In fairness, though, we should add that it owes much of its success in recent years to the astounding computational powers of super-computers.

Consider first a photon going from one point in space–time to another (Fig. 4.27a). The probability amplitude, Π, for this is calculated from a formula that depends only on the inverse of the space–time interval, I, between the

[20]The probability amplitudes are combined and interpreted according to the general rules laid down in Section 2.4.3.

points.[21] In a regular Minkowski space–time diagram (Fig. 1.32), a photon's motion between two points is represented by a world-line which has a slope of 45°, corresponding to an *in vacuo* speed of c. Over large distances this is quite correct and satisfactory. However, when the distances are short, as they are in particle–photon interactions such as those that occur, say, in the scattering of light by a material medium, there is a non-zero probability for a photon to appear at any point after the event. In such instances the photon's world-line will not necessarily have a slope of 45°, it may be aligned at any angle and its speed may be more or less than c (Fig. 4.28). The small contributions made by other intervals, $I < 0$ and $I > 0$, that correspond respectively to speeds slower and faster than c, cannot be totally ignored, although it is the 45° world-line, corresponding to a speed c, that makes by far the greatest contribution to the probability amplitude.

Turning to the electrons, we recall that quantum mechanics prohibits the assigning of a specific trajectory to their individual motions. This is equivalent to saying that every conceivable trajectory is a possibility, though not all may be equally probable. There is, likewise, a non-zero probability for the electron to appear at any point after the event. Consequently, when considering electron–photon interactions, even possibilities such as an electron emitting a photon, travelling backwards in time to absorb a photon and then proceeding forwards in time again cannot be ruled out (Fig. 4.28c). The path of such a 'backwards-moving' electron can be long enough to appear real and detectable. When viewed with time moving forward, a backward-moving electron appears to have all the familiar properties of ordinary electrons except that it is attracted and not repelled by them; it is as though it were positively charged. Such a backward-moving electron is a 'positron', the anti-particle of the electron. In general, every particle has an amplitude to move backwards in time and, therefore, has a corresponding anti-particle.

The motion of an electron from one point in space–time to another can be regarded as taking place in an infinite number of different ways; it can be envisaged as taking place in a straight path involving one hop (Fig. 4.27b) or by an infinity of alternate paths each involving more than a single hop. All these possibilities must be taken into account when computing the overall probability amplitude, Φ_{AB}, for an electron to go from, say, point A to point B.

The formula for the probability amplitude of each discrete hop is the same as that for a photon making the same move, *viz.* $\Phi_{Hop} = \Pi_{Hop}$. However, those paths that comprise more than a single hop also involve changes in the electron's direction. This must also be taken into account in calculating the

[21] See Eq. (1.77) Section 1.4.7. Feynman has called this formula 'one of the great laws of Nature.'

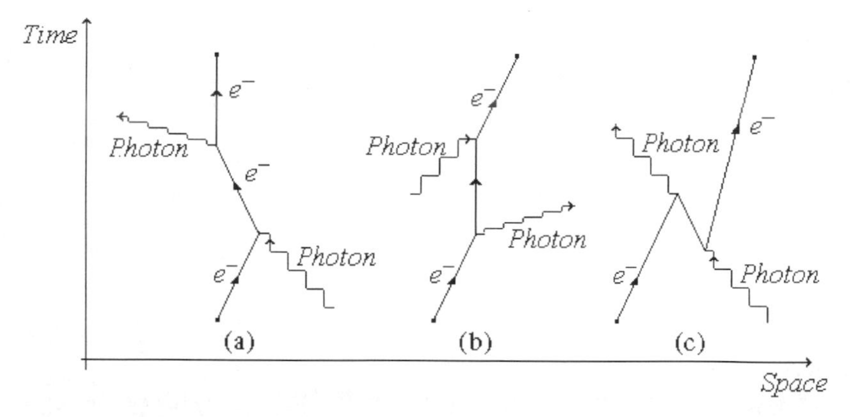

Fig. 4.28. The scattering of light involves the absorption and emission of a photon by an electron – not necessarily in that order. In (a) the photon is first absorbed and another is later emitted, as in the Compton effect (Section 2.3.2). In (b) the electron first emits a photon and later absorbs another identical one (though not necessarily the same one). In (c) an electron emits a photon, rushes backwards in time to absorb a photon, and then continues forward in time. All these alternative scattering scenarios, as well as a multitude of others, contribute to the overall probability amplitude for the phenomenon.

probability amplitude for its overall displacement. This is done by assigning a probability amplitude to the change in direction. This probability amplitude is the square, k^2, of a number k whose value is associated with the particle's mass.[22] An electron can change direction any number of times and so the number of points at which this can occur can be anywhere from zero to infinity (Fig. 4.29).[23]

In accordance with the rules for the combination of probability amplitudes (wave functions),[24] the probability amplitude, Φ_{AB}, for the displacement of an electron from point A to point B will be given by the sum of a series of terms each of which is the probability amplitude for one of the possible paths the electron could take:

$$\begin{aligned}\Phi_{\text{AB}} \propto\ & \Pi_{\text{AB}} \\ & + \Pi_{\text{AC}} \cdot k^2 \cdot \Pi_{\text{CB}} \\ & + \Pi_{\text{AD}} \cdot k^2 \cdot \Pi_{\text{DE}} \cdot k^2 \cdot \Pi_{\text{EB}} \\ & + \Pi_{\text{AF}} \cdot k^2 \cdot \Pi_{\text{FG}} \cdot k^2 \cdot \Pi_{\text{GH}} \cdot k^2 \cdot \Pi_{\text{HB}} \\ & + \cdots .\end{aligned} \tag{4.36}$$

[22] In the case of a photon the value of k is zero.

[23] For the sake of simplicity, we are ignoring electron spin and imagining the electron to be a 'spin-zero' particle.

[24] See Section 2.4.4.

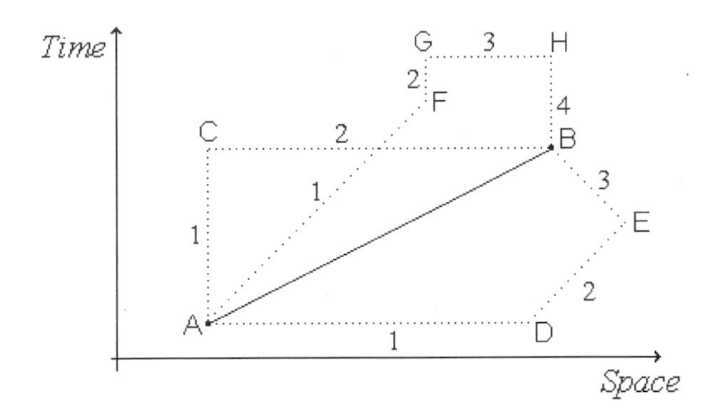

Fig. 4.29. The displacement of a particle from point A to point B can be envisaged as taking place in a single hop or by an infinity of alternate paths each involving more than a single hop. Three such alternative paths are shown: ACB – two hops; ADEB – three hops; AFGHB – four hops. Note that even hops that show the particle going backwards in time such as the HB hop must be taken into account.

Limiting ourselves to the three alternatives depicted in Fig. 4.30, we can now show how to compute the probability amplitude, Ψ, for two electrons at points A and C to become two electrons at points B and D.

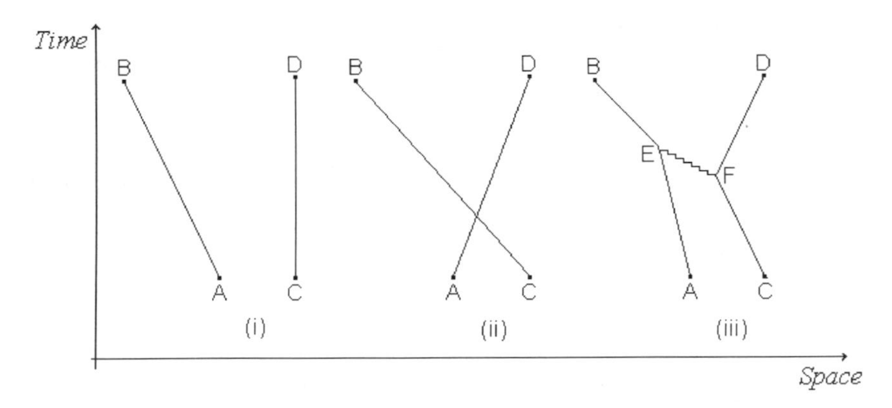

Fig. 4.30. Three alternative ways by which a system of two electrons at points A and C can become a system of two electrons at points B and D. In the first the electrons move directly; one goes from A directly to B, the other from C directly to D. In the second they 'cross-over'. In the third a photon is exchanged between the electrons at some point during their passage from the initial to final points, say between the points E and F. Such a photon which appears in neither the initial nor the final conditions is usually called a 'virtual' photon.

The first step is to specify the probability amplitude of each of the first two alternatives, Figs 4.30(i) and 4.30(ii).

1. The two sub-events that comprise the first alternative occur concomitantly, and so the probability amplitude, $\Psi(\text{i})$, for this event is the product of the probability amplitudes, Φ_{AB} and Φ_{CD}, of the two sub-events:

$$\Psi(\text{i}) = \Phi_{\text{AB}} \cdot \Phi_{\text{CD}}. \tag{4.37}$$

2. Similarly, the two sub-events that comprise the second alternative occur concomitantly, and so the probability amplitude, $\Psi(\text{ii})$, for this event is the product of the probability amplitudes, Φ_{AD} and Φ_{CB}, of the two sub-events:

$$\Psi(\text{ii}) = \Phi_{\text{AD}} \cdot \Phi_{\text{CB}}. \tag{4.38}$$

Combining these two alternatives gives a first fair approximation for the probability amplitude of the process:

$$\Psi = \Psi(\text{i}) - \Psi(\text{ii}). \tag{4.39}$$

The probability amplitudes are subtracted because the electrons are fermions.

So far, no electron–photon interactions have been considered in the calculation of the probability amplitude. However, the first approximation we have obtained can be improved by taking the third alternative shown in Fig. 4.31 into account, namely, the exchange of a 'virtual' photon between the two electrons at some point in space–time. The probability amplitude for this third option is:

$$\Psi(\text{iii}) = \Phi_{\text{AE}} \cdot j \cdot \Phi_{\text{EB}} \cdot \Phi_{\text{CF}} \cdot j \cdot \Phi_{\text{FD}} \cdot \Pi_{\text{EF}}, \tag{4.40}$$

where Φ_{AE} and Φ_{EB} are the probability amplitudes for the hops from A to E and from E to B respectively; Φ_{CF} and Φ_{FD} are the probability amplitudes for the hops from C to F and from F to D respectively; j is the probability amplitude for the emission/absorption of a photon and Π_{EF} is the probability amplitude for the exchange of the photon. However, this is only one of the possible probability amplitudes for alternative (iii). As we stated above, there are no constraints on where a particle can go, and so the points E and F could be anywhere in space–time. The probability amplitudes for all these possibilities must be added when calculating the overall probability amplitude, $\Psi(\text{iii})_{\text{Overall}}$, for this alternative:

$$\Psi(\text{iii})_{\text{Overall}} = \sum \Psi(\text{iii}) = \Psi(\text{iii})_1 + \Psi(\text{iii})_2 + \Psi(\text{iii})_3 + \ldots + \Psi(\text{iii})_\infty. \tag{4.41}$$

Taking this summation into account gives the following expression for the probability amplitude, Ψ, for two electrons at points A and C becoming two electrons at points B and D:

$$\Psi = \Psi(\text{i}) - \Psi(\text{ii}) + \Psi(\text{iii})_{\text{Overall}}. \tag{4.42}$$

Up to this point, we have considered just three ways in which the event could be considered to occur. However, more complex options that include quantum-mechanical effects such as photons being emitted and reabsorbed by the same electron and the morphing of a photon into an electron–positron pair that subsequently annihilates are also possibilities (Fig. 4.31). Better approximations for the overall probability amplitude are obtained when these are also taken into account.

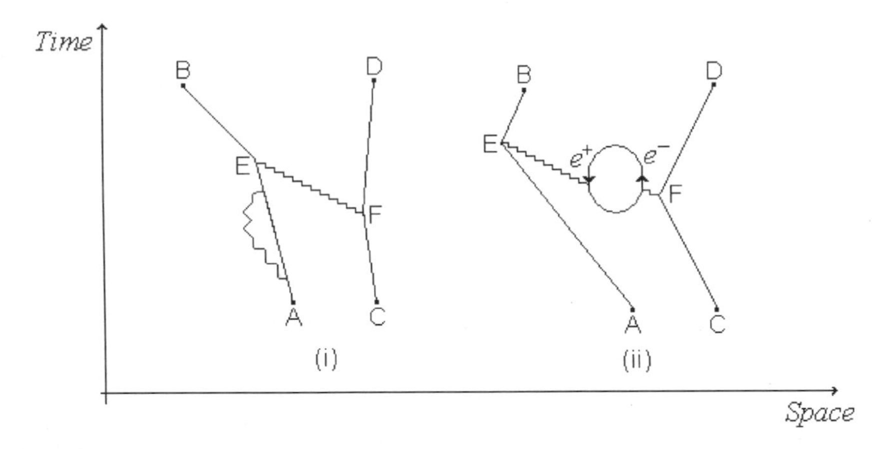

Fig. 4.31. Two further ways in which two electrons at points A and C can become two electrons at points B and D. In (i), the electron from A emits and reabsorbs a photon before it reaches point E where it exchanges a virtual photon with the other electron. In (ii) the second, an exchanged virtual photon morphs into an electron–positron pair which later annihilates.

Feynman diagrams can be divided into two types, 'trees' and 'loops', on the basis of their topology. Tree diagrams only have branches and do not take account of quantum-mechanical effects; they describe processes such as scattering (Fig. 4.28). Loop diagrams, as their name suggests, involve closed loops (Fig. 4.31). These diagrams involve what are termed 'off mass-shell' virtual particles that can appear and disappear, violating the rules of energy and momentum conservation (they do not satisfy the relationship $E^2 = p^2c^2 + m^2c^4$), so long as the uncertainty principle is satisfied.

Although, theoretically, there are always an infinite number of ways in which the basic processes can be construed, in practice, only a relatively small number are actually taken into account. The reason for this is that, in most instances, the more complex the Feynman diagram, the less significant is its contribution to the overall probability amplitude. Accordingly, we only need to consider alternatives up to a complexity consistent with our purposes and with the precision of the experimental techniques available to us, those by which the predictions will be tested.

4.4.3 Renormalisation

Although the two numbers, j and k, needed for calculating the probability amplitudes, are both associated with known properties of the electron – its electrical charge and mass, respectively – there is no definite mathematical connection between j and k, on the one hand, and e and m on the other. If electrons moved through space–time only by the direct one-hop path and did not emit and absorb virtual photons on the way, i.e. they were 'ideal' or 'bare' electrons, the value of j would simply be their observed electrical charge and of k their observed rest-mass. But no such electrons exist. Real electrons cannot be separated from the quantum-mechanical effects that encompass them. So how, without knowing the values of k and j, can computations be made?

It is here that the Feynman diagrams really come into their own. The information we lack in order to reconcile, say, j with e, is the contribution made by the quantum-mechanical effects depicted by these diagrams to the observed value of the electron's charge, namely, the 'corrections' that need to be made to the observed charge in order to obtain the 'bare' electron's charge. Similarly, for k and m.

In the absence of any definite mathematical functions to guide us, we turn to making approximations; the necessary technique for this is provided by *perturbation theory*. The particular physical parameter we wish to calculate is expressed as the sum of a power series of diminishing terms in the hope that, in the limit, we will end up with the observed experimental value of the parameter. The terms are ordered according to the number of electron–photon couplings that appear in the various Feynman diagrams by which the phenomenon or event can be represented. A first estimate, a *first order approximation*, is made on the assumption that we are dealing with an ideal electron; those diagrams depict no couplings (Fig. 4.32a). To this is added a *second order correction*, which takes two electron–photon couplings, j^2, into account; this gives a *second order approximation* (Fig. 4.32b). A *third order correction*, which takes four couplings, j^4, into account is calculated and added to provide a *third order approximation* (Fig. 4.32c); and so on.

In theory, the infinity of points where couplings could possibly occur, even down to where the distance between them is zero, should be taken into

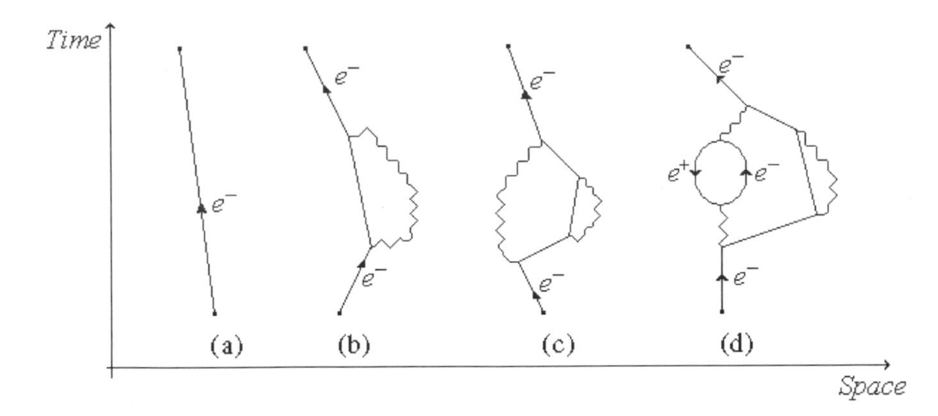

Fig. 4.32. If electrons were ideal and went from point to point only by the direct path shown in (a) there would be no couplings. However, because of the quantum effects other alternatives must also be considered, such as the emission and re-absorption of photons as in the second term, (b), which has two couplings and the third term, (c), which has four couplings. One of the photons could also morph into an electron–positron pair which subsequently annihilates as depicted in (d).

account in calculating the terms. But when this is done, impossible infinities start to appear in the results from the second order corrections on. This is not really surprising. Taking every conceivable point into account is tantamount to assuming that the interacting particles are point-sized; only a point-sized particle could change its direction or emit/absorb a photon at every conceivable point in space–time. But, as we noted above, the magnitude of the electric field close to a point-sized electron would rise asymptotically to infinity, which would mean that the electron had an infinite mass; this is obviously nonsensical. Evidently, electrons cannot be treated as mathematical points.

The problem of the appearance of infinities or *divergences* such as these was not new; they had also arisen in the classical theory of the electron too. However, in view of that theory's success in explaining such a large variety of phenomena at the time, they were ignored as being no more than mathematical inconveniencies. But when they first arose in quantum electrodynamics, it led many to believe that the theory was fatally flawed. If the theory was to survive, a way had to be found to *renormalise* the values for the parameters given by the theory such that they agreed with those obtained experimentally. This required finding a method of removing the infinities.

After some 20 years of fruitless effort, it was realised that by stopping the calculation of the higher order terms when the distance between the coupling points is very small, say 10^{-30} m or 10^{-40} m, both billions of times smaller than anything experimentally observable, the divergences are eliminated and renormalisation is achieved. The use of such *cut-offs* and of other such devices for obtaining finite results was first elaborated by Schwinger,[25] Tomonaga[26] and Feynman in the late 1940s; the three subsequently shared the 1965 Nobel prize for physics.

It transpires that the values that must be assigned to j and k in order that the computations give rise to predictions that accord with actual experiments, depend on the level at which the computations are ceased; different values are needed if all hops shorter than 10^{-30} m are disregarded than if the limit is set at, say, 10^{-40} m. However, if the values of j and k obtained at a particular limit from the analysis of one phenomenon are used for the analysis of a different phenomenon down to the same limit, the new predictions are also found to agree with experiment. Thus, it appears that the only factors that depend on the minimum length set for the hops are the values of the theoretical numbers j and k. Everything else about the phenomena, and in particular their observables, seems to be unaffected by the limit chosen.

The numbers j and k are functions of just space and time. Their effect is to change the phase of the particle's probability amplitude (wave function) at each change in direction or emission/absorption of a photon whilst leaving the probability density, $|\Psi|^2$, which is an observable quantity, unchanged. Theories in which all measurable quantities (observables) remain unchanged when the phases of the fields are altered by an amount that is a function of space and time are called *gauge theories.* The observables are said to be *phase symmetrical.* Quantum electrodynamics is a gauge theory. It is believed that gauge theories can provide the basis for a description of all elementary particle interactions.[27]

4.4.4 Quantum Electrodynamics: Reality or Fancy

When Dirac first elucidated electron spin, he predicted an absolute value of two for the electron's spin gyromagnetic ratio, γ_S.[28] In making this calculation, he presumed that the only interaction that had to be taken into account was that between the electron and the external magnet field (Fig. 4.33a). He

[25] Julian S. Schwinger, American physicist who was awarded the Nobel prize in 1965 for his work on quantum electrodynamics.
[26] Sin-Itiro Tomonaga 1906–1979, Japanese physicist who shared the 1965 Nobel prize in physics for his work on quantum electrodynamics.
[27] See Section 6.5.4 for a further discussion of gauge theories.
[28] See Sections 3.4.1 and 6.3.1.

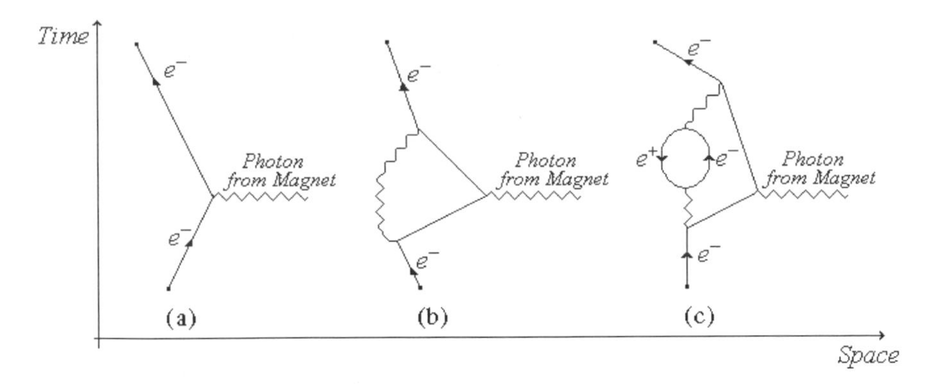

Fig. 4.33. Three of the indistinguishable ways, of many, by which an electron, starting at one point in a magnetic field, can end up at another. Alternative (b) involves the emission and subsequent absorption of a photon by the electron; in alternative (c) the photon emitted by the electron disintegrates into a positron–electron pair which, subsequently, mutually annihilate to form a new photon which, in turn, is absorbed by the electron.

likewise presumed that the appropriate values for the mass and electrical charge of the electron to employ in his calculations were those observed experimentally. However, subsequent experiments found an absolute value of 2.00232 for the electron's spin gyromagnetic ratio, about 0.1% outside Dirac's prediction. This discrepancy became known as the *electron's anomalous magnetic moment.*

With the passage of time and a fuller appreciation of the implications of the quantum-mechanical effects such as those shown in Figs 4.33b and 4.33c, it began to be appreciated that the simple model pictured by Dirac was not the whole story. It was a fair first approximation but needed the addition of one or more correction terms. The first of these *radiative corrections*, as they were termed, which corresponded to Fig. 4.33b, was found by Schwinger in 1948. He showed that the necessary correction was given by the factor $\alpha/2\pi$ such that

$$\gamma_S = 2\left(1 + \frac{\alpha}{2\pi}\right), \tag{4.43}$$

where α is the dimensionless fine structure constant. Substituting the estimated value of the fine structure constant, $\alpha = \frac{e^2}{2\varepsilon_0 hc} \approx \frac{1}{137}$, in this equation gave a value for the electron's spin gyromagnetic ratio (or constant) that agreed with the then available experimental value.

However, measurements of the electron's gyromagnetic ratio to an accuracy of over one part in ten trillion have since been made. In the latest (2008),

Gabrielse's group at Harvard University measured an experimental value of

$$\gamma_S = -2.002\ 319\ 304\ 361\ 460\ (56).$$

Expressed in terms of the fine structure constant, as is the usual practice in quantum electrodynamics, the power series which, according to quantum mechanics (and the standard model too), gives the absolute value of the spin gyromagnetic constant is

$$\gamma_S = 2\left[1 + C_2\left(\frac{\alpha}{\pi}\right) + C_4\left(\frac{\alpha}{\pi}\right)^2 + C_6\left(\frac{\alpha}{\pi}\right)^3 + C_8\left(\frac{\alpha}{\pi}\right)^4 + C_{10}\left(\frac{\alpha}{\pi}\right)^5 + \ldots\right]. \quad (4.44)$$

By reason of the smallness of the fine structure constant, α, successive terms would be expected to have diminishing values, as required by perturbation theory:

$$\alpha \approx \frac{1}{137} = 0.0073 \quad \alpha^2 \approx \left(\frac{1}{137}\right)^2 = 0.00053 \quad \alpha^3 \approx \left(\frac{1}{137}\right)^3 = 0.00000039.$$

From Schwinger's result, the value of the coefficient C_2 is seen to be exactly 0.5. However, his calculation took only the first two terms of the power series into account, i.e., just one Feynman diagram (Fig. 4.33a). Subsequent calculations of the spin gyromagnetic constant have included more correction terms corresponding to other Feynman diagrams such as Figs 4.33b and 4.33c. As of November 2010, the value of C_4 has been calculated analytically on the basis of seven Feynman diagrams; that of C_6, analytically and numerically, on the basis of 72 diagrams; and that of C_8, just numerically, on the basis of 891 diagrams:

$$C_4 = -0.328\,478\,444\,00 \quad C_6 = +1.181\,234\,017 \quad C_8 = -1.914\,4(35).$$

A calculation of C_{10}, on the basis of 12,672 Feynman diagrams, is currently in progress. It will be appreciated that these computations require the use of super-computers and, even so, take years to complete.

Substituting the experimental value of the spin gyromagnetic constant, γ_S, found by Gabrielse's in the power series gives a value for the fine structure constant of:

$$\alpha^{-1} = 137.035\,999\,085(37).$$

Tests of Quantum Electrodynamics

Tests of a theory usually involve comparing its *a priori* predictions to actual experimental results. However, since there is no way of calculating the fine structure constant from first principles, the only way we can test the correctness of the value obtained from measurements of the electron's spin gyromagnetic constant and, by implication, of the validity of the processes by which it was derived – quantum electrodynamics and perturbation theory – is by comparing it to values obtained from other experimental procedures.

The fine structure constant, α, can be derived indirectly from the Rydberg constant,[29]

$$R_0 = \frac{\alpha^2 m_e c}{4\pi\hbar}, \tag{4.45}$$

which is known to a precision of seven parts in a trillion, and precise measurements of the mass of the electron, m_e. The mass of the electron relative to that of rubidium and caesium atoms is known with exceptional accuracy. Thus, its mass can be deduced from measurements of the recoil speed of, say, a ^{87}Rb atom after it emits a photon of known frequency in an atomic transition.

This method for determining α, which is considered the next most precise determination after that based on measurements of the electron's spin magnetic constant, gives a value for the fine structure constant of

$$\alpha^{-1} = 137.035\,998\,78(91).$$

Other experimental procedures have produced similar values.

To what extent this result, so close to that obtained from measurements of the electron's spin gyromagnetic constant, can be accepted as confirmation of the validity of quantum electrodynamics is still the subject of debate. Perhaps the final word in the matter should be left to Freeman Dyson,[30] one of the originators and first promoters of quantum electrodynamics, in particular, of the use of Feynman diagrams.

> '*... your statement that QED is tested more stringently than its inventors could have imagined is correct...we thought of QED in 1949 as a temporary jerry-built structure, with mathematical inconsistencies and renormalized infinities swept under the rug. We did not expect it to last more than ten years ... We expected and hoped that some new experiments would reveal discrepancies that would point the way to a better theory ... it remains perpetually amazing that Nature dances to the tune we scribbled so carelessly ... and it is amazing that experimenters can measure her dance to one part per trillion and find her still following our beat.*'[31]

---

[29]See Section 3.2.1.

[30]Freeman John Dyson 1923–, British/American physicist.

[31]From a letter to Gabrielse quoted in *Physics Today* (August 2006) p. 15.

Appendices to Part Four

4.5.1 Rayleigh Scattering

Consider a beam of unpolarised radiation of wavelength λ and intensity I_0, that is scattered by a suspension of particles. Rayleigh showed that the intensity per solid angle, $I(\theta)$, of the scattered beam at an angle θ relative to the direction of the incident beam is given by

$$I(\theta) = 9\pi^2 I_0 \frac{\left(n_1^2 - n_0^2\right)^2}{2\left(n_1^2 + 2n_0^2\right)} \cdot \frac{NV^2}{\lambda^4} \cdot (1 + \cos^2\theta), \tag{4.46}$$

where N is the number of scattering particles in each unit volume, V is the volume of each scattering particle, n_0 is the refractive index of the scattering material and n_1 is the refractive index of the suspension. So long as the dimensions of the particles are smaller than the wavelength of the incident radiation, this expression corresponds very well to the results obtained from experiments.

The scattering coefficient of attenuation, $\mu_{Scattering}$, can be estimated from the integration of $I(\theta)$ over all angles, i.e. from the integration $2\pi \int_0^{\pi} I(\theta)\sin\theta.d\theta$.

4.5.2 Moment of Inertia of a Diatomic Molecule

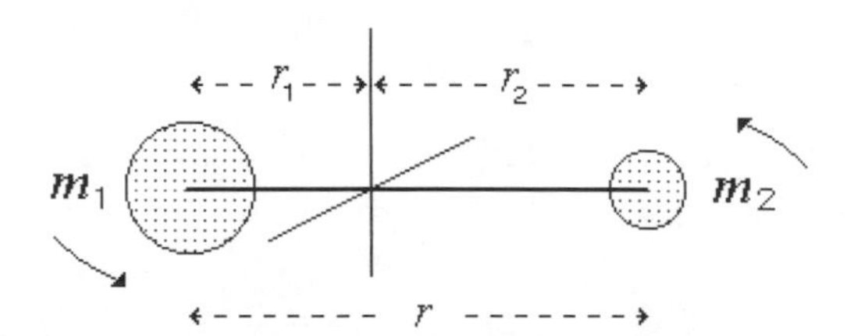

Fig. 4.34. Schematic representation of a diatomic molecule such as HCl rotating around an axis through its centre of mass.

Balancing moments around the centre of mass of the diatomic molecule in Fig. 4.34 gives:

$$m_1 r_1 = m_2 r_2. \tag{4.47}$$

The distance, r, between the atomic centres is

$$r = r_1 + r_2. \tag{4.48}$$

Eliminating r_2 from these two equations gives

$$r_1 = \frac{rm_2}{m_1 + m_2}, \tag{4.49}$$

and eliminating r_1 gives

$$r_2 = \frac{rm_1}{m_1 + m_2}. \tag{4.50}$$

By definition, the moment of inertia, I, of the molecule around its centre of mass is given by

$$I = m_1 r_1^2 + m_2 r_2^2. \tag{4.51}$$

Substituting the expressions for r_1 and r_2, Eqs (4.49) and (4.50) respectively, gives

$$I = r^2 \left(\frac{m_1 m_2}{m_1 + m_2} \right) = r^2 m', \tag{4.52}$$

where m', the molecule's reduced mass, is

$$m' = \left(\frac{m_1 m_2}{m_1 + m_2} \right) \tag{4.53}$$

or

$$\frac{1}{m'} = \frac{1}{m_1} + \frac{1}{m_2}. \tag{4.54}$$

In its oscillations, a two-body system such as a diatomic molecule behaves like an ordinary harmonic oscillator having the same force constant, k, but with the reduced mass m' (Fig. 4.35).

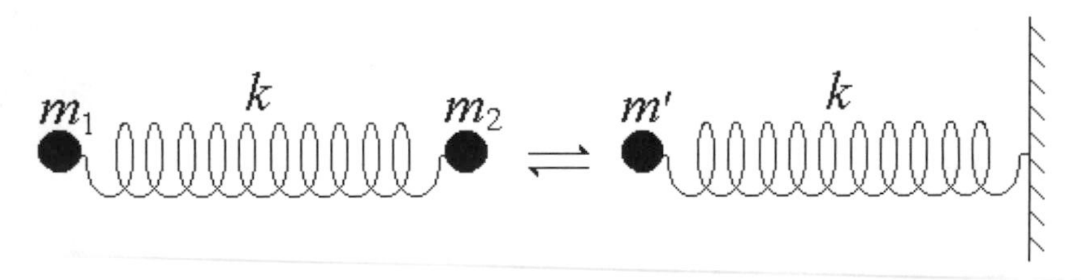

Fig. 4.35. The equivalence of a two-body oscillator and an ordinary oscillator with the same force constant, k.

Part Four: Questions, Exercises and Problems

Questions

1. Estimate the number of rotational levels between the first two vibrational levels in HCl.
2. How does a microwave oven work?
3. Raising the temperature of a diatomic molecule increases its speed of rotation. This strengthens the centrifugal force acting on the atoms. As a result, the bond between them lengthens. How would this affect the rotational spectrum?
4. Why does microwave radiation not cause diatomic molecules to rotate around the axis (the bond) joining the atoms?
5. Explain how the presence of isotopes affects a substance's rotational spectrum.
6. Explain the use of 'phosphors' – phosphorescent materials – such as zinc sulphide in the screens of cathode-ray and TV tubes.
7. A single carbon–carbon bond vibrates at about $3.3 \cdot 10^{13}$ Hz, a double carbon–carbon bond at about $5.0 \cdot 10^{13}$ Hz and a triple bond at about $6.7 \cdot 10^{13}$ Hz. What does this indicate about the relative strengths of the bonds?
8. Pure vibrational spectra are observed only in liquids and solids. The spectra of molecular gases do not show isolated lines corresponding to each vibrational transition. Explain.
9. 'One of the most powerful tools for determining molecular structure is the analysis of molecular spectra.' Discuss.

Exercises

1. If the absorbance of a sample of thickness 2 cm is 0.5, what will be the absorbance of a sample of the same material of thickness 0.2 cm?
2. If the optical density (absorbance) of a solution of concentration 0.032 M is 0.1, calculate the absorbance of a solution made with the same constituents but at a concentration 0.16 M.
3. Given that the bond length is 0.075 nm, calculate the four lowest rotational energy eigenvalues of the H_2 and D_2 molecules.
4. The length of the bond in carbon monoxide, CO, is 0.113 nm. Calculate the rotational energy and angular velocity of a carbon monoxide molecule in the ground state.
5. The spacing between the lines in the pure rotational spectrum of $^{35}Cl^{19}F$ is $1.12 \cdot 10^{12}$ Hz. Calculate the bond length in this molecule.
6. A series of lines spaced at 16.94 cm^{-1} from one another appears in the microwave spectrum of HBr. Calculate the moment of inertia of this molecule and its bond length.
7. Hydrogen iodide, HI, strongly absorbs radiation of wavelength 4.33 μm. Calculate the force constant in this molecule.
8. Using the Boltzmann equation, $(N_{v+1}/N_v) = e^{-(\Delta E/kT)}$, and the data in the previous question, find the ratio between the number of HI molecules occupying the excited $v = 1$ vibrational state and the number occupying the vibrational ground state, $v = 0$, at a temperature of (i) 300 K; (ii) 1300 K.

9. Calculate the force constant and the zero-point energy of HCl on the assumption that its fundamental vibrational frequency is $9 \cdot 10^{13}$ Hz.

Problems

1. Calculate the frequency of rotation of an HCl molecule that is in the $J = 9$ state.
2. The first three vibrational frequencies of the carbon dioxide molecule are 667 cm^{-1}, 1,388 cm^{-1} and 2,349 cm^{-1}. Sketch a diagram of the first few vibrational energy levels of this molecule.
3. The force constant of the bond in HCl is 516 Nm^{-1}. Assuming the value of the force constant does not depend on the mass of the atoms in the molecule, calculate the fundamental vibrational frequency of each of the following: $^{1}H^{35}Cl$, $^{1}H^{37}Cl$ and $^{2}H^{35}Cl$.
4. When monochromatic light of wavelength 632 nm is incident on a sample of HCl, wavelengths of 532.1 nm and 780.5 nm are observed in the scattered light. These wavelengths are attributed to Raman scattering and are associated with vibrational transitions. Calculate the vibrational frequency of HCl.
5. β-carotene strongly absorbs radiation of wavelength 450 nm. The β-carotene molecule is a conjugated linear system comprising 22 π electrons. Estimate the length of the β-carotene molecule.

Part Five Nuclear Physics

With a surface temperature of approximately 6,000 K, the sun must be emitting heat radiation at a rate of about $5 \cdot 10^{23}$ kW.[1] The mechanism by which the sun produces this immense amount of energy baffled scientists and astronomers for many years. The great age attributed to the sun and its planets by cosmology, geology and the theory of evolution, precluded the possibility that it was either a chemical reaction such as combustion or that it resulted from gravitational collapse – the sun contracting under its own weight. Calculations showed that neither of these processes could have sustained the required rate of energy production for the imputed æons of deep time.

Becquerel's[2] discovery in 1896 of the penetrating ionising radiation emitted by substances containing the element uranium presented a similar problem. Typically, the energies associated with this radiation are some million times greater than those associated with chemical reactions. At first, Becquerel thought that the radiation was a form of phosphorescence and that its emission required the prior exposure of the uranium-bearing substances to sunlight. However, he subsequently found that

> '*... the radiations are emitted not only when the salts are exposed to light but also when they are kept in the dark, and for two months the same salts continued to emit these new radiations, without noticeable decrease in amount.*'[3]

Further investigations revealed that other heavy elements, such as thorium and radium, also emitted these radiations; they too are *radioactive*. Becquerel remarked on the similarity between these radiations and the X-rays that had been recently discovered by Röntgen. But the actual mechanism by which these radiations are generated and emitted from atoms remained an enigma for many years.

[1] This follows from the Stefan–Boltzmann law of black-body radiation, $E = \sigma T^4$. In order to generate this immense amount of energy, mass–energy equivalence, $E = mc^2$, requires that the sun's mass must be decreasing by about 5,000,000 tons each second.

[2] Antoine Henri Becquerel 1852–1908, French physicist who was awarded the Nobel prize in 1903 together with Pierre and Marie Curie for the discovery of radioactivity.

[3] (1896) *Comptes Rendus*, 122, p. 1086, translation in M.H. Shamos (ed.) (1959) *Great Experiments in Physics*, New York: Dover Publications, p. 8.

The nuclear model of the atom, proposed by Rutherford in 1913 to account for the scattering of α-rays by metal foils, had presented two fundamental problems: the stability of the atomic electrons and the incredibly great density it posited for the atom's nucleus. The former, as we have seen, was subsequently explained by applying quantum mechanics to atomic structure. The search for an understanding of the latter would not only resolve the enigma of the source of the sun's energy and of that emitted by radioactive substances, but would open up a whole new field of physics – the world of sub-atomic and high-energy particles – whose existence was unforeseen by either relativity or quantum mechanics.

Tragically, almost the first practical use made of this new knowledge was in the manufacture of weapons – *nuclear weapons*. These horrific devices have presented mankind with the awesome means for its own almost total and instantaneous obliteration.

'*Now I am become Death, the destroyer of worlds.*'[4]

However, wisely employed, this knowledge has the potential for developing virtually inexhaustible sources of useful energy: micro-suns working under human control, that could supply all of mankind's energy needs for as far into the future as anyone can possibly hope to plan.

[4] A line from the Hindu scripture, the *Bhagavad-Gita*, recalled by Robert Oppenheimer (1904–1967), head of the laboratories at Los Alamos, New Mexico where nuclear weapons were first produced, as he watched the test-explosion (code named 'Trinity') of the first atomic bomb in 1945.

Chapter 5.1

The Structure of the Nucleus

The proportions of atomic nuclei were first appraised by Rutherford in 1910 when he proposed his nuclear model of the atom. His estimates were based on the results of the α particle scattering experiments he had initiated. More recently, nuclear sizes have been measured by Hofstader.[5] The methods he used were based on the scattering of high-energy electrons. When the De Broglie wavelength of the electrons is of the order of the nuclear diameter, the scattered electrons produce a diffraction pattern from which the dimensions of the nucleus can be computed.

Nuclei are several orders of magnitude smaller than the atom. Whereas atomic dimensions are of the order 10^{-10} m, those of nuclei are of the order 10^{-15} m to 10^{-14} m. Nuclear dimensions are usually expressed in terms of the unit known as the *femtometer* or *fermi*:

$$1 \text{ femtometer (fm)} = 10^{-15} \text{ m}.$$

In these units, $R_{\text{Nucleus}} = 1 - 10\,\text{fm}$ where R represents the radius of a spherical particle.

Atomic nuclei are generally spherical in shape and of almost uniform density. The positive charge of the atom is localised in its nucleus, together with almost all its mass. Most nuclei possess both an intrinsic angular momentum, called *nuclear spin*, and a magnetic moment. Typically, the magnitude of a nuclear magnetic moment is about a thousandth that of the electron. However, there is no simple relationship between the spin of a nucleus and its magnetic moment; in some nuclei the magnetic moment is in the same direction as the spin and in others it is in the opposite direction.

Quantum mechanics must always be used in the description of nuclei and their interactions. Accordingly, atomic nuclei occupy discrete quantum states and they absorb and emit energy by means of transitions between their energy levels.

[5] Robert Hofstader 1915–1990, American physicist who was awarded the Nobel prize in 1961 for his research on nuclear structure.

5.1.1 Nucleons

Atomic nuclei are not elementary particles; they are themselves composed of more fundamental particles. The particles that comprise atomic nuclei are termed *nucleons*; the term indicates a proton or a neutron.

Protons carry the unit positive charge, $+e$, and have a rest-mass some 1,836 times greater than that of the electron. Neutrons carry no net electrical charge; their rest-mass, 1,839 times that of an electron, is marginally greater than that of the proton.

Protons and neutrons are both fermions; each has a spin of a half. They also both possess a magnetic moment, the magnitude of which is about one thousandth that of the electron's magnetic moment (Table 5.1). However, whereas that of the proton is in the same direction as its spin, that of the neutron is in the opposite direction. The existence of a magnetic moment in neutrons, despite the fact that they carry no net electrical charge, attests to an internal structure. Neutrons are not electrically homogeneous. Experiments on the scattering of neutrons have shown that they comprise an inner core of positive charge overlaid by an outer coating of negative charge.

Table 5.1. Properties of the electron and the nucleons.

	Electron	Proton	Neutron
Rest-mass	$m_e = 9.11 \cdot 10^{-31}$ kg	$m_p = 1836 m_e$	$m_n = 1839 m_e$
	$= 0.000055$u	$= 1.00726$u	$= 1.00867$u
Radius	point	about 1 fm	about 1 fm
Charge	$-e = 1.6 \cdot 10^{-19}$ C	$+e = 1.6 \cdot 10^{-19}$ C	0
Spin	$\frac{1}{2}$	$\frac{1}{2}$	$\frac{1}{2}$
Magnetic moment	$-e\hbar/2m_e$	$+2.79 e\hbar/2m_p$	$-1.91 e\hbar/2m_p$

Like all minute particles confined to a particular region – a potential well – the nucleons in an atomic nucleus occupy discrete energy levels. Two distinct, albeit similar, sets of nuclear energy levels exist, one for the protons the other for the neutrons. Since the protons and neutrons are both fermions, Pauli's exclusion principle pertains to them. Consequently, each nuclear energy level can hold at most a pair of protons or a pair of neutrons, with their spins opposed (Fig. 5.1).

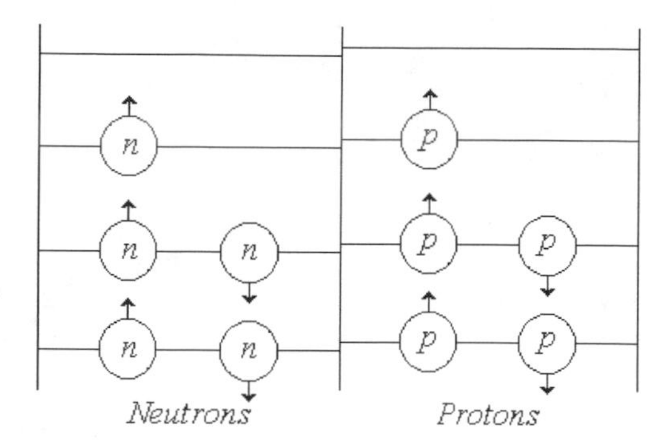

Fig. 5.1. Nuclear energy levels. The proton energy levels are slightly higher than those of the neutrons because of their mutual electrostatic repulsion. Each energy level is occupied by no more than two particles with their spins opposed.

5.1.2 Nuclear Nomenclature

1. Mass number

The total number, A, of nucleons in an atomic nucleus, i.e. the number of protons plus the number of neutrons, is called its *mass number*. The mass numbers of the naturally occurring nuclei range from 1 (hydrogen) to 238 (uranium).

It has been found, empirically, that the number of nucleons in each unit volume of an atomic nucleus is ~ 0.15 nucleons/fm^3, irrespective of its mass number. The following conclusions can be inferred from this observation:

a. nucleons can be regarded as having a fixed volume;
b. the volume of a nucleus is directly proportional to its mass number;
c. the nuclear radius, R, can be estimated from the formula

$$R = R_0 \sqrt[3]{A}, \tag{5.1}$$

where $R_0 = \sqrt[3]{\frac{3}{4} \cdot \pi \cdot 0.15} \approx 1.2\,\text{fm}$.

2. Atomic number

The number of protons, Z, in an nucleus is called its *atomic number*. The atomic numbers of the naturally occurring nuclei range from 1 (hydrogen) to 92 (uranium). The nuclear charge is equal to $+Ze$; in a neutral atom the number of electrons around the nucleus also equals Z.

The interaction of an atom with its surroundings – its chemical properties – depends on its electronic structure. This, in turn, depends *inter alia* on the number of protons in its nucleus. Thus, all atoms having the same atomic number are atoms of the same chemical element. For example, the atomic number of the element carbon is 6; *ergo*, all carbon nuclei contain six protons and possess a nuclear charge $+6e$.

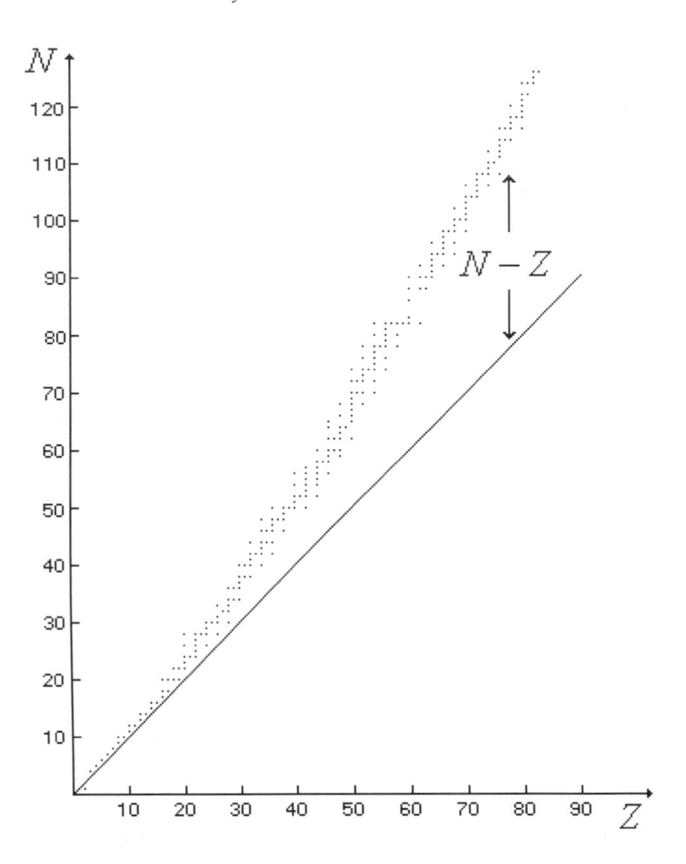

Fig. 5.2. A plot of the number of neutrons, N, against the atomic number (number of protons), Z, in the stable nuclei. The neutron excess, $N - Z$ increases as the size of the nucleus grows.

3. Neutron excess

The number of neutrons, N, in an atomic nucleus is given by the formula

$$N = A - Z. \tag{5.2}$$

The difference, $N - Z$, between the number of neutrons and the number of protons in a nucleus is called the *neutron excess.* In stable nuclei, the general rule is that the larger the nucleus the greater the neutron excess (Fig. 5.2).

4. Nuclear composition

The composition of a nucleus of the element whose chemical symbol is X, is designated by the symbol ${}^{A}_{Z}\mathrm{X}$. Thus, for example, the symbol ${}^{12}_{6}\mathrm{C}$ designates a carbon nucleus containing 12 nucleons, 6 of which are protons, i.e. $A = 12$; $Z = 6$. From Eq. (5.2), $N = A - Z$, it follows that this nucleus also contains 6 neutrons.

Alternative nomenclatures designate a nucleus by the symbols $X - A$ and ${}^{A}\mathrm{X}$. These give $\mathrm{C} - 12$ and ${}^{12}\mathrm{C}$, respectively, for the above carbon nucleus.

These nomenclatures omit the atomic number, Z, because it is implicit in the element's symbol.

5.1.3 Nuclear Masses; Isotopes

Nuclear masses can be measured with great accuracy with the help of techniques such as mass spectography. They are variously expressed in terms of different units (Table 5.2):

* the kilogram, kg;
* the atomic mass unit, u:
* the relativistic unit, MeV/c^2.

The *atomic mass unit* is defined as one twelfth of the mass of the carbon atom that contains 6 protons, 6 neutrons and 6 electrons, i.e. the isotope C−12. Given that, by definition, the mass of a mole of such carbon atoms is 0.012kg, this gives:

$$1\mathrm{u} = \frac{0.012}{12 \cdot 6.022 \cdot 10^{23}} = 1.661 \cdot 10^{-27}\,\mathrm{kg}. \tag{5.3}$$

From the mass–energy relationship $E = mc^2$, the energy equivalent, $E_{1\mathrm{u}}$, of one atomic mass unit is:

$$E_{1\mathrm{u}} = m_{1\mathrm{u}}c^2 = 1.66 \cdot 10^{-27} \cdot 9 \cdot 10^{16} = 1.49 \cdot 10^{-10}\,\mathrm{J}. \tag{5.4}$$

Or, alternatively, in units of millions of electron-volts:

$$E_{1\mathrm{u}} = \frac{1.49 \cdot 10^{-10}}{1.6 \cdot 10^{-19} \cdot 10^{6}} = 931\mathrm{MeV}. \tag{5.5}$$

Table 5.2. The rest-masses of the electron and the nucleons in various units.

	kg	u	$\mathrm{MeV/c^2}$
Proton	$1.6726 \cdot 10^{-27}$	1.00726u	938.3
Neutron	$1.6750 \cdot 10^{-27}$	1.00867u	939.6
Electron	$9.109 \cdot 10^{-31}$	0.00055u	0.511

Though all the nuclei of the atoms of a particular element contain the same number of protons, they do not necessarily contain the same number of neutrons, i.e. nuclei of the same element can have different mass numbers. Atoms of the same element that possess different mass numbers are called *isotopes.* For example, the element uranium has a number of isotopes, the principal ones being U−235 and U − 238. Both have 92 protons in their nuclei. But, whereas the U−235 isotope has 143 neutrons in its nucleus, that of the U−238 isotope contains 146 neutrons.

The isotope of hydrogen with mass number $A = 2$ is called *deuterium* or *heavy hydrogen.* The nucleus of this isotope is called a *deuteron* and is designated ^{2}D or ^{2}H; it is composed of one proton and one neutron. *Heavy water*, which constitutes about 0.015% of naturally occurring water, embodies this isotope of hydrogen (Table 5.3).

Table 5.3. The naturally occurring isotopes of Hydrogen and Chlorine.

	Atomic Number	Symbol	Number of protons (Z)	Number of Neutrons (N)	Mass Number (A)	Atomic Mass	Relative Abundance (%)
Hydrogen	1	^{1}H	1	0	1	1.008u	99.99
		^{2}H	1	1	2	2.014u	0.01
		^{3}H	1	2	3	3.016u	< 0.001
Chlorine	17	^{35}Cl	17	18	35	34.97u	75.53
		^{37}Cl	17	20	37	36.97u	24.47

Because all the isotopes of a given element exhibit the same chemical properties, they cannot be separated by chemical means. The traditional methods of separating isotopes exploit the differences in their masses. These methods have included mass spectography, centrifuging and processes that depend on the different rates at which heavy and light particles diffuse. More recent techniques using tuneable lasers rely on the different nuclear excitation energies of the isotopes.

Example 5.1: The Density of Nuclear Material

The mass of the material that constitutes the nucleus of the most abundant naturally occurring isotope of uranium, ^{238}U, is almost exactly 238u. From Eq. (5.1), the radius of this nucleus will be $R = 1.2 \cdot \sqrt[3]{238} = 7.44\,\text{fm}$ and so, assuming it is spherical, its volume will be $V = 4\pi R^3/3 = 1.73 \cdot 10^{-42}\,\text{m}^3$. The usual definition of density gives for its density:

$$\rho_{nucleus} = \frac{\text{mass of the nucleus}}{\text{volume of the nucleus}} = \frac{238 \cdot 1.66 \cdot 10^{-27}}{1.73 \cdot 10^{-42}} = 23 \cdot 10^{17}\,\text{kg/m}^3.$$

Thus, the density of nuclear material is more than $2 \cdot 10^{14}$ times greater than that of water!

Such immense densities are also found in neutron stars. A neutron star can be regarded as a giant nucleus with a radius of approximately 10,000 m and mass number

10^{57}, that rotates at high speed. In 1967, neutron stars were identified as the sources of the high intensity radio waves emitted at a fixed frequency from the stellar bodies called *pulsars*.

5.1.4 Nuclear Binding Energy

An examination of nuclear masses shows that the mass of a nucleus is always less than that of its constituent nucleons. For example, the mass of a ${}^{12}_{6}\mathrm{C}$ nucleus, $m_{{}^{12}_{6}\mathrm{C}} = 11.997\mathrm{u}$, is less than that of six free protons and six free neutrons:

$$6m_p + 6m_n = (6 \cdot 1.00726\mathrm{u}) + (6 \cdot 1.00867\mathrm{u}) = 12.0956\mathrm{u}.$$

The difference,

$$\Delta m = 6m_p + 6m_n - m_{{}^{12}_{6}\mathrm{C}} = 0.0986\mathrm{u},$$

between the mass of the nucleus and that of its constituent nucleons is called the *mass defect* of the nucleus. Alternatively, the mass defect, Δm, can be expressed as an *energy defect*, E_B, such that where Δm is expressed in atomic mass units

$$E_B = 931\Delta m\,\mathrm{MeV}. \tag{5.6}$$

In the case of the ${}^{12}_{6}\mathrm{C}$ nucleus: this gives

$$E_B = 931 \cdot 0.0986 = 91.8\,\mathrm{MeV}.$$

This quantity is also termed the *binding energy* of the nucleus. It is the amount of energy that would be emitted during the formation of a nucleus from its constituent nucleons. Alternatively, it can be defined as the quantity of energy that must be invested in order to break up a nucleus into its free constituent nucleons.

A key parameter in the study of atomic nuclei is their *average binding energy per nucleon*, E_B/A. In the case of the ${}^{12}_{6}\mathrm{C}$ nucleus, its value is

$$\frac{E_B}{A} = \frac{91.8}{12} = 7.65\,\mathrm{MeV}. \tag{5.7}$$

The average binding energy per nucleon varies from one nucleus to another. It reaches its maximum value, almost 9 MeV, in the nuclei of the elements situated just before the middle of the periodic table, i.e. for values of A between 40 and 100 (Fig. 5.3).

Example 5.2: Binding Energy Per Nucleon

The experimentally measured mass of a neutral atom of O$-$16 is 15.99492 u. What is the average binding energy per nucleon in the nucleus of this oxygen isotope?

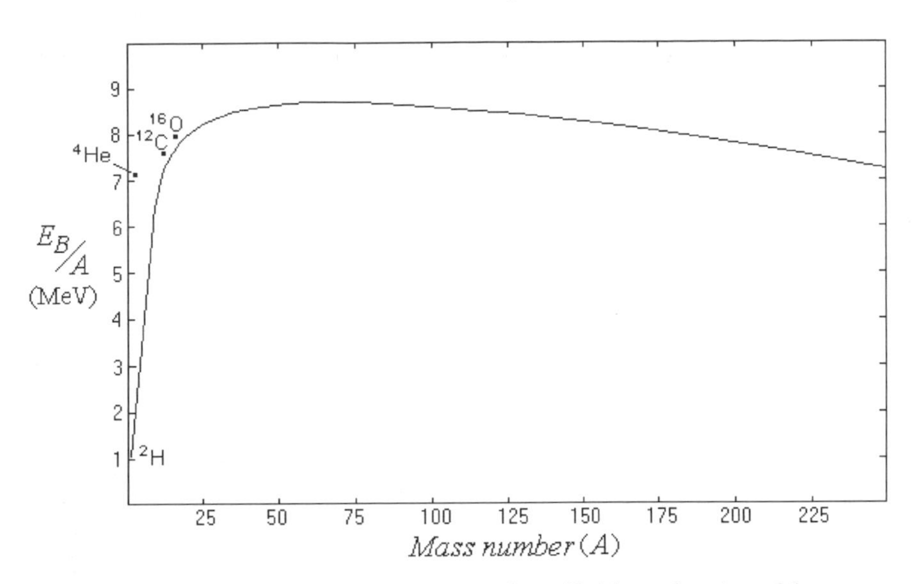

Fig. 5.3. The average binding energy per nucleon, E_B/A, as a function of the mass number, A. It reaches its maximum in the range $A = 40$ to 100.

Solution: The measured mass of the neutral O−16 atom includes the mass of its eight electrons as well as that of its nucleus. Thus, it may be considered as compounded from eight hydrogen atoms, each containing a proton and an electron, and eight free neutrons. The mass of a hydrogen atom is

$$\begin{aligned} m_{H-atom} &= m_p + m_e \\ &= 1.00726 + 0.00055 \\ &= 1.00781\text{u}. \end{aligned}$$

Accordingly, the mass of the O − 16 atom's components is

$$\begin{aligned} 8(m_{H\text{-}atom} + m_n) &= 8(1.00781 + 1.00867)\text{u} \\ &= 16.13184\text{u}. \end{aligned}$$

However, its actual experimental mass is 15.99492 u. This gives a mass defect of

$$16.1318\text{u} - 15.99492\text{u} = 0.13692\text{u}.$$

The nucleus contains 16 nucleons, eight protons and eight neutrons. Hence, the average binding energy per nucleon is

$$\frac{E_B}{A} = \frac{0.13692 \cdot 931}{16} = 7.97\,\text{MeV}.$$

Example 5.3: Nuclear Magic Numbers

Nuclei that contain certain *magic numbers* of protons or neutrons show special properties. For example, nuclei having 2, 8, 20, 28, 50, 82 or 126 protons or neutrons are more abundant in nature than those having similar mass numbers but different compositions. In addition, whereas most other nuclei are distorted, these nuclei appear to be perfectly spherical in shape; they have zero quadrupole electric moments. Nuclei with these 'magic numbers' of nucleons are also relatively more stable. The nuclei ^{4}He, ^{16}O, ^{40}Ca, and ^{208}Pb (82 protons and 126 neutrons) that contain magic numbers of both neutrons and protons are particularly stable. The average binding energy per nucleon is also relatively higher in the nuclei ^{4}He, ^{12}C, ^{16}O and ^{40}Ca, nuclei in which the number of protons equals the number of neutrons.

All these properties have been attributed to the exceptionally stable nucleonic configuration in these nuclei. The relative stability of these 'magic number' nuclei is reminiscent of that of inert gas atoms. These chemically inactive atoms contain specific numbers of electrons (2, 10, 18, ...) all of which occupy closed shells (K, L, M, ...). Evidence of the stability of these closed electronic configurations is the very high first ionisation energies of these atoms. On the other hand, elements such as sodium, $Z = 11$, that have one electron more than an inert gas, have low first ionisation energies; they forego their single valence electron with just a small investment of energy, thereby acquiring the more stable arrangement of closed shells. In a similar manner, nuclei that contain one proton or one neutron more than a magic number, forego this extra nucleon relatively easily. Thus, for example, whereas only 5.8 MeV are required to remove a proton from the nucleus ^{121}Sb that contains 51 protons ($Z = 51$), i.e. one more than a magic number, 11 MeV are required to remove a proton from the nucleus ^{120}Sn that contains the magic number of 50 protons ($Z = 50$). It is postulated that nucleons are also arranged inside nuclei in shells, the nucleonic configurations corresponding to the magic numbers being particularly stable.

	Number of neutrons (N)	Number of protons (Z)	Atomic mass
^{209}Bi	126	83	208.9804u
^{209}Pb	127	82	208.9811u
^{208}Pb	126	82	207.9767u
^{207}Pb	125	82	206.9759u
^{207}Tl	126	81	206.9774u

The nucleus ^{208}Pb has double magic numbers: $Z = 82$, $N = 126$. Using the data in the table above:

1. compare the energy required to remove a proton from the nucleus of an atom of ^{209}Bi with that required for the removal of a proton from an atom of ^{208}Pb;
2. compare the energy required to remove a neutron from the nucleus of an atom of ^{209}Pb with that required for the removal of a neutron from an atom of ^{208}Pb.

Solution: The energy required to remove a nucleon from an atom is given by the energy equivalent of the difference between the mass of the atom from which the nucleon is extracted, and the sum of the mass of the atom produced in the process and that of the particles released from the original atom.

a. The nuclei ^{209}Bi and ^{208}Pb have the same number of neutrons. The removal of a proton from the nucleus is accompanied by the removal of an electron from the atom. The equation for the first process is

$$^{209}\,\text{Bi} \to {}^{208}\,\text{Pb} + p + e$$

$$\begin{aligned}\Delta m &= (m_{\text{Pb}-208} + m_p + m_e) - m_{\text{Bi}-209}\\ &= (207.9767 + 1.007276 + 0.0005486) - 208.9804\\ &= 0.0041246\text{u}\\ &= 3.84\,\text{MeV}.\end{aligned}$$

The equation for the second process is

$$^{208}\text{Pb} \to {}^{207}\text{Tl} + p + e$$

$$\begin{aligned}\Delta m &= (m_{\text{Tl}-207} + m_p + m_e) - m_{\text{Pb}-208}\\ &= (206.9774 + 1.007276 + 0.0005486) - 207.9767\\ &= 0.0085246\text{u}\\ &= 7.94\,\text{MeV}.\end{aligned}$$

b. The nuclei ^{208}Pb and ^{209}Pb have the same number of protons. The removal of a neutron from the nucleus does not change the atomic number; the element is unchanged. The equation for the first process is

$$^{209}\text{Pb} \to {}^{208}\text{Pb} + n$$

$$\begin{aligned}\Delta m &= (m_{\text{Pb}-208} + n) - m_{\text{Pb}-209}\\ &= (207.9767 + 1.008665) - 208.9811\\ &= 0.004265\text{u}\\ &= 3.97\,\text{MeV}.\end{aligned}$$

The equation for the second process is

$$^{208}\text{Pb} \to {}^{207}\text{Pb} + n$$

$$\begin{aligned}\Delta m &= (m_{\text{Pb}-207} + n) - m_{\text{Pb}-208}\\ &= (206.9759 + 1.008665) - 207.9767\\ &= 0.007865\text{u}\\ &= 7.33\text{MeV}.\end{aligned}$$

In both cases, more energy is required for the removal of the nucleon from the nucleus ^{208}Pb, that has a double magic number of nucleons. This is indicative of the greater stability of this nucleus.

5.1.5 The Nuclear ('Strong') Force

Atomic nuclei are very small; their diameters range from about 1 fm to 8 fm.[6] Furthermore, all atomic nuclei, except for those of hydrogen, contain two or more protons. At such close distances, the electrostatic repulsive force that acts between two protons is of macroscopic proportions – tens or even hundreds of newtons. This force, acting between the closely packed protons, should cause the nuclei to blow themselves apart. The fact that atomic nuclei do not spontaneously explode, despite the presence of these potentially destructive forces within them, requires an explanation.

In order to account for the existence of atomic nuclei despite the mutual repulsion of their constituent protons, it is postulated that a short-range attractive force acts between nucleons: a cohesive force that is substantially stronger than the repulsive electrostatic force between the protons. This second force, called the *nuclear force* or the *strong force*, binds the nucleons in a nucleus together. At very close distances, this tremendous force of attraction acts between all the nucleons: protons attract protons, neutrons attract neutrons and protons attract neutrons.

The strong force is essentially charge independent: the forces that act between two protons, between two neutrons or between a proton and a neutron have virtually the same strength. Evidence for this is provided by the binding energies of 'mirror nuclei'. Pairs of nuclei, such as ^{3}He and ^{3}H, ^{7}Be and ^{7}Li or ^{35}Cl and ^{35}Ar, that have the same mass number, A, and can be obtained from one another by the conversion of a neutron into a proton, or *vice versa*, are called *mirror nuclei*. Such pairs of nuclei are found to have almost identical binding energies. Evidently, the electrical charge carried by the nucleons has no significant effect on their binding energies. The slight differences that are found are attributed to the energy of the electrostatic repulsion that acts between the protons in the nucleus.

The nuclear force is very short-range. Plainly, its reach does not extend to distances of the order of magnitude of the atom, $\sim 10^{-10}$ m. If it did, the isotopes of a given element would not all exhibit the same chemical properties.

---

[6]The ratio between the size of an atomic nucleus and that of the atom is comparable to that between the size of a football and that of a football stadium.

In practice, at distances beyond 3 fm it fades away rapidly. Thus, although an attractive force of some 2,000 newtons acts between two nucleons at a distance of 2 fm, nucleons on one side of a large nucleus such as the uranium nucleus, are not directly affected by the nuclear force exerted by the nucleons on the opposite side.

Experimental evidence of the existence and nature of the nuclear force has been obtained from nuclear scattering experiments of the kind carried out by Rutherford. When such experiments are carried out with very high-energy protons or α particles, qualitatively different results from those he encountered are obtained. At particle energies of just a few MeV, such as those Rutherford employed, the scattered protons or α particles are unaffected by the nuclear force; even their closest encounters with the nucleus do not bring them within its range. However, at higher energies the particles can overcome the electrostatic forces of repulsion acting between them and the nucleus, and can penetrate so deep into the atom that they 'touch' the nucleus and enter the range of the nuclear force.

The increase in the neutron excess, $N - Z$, in stable nuclei as their size increases (Fig. 5.1) and the variation of the average binding energy per nucleon, E_B/A, with mass number (Fig. 5.3) can both be qualitatively explained in terms of the respective ranges and strengths of the nuclear and electrostatic forces, as follows.

* In the smallest nuclei, those with a mass number of ten or less, $A \leq 10$, all the nucleons are within range of the nuclear force exerted by all of the other nucleons. The addition of a nucleon to such a nucleus enhances the overall cohesive force binding the nucleons. As a result, the average binding energy per nucleon rises sharply with each increase in mass number (Fig. 5.3). In these small nuclei, the ratio between the number of neutrons to the number of protons is close to one (Fig. 5.2), and the effect of the repulsion between the protons is negligible.
* As the mass number, A, rises above ten, each additional nucleon produces a diminishing return in the amount by which the average binding energy per nucleon increases. This is because the additional nucleons are only attracted by those closest to them, not by all the nucleons in the nucleus. Moreover, as the mass number increases and the nucleus grows larger, the number of nucleons 'in touch' with each additional nucleon drops lower still. The average binding energy per nucleon reaches its maximum value at around $A \approx 60$.
* Above $A \approx 60$, the average binding energy per nucleon starts to decline. This is attributed to the repulsive forces between the nuclear protons. Although the number of nucleons 'in touch' with each additional nucleon ceases to

rise, all the extra protons are in range of the electrostatic forces of repulsion exerted by all the others. As a result, whereas the average cohesive force remains unchanged, the repulsive forces within the nuclei continue to grow. This is moderated to some extent by the increase in the neutron excess, $N - Z$, as the nuclei become larger; the extra neutrons add cohesion without contributing any repulsion and by stretching the average distance between the protons they weaken the repulsive forces between them. Thus, stable heavy nuclei can exist despite the large number of protons they contain. The heaviest stable nucleus is ^{209}Bi; the ratio between the number of its neutrons and protons being $N/Z = 1.52$. Heavier naturally occurring nuclei do exist but they are all unstable, i.e. radioactive.

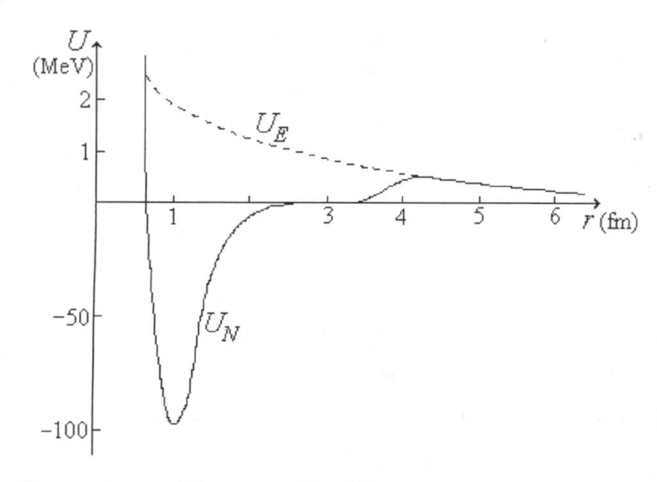

Fig. 5.4. The total potential energy, U, of the interaction between two protons as a function of the distance, r, between them. U_E is the electrostatic potential energy and U_N is the nuclear potential energy; $U = U_E + U_N$. Note that the energy scale is different for positive and negative energy values. Where $r > 4$ fm the total potential energy curve, U, merges with that of the electrostatic potential energy, U_E, and where $r < 3$ fm it merges with that of the nuclear potential energy, U_N.

In 1935, Yukawa[7] suggested that the interaction between nucleons could be described in terms of a potential energy function, U_N, such that the total potential energy of the interaction between two protons is given by the sum, $U = U_N + U_E$, of the nuclear and electrostatic interactions (Fig. 5.4).

Since nucleons can be regarded as particles of fixed volume, when they are closer than 0.8 fm to one another, a repulsive force, greater than the attractive nuclear force, acts between them; this appears as positive energy in the potential energy curve. At separations greater than 4 fm, the nuclear force is

[7]Hideki Yukawa 1907–1981, Japanese physicist who was awarded the Nobel prize in 1949 for predicting the existence of mesons and for his work on nuclear forces.

out of range; only the electrostatic repulsion between the protons operates and so the total potential energy is positive. At separations between 0.8 fm to 3 fm, the cohesive nuclear force between the protons overwhelms the electrostatic repulsion between them; the net force between them is attractive and so the total potential energy is negative.

However, in reality, the nuclear force is too complex to be summarised in terms of a simple function of the distance between the nucleons. Experiments show that the nuclear force also depends on the spin of the nucleons and does not necessarily act along the axis joining the particles. Despite all the knowledge we have about nuclear forces, the function that correctly describes nuclear interaction energy has yet to be discovered.

5.1.6 Nuclear Models

In the attempt to understand the structure and stability of nuclei, three models have been proposed: the liquid drop model, the shell model and the collective model.

The liquid drop model, first proposed by Niels Bohr, views the nucleus as a droplet of nuclear fluid. It assumes that each nucleon interacts only with its closest neighbours in the nucleus, like the molecules in a liquid. By analogy with liquid drops, the model predicts the generally spherical shape of nuclei and provides a good qualitative description of the fission of large nuclei. Quantitatively, it provides a function for the binding energy per nucleon in terms of the mass number that closely fits the empirical data.

The shell model attempts to apply the same quantum tools to nuclear structure as were successfully applied to the elucidation of the electronic structure of atoms. It regards each nucleon as interacting with the overall force field created by the other nucleons comprising the nucleus. Its outstanding achievement is a computation of the nucleonic energy levels that gives a rationale to the postulate that the nucleons are configured in shells inside nuclei. In making the calculation it was assumed that the nucleons possess an orbital angular momentum, **L**, in addition to their spin, **S**, the two being coupled to form a total (intrinsic) nuclear angular momentum, **I**, whose quantisation prescribes the nuclear energy levels and, hence, the nucleonic configuration (see Section 6.3.2).

The collective model is eclectic, uniting the best features of the other two. Not surprisingly, it has been the most successful of the three. Nevertheless, it still fails to give a comprehensive picture of the nucleus. The collective model views the nucleus as comprising a stable core of closed shells, each containing a magic number of nucleons, with which the 'extra' nucleons in the nucleus, i.e. those that are not constituents of the core, interact. These 'extra' nucleons occupy quantum states whose energies are determined by their interaction with the core. The core, in turn, is agitated by this interaction. As a result, rotational and vibrational motions are set up in the nucleus that are analogous to those that occur in droplets.

5.1.7 The Elementary Particles of Matter

The notion that the atomic nucleus comprises yet smaller particles, begs the question of whether these particles – the nucleons – might not themselves be composites of still smaller particles. And if so, where does the sequence (the 'pecking order') end? We shall conclude our discussion of the structure of the nucleus by briefly considering some of the basic building blocks – *elementary* or *fundamental particles* – from which matter is believed to be constructed.

1. Pions

As well as proposing a mathematical function to describe the nuclear force, Yukawa also suggested a mechanism for its action. He posited that the strong nuclear force that acts between nucleons arises from an exchange of particles between them. Originally called *π mesons*[8] these particles are now called *pions*. The mechanism by which they mediate the strong interaction between nucleons is illustrated in the Feynman diagrams in Fig. 5.5.

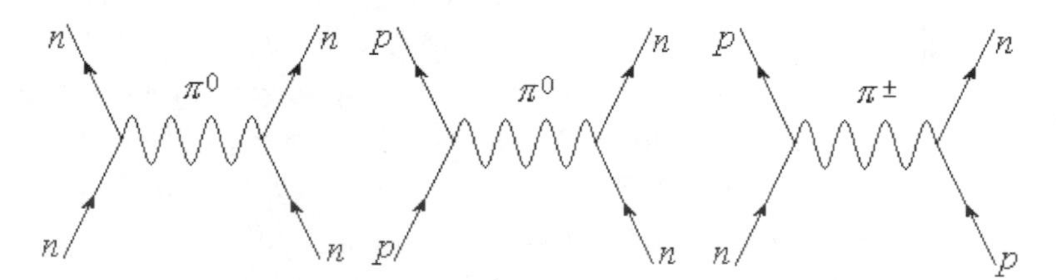

Fig. 5.5. Feynman diagrams showing the strong interaction between nucleons mediated by virtual pions. The neutral pion, π^0, mediates interactions in which the nucleons retain their identity. The charged pions, $\pi^\pm$, mediate those in which protons become neutrons or neutrons become protons.

The notion that particle–particle interactions are mediated by an exchange of other particles is not new to us. It is a basic concept of the quantum theory of fields. The mediating particles can be *actual* particles, as when two atoms form a covalent bond by the exchange of electrons, or they can be *virtual* particles, as in the interaction between stationary charged particles. The spontaneous generation of a virtual particle, in the absence of any external source of energy, is allowed for by Heisenberg's uncertainty principle: the particle can appear on condition that it exists for a period no longer than $\Delta t \approx \hbar/\Delta E$, where

[8]From the Greek word *mesos* – $\mu\varepsilon\sigma o\zeta$ – meaning 'middle'.

ΔE is its total energy. If it is to mediate the interaction between two particles, the ephemeral particle must manage to move from one to another during its fleeting lifetime. Evidently, the less time it has to achieve this, the more energy must be 'borrowed', i.e. the heavier must be the particle.

Given that the equilibrium distance between two nucleons is typically $r \approx 1.7\,\mathrm{fm}$, and assuming that the virtual particles that mediate the strong interaction between them move at close to the speed of light, c, their lifetime can be estimated to be

$$\Delta t = \frac{r}{c} = \frac{1.7 \cdot 10^{-15}}{3 \cdot 10^{8}} = 0.57 \cdot 10^{-23}\,\mathrm{s}. \tag{5.8}$$

This gives for the particle's total energy:

$$\Delta E = \frac{\hbar}{\Delta t} = \frac{6.58 \cdot 10^{-22}}{0.57 \cdot 10^{-23}} = 115.4\,\mathrm{MeV}. \tag{5.9}$$

This is equivalent to a rest-mass of $0.124\,\mathrm{u}$, i.e. somewhere between the mass of an electron and that of a nucleon.

Virtual pions cannot be directly observed. In order for them to become observable, an external supply of energy must be provided. According to the above calculation, the realisation of free pions requires the investment of considerable energy; at least 115.4 MeV. To furnish a stationary target particle with this much energy in a collision, the incident particle must possess considerably more. Before the construction of powerful particle accelerators in the 1950s the only source of such particle energies was cosmic rays.[9] Free pions were first observed by Powell[10] in 1947 in the microscopic tracks they left in nuclear emulsions exposed at high altitudes to the bombardment of cosmic rays. Three different pions were discovered: a neutral pion, π^0, and two charged pions, π^+ and π^-. All three pions are unstable: the neutral pion decays after a lifetime of $\sim 10^{-17}$ s into a pair of high-energy photons – γ-rays; the charged pions, π^+ and π^-, decay after a lifetime of $\sim 10^{-8}$ s, by stages, into a positron, e^+, and an electron, e^-, respectively.

2. Particles and anti-particles

The existence of the positron or *anti-electron* was predicted by Dirac in 1928. It arose as a possible solution of the wave function of the electron when relativistic

[9]Cosmic rays are high-energy particles, $10^2 - 10^{14}$ MeV, primarily protons, from an unidentified source in outer space that continuously bombard the earth. Most do not reach the earth's surface because they are either deflected by the earth's magnetic field or absorbed in the atmosphere. They are only easily detected high in the atmosphere.

[10]Cecil F. Powell 1903–1969, English physicist who was awarded the Nobel prize in 1950 for his discovery of the pion.

considerations were incorporated into its formulation. The solution to the relativistic wave equation required the existence of negative energy states. But if such states existed and were unoccupied, free electrons possessing positive energy would be expected to fall into them, emitting a photon in the process. Why is this not generally observed?

Dirac postulated that the negative states are all generally occupied, even in a vacuum. However, should one be unoccupied for any reason, a *hole* would be left in the 'sea' of otherwise filled negative energy states. This hole, which behaves like a particle with positive charge, appears to us as a positron.[11]

The actual positron, e^+, was first identified by Anderson[12] in 1932 in cosmic rays. Positrons and electrons are identical except for their electric charge; the positron is the *anti-particle* of the electron. The charged pions are also identical except for the sign of the electric charge they carry; the negatively charged pion is the anti-particle of the positively charged one.

It transpires that for every particle there exists a corresponding anti-particle. In general, the particle and its anti-particle have the same mass, spin and lifetime, if they are unstable, but opposite electromagnetic properties. If the one is positively charged the other will be negatively charged; if the one has its magnetic moment in the direction of its spin, that of the other will be in the opposite direction. Chargeless particles like the neutral pion and the photon are their own anti-particle.[13] The anti-proton, p^- or $\bar{p}$, was discovered in 1955 in the shower of particles produced when a target was bombarded with protons that had been accelerated to an energy of 6.2 GeV ($6.2 \cdot 10^3$ MeV).

There is no obvious reason why matter composed entirely of anti-particles should not exist. Such 'anti-matter' would behave exactly like ordinary matter. However, if matter and anti-matter were to meet, they would mutually annihilate with the emission of a flood of very high energy photons, γ-rays. No such emissions have been observed on earth or anywhere in space. For whatever reason, there does not appear to be much anti-matter around.

3. Particle physics

The discovery of the free pions in cosmic ray tracks in 1946 marks the birth of modern particle physics. Subsequent investigations have revealed that pions and anti-protons are not the only species produced when high energy

---

[11] The holes in the 'Dirac sea' of negative energy states are analogous to the positive holes in the valence band of a semiconductor. See Section 6.4.6 below.

[12] Carl Anderson 1905–1984, American physicist who was awarded the Nobel prize in 1936 for the discovery of the positron.

[13] Anti-particles are usually designated by adding a bar over the symbol for the particle: the anti-proton by $\bar{p}$; the anti-neutron by $\bar{n}$.

particles are slammed into stationary targets or into one another. A plethora of unstable and short-lived[14] particles (more than 300) has been observed in the debris produced when protons or electrons accelerated up to energies of 1,000 GeV (10^6 MeV $= 1$ TeV) are smashed into stationary targets or when two such high-energy particles, or a particle and its anti-particle, collide head on.

The early particle accelerators were linear. A supply of charged particles was accelerated along the length of an evacuated tube by the electric field produced by large static electric charges. In more advanced versions currently employed, the particles are accelerated repeatedly as they pass through a linear array of perforated plates. An alternating electrical potential is applied to the plates such that a particle is attracted as it approaches a plate and is repelled after passing through a perforation in it on its way to the next plate. Particle accelerators of this type, operating at energies of up to 30 MeV, comprise almost all the 30,000 units currently used in medicine and industry; in the field of radiotherapy they have largely supplanted the use of Co-60 sources.

Most modern particle accelerators used in high energy physics ($>$ 1 GeV) are descended from the cyclotron invented by Lawrence[15] in 1931. This ingenious device employs magnetic fields to keep the charged particles moving in circular paths whilst an alternating electric field of moderate strength accelerates them incrementally (Fig. 5.6). However, because of relativistic effects – the rise in the mass of the particle as its energy increases – the frequency of their orbits drops out of synch with the alternating electric field at high energies. Consequently, simple cyclotrons can only accelerate protons to energies of about 15 MeV corresponding to a speed of $\sim$0.1 c.

To achieve higher energies, an accelerator in which the particles move in a constant circular path, usually inside a narrow pipe torus (tube) just a few centimetres in cross-section, is employed. This arrangement requires the synchronous adjustment of the electric field, which accelerates the particles, with the magnetic fields that keep them in the fixed orbit as their momentum increases; hence the name *synchrotron.* The magnetic field only has to be applied where the particle actually orbits, and not across the whole diameter of the cyclotron 'Dees'; this does away with the single huge cyclotron magnet. Instead, a series of tightly focusing bending magnets surround the torus at appropriate points along its length (Fig. 5.7).

---

[14] If a particle exists for only 10^{-23} s its track is unobservable, since even if it travels at the speed of light it will move at most 10^{-15} m before decaying, *viz.* about one proton diameter. Such a short track is totally unobservable. On the other hand, particles that last as long as 10^{-10} s are considered 'stable'. They can leave tracks a few centimetres long and may even last long enough to be assembled into a beam that can be aimed at another target.

[15] Ernest Orlando Lawrence 1901–1958, American physicist who was awarded the Nobel prize in 1939 for the development of the cyclotron.

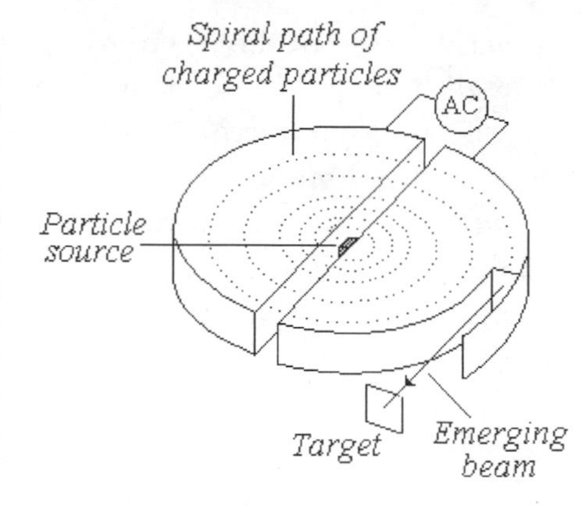

Fig. 5.6. The cyclotron. A strong vertical magnetic field holds the charged particles (ions) on a circular path within two flat horizontal semicircular metal vessels – 'Dees'. Applying an alternating potential of the same frequency as their orbit across the gap between the Dees gives the charged particles a push every time they cross the gap, accelerating them incrementally to ever higher speeds. As their speed increases, the particles spiral outwards until they emerge and strike the target.

Accelerated charged particles emit photons (electromagnetic radiation), which sets the upper limit on the operation of this device. The limiting beam energy is reached when the energy lost equals the energy added each cycle. Lighter particles (such as electrons) lose a larger fraction of their energy and the energy of electron/positron accelerators is limited by this radiation loss; however, it does not play such a significant role in proton or ion accelerators. In many installations use is made of the *synchrotron radiation* produced by the

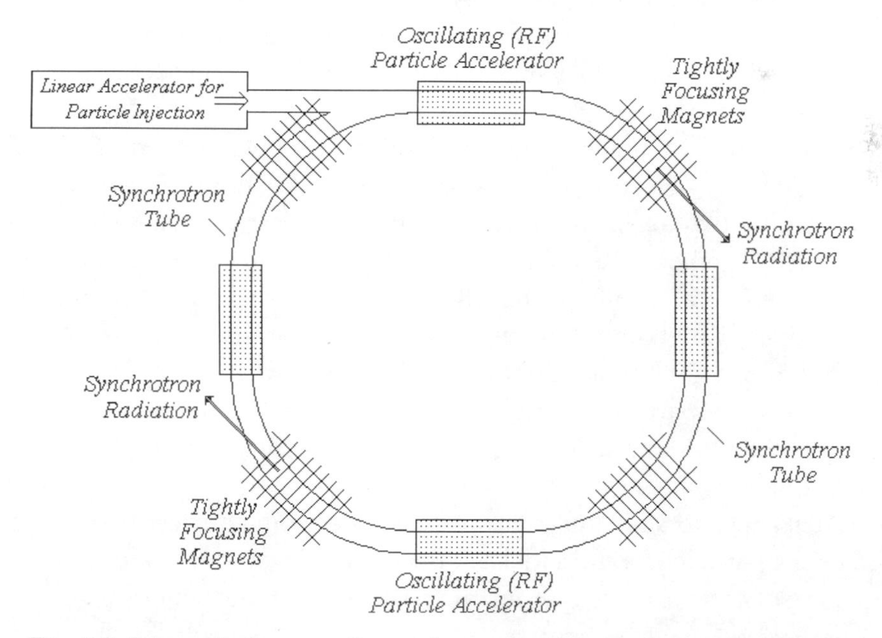

Fig. 5.7. Schematic diagram of a synchrotron particle accelerator. Particles are injected into the ring at substantial energies from a linear accelerator. In practice, it is easier to make the torus in the shape of a round-cornered polygon, the sections between the bending magnets where the radio frequency (RF) accelerators are situated being straight. The beams of synchrotron radiation emerge at a tangent to the curved sections of the synchrotron tube.

radial acceleration of the electrons. This radiation is of high beam intensity and low divergence, has a characteristic polarisation and can be adjusted to any frequency from microwaves to X-rays; the latter are employed for such purposes as the X-ray spectroscopy of weakly scattering crystals, such as those of proteins and similar biological macro-molecules.

At the very high particle velocities achieved in these devices, allowance must be made for relativistic effects such as the greatly increased mass of the particles as they approach the speed of light.[16] As a result, in order to reach the tremendous energies the researchers wish to investigate, gigantic and expensive machines are required. The *Tevatron* proton accelerator at the Fermilab near Chicago (shut down in 2011) had a circumference of 6.3 km and produced beams with particle energies of 1 TeV (a speed of 0.9999995 c).[17] The largest accelerator installation at the Centre Européen de la Recherche Nucléaire (CERN) comprises a tunnel of circumference 27 km buried about 100 m underground. Until the year 2000, it housed the Large Electron–Positron Collider (LEP) in which beams of these particles circulating in opposite directions met head-on with a combined energy in excess of 200 GeV. The results of these experiments gave precise values for many of the parameters of the *standard model* of particle physics. This theory, which was first postulated in the 1960s, seeks to unite the electromagnetic, weak and strong interactions that mediate the dynamics of the elementary particles into a single comprehensive framework.[18]

In 2001, the LEP collider was dismantled and construction of the Large Hadron Collider (LHC) began. Now fully operational, this technological colossus accelerates two beams of protons in opposite directions such that the particles smash together with a combined energy of 14 TeV per collision, equivalent to a temperature of 10^{16} K. The beams are controlled by some 1,800 superconducting niobium-titanium magnets operating at a temperature of 1.9 K (−271 °C). Key components of both the LEP and LHC are the instrumentation that controls the particle beams and the detectors that register the properties and paths of the particles in the debris that results from the high-energy collisions.

Modern detectors are huge purpose-built electronic devices that signal the passage of a particle in a form that can be fed into a computer and from which the pattern of its track can be reconstructed and displayed. Gone are the bubble

[16] The Stanford Linear Accelerator produces a beam of electrons with a speed that is within 8 cm/s of the speed of light.
[17] 1 TeV = 10^{12} eV is roughly the energy of a flying mosquito.
[18] See Section 5.3.7 below.

chambers that provided photographic images of the tracks left by particles, replaced by arrays of silicon-based semiconductors. A particle's momentum can be calculated from the curvature of its path through a magnetic field and its energy from its penetration into layers of 'passive' or 'absorbing' high-density materials (lead for instance) interleaved with layers of 'active' medium such as solid lead-glass or liquid argon. Particle velocities can be derived from the *Çerenkov radiation* produced when they move through a material medium at speeds greater than the speed of light in the medium or, alternatively, from the *transition radiation* emitted when high-energy charged particles cross the boundary between two insulating media of differing dielectric constants. Combined with measurements of its momentum, the velocity thus obtained can be used to determine a particle's mass and hence its identity.

So great is the quantity of data produced by the experiments conducted at the LHC, that a new networking technology called the GRID is being developed at CERN that will link computers world-wide into a vast global computing resource that will hopefully cope with it all; certainly an apt initiative for the institution where the World Wide Web (WWW), familiar to all internet users, was invented.

The quark hypothesis

Attempts to make sense of the bewildering host of unstable particles produced from these brute-force high-energy collisions and of the processes by which the particles interact and decay, have centred on the notion that particles such as the protons and neutrons that constitute atomic nuclei, as well as the pions that mediate between them, are themselves composed of still more fundamental particles. The evidence for the existence of an internal structure in these particles is well established.

In experiments similar in their design to that of the Franck–Hertz experiment, evidence has been found for the existence of excited nucleon states, i.e. well-defined higher states to which a nucleon can be raised by the absorption of specific amounts of energy. In these experiments, a beam of accelerated protons is aimed at a liquid-hydrogen target and the distribution of the energies of the scattered protons is analysed. The distribution indicates that although most of the incident protons collide elastically with those in the target, some do so inelastically. In these inelastic proton–proton collisions, specific amounts of energy are taken up by the target protons.

Similar results are obtained when protons are bombarded by energetic π^+ pions. The result of the interaction between the π^+ pion and the proton is the creation of three new pions

$$\pi^{+^*} + p \rightarrow \pi^+ + p + (\pi^+ + \pi^- + \pi^0).$$

In this case, too, an analysis of the distribution of the energies of the three new pions indicates that the reaction can proceed via an intermediate stage, usually designated η_0, that is characterised by a specific energy, ~550 MeV. This intermediate stage, which is usually termed a *resonance*, can be regarded as being a very short-lived particle, $\sim 10^{-20}$ s, such that overall the reactions can be described as occurring in two steps:

$$\pi^{+^*} + p \rightarrow \pi^+ + p + \eta_0$$
$$\eta_0 \rightarrow \pi^+ + \pi^- + \pi^0.$$

At different times, various models have been suggested for the inner structure of the nucleons. For many years, they were considered to comprise an inner core surrounded by a cloud of pions; a sort of microcosm of the nuclear model of the atom. Subsequently, it was suggested that they are composed of two different fundamental components called the *up quark* and the *down quark*.[19] According to this theory, protons contain two up quarks and one down quark; neutrons, two down quarks and one up quark (Fig. 5.8).

It is postulated that the quarks carry fractional electric charges. The up quark carries a charge of $+\frac{2}{3}e$; the down quark a charge of $-\frac{1}{3}e$. Corresponding to each quark, there is an anti-quark that carries an electric charge of the opposite sign. They were presumed to be truly elementary particles, i.e. essentially point particles that, like electrons, possess no internal structure and are likewise, fermions. However, there the similarity ends, for, unlike the quarks, electrons do not participate in strong interactions (Table 5.4).

Table 5.4. The component elementary particles of ordinary matter.

Particle	Type	Rest-mass[20] (MeV/c^2)	Spin	Charge (e)
Up	quark	$2-4$	$\frac{1}{2}$	$+\frac{2}{3}$
Down	quark	$4-6$	$\frac{1}{2}$	$-\frac{1}{3}$
Electron	lepton	0.511	$\frac{1}{2}$	-1

[19]The name was coined by the American physicist Murray Gell-Mann 1929–, who was awarded the Nobel prize in 1969 for his work on the classification of elementary particles. The term comes from *Finnegan's Wake* by James Joyce: 'Three quarks for Muster Mark!' Quarks were independently postulated by George Zweig.

[20]The values given in this table are of the quantity known as *current quark mass* and is the mass of a quark by itself; they are inferred values. Since free quarks have never been found their mass has never been directly measured.

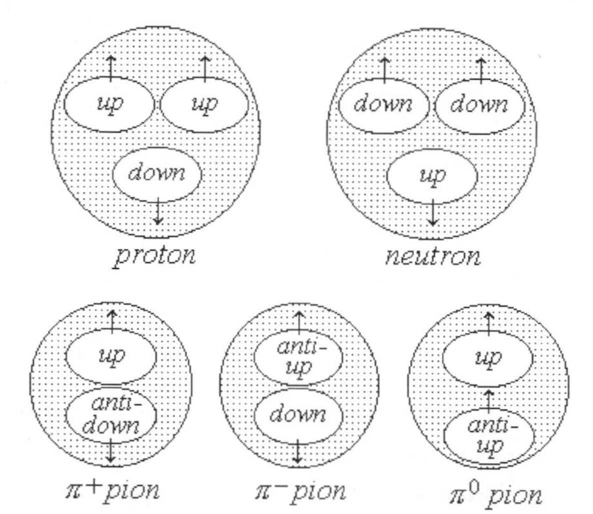

Fig. 5.8. The quark configurations of the nucleons and pions. The arrows indicate the spins. The composite particle's spin and charge is the sum of those of its components.

Whether or not they participate in the strong interaction is the basis of the division of elementary particles into two distinct groups: *leptons* and *hadrons*:

Particles that do not take part in strong interactions are called leptons: electrons are leptons.

Particles that do participate in strong interactions are called hadrons: protons and neutrons are hadrons.

Hadrons are composed of quarks and are further sub-divided into *baryons* and *mesons*, according to the number they contain (Fig. 5.8).

Protons and neutrons each contain three quarks; they belong to the family of particles called the baryons, each member of which is a fermion.

Pions are composed of a quark and an anti-quark; they belong to the family of particles called the mesons, each member of which is a boson.

Together, the baryon and meson families make up the clan of the hadrons.

Unlike the protons, neutrons and electrons that comprise atoms and nuclei, free quarks have never been observed; all quarks appear to be locked inside the particles they comprise. They are confined by the strong force fields and their mass can only be inferred; it appears to be much less than that required to account for the mass of the nucleons. For example, compare the actual mass of a proton ($938\,\mathrm{MeV}/c^2$) to the sum of the inferred masses of two free up quarks and one free down quark ($\sim 11\mathrm{MeV}/c^2$). Evidently, the mass of the *constitutent quarks* of a hadron, as distinct from that of free quarks, is greatly enhanced; this additional mass is attributed to the binding energy that holds the constituent quarks together in the hadron.

Remarkably, the bewildering host of unstable particles produced from the brute-force high-energy collisions that provide the experimental data of particle physics and from the processes by which the particles interact and decay, can be systematised. This requires postulating the existence of additional types (flavours)[21] of quarks and anti-quarks. To date, the existence of six quarks with their corresponding anti-quarks has been postulated. They are divided into three 'generations'. In addition to their electric charge, rest-mass and intrinsic angular momentum (spin), each of these new quarks and its corresponding anti-quark is characterised by a previously unknown unique physical property. These novel properties have been assigned whimsical names such as *strangeness*, *charm*, *bottomness* and *topness*. The allowed values of these physical quantities are characterised by integer quantum numbers (Table 5.5).

Table 5.5. The parameters of the six quarks.

	Quark flavour	Symbol	Unique property	Rest-mass $(\mathrm{MeV}/c^2)^{21}$	Spin	Charge (e)
First Generation	Up	u		$2-4$	$\frac{1}{2}$	$+\frac{2}{3}$
	Down	d		$4-6$	$\frac{1}{2}$	$-\frac{1}{3}$
Second Generation	Strange	s	Strangeness	~ 100	$\frac{1}{2}$	$-\frac{1}{3}$
	Charm	c	Charm	~ 1300	$\frac{1}{2}$	$+\frac{2}{3}$
Third Generation	Bottom	b	Bottomness	~ 5000	$\frac{1}{2}$	$-\frac{1}{3}$
	Top	t	Topness	$> 100,000$	$\frac{1}{2}$	$+\frac{2}{3}$

The existence of hadrons containing three identical quarks, all in the ground state and with their spins parallel, would appear to be forbidden by the exclusion principle; the occupation of the same orbital and spin state by three identical fermions is prohibited. However, such a hadron, the relatively stable *omega baryon*, Ω^-, has been observed; it comprises three strange quarks with their spins parallel. To get round this difficulty, it is simply postulated that the strange quark, as well as each of the other five quarks, can come in one of three different *colours* – red, blue and green – making a total of 18 distinctive quarks each with its corresponding anti-quark. It is postulated that each of the three strange quarks in the omega particle is of a different colour. Thus, they are not truly

[21] Imputed values for the curernt (free) quark mass.

identical and so the exclusion principle is not violated. This is the essence of the theory called *quantum chromodynamics.*

Colour is an elusive quality of matter, but it lies at the heart of the quark hypothesis; it is the property that is the source of the strong force just as mass is the source of gravity and charge the source of electromagnetism. And just as electromagnetic interactions can be interpreted in terms of an exchange of photons, so can both colour and gravitational interactions. The photon-like particles that mediate the former are the *gluons*; those that might be the mediators of gravity are called *gravitons.*

The basic interaction that holds hadrons together is the *colour force* that acts between their constituent quarks by virtue of their colour. All normal hadrons are colourless; they contain counterbalancing amounts of colour.[22] On this view, the strong force that acts between nucleons is a product of the colour force. Thus, it may be considered analogous to the familiar van der Waals force that acts between atoms and molecules as a consequence of the electromagnetic force and, likewise, it falls off rapidly with distance. However, here the similarity ends. For, unlike the electromagnetic force, the underlying colour force does not decrease with distance; it may even increase as the distance between the quarks grows and it is this property that prevents the observation of free quarks. The quarks can be pictured as joined to one another by an elastic string. So long as the distance between them is less than the unstretched length of the string, they can move freely. But should they move apart beyond this distance, the elasticity of the stretched string brings them back together. And should the string joining two quarks snap, the energy invested in the process appears as a pair of new quarks, one at each of the loose ends of the snapped string. Thus, free unpaired quarks can never be produced.

Evidence for the presence of hard point-like constituents in hadrons (nucleons) has been obtained from the scattering of very high-energy electrons (short De Broglie wavelength, $\sim 10^{-17}$ m) by protons and neutrons. The results indicate that these constituents have fractional charges and spin $\frac{1}{2}$. However, they also indicate that only part of the nucleon's momentum is stored in its quarks; part is also stored in electrically neutral particles within the nucleon, possibly

[22]Normal particles are 'white' or colourless. If they are baryons, each of the three component quarks has a different colour: red, green or blue. Together these three colours make white. If they are mesons, i.e. they comprise a quark and an anti-quark, the colour of the quark is neutralised by the anti-colour of the anti-quark. Since they do not participate in the colour force, leptons too are colourless.

the gluons. However, these photon-like particles which, it is postulated, mediate the interactions between the quarks have never been observed experimentally. Notwithstanding, the quark hypothesis remains central to the *standard model* of particle physics and has proved both useful and effective in explaining nuclear processes and the products of collisions between high-energy particles (see Section 5.3.7 below).

Chapter 5.2

Nuclear Radiations

The true origin of the *ionising radiations* first discovered by Becquerel remained an enigma until the nuclear model of the atom was postulated by Rutherford. Subsequent research showed that these high-energy radiations originated in the atomic nucleus and they were thus given the name *nuclear radiations.*

5.2.1 The Nature of the Nuclear Radiations

Initially, the radiations were classified on the basis of their penetrating power, i.e. in terms of the thickness of material required to attenuate a beam of the radiation. In this way, three different types of radiation were identified: they were designated α, β and γ-rays (Table 5.6).

Table 5.6. The penetrating power of nuclear radiations.

Type	Penetrating Power
α-rays	Absorbed by a sheet of thin paper or cardboard.
β-rays	Absorbed by a 2 mm sheet of lead.
γ-rays	Penetrate several centimetres of lead sheets before the beam is significantly attenuated.

More information about the radiations was obtained by passing them through electric and magnetic fields. It was found that γ-rays are not deflected by either electric or magnetic fields; they carry no net charge. However, both α-rays and β-rays are deflected, each in the opposite direction; α-rays are positively charged and β-rays negatively charged (Fig. 5.9).

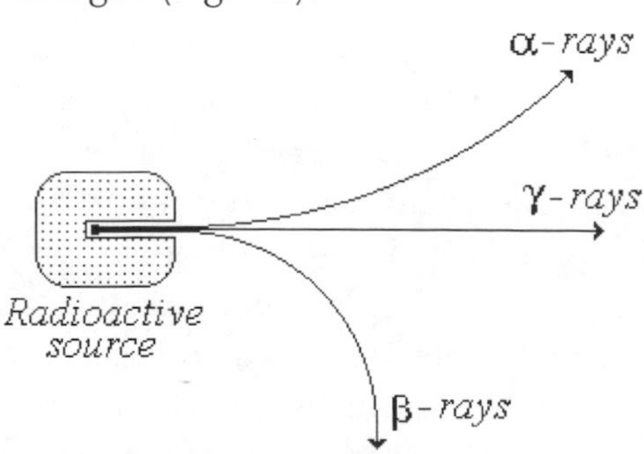

Fig. 5.9. The deflection of α, β and γ-rays by a magnetic field whose direction is into the plane of the page.

The actual nature of the radiations emitted by radioactive substances was first elucidated by Rutherford in 1899, just three years after their discovery. The ratio, q/m, between the charge and the mass of the β-rays was found to be the same as that of the electron. This, together with their negative charge, sufficed to identify the β-rays emitted from the then known radioactive substances as electrons. With the discovery and manufacture in the 1930s of radioactive isotopes of light elements, positively-charged β-rays were detected. These were found to be high-energy positrons (anti-electrons).

The charge to mass ratio found for the α-rays corresponded to that of both the H_2^+ ion and the He^{2+} ion (helium nucleus). In order to decide between these two alternatives, Rutherford collected α-rays in an evacuated discharge tube whose walls were too thick to allow their escape (Fig. 5.10). He found that when a high voltage was applied between the electrodes of the discharge tube, light that exhibited the typical emission spectrum of helium filled the tube. This identified the α-rays as helium nuclei.

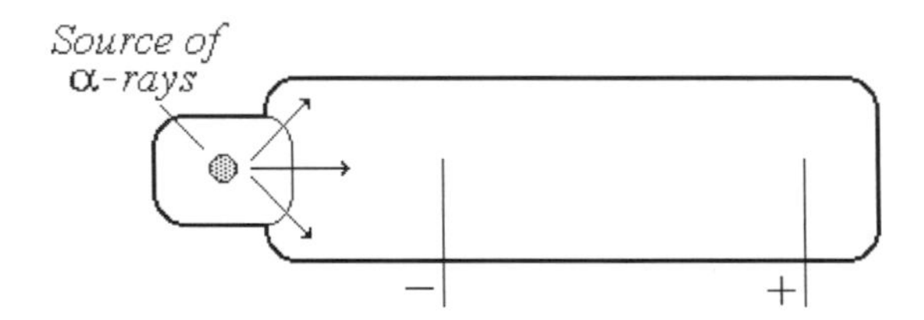

Fig. 5.10. Rutherford's identification of α-rays. The α-rays cannot penetrate the thick walls of the compartment containing the radioactive source or those of the discharge tube. However, they can pass from the former to the latter through the thin wall between them. Sufficient time was allowed for a significant number of α particles to accumulate in the discharge tube. A high potential was then applied between the electrodes and light that exhibited the typical emission spectrum of helium filled the tube.

The more important features of the various nuclear radiations are summarised in Table 5.7.

With their identities established, the kinetic energies of the particles could now be determined; this required measuring their velocities. Beams of α or β particles were passed through crossed electric and magnetic fields and the field strengths were adjusted until the particle beams passed through the crossed fields undeflected. The particle velocities could then be calculated from the ratio, $v = E/B$, where E and B are the magnitudes of the respective fields. It was found that the velocity of α particles is significantly lower than that of β particles. However, both the helium nuclei and the electrons that, respectively, comprise the α and β-rays were found to possess considerable amounts of kinetic energy, of the order of millions of electron-volts.

Table 5.7. The properties of nuclear radiation.

	α-rays	β-rays: β^+	β-rays: β^-	γ-rays
Identity	Helium nuclei He^{2+}	Positrons	Electrons	Photons
Charge	$+2e$	$+e$	$-e$	0
Rest-mass	4.0015 u	0.00055 u		0
Velocity	~0.06c	up to 0.98c		c
Energy	~6 MeV	up to 1 MeV		≥1 MeV
Penetrating power	~10^5 ion-pairs per cm of air	~10^3 ion-pairs per cm of air		~10 ion-pairs per cm of air

5.2.2 Mechanisms of Nuclear Radiation Attenuation

As they pass through a material medium, the α or β particles emitted by radioactive sources repeatedly collide with and ionise the atoms and molecules comprising the medium. In each such ionising collision a pair of charged particles is produced, an electron and a positive ion, and the incident α or β particle loses about 30 eV of its kinetic energy, slowing it down accordingly. A typical α particle, with an initial energy of 3 MeV, will produce about 100,000 ion-pairs before completely stopping (Table 5.8). In general, the penetrating power of ionising radiation is inversely proportional to the number of such ion-pairs produced in each unit layer of material through which they pass. The more ion-pairs produced in each unit layer, the thinner the thickness through which the ionising α or β particles must travel before being stopped.

Table 5.8. The stopping distances of α and β particles.

		Stopping distance (cm) in:		
Particle	Energy (MeV)	Air	Living Tissue	Aluminium
α	1	0.55	0.003	0.0003
	2	1.0	0.006	0.0006
	4	2.6	0.015	0.0015
β	0.01	12	0.015	0.004
	0.5	150	0.2	0.5
	1	420	0.5	0.15
	2	840	1.0	0.3

The relative penetrating powers of α and β radiation can be explained by the difference in the velocities of their component particles. The massive and slower α particles (helium nuclei) are much more efficient ionisers than the lighter and faster β particles (electrons); they ionise almost every atom with which they collide, losing some energy each time. Consequently, α particles lose all their energy after passing through just a thin layer of matter (Fig. 5.11).

However, the lighter and fleeter β particles are more easily deflected from their paths by the atoms and molecules of the material medium than are the heavier and more dogged α particles. As a result, β particles do not ionise every atom with which they collide and so a thicker layer of material is required to stop them. Moreover, the deceleration of the β electrons (or positrons) that accompanies these non-ionising collisions generates X-rays (bremstrahlung). The penetrating power of these X-rays exceeds that of the β particles and so even after all the β particles have been stopped by the material, these secondary X-rays may still be advancing through it.

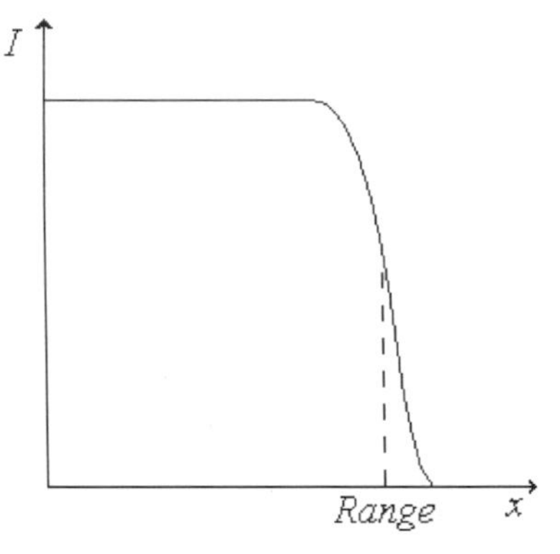

Fig. 5.11. The intensity, I, of a beam of α radiation as a function of the thickness, x, of the material through which it passes. The intensity falls off sharply at a certain thickness, called the *range*, indicating that the particles all had the same initial energy. The range is approximately proportional to the density of the material.

The third type of nuclear radiation, the γ-rays, are high energy photons: $f > 10^{19}$ Hz. Like other forms of electromagnetic radiation, the intensity, I, of a beam of γ-rays is attenuated exponentially when passed through matter[23]

$$I(x) = I_0 e^{-\mu x}. \tag{5.10}$$

In the case of γ-rays, the coefficient μ is called the *linear attenuation coefficient* or the *macroscopic cross-section*. The attenuation is usefully expressed in terms of the thickness of the material required to reduce the intensity of a beam by a half (Table 5.9).

Table 5.9. The thickness required to reduce the intensity of a beam of γ rays by a half.

Energy of the γ-ray photons (MeV)	Thickness of material (cm)	
	Living Tissue	Lead Metal
0.01	0.13	0.0008
0.1	4.0	0.01
1	9.8	0.9
5	23	1.5

[23] See Section 4.1.1.

The attenuation of high energy X-rays and γ-rays involves three different mechanisms.

1. The photoelectric effect

The photons in the incident high energy radiation release electrons from the atoms and molecules. The photons in the beam of radiation are annihilated, resulting in a corresponding reduction in its intensity.[24]

2. The Compton effect

When a beam of high-frequency electromagnetic radiation passes through a region containing free electrons, an interaction takes place between the incident high-frequency radiation and the free electrons in which energy is transferred from the radiation to the electrons. Relative to the high energies of the X-rays and γ-rays the outermost electrons of any atom are essentially free. As a result of the transfer of energy to the electrons the intensity of the radiation is reduced.[25]

3. Pair production

When a positron and electron collide they mutually destruct, releasing a burst of electromagnetic energy – photons. Conservation requires that the energy of these photons equal that of the annihilated particles. The rest-energy of an electron or positron is 0.51 MeV and so the energy of these photons must be at least 1.02 MeV. The reverse process also occurs. When passing close to a nucleus, photons possessing an energy greater than 1.02 MeV can transmute into an electron–positron pair (Fig. 5.12). Charge, linear and angular momentum are all conserved in the process. This is an example of the ***pair-production*** postulated by Dirac (Section 5.1.7).

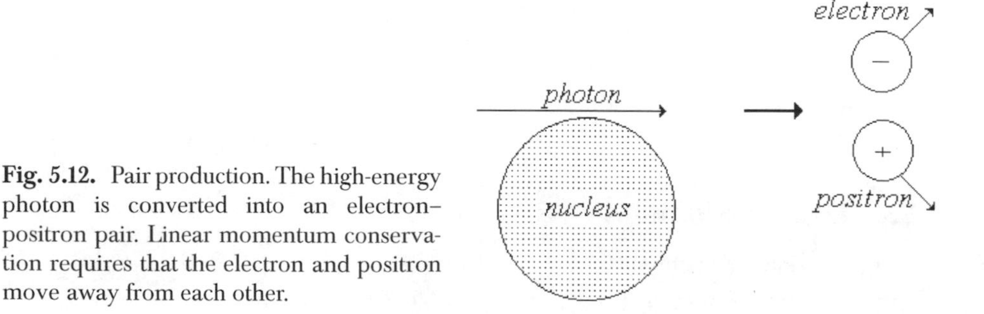

Fig. 5.12. Pair production. The high-energy photon is converted into an electron–positron pair. Linear momentum conservation requires that the electron and positron move away from each other.

[24]The phenomenon is comprehensively dealt with in Chapter 2.2.

[25]The phenomenon is comprehensively dealt with in Section 2.3.2.

The overall linear attenuation coefficient, μ, is the sum of the coefficients for the three mechanisms:

$$\mu = \mu_{Photoelectric} + \mu_{Compton} + \mu_{Pair-production}. \tag{5.11}$$

For photons with an energy of between 1 MeV and 5 MeV the Compton effect is the primary absorption mechanism. At higher energies, pair production becomes more prevalent (Fig. 5.13).

Although not emitted by naturally occurring sources of nuclear radiation, neutrons of both high and low energy are emitted from man-made sources such as atomic reactors. The penetrating power of neutrons is very high and so atomic reactors must be encased in thick-walled containers. Possessing no net charge, neutrons can quite easily reach into the nuclei of the material through which they are passing, disturbing their structure and thereby setting up secondary radiations. Certain materials such as boron and cadmium absorb neutrons without setting up these secondary radiations and are therefore used as neutron absorbers.[26]

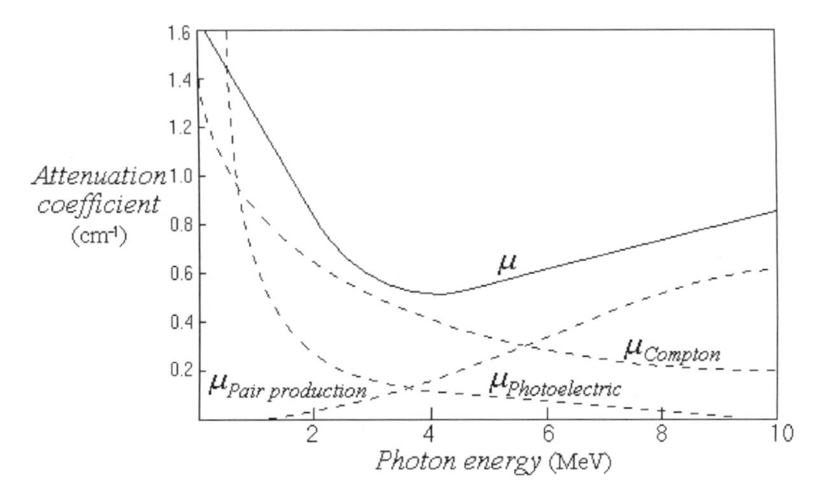

Fig. 5.13. The linear attenuation coefficients of the three mechanisms of X-ray and γ-ray absorption as a function of the photon energy.

5.2.3 Detectors of Ionising Radiation

Ionising radiation is detected by the ionisation it causes. Some of the most common devices used for their detection are briefly described here.

[26]The topic of neutron reactions is dealt with exhaustively in Chapter 5.4.

1. The cloud-chamber

The ions produced as an ionising particle passes through clean air that is saturated with water or alcohol vapour act as centres of condensation upon which tiny drops of water or alcohol are formed. These drops are formed all along the particle's path and so a visible track appears in the air. The cloud-chamber is a device for producing such a saturated atmosphere and for viewing the tracks made by ionising particles. In the version invented by Wilson in 1912, the saturated atmosphere is produced by a sudden adiabatic expansion (Fig. 5.14). The tracks formed by α-rays are thick and straight; those formed by β-rays are thin and curved. γ-rays eject electrons from air molecules which, in turn, produce a tangle of β-ray tracks.

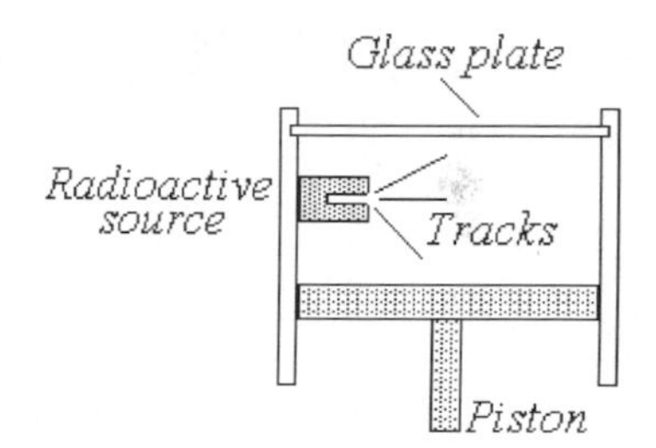

Fig. 5.14. The Wilson cloud-chamber. The air enclosed in the space above the piston is saturated with water vapour. Rapid withdrawal of the piston causes the air to cool and condensation results. The condensation takes place on the centres provided by the ions produced by the radiation emitted from the radioactive source and appears as tracks that can be viewed through the glass plate.

2. The bubble chamber

A liquid can be heated – superheated – above the temperature at which it normally boils if there are no centres in the liquid bulk upon which bubbles can form. In the bubble chamber a liquid, typically liquid hydrogen, is heated under pressure. Suddenly reducing the pressure leaves the liquid in a superheated state. An ionising particle passing through the superheated liquid generates ions which act as ebullition centres. The tiny bubbles appear as a visible track in the liquid.

3. Nuclear emulsions

Ionising radiation can be detected by the fine tracks it leaves in sensitive photographic transparencies called *nuclear emulsions.* After developing, the tracks are revealed by examining the transparencies through a microscope. Typically, the tracks are just a few millimetres in length. Each type of ionising particles leaves a characteristic track by which it can be identified and its energy estimated. The emulsions are rugged and can be used where more delicate detectors would be inappropriate. For example, they can be taken aloft in unmanned balloons for the investigation of cosmic rays in the upper atmosphere.

4. The ionisation chamber – Geiger counter

The ionisation chamber is a gas-filled tube inside which a pair of electrodes have been installed. The gas, typically a mixture of argon and a little ethyl alcohol vapour at low pressure (~0.1 atmospheres), is not an electric conductor under normal conditions. A potential difference is maintained between the electrodes (Fig. 5.15).

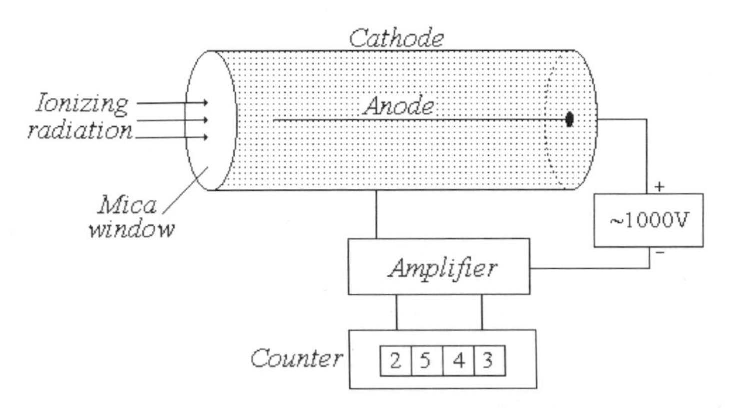

Fig. 5.15. Geiger counter. Ionising radiation enters the ionising chamber through the thin mica window. The walls of the chamber are the cathode; the anode is a fine wire down its centre.

The positive ions, Ar^+, and electrons liberated in the gas-filled tube when an ionising particle enters the tube are attracted to the cathode and anode, respectively. With a relatively high potential difference between the anode and cathode, ~1,000 V, the electron produced in the initial event collides with and ionises other argon atoms on its way to the anode, producing more argon ions and free electrons. The latter, in turn, generate still more ion-pairs inside the tube. A chain reaction takes place resulting in an avalanche of ionisation in the gas. The arrival of the free electrons at the anode is signalled by a detectable pulse of anode current. Since each pulse of anode current corresponds to the entry of an ionising particle into the tube, the number of particles can be determined by counting the pulses.

For the ionisation chamber to detect the next ionising particle, it must first be restored to its neutral state. This requires the neutralisation of the positive argon ions remaining in the gas. This is done by the ethyl alcohol molecules. Before reaching the cathode, where they might release electrons that would blur the pulses of anode current, the energetic argon ions collide with the alcohol molecules giving up their energy to them. Only positive alcohol ions reach the cathode where any excess energy they possess results in their decomposition rather than in the release of electrons.

Preparing the tube to receive the next ionising particle takes time, about a millisecond. During this 'dead time', as the argon atoms are being neutralised, the entry of an ionising particle would go undetected. Consequently, Geiger counters cannot be used for particle rates greater than a few hundred per second. They also give no indication of the energy of the ionising particles. They are employed mainly for counting β particles. They are inefficient counters of γ radiation or of high-energy neutrons since not all of the photons or neutrons entering the tube induce the avalanche of ionisation in the gas.

5. Scintillation counter

In the famous experiment that demonstrated the existence of atomic nuclei, Geiger and Marsden observed the scattering of α particles by laboriously counting the light scintillations produced as each particle struck a zinc sulphide coated plate. In a modern scintillation counter this is done automatically by use of a *photomultiplier tube.*

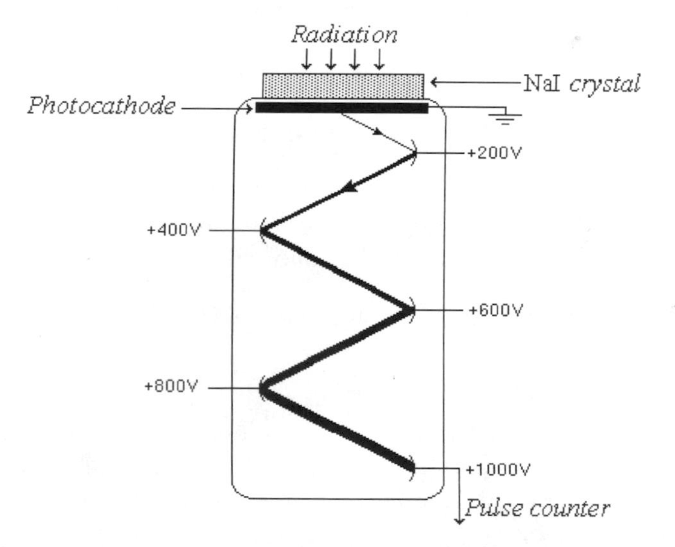

Fig. 5.16. Scintillation counter. The entire device is kept in a light-proof container. It is also shielded from external magnetic fields. The number of electrons in the beam is amplified at each electrode.

When an ionising particle strikes a sodium iodide crystal a flash of light is produced. Some of this light enters the photomultiplier tube where it strikes the photocathode and releases a number of electrons from it by means of the photoelectric effect. These electrons are accelerated towards a second electrode that is held at a higher potential than the photocathode (Fig. 5.16). Each electron that strikes this second electrode ejects a number of electrons from it, increasing the total number of electrons in the pulse. The shower of electrons is now accelerated towards a third electrode held at a still higher potential where the amplification process occurs once again. This is repeated a number of times until an electron pulse of significant strength is produced. Finally, the pulses are counted.

Because the number of electrons emitted from the final electrode depends on the number incident on the first, each pulse recorded corresponds to a single ionising particle. Moreover, since the intensity of the light incident on the photocathode is determined by the energy of the original ionising particle, the strength of the pulse reflects the particle's energy. Thus, the scintillation counter not only counts the ionising particles but also measures their energy. The dead-time of a good scintillation counter is short, about one microsecond. It is also an efficient counter of both β and γ radiation.

By using light-emitting substances other than sodium iodide and by adjusting the potential differences between the electrodes, scintillation counters can be used to detect and measure ionising radiations over a wide range of energies and from different types of sources.[27]

5.2.4 The Biological Effects of Nuclear Radiation

Since the major constituent of living tissue is water, the most probable initial result of the passage of an ionising particle through living tissue is the ionisation of a water molecule in reactions of the following type:

$$\begin{aligned} H_2O + \text{ionizing radiation} &\rightarrow H_2O^+ + e^- \\ H_2O + e^- &\rightarrow H_2O^-. \end{aligned} \tag{5.12}$$

The products of these reactions, H_2O^+ and H_2O^-, are very active species and can react with and cause damage to other molecules in the tissue. In addition to the gross tissue damage they cause, these species can also induce genetic mutations. Most of the biological damage that results from exposure to nuclear radiation arises from these secondary reactions.

The extent of the physical damage caused to living tissues by nuclear radiation is a function of four factors:

1. the amount of radiation absorbed;
2. the nature of the ionising radiation;
3. the tissue or body part exposed to the radiation;
4. the rate at which radiation is absorbed.

[27]The radiation detectors based on semiconductors are described in Section 6.4.9.

We will consider each of these factors, one by one.

1. The amount of radiation absorbed
The capacity of a source of radiation is often reported in a unit called the *röntgen* whose symbol is R. For instance, the X-ray beam from a typical dental machine may have a rating of 250 mR/s.[28] But this says nothing about the radiation actually absorbed by the patient; the machine has the same rating whether anyone is sitting in the dentist's chair or not.

Living tissue is only affected by the radiation energy that it actually absorbs; it is unaffected by radiation that passes through it unabsorbed. The unit used to express the amount of radiation energy absorbed by living tissue is the *gray* (Gy).

> An absorbed dose of 1 gray equals the absorption of 1 Joule of radiation energy by 1 kg of tissue.

An alternative unit, used widely in the past, is the RAD (acronym of *radiation absorbed dose*):

$$1\,\text{RAD} = 10^{-2}\,\text{Gy} = 10^{-2}\,\text{J/kg}.$$

The dose received from a medical X-ray can range from 10 μGy to 1,000 μGy.

2. The nature of the ionising radiation
The damage caused to living tissue varies according to the type of radiation to which it is exposed and how localised is the energy absorption. Densely ionising particles such as α particles and slow β particles, whose ranges in biological tissue are relatively short, are more damaging than the high-energy photons in X or γ-rays. The relative biological effect of each type of ionising radiation is indicated by its *quality factor* (QF) or *relative biological effectiveness* (RBE) (Table 5.10).

The effective absorbed dose, i.e. the biological effect of a dose of radiation, is the product of the absorbed dose (measured in grays or RADs) and the quality factor (or RBE) of the particular radiation. The effective absorbed dose is measured in units called *sieverts* (Sv). For example, the biological effect of a dose of 10 Gy of γ or X-rays whose quality factor is equal to unity is $10 \cdot 1 = 10$ Sv. A dose of 5 Gy of slow β particles or of 0.5 Gy of α particles would have the same biological effect.

[28] One röntgen is defined as the capacity of a γ or X-ray source that delivers 8.78 mJ of energy to 1 kg of dry air.

Table 5.10. The quality factor (relative biological effectiveness) of various ionising radiations.

Radiation	QF or RBE
γ or X-rays	1
Fast β particles	1
Slow β particles	2
Fast neutrons and protons	10
Slow neutrons	3–5
α particles	10–20

An alternative unit, frequently used in the past to measure the biological effect of radiation, is the REM (acronym of *RAD equivalent mammal*[29]):

$$100\,\text{REM} = 1\,\text{Sv}.$$

3. The tissue or body part exposed to the radiation
Because of their greater density, bones absorb relatively more radiation than soft tissue. The cells most sensitive to radiation damage are those which have the highest rate of division, and are the least differentiated. Cancer cells exhibit these characteristics and so are more susceptible to radiation damage than other cells. This is the basis of the treatment of cancer by high-energy radiations. On the other hand, these characteristics are also typical of zygotes and young embryos and so great care must be taken in the irradiation of women of child-bearing age. A dose of 40 Sv to the gonads produces sterility in men; a dose of just 10 Sv renders women sterile.

Red blood cells are more resistant to radiation damage than white blood cells. Changes in the ratio between the number of red and white cells in the blood can be used to measure the doses to which an individual has been exposed. Whole body doses as low as 0.25 Sv can affect the ratio between the cells. Apart from the gross tissue damage inflicted by an exposure to ionising radiation, mutations in the organism's genetic material may also occur. Whether genetic damage occurs only above a certain minimum threshold of radiation or at any level, however low, is still unresolved.

Ingesting radioactive materials is particularly dangerous. The fall-out from the explosion of a typical 1 megaton nuclear weapon contains about 0.5 kg of the relatively long-lived radioactive strontium isotope Sr–90.[30] Chemically, strontium is similar to calcium and it was found during the 1950s that radioactive Sr–90, released by the atmospheric testing of nuclear weapons, was accumulating in the bones of children fed the milk of cows that had grazed

[29] Or *röntgen equivalent in man.*
[30] The half-life of Sr–90 is 29 years.

in regions where the fall-out came to earth. The long-term implications of this were considered so dangerous that the nuclear powers agreed to cease the atmospheric testing of nuclear weapons. After the Chernobyl disaster in 1986, the meat of lambs that had grazed in wet areas of the UK was taken off the market, for fear that they had become contaminated by the radioactive fall-out the rain had brought down in the days and weeks after the explosion.

4. The rate at which the radiation is absorbed

Whereas a single dose of 10 Sv can be lethal, certain cancer patients may receive daily doses of 1 Sv for a period of a fortnight, making a total of 14 Sv for the period. Evidently, the rate at which the radiation is received is an important factor in determining its biological effect.

Skin reddening, *erythema*, is one of the manifestations of tissue damage caused by high-energy radiation and it has been used to estimate the effects of radiation. Table 5.11 shows the effect of different radiation rates on skin. Clearly, substantial doses can be tolerated if they are given slowly and the tissue is allowed time to repair.

Table 5.11. Doses of X-ray irradiation that produce the same degree of erythema.

Total amount of radiation (Gy)	Irradiation time (mins)	Rate of radiation (Gy/min)
5.0	1	5
7.8	15	0.50
13.0	260	0.05
22.5	4500	0.005

The whole body dose that kills 50% of a population of organisms is designated LD_{50}. A burst of 30 Sv of X-rays will kill 50% of a newt population; the LD_{50} for guinea pigs is just 2.5 Sv. On the basis of the mortality rates recorded in Hiroshima and Nagasaki after their devastation by nuclear weapons in 1945, it has been estimated that the human LD_{50} is about 6.5 Sv. Protecting the more sensitive organs such as the spleen increases the chances of survival.

We are all unavoidably exposed to the natural sources of ionising radiation: the cosmic rays entering the atmosphere from outer space and the emissions from radioactive sources such as the isotope ^{40}K present in our own bodies and the radon gas emitted by rocks and stone-based building materials (Table 5.12). Thus, total avoidance of high-energy radiation is impossible. In considering the risks implicit in the man-made radiation to which we are exposed, which accounts on average for less than 20% of our total yearly

Table 5.12. Annual individual exposure to ionising radiation in the UK.[31]

Radiation Source	Dose (μSv)	
Natural Sources		
Cosmic rays	330	
Terrestrial γ sources	350	
Radioactive isotopes in the human body	250	
Radon	700	
		2230
Artificial Sources		
Medical	410	
Fall-out from weapon testing	6	
Occupational	6	
Miscellaneous	1	
		419
Total		2,700

exposure, the undoubted benefits of procedures such as medical X-rays must also be taken into account.

The International Commission on Radiological Protection (ICRP) has set guidelines for 'acceptable' individual levels of radiation exposure, many of which have been incorporated into national regulatory requirements. The primary aim of the Commission was 'to contribute an appropriate level of protection for people and the environment against the detrimental effects of radiation exposure without unduly limiting the desirable human actions that may be associated with such exposure.'

In establishing the dose limits, the conservative assumption made was that there is no safe level of exposure; even the slightest exposure may have a stochastic effect, cancerous or otherwise. Based on this assumption, it was recommended that all exposure be kept 'as low as reasonable achievable (ALARA)'. After examining the best available evidence, the ICRP settled on a dose of 20 mSv (20,000 μSv) per year, averaged over five years (i.e. no more than 100 mSv in any five years), as a tolerable limit for occupational exposure (for radiologists, workers in nuclear industries, etc.) with the further provision that the dose not exceed 50 mSv (50,000 μSv) in any one year. For the general public, the limit was set at 1 mSv (1,000 μSv) per year over and above the ionising radiation to which we are all exposed from natural and medical sources; individual annual doses from practices other than medical procedures are generally much less than this limit.

[31]UK Health Protection Agency, *Ionising Radiation Exposure of the UK Population: 2005 Review*, HPA-RPD-001.

It must be emphasised that these limits are only guidelines; they take no account of the particular body-part exposed to the radiation (the sensitive eye lens, the skin or the extremities) and the age or condition of the individual (minors, pregnancy, etc.). Whether these doses are as safe genetically in the long term is also impossible to gauge at present.

Whilst there is probably no completely safe dose of ionising radiation, neither is there an entirely safe way of crossing the road; caution is required in both cases. However, the experience of the more than 100 since the discovery of radioactivity and X-rays has shown that in properly administered small and intermittent doses, ionising radiation is not harmful and certainly not life-threatening.

Chapter 5.3

Radioactivity, Neutrinos and the Standard Model

Of the more than 1,400 known atomic nuclei, only about 280 are stable. All the stable nuclei occur naturally. Some of the unstable ones, particularly those of the heavy elements, are also found naturally. All the other unstable nuclei are manufactured from naturally-occurring nuclei by nuclear reactions.[32] These are usually referred to as *artificial nuclei* and the atoms they comprise as *artificial isotopes.*

5.3.1 The Disintegration of Unstable Nuclei

In general, unstable nuclei disintegrate spontaneously by the emission of nuclear radiation; they are *radioactive.* When describing nuclear decays, the disintegrating nucleus is usually referred to as the *parent nucleus* and the nucleus remaining after the event as the *daughter nucleus.*

The total rest-mass of the daughter nucleus and of the nuclear radiation released in a radioactive disintegration, $m_{Daughter} + m_{Radiation}$, is always less than that of the parent nucleus, m_{Parent}. The mass difference,

$$\Delta m = (m_{Daughter} + m_{Radiation}) - m_{Parent}$$

appears as the *disintegration energy*, $Q = \Delta mc^2$, liberated in the process.

The mass number, A, of the daughter nucleus produced when an α particle is emitted from a radioactive nucleus is always four units less than that of the parent nucleus and its atomic number, Z, is always two less. For example:

$$\underset{\substack{\textit{Parent}\\ \textit{Nucleus}}}{{}^{226}_{88}\mathrm{Ra}} \rightarrow \underset{\substack{\textit{Daughter}\\ \textit{Nucleus}}}{{}^{226}_{88}\mathrm{Rn}} + \underset{\substack{\textit{Alpha}\\ \textit{Particle}}}{{}^{4}_{2}\mathrm{He}} + \underset{\substack{\textit{Disintegration}\\ \textit{Energy}}}{4.78\,\mathrm{Mev}} \tag{5.13}$$

The emission of a β particle, either an electron, β^-, or a positron, β^+, changes the atomic number of the nucleus without affecting its mass number. In β^-

[32] The production of artificial isotopes is described in Chapter 5.4.

emission, a neutron in the parent nucleus becomes a proton in the daughter nucleus; this raises the atomic number by one

$$\underset{\substack{\textit{Parent}\\\textit{Nucleus}}}{{}^{214}_{83}\mathrm{Bi}} \rightarrow \underset{\substack{\textit{Daughter}\\\textit{Nucleus}}}{{}^{214}_{84}\mathrm{Po}} + \underset{\substack{\textit{Beta}\\\textit{Particle}}}{e^-} + \underset{\substack{\textit{Disintegration}\\\textit{Energy}}}{3.26\,\mathrm{Mev}} \tag{5.14}$$

In β^+ emission, a proton in the parent nucleus becomes a neutron in the daughter nucleus; this reduces the atomic number by one

$$\underset{\substack{\textit{Parent}\\\textit{Nucleus}}}{{}^{25}_{13}\mathrm{Al}} \rightarrow \underset{\substack{\textit{Daughter}\\\textit{Nucleus}}}{{}^{25}_{12}\mathrm{Mg}} + \underset{\substack{\textit{Beta}\\\textit{Particle}}}{e^-} + \underset{\substack{\textit{Disintegration}\\\textit{Energy}}}{3.26\,\mathrm{Mev}} \tag{5.15}$$

A nuclear proton can also be converted into a nuclear neutron by the capture of an electron. This process competes with β^+ emission

$$\underset{\substack{\textit{Parent}\\\textit{Nucleus}}}{{}^{15}_{8}\mathrm{O}} \rightarrow \underset{\substack{\textit{Beta}\\\textit{Particle}}}{e^-} + \underset{\substack{\textit{Daughter}\\\textit{Nucleus}}}{{}^{15}_{7}\mathrm{N}} + \underset{\substack{\textit{Disintegration}\\\textit{Energy}}}{2.75\,\mathrm{Mev}}\,. \tag{5.16}$$

After an α or β emission, the daughter nucleus is often left in an excited energy state. It subsequently reaches its ground state by emitting high-energy photons, γ-rays

$$\underset{\substack{\textit{Parent}\\\textit{Nucleus}}}{{}^{12}_{5}\mathrm{B}} \rightarrow \underset{\substack{\textit{Excited Daughter}\\\textit{Nucleus}}}{{}^{12}_{6}\mathrm{C}^*} + \underset{\substack{\textit{Beta}\\\textit{Particle}}}{e^-}$$

$$\underset{\substack{\textit{Excited Daughter}\\\textit{Nucleus}}}{{}^{12}_{6}\mathrm{C}^*} \rightarrow \underset{\substack{\textit{Ground-state Daughter}\\\textit{Nucleus}}}{{}^{12}_{6}\mathrm{C}} + \underset{\substack{\textit{Gamma Particle-}\\\textit{Photon}}}{\gamma} \tag{5.17}$$

The emission of a high-energy photon affects neither the mass number nor the atomic number of the nucleus. However, it does reduce its rest-mass by an amount, $\Delta m = hf/c^2$, equivalent to that of the photon's energy. A typical nuclear energy level diagram illustrating this process is shown in Fig. 5.17.

Certain excited nuclei, instead of emitting high-energy photons, make the transition to the lower energy state by transferring energy to one of their atomic electrons, which is then emitted at high energy.

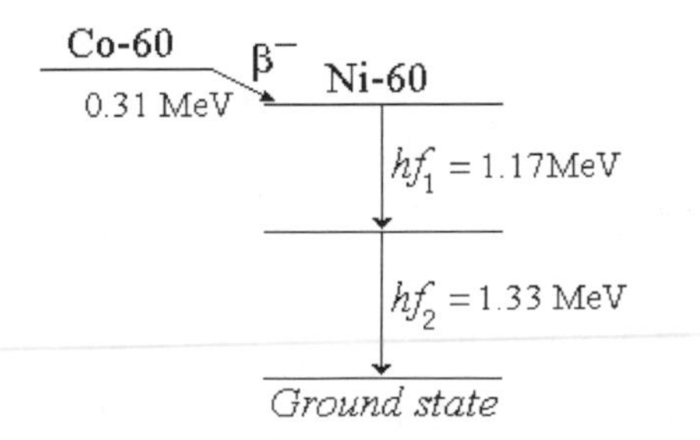

Fig. 5.17. The Ni-60 nucleus produced from the emission of a β^- particle from Co-60 is in an excited nuclear energy state. It falls to its ground state in two stages, emitting first a photon of energy 1.17 MeV and then another of 1.33 MeV.

The daughter nucleus produced from the decay of a radioactive nucleus is often itself radioactive. This can result in a sequence of decays. For example:

$$\begin{array}{ccccccccc} {}^{238}_{92}\mathrm{U} & \to & {}^{234}_{90}\mathrm{Th} & \to & {}^{234}_{91}\mathrm{Pa} & \to & {}^{234}_{92}\mathrm{U} & \to & {}^{230}_{90}\mathrm{Th} \to \text{etc.} \\ & & +\alpha & & +\beta^- & & +\beta^- & & +\alpha \end{array}$$

Almost all nuclei with an atomic number greater than 82 – the atomic number of the element lead, Pb – are radioactive. Since the only change in the mass number of a nucleus that can occur as a result of radioactive decay (α emission) is -4, all these unstable heavy nuclei can be fitted into one of four decay sequences or series, each characterised by the mass number, A, of its component nuclei: $4n, 4n + 1, 4n + 2$ and $4n + 3$, where n is an integer between 51 and 59. These are known, respectively, as the thorium, neptunium, uranium and actinium series (Table 5.13):

Table 5.13. The four radioactive decay series.

Series	Parent Nucleus	Mass numbers
Thorium series	${}^{232}_{90}\mathrm{Th}$	$A = 4n$
Neptunium series	${}^{237}_{93}\mathrm{Np}$	$A = 4n + 1$
Uranium series	${}^{238}_{92}\mathrm{U}$	$A = 4n + 2$
Actinium series	${}^{235}_{92}\mathrm{Ac}$	$A = 4n + 3$

All the intermediate nuclei of a decay series are less stable than the parent nucleus. The thorium, uranium and actinium series, which occur naturally, all end with a stable isotope of the element lead (Fig. 5.18).

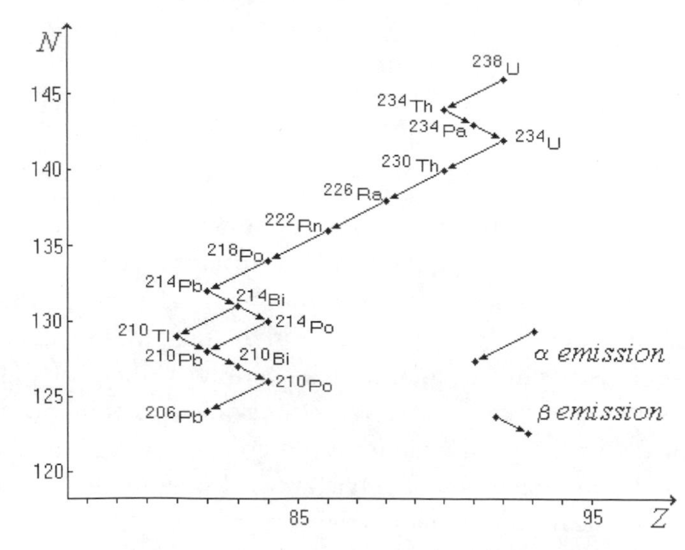

Fig. 5.18. The uranium series. The Bi–214 nucleus can decay either by α emission or by β^- emission. However, both routes lead ultimately to the Pb–210 nucleus.

The parent nucleus of the neptunium series is not found in nature, though it can be manufactured. Compared to the parent nuclei of the other series it is very unstable. Thus, within the estimated age of the solar system, it and any of the intermediate nuclei of its series that might once have existed have all decayed into the stable nucleus, Bi–209, with which the series ends.

Example 5.4: Disintegration Energy

Calculate the amount of energy liberated in the radioactive disintegration (β^- emission) of P–32:

$$^{32}_{15}\mathrm{P} \rightarrow {}^{32}_{16}\mathrm{S} + e^-.$$

The rest-mass of a P–32 atom is 31.97391 u and that of an S–32 atom is 31.97207 u.

Calculation: The energy liberated in the disintegration is given by:

$$Q = \Delta m \cdot c^2,$$

where Δm is the difference between the rest-mass, m_{P-32}, of the parent nucleus and the sum, $m_{S-32} + m_e$, of the rest-masses of the daughter nucleus and the emitted electron:

$$\Delta m = m_{\mathrm{P}-32} - (m_{\mathrm{S}-32} + m_e).$$

A neutral sulphur atom contains one electron more than a neutral phosphorus atom. It follows, therefore, that in this case the mass difference Δm is also given by the difference between the rest-mass of a neutral P–32 atom and that of a neutral S–32 atom, i.e.

$$\Delta m = 31.97391\mathrm{u} - 31.97207\mathrm{u} = 0.00184\mathrm{u}.$$

From which

$$Q = 0.00184 \cdot 931 = 1.71\,\mathrm{MeV}.$$

5.3.2 The Kinetics of Radioactive Disintegration

The disintegration of a radioactive nucleus is a random process; there is no way of predicting precisely when a particular unstable nucleus will disintegrate. 'Old' radioactive nuclei that have been around for a long time are no more or less likely to disintegrate in the next minute than freshly produced but otherwise identical 'young' nuclei. Similarly, the probability that a nucleus will disintegrate is unaffected by its environment, be it hot or cold, high pressure or low pressure, or whether it is a component of one chemical compound or another. It follows that the rate at which a sample of radioactive matter decays and emits nuclear radiation depends only on the number of radioactive nuclei it contains.

The rate of decay, dN/dt, of a sample containing N radioactive nuclei is given by:

$$\frac{dN}{dt} = -\lambda N. \tag{5.18}$$

The rate of decay is negative because the number of radioactive nuclei decreases with time. The parameter, λ, called the *decay constant*, is a characteristic of the unstable nuclei.

The absolute value, $|dN/dt|$, of its rate of decay is called the sample's *activity*. The units in which activities are expressed are the *Becquerel* whose symbol is Bq or the *Curie*[33] which is designated Ci.

1 Bq is an activity of one disintegration per second: $1\ \text{Ci} = 37 \cdot 10^9\ \text{Bq}$.

Radioactive sources used in cancer treatment, such as those containing the isotope Co−60, have activities of about 10^{14} Bq. In the explosion of the nuclear reactor and power station at Chernobyl in 1986, radioactive materials with an activity of at least $90 \cdot 10^6$ Ci were liberated into the atmosphere. This is equivalent to at least 6,750 Bq on each square meter of the earth's surface, an amount of radioactive fall-out comparable to that it is estimated would be released in a medium-scale tactical nuclear strike.[34] Radioactive sources used in student laboratories have activities of 10^5 Bq or less. Antique luminous watches whose numerals are painted with uranium-based paints have activities of about 20,000 Bq.

The solution to Eq. (5.18) is:

$$N_t = N_0 e^{-\lambda t}, \tag{5.19}$$

where N_0 is the number of unstable nuclei in the sample at $t = 0$. This exponential law of radioactive decay was first discovered by Rutherford and Soddy[35] in 1902 (Fig. 5.19).

[33]Named after Marie Curie 1867–1934 and her husband Pierre Curie 1859–1906, French physicists who were awarded the Nobel prize in 1903 for their work on radioactivity and the discovery of the radioactive elements radium and polonium. The latter element was named after Marie's land of birth, Poland. In 1911, Marie Curie was awarded a second Nobel prize for her further work on radium.

[34]Yuri M. Scherbak (1996) 'Ten Years of the Chornobyl Era', *Scientific American*, 274, pp. 44–54.

[35]Frederick Soddy 1877–1956, British chemist who was awarded the Nobel prize in 1921 for predicting the existence of isotopes.

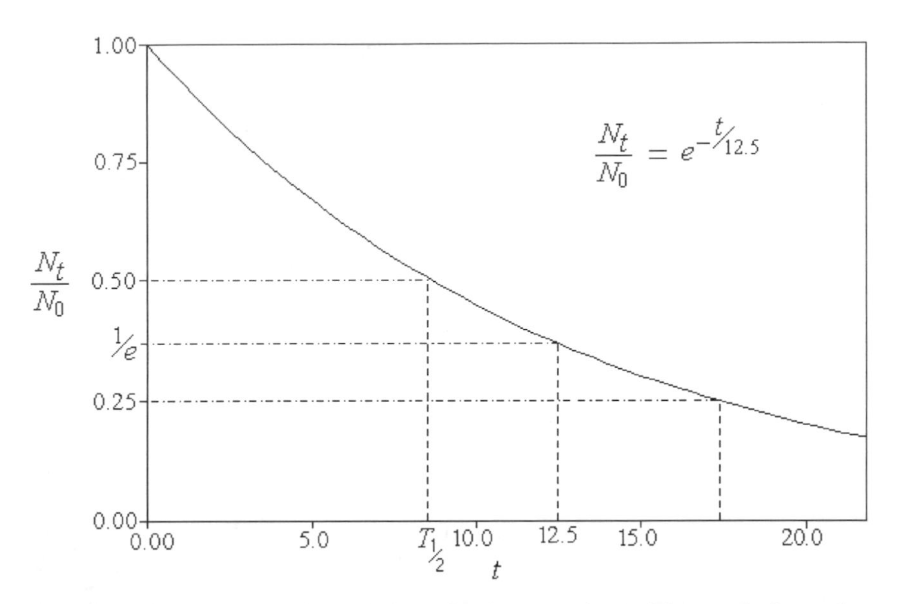

Fig. 5.19. The Rutherford–Soddy law of radioactive decay. The graph shows the quotient N_t/N_0 of the number of unstable nuclei in a sample as a function of the time, t, where $\tau = 12.5$ time units. A graph of the sample's activity, $|dN/dt|$, as a function of time would have exactly the same form.

The disintegration (decay) law can also be expressed in terms of the parameter $\tau = 1/\lambda$, the *average lifetime* of the unstable nucleus[36]

$$N_t = N_0 e^{-1/\tau}. \tag{5.20}$$

The most commonly quoted parameter of an unstable nucleus is its *half-life*, $T_{\frac{1}{2}}$. It is the length of time after which half the nuclei of its type will have disintegrated in any given sample of them. In a sample containing N_0 unstable nuclei at any given moment, it is the time after which $N_t = N_0/2$. Substitution in Eq. (5.19) gives:

$$\frac{N_0}{2} = N_0 e^{-\lambda T_{\frac{1}{2}}}, \tag{5.21}$$

which reduces to

$$2 = e^{\lambda T_{\frac{1}{2}}} \tag{5.22}$$

or

$$T_{\frac{1}{2}} = \frac{\ln 2}{\lambda} = \frac{0.693}{\lambda}. \tag{5.23}$$

[36]The proof of this definition of τ is given in Appendix 5.5.1.

Example 5.5: Radioactive Disintegration

Two hours after it was prepared, a sample of the radioactive isotope Pb–214 was found to contain one hundred million Pb–214 nuclei. The half-life of this isotope is 27 minutes. How many Pb–214 nuclei were in the sample when it was prepared?

Calculation: From Eq. (5.23) the decay constant of the Pb–214 isotope is

$$\lambda_{\mathrm{Pb}-214} = \frac{0.693}{27} = 0.0257 \mathrm{min}^{-1}.$$

At time $t = 120\mathrm{m}$ the number of radioactive Pb–214 nuclei left in the sample was $N_t = 10^8$. Substitution in Eq. (5.19) gives

$$10^8 = N_0 e^{-0.0257 \cdot 120}.$$

From which

$$N_0 = 21.8 \cdot 10^8.$$

5.3.3 Age Determination with Radioisotopes

Because the rate at which radioactive nuclei disintegrate is unaffected by their environment, it can be used to estimate the age of any material sample or object which contains a radioactive isotope (radioisotope). In addition to a knowledge of the radioisotope's half-life and of its present abundance and that of its decay products in the sample, this dating method also requires making certain assumptions about their original abundance.

Thus, for example, the age of the earth's crust can be estimated from the ratio between the amounts of U–238 and Pb–206 found in geological specimens. These two isotopes are respectively the first and last members of the uranium decay series, all the intermediate members of which are so unstable that the decay of the parent radioactive uranium can be considered to lead directly to the stable lead isotope. The assumption made is that all the Pb–206 nuclei found in the specimen today were originally U–238 nuclei, i.e. that at the crust's formation the specimen contained no Pb–206 nuclei. If no other lead isotopes are found in the specimen, this is a reasonable assumption. Typically, geological samples contain equal numbers of U–238 nuclei and Pb–206 nuclei, i.e. half of the uranium nuclei they originally contained have transmuted into lead. Thus, the age of the earth's crust approximately equals the half-life of U–238, *viz.* $4.5 \cdot 10^9$ years.

The carbon radioisotope C–14 is frequently used for dating archaeological finds. This isotope is formed in the upper atmosphere from the collision of high-energy neutrons with nitrogen atoms in the air

$$^{14}_{7}\mathrm{N} + n \rightarrow {}^{14}_{6}\mathrm{C} + p.$$

The high-energy neutrons are a constituent of the shower of particles produced when cosmic ray protons enter the atmosphere and collide with the atoms in the air molecules. The half-life of C–14 is 5,730 years.

The stable C–12 isotope is by far the most common carbon isotope in atmospheric carbon dioxide. On average, just one out of every $1.3 \cdot 10^{12}$ carbon atoms in the atmosphere is a C–14 atom. As a result of photosynthesis, the same ratio is found in green plants and subsequently in the food chain of living organisms. However, once an organism dies it ceases to absorb any more C–14. Assuming that the incidence of both carbon isotopes in the atmosphere has remained unchanged over the period, the age of the remains of a once living object can be determined by comparing the ratio of C–14 to C–12 atoms found in its remains to that in the atmosphere. The method enables datings to be made up to about 20,000 years ago with an accuracy of about ±100 years.

Example 5.6: Carbon–14 Dating

The activity of a piece of charcoal found in an archaeological dig was found to be 1 Bq. The piece contained six grams of carbon. Estimate the charcoal's age.

Calculation: Since the molecular weight of carbon is 12 grams, the piece of charcoal contains half a mole of carbon atoms, i.e. $3.0 \cdot 10^{23}$ atoms. The number, N_0, of C–14 atoms the sample originally contained is, therefore:

$$N_0 = 3.0 \cdot 10^{23} \cdot 1.3 \cdot 10^{-12} = 3.9 \cdot 10^{11}.$$

The sample's activity, $|dN/dt| = 1$, is directly proportional to the number, N, of C–14 atoms it contains.

$$\left|\frac{dN}{dt}\right| = \lambda N = \frac{0.693N}{T_{\frac{1}{2}}}.$$

The half-life of C–14 is 5,730 years, i.e. $1.807 \cdot 10^{11}$s. Substitution gives:

$$N = \frac{1.807 \cdot 10^{11}}{0.693} = 2.607 \cdot 10^{11}.$$

The age, t, of the charcoal is obtained by substituting $T_{\frac{1}{2}} = 5730$ years and these values of N_0 and N and in Eq. (5.19), $N = N_0 e^{-0.693t/T_{\frac{1}{2}}}$:

$$t = \frac{5730}{0.693} \ln\left(\frac{3.9 \cdot 10^{11}}{2.61 \cdot 10^{11}}\right) = 3320\,\text{years}.$$

5.3.4 Uses of Radioisotopes

Apart from their usefulness in dating, radioisotopes are widely employed both in research and industry. Because (1) radioisotopes are chemically identical to the stable isotopes of the same element and (2) the radiation they emit can be detected with relative ease, they are most frequently used as *tracers.* For instance, the transport of sodium and potassium ions across cell membranes can be followed with the radioisotopes Na−22 and K−42. Similarly, binding to membranes, proteins and DNA molecules can be tracked with radio-tracers such as Ca−45. The take up of fertilisers by plants is studied by labelling the fertilisers with a minute amount of the radioisotope P−32 and the activity of the thyroid gland can be examined by giving the patient a dose of the radioisotope I−131 (Table 5.14).

Table 5.14. The half-lives and radiations emitted from a number of radioactive isotopes.

Isotope	Half-life	Radiation	Energy (MeV)
H−3	12 years	β^-	0.018
C−14	5,730 years	β^-	0.160
Na−22	2.6 years	β^+	0.51
		γ	1.28
P−32	14 days	β^-	1.7
S−35	87 days	β^-	0.17
K−40	$1.3 \cdot 10^9$ years	β^-	1.3
		γ	1.5
K−42	12 hours	β^-	2.0
		γ	3.6
I−131	8 days	β^-	0.610
		γ	0.360

The high-energy penetrating energy emitted by radioisotopes has been used for the treatment of deep tumours, for the continuous measurement of the thickness of metal sheeting and the sterilisation of heat-sensitive disposable medical equipment such as plastic syringes.

In the technique called *positron emission tomography* (*PET*), real-time images of the functioning of body organs including the brain have been obtained. The person or animal under examination receives an injection of a short-lived β^+ emitting radioisotope such as ^{15}O. The moment they are emitted, the positrons released by the isotope react with an electron in the surrounding matter producing two γ photons. These photons are detected and located by a bank of scintillation counters. Images of the movement of the radioisotope can be obtained by feeding the locations into a computer. The technique has been used to detect changes in blood flow in the heart and brain as a conscious subject performs various tasks or is stimulated.

Example 5.7: Dosimetry

Co–60 decays by β^- emission with a half-life $T_{\frac{1}{2}} = 5.26$ years into an excited Ni–60 nucleus. This excited nucleus immediately releases two γ photons, one of energy 1.17 MeV and the other 1.33 MeV. Calculate the radiation dose received by a patient of mass 60 kg who is given a 30-second whole-body exposure to the γ-rays emitted by an 8 gm Co–60 source. Assume that 1% of the radiation released by the source can be localised on the patient.

Calculation: The 8 gm source contains 8/60 mole of cobalt atoms, i.e. $N_{\text{Co}-60} = 8/60 \cdot 6.02 \cdot 10^{23}$ atoms. Substitution in Eq. (5.18) gives for the activity, $|dN/dt|$, of the source:

$$\left|\frac{dN}{dt}\right| = \lambda N_{\text{Co}-60} = \frac{0.693}{T_{\frac{1}{2}}} \cdot \left(\frac{8}{60} \cdot 6.02 \cdot 10^{23}\right),$$

where $T_{\frac{1}{2}} = 5.26 \cdot 365 \cdot 24 \cdot 3600 = 1.66 \cdot 10^8$ s.

During the 30-second exposure, $30 \cdot |dN/dt|$ radioactive Co–60 atoms disintegrate. Thus, the total energy, E_γ, released as γ radiation during the 30 seconds will be:

$$\begin{aligned} E_\gamma &= 30 \cdot \left|\frac{dN}{dt}\right| \cdot (1.17 + 1.33) \\ &= 30 \cdot \left(\frac{0.693 \cdot \left(\frac{8}{60} \cdot 6.02 \cdot 10^{23}\right)}{1.66 \cdot 10^8}\right) \cdot 2.5 = 25.13 \cdot 10^{15}\,\text{MeV}. \end{aligned}$$

Of this amount, just 1% is actually received by the patient.

$$E_{Patient} = \frac{25.13 \cdot 10^{21}}{100} \cdot 1.6 \cdot 10^{-19} = 40.2\,\text{J}.$$

Not all of this radiation is absorbed by the body. According to the data in Table 5.8, each layer of 9.8 cm of living tissue absorbs 50% of the 1 MeV γ radiation passing through it. Assuming that the patient's body is, on average, about 10 cm thick, the radiation energy absorbed per kg is:

$$E = \frac{40.2}{2 \cdot 60} = 0.33\,\text{Gy}.$$

Since the quality factor (QF) of γ-rays equals unity, the effective radiation dose received by the patient is 0.33 Sv.

5.3.5 The Factors Affecting Nuclear Stability

Although no comprehensive nuclear theory has yet been developed, certain factors that appear to affect nuclear stability can be clearly identified.

1. The ratio between the number of neutrons and protons

A plot of the number of neutrons, N, in the 280 stable nuclei versus the number of protons, Z, suggests that they constitute a band of stability (Fig. 5.20). Apparently, the ratio, N/Z, is a factor in their stability. Whereas light nuclei ($A < 20$) contain approximately equal numbers of neutrons and protons, the ratio between the number of neutrons and protons rises as the total number of nucleons in the nucleus increases.

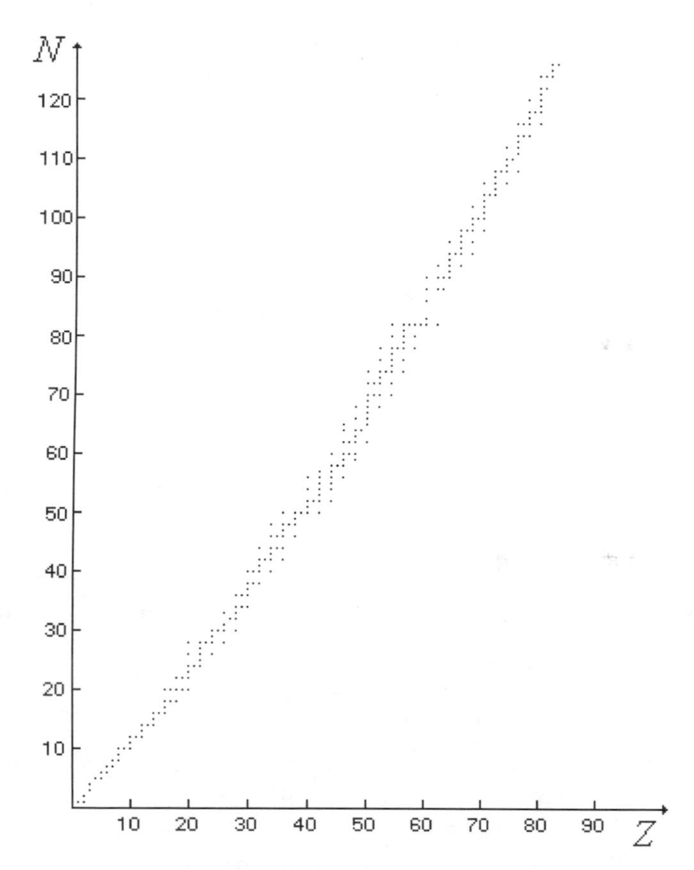

Fig. 5.20. The band of stable nuclei. There are no stable nuclei of atomic number 43 or 61; with 19, 21, 35, 39, 45, 61, 89, 115 or 123 neutrons; or containing a total of 5 or 8 nucleons. All nuclei with atomic number greater than 83, or those containing more than 126 neutrons or 209 nucleons, are unstable (radioactive).

2. The size of the nucleus

All nuclei containing more than 210 nucleons are unstable. Evidently, size, too, affects nuclear stability. This is attributed to the very short range over which the cohesive nuclear force acts. As the size of the nucleus increases, the long-range electrostatic repulsive interaction between its protons becomes more dominant and even the substantial neutron excess found in these large nuclei cannot overcome it.

3. The pairing of the nucleons

Except for just five nuclei, ${}^{2}_{1}H$, ${}^{6}_{3}Li$, ${}^{10}_{5}Be$, ${}^{14}_{7}N$ and ${}^{180}_{73}Ta$, all the stable nuclei contain an even number of protons or an even number of neutrons. Sixty percent contain even numbers of both protons and neutrons. Thus, the pairing of its nucleons appears to be a factor in the stability of a nucleus. This is consistent with the presence of nuclear energy levels each of which, in compliance with the exclusion principle, can be occupied by no more than two protons or two neutrons.

4. Magic numbers of nucleons

Nuclei having 2, 8, 20, 28, 50, 82 or 126 protons or neutrons are more abundant in nature than those having similar mass numbers but different compositions. Nuclei with these magic numbers of nucleons are also relatively more stable. This suggests that nucleonic configuration is also a stability factor.

On the basis of these four factors, the particular type of radiation emitted by different unstable nuclei, in their search for stability, can be explained.

1. Nuclei having a low neutron/proton ratio

If its instability arises from too low a neutron/proton ratio, a radioactive nucleus can improve the situation by β^+ emission, i.e. by emitting a positron or capturing an electron. In the process, one of its protons is converted into a neutron, thereby raising the ratio. Typically, such radioactive nuclei lie to the right of the stability band. If, on the other hand, its instability arises from too large a neutron/proton ratio, stability will be gained by β^- emission in which a neutron is converted into a proton. Typically, such nuclei lie above and to the left of the band.

2. Nuclei with unstable configurations

The instability of the radioactive nuclei that occupy the empty spaces in the stability band is associated with their nucleonic configurations. In many instances, these nuclei embody unpaired nucleons. In such cases, the stability of the nuclei can be improved by β radiation. For example, if an unpaired neutron in the nucleus decays by β^- emission into a proton, and the proton into which it transmutes pairs up with another previously unpaired proton, the total energy of the nucleus will generally be reduced. Overall, two previously unpaired nucleons will have become paired.

3. Nuclei containing more than 200 nucleons

Nuclei that are unstable because they contain too many nucleons cannot gain stability by β emission since this leaves their mass number unchanged. In principle, such nuclei could reduce their nucleon excess by the emission of free nucleons (protons or neutrons), deuterons ($^2H^+$) or helium nuclei ($^3He^{++}$ or $^4He^{++}$). However, as the calculations in Example 5.8 show, the emission of an α particle ($^4He^{++}$) is the only one of these disintegration modes that is energetically possible for heavy nuclei. As a result, most unstable heavy nuclei with a mass number $A > 200$, are α emitters. The key to this is the considerable binding energy, 28 MeV, that the $^4He^{++}$ nucleus can make available to the disintegrating nucleus. None of the other disintegration modes have anything like this amount of energy available to them (Table 5.15).

Table 5.15. The binding energies of helium nuclei, deuterons and a free nucleon.

Particle	Binding Energy
α particle ($^4He^{++}$)	28 MeV
$^3He^{++}$	7 MeV
deuteron	2 MeV
nucleon	0 MeV

Having identified the factors that determine nuclear stability and the choice of the decay mode by which unstable nuclei can attain stability, we must now address a more fundamental question: Why do unstable nuclei still exist? Why haven't all the naturally occurring radioactive nuclei decayed long ago into stable nuclei? If these entities are inherently unstable, what mechanism prevents them from instantly disintegrating? This is a key question that must be addressed by the theories postulated to explain α and β decay.

Example 5.8: Possible Disintegration Modes of Heavy Nuclei

In principle, heavy nuclei ($A > 200$) could reduce their nucleon excess by the emission of free nucleons (protons or neutrons), deuterons ($^2H^+$) or helium nuclei ($^3He^{++}$ or $^4He^{++}$). Which of these modes is energetically possible?

1. Emission of a single nucleon

The average binding energy per nucleon in heavy nuclei ($A > 200$) is about 7.5 MeV and it increases by about 0.0075 MeV with each unit reduction in mass number. Hence, the energy stored in an Ra–226 nucleus is smaller than that in the daughter nucleus, Ra–225 or Fr–225, that would result from the emission of a nucleon by the amount

$$(226 \cdot 7.5) - (225 \cdot 7.5075) = 6\,\text{MeV}.$$

In order for the Ra–226 nucleus to decay by emitting a single nucleon, it would first have to be supplied with this amount of energy. There being no source for this energy, this decay mode cannot occur.

2. Emission of a deuteron or the $^3He^{++}$ ion
The emission of a deuteron by the Ra–226 nucleus to give a nucleus with mass number $A = 224$ would result in a loss of

$$(226 \cdot 7.5) - (224 \cdot 7.515) = 11.6\,\text{MeV}$$

in the binding energy of the nucleus. The 2 MeV of binding energy possessed by the deuteron is insufficient to make up for this. Even the 7 MeV of binding energy possessed by the $^3He^{++}$ ion cannot make up for the

$$(226 \cdot 7.5) - (223 \cdot 7.5225) = 17.5\,\text{MeV}$$

of binding energy that would be lost were it to be the particle emitted in the decay.

3. Emission of an α particle
The only process that is energetically possible is the emission of an α particle. Its 28 MeV of binding energy exceeds the amount lost by the nucleus as its mass number falls from 226 to 222,

$$(226 \cdot 7.5) - (222 \cdot 7.53) = 23.4\,\text{MeV},$$

by about $28 - 23.4 = 4.6$ MeV. Typically, in the α decay of heavy nuclei about 4–9 MeV is available for the kinetic energy of the emitted α particle. Furthermore, as they are all produced by the same process, the α particles emitted from a particular isotope generally have the same energy.[37]

5.3.6 The Mechanism of α Decay

Inside the parent nucleus, the potential energy of the component nucleons of an α particle – the two protons and two neutrons – must be negative since they are bound by the strong force to the rest of the nucleus. However, the moment it finds itself outside the nucleus, the α particle has a positive potential energy due to its electrostatic repulsive interaction with the daughter nucleus.

Consider, for example, the emission of an α particle from the $^{226}_{88}$Ra nucleus. Assuming the nucleus to be a sphere of radius $R = 1.2 \cdot \sqrt[3]{222} = 7.3$ fm, the electrostatic potential energy, U_E, of the α particle on the surface of the daughter

[37]Some isotopes do emit α particles with different energies. For example, Ra–222 emits α particles with energies of 4.78 MeV and 4.6 MeV. The higher values correspond to the instances in which the parent nucleus was in an excited state and/or the daughter nucleus was produced in its ground state; the smaller value to the instances in which the parent nucleus was in its ground state and/or the daughter nucleus was produced in an excited state.

nucleus, $^{222}_{86}$Rn, is, according to Coulomb's law:

$$U_E = \frac{kq_\alpha q_{\text{Rn}-222}}{R} = \frac{9 \cdot 10^9 \cdot 2 \cdot 86 \cdot (1.6 \cdot 10^{-19})^2}{7.3 \cdot 10^{-15} \cdot 10^6 \cdot 1.6 \cdot 10^{-19}} = 34\,\text{MeV}. \qquad (5.24)$$

The energy, ~4.6 MeV, made available to the α particle by the decay of the parent nucleus, is almost an order of magnitude less than this. This situation is illustrated in the potential energy diagram in Fig. 5.21.

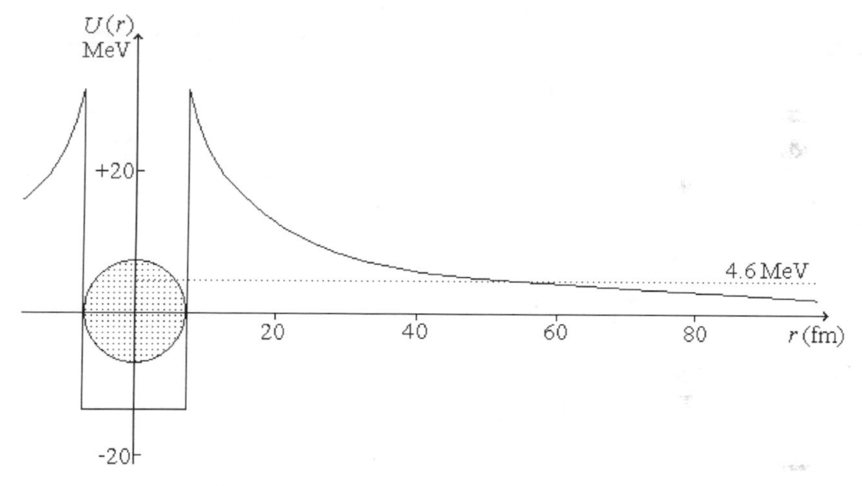

Fig. 5.21. The potential energy curve of an α particle and an Rn-222 nucleus. The radius of the nucleus is 7.3 fm and the potential energy at its surface is 34 MeV. The energy of the α particle, 4.6 MeV, is substantially below the maximum height of the potential well enclosing the nucleus.

It would appear that the question to be answered is not why haven't all the α emitting nuclei decayed long ago, but how do any of them manage to emit an α particle at all!

The answer is provided by quantum mechanics. As we saw in Section 2.4.8, a particle can escape from a finite potential well even though it does not possess sufficient energy to mount its walls; it may simply 'tunnel' through the energy barrier. The probability of the particle's escape depends principally on three factors: (1) the height of the energy barrier; (2) its shape; (3) the energy of the particle.

The emission of an α particle from a nucleus is an example of quantum tunnelling. Thus, the stability of an α emitting nucleus is a function of the escape probability of the α particle. Generally, this probability is extremely low. This is why alpha emitting nuclei still occur naturally. The probability of their decay is so low, that even the æons of time since their creation have not sufficed for all of them to disintegrate.

The stability of a radioactive isotope is reflected by its half-life. The longer the half-life of an α emitting isotope, the more stable are its nuclei and the lower is the escape probability of the α particles. The empirical dependence of the half-life of the α emitting isotopes of thorium on the energy, E_α, of the α particle is illustrated in Fig. 5.22.

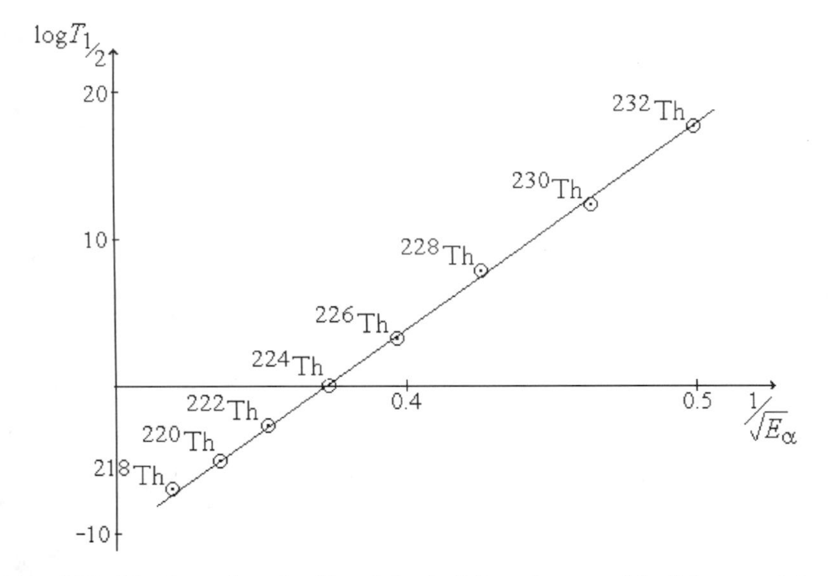

Fig. 5.22. The logarithm, $\log T_{\frac{1}{2}}$, of the half-lives (in seconds) of the α emitting isotopes of thorium as a function of the reciprocal, $1/\sqrt{E_\alpha}$, of the square root of the α particle's energy.

To a close approximation, the escape probability, P_α, of the α particle from a nucleus is an exponential function of the reciprocal of the square root of the α particle's energy:

$$P_\alpha \propto e^{1/\sqrt{E_\alpha}}. \tag{5.25}$$

Consequently, small differences in particle energy produce great differences in the escape probability. This rule was first derived from quantum mechanical considerations by assuming that α particles exist as such inside the nucleus and that they bounce back and forth between the walls of the potential barrier impacting repeatedly on the energy barrier. Each impact is considered an attempt to escape. The escape probability is then expressed in terms of the number of times the particle must try before succeeding. Though, on this view, an α particle typically knocks at its confining wall 10^{21} times every second, it may have to wait millions of years before it escapes.

5.3.7 The Mechanism of β Decay — Weak Charge

Few natural phenomena have generated as many difficulties or have required as many apparently contrived and *ad hoc* postulates for their elucidation as has the phenomenon of β decay. On the other hand, few have ultimately proven as fruitful a source of new knowledge and ideas.

An examination of the disintegration energies, Q, of the three basic processes underlying β decay reveals that, without an additional source of energy, neither β^+ emission nor electron capture can occur spontaneously.

$$\begin{array}{lrl} \beta^- \text{ emission:} & n \rightarrow p + e^- & Q_{\beta^-} = +0.78\,\text{MeV} \\ \beta^+ \text{ emission:} & p \rightarrow n + e^+ & Q_{\beta^+} = -1.80\,\text{MeV}. \\ \text{electron capture:} & p + e^- \rightarrow n & Q_{EC} = -0.78\,\text{MeV} \end{array} \quad (5.26)$$

The endothermicity of β^+ emission accounts for the stability of free protons and that of electron capture for the existence of the hydrogen atom; the spontaneous fusion of the latter's proton and electron to produce a neutron is energetically forbidden. It follows that β^+ emission and electron capture can only take place if the re-arrangement of the nucleons in the daughter nucleus increases the total binding energy at least by the amount needed to make up for the energy lost in the conversion of the proton into a neutron.

The exothermicity of β^- emission indicates that the spontaneous disintegration of free neutrons is energetically possible. In fact, free neutrons are unstable and disintegrate by β^- emission after an average lifetime of about 15 minutes. The neutrons inside a nucleus are only prevented from disintegrating by the absence of available energy levels into which the protons that would be produced from their decay could be placed. The lower proton energy levels are already fully occupied by the existing protons in the nucleus and the new protons that would be produced lack the necessary energy to occupy the vacant higher levels.

Neither the electrons nor the positrons involved in the three modes of β decay are themselves nuclear components. Nor, since protons, neutrons, positrons and electrons are all fermions with a spin of $\frac{1}{2}$, can any one of them be a combination of any two of the others. The neutron cannot comprise a proton and an electron as was once thought, since it would then have spin of $\frac{1}{2} + \frac{1}{2} = 1$ or $\frac{1}{2} + (-\frac{1}{2}) = 0$.

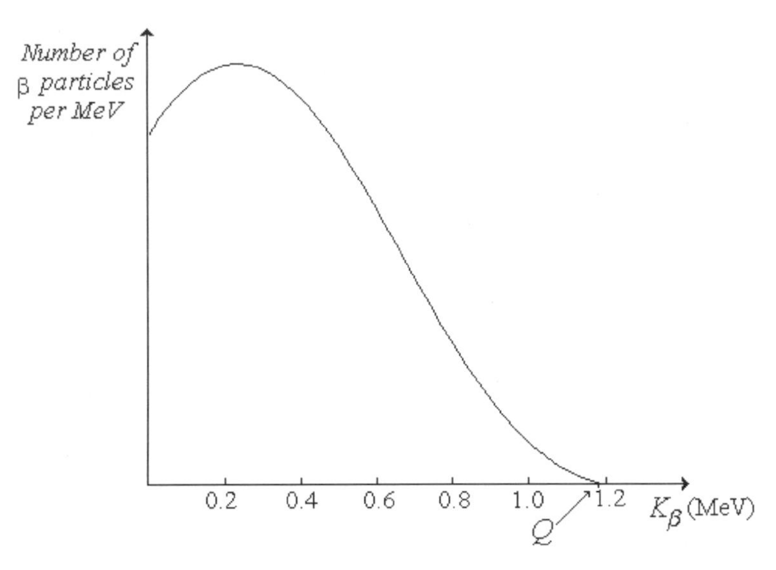

Fig. 5.23. Energy spectrum of the electrons emitted from a typical β emitter, Bi-210. Most of the electrons are emitted with energies, K_β, less than half that of the disintegration energy, Q.

By the same reasoning, each of the formulations that appear in Eq. (5.26) for the three basic processes underlying β decay violates angular momentum conservation. Taking β^- emission as the example: whereas the neutron has a spin of $\frac{1}{2}$, the sum of the spins of the products of its disintegration, a proton and an electron, must be $\frac{1}{2} + \frac{1}{2} = 1$ or $\frac{1}{2} + \left(-\frac{1}{2}\right) = 0$.

The problem is compounded by an examination of the energies of the electrons and positrons emitted in β decay (Fig. 5.23). Whereas α particles are all emitted with energies equal to or close to the disintegration energy of their decay process, β particles are emitted with a range of energies from almost zero up to a maximum equal to the disintegration energy, Q. The recoil energy of the nucleus in β decay is negligible, so what happens to the excess energy? Is energy not conserved in β decay?

In 1931, as a 'desperate remedy' and in an act of faith based on no more than his deep belief in the laws of conservation, Pauli proposed that another undetected particle was involved in β decay. Fermi[38] later christened the particle the *neutrino.* However, another 20 years were to pass before experimental evidence for the existence of neutrinos was obtained.

[38]Enrico Fermi 1901–1958, Italian physicist who was awarded the Nobel prize in 1938 for his researches on nuclear reactions with slow neutrons.

Neutrinos are stable, uncharged and virtually massless ($m_{neutrino} < 0.001\ m_e$) fermions. They belong to the family of leptons and are the fourth stable fundamental component of ordinary matter.[39] Both neutrinos, ν, and anti-neutrinos, $\bar{\nu}$, are involved in the three types of β decay.[40]

$$\begin{array}{ll} \beta^{-}\text{emission:} & n \rightarrow p + e^{-} + \bar{\nu} \\ \beta^{+}\text{ emission:} & p \rightarrow n + e^{+} + \nu. \\ \text{electron capture:} & p + e^{-} \rightarrow n + \nu \end{array} \tag{5.27}$$

The neutrinos and anti-neutrinos provide the spin, $\frac{1}{2}$, necessary for angular momentum conservation in β decay. Moving with a velocity close to that of the speed of light, they likewise account for the energy and linear momentum produced in the decay but not taken up by the other particles. In terms of the quark hypothesis, each of the three types of β decay involves the transmutation of one first generation quark into another. In each case, two leptons are also involved, an electron or positron and an anti-neutrino or neutrino, respectively.

$$\begin{array}{lrcl} \beta^{-}\text{ emission:} & \text{down quark} & \rightarrow & \text{up quark } + e^{-} + \bar{\nu} \\ \beta^{+}\text{ emission:} & \text{up quark} & \rightarrow & \text{down quark } + e^{+} + \nu. \\ \text{electron capture:} & \text{up quark} + e^{-} & \rightarrow & \text{down quark } + \nu \end{array} \tag{5.28}$$

However, since leptons do not participate in the colour/strong force, the interaction underlying these transmutations cannot be the colour/strong interaction. In order to account for β decay, a fourth fundamental force, in addition to gravity, electromagnetism and the colour (strong) force, must exist in nature. This is known as the *weak force* or *weak interaction*; it is weak only relative to the strong force. It is a very short-range interaction, operating only at distances of up to 10^{-18} m.

Because they participate in neither strong nor electromagnetic interactions, neutrinos pass through both space and matter unhindered. As a result, they generally go undetected. Notwithstanding, the existence of neutrinos and anti-neutrinos was ultimately confirmed experimentally when it was shown that the following two *inverse beta decay* reactions occur:

$$\begin{array}{c} p + \bar{\nu} \rightarrow n + e^{+} \\ n + \nu \rightarrow p + e^{-} \end{array} \tag{5.29}$$

In order to accomplish the first of these reactions, the intense flux of anti-neutrinos generated by the large-scale beta decay that takes place in a nuclear reactor was aimed into a tank of water. Protons in the water molecules react

[39] The other three are the up and down quark and the electron.

[40] Strictly speaking, *electron* neutrinos and *electron* antineutrinos, so called to distinguish them from the other two types, the muon and tau neutrinos and antineutrinos (see below).

with the anti-neutrinos and are converted into neutrons and positrons. The positrons collide almost at once with electrons in the water. The positron–electron pairs mutually destruct with the production of a pair of γ-ray photons. The brief existence of the positrons, and, hence, the occurrence of the inverse beta decay reaction, was inferred from the detection of these photons. In 1987 a shower of neutrinos (about 20 altogether) was similarly detected in the radiation reaching the earth from an exploding star – a supernova – observed in the Large Magellanic Cloud, one of the closest galaxies to Earth, just 170,000 light years away.

1. Neutrinos and parity

A particularly intriguing feature of neutrinos and anti-neutrinos is the distinction between them. Being uncharged particles, the difference between them cannot be one of electric charge. Instead, it is their *helicity* that distinguishes between them. Relative to their direction of motion (their linear momentum), each spins in a different manner. Whereas the neutrino spins in a counter-clockwise direction – it has *negative helicity*, the anti-neutrino spins in a clockwise direction – *positive helicity*. The significance of this distinction is that the neutrino is the mirror image of the anti-neutrino, and *vice versa* (Fig. 5.24).

According to classical thinking, there is no way of telling, *a priori*, whether what we are looking at is an object or its mirror image. A physical object

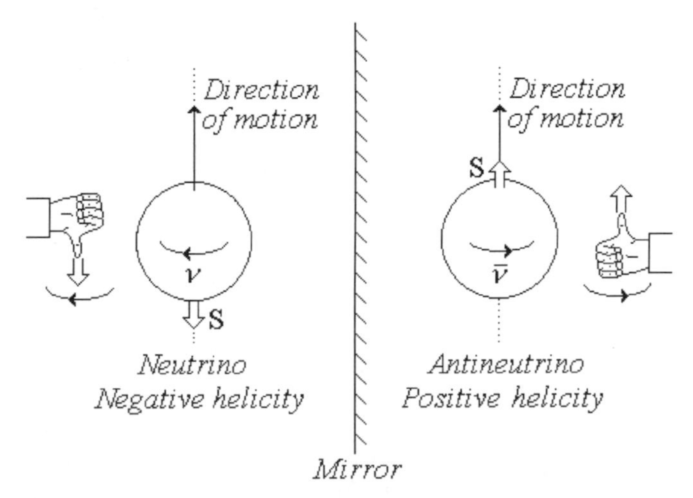

Fig. 5.24. The neutrino and anti-neutrino are mirror images of one another. Whereas the neutrino advances through space like a left-handed screw (viewed from behind as spinning in a counter-clockwise direction), the anti-neutrino moves like a right-handed screw (spinning in a clockwise direction). The direction of the axial vector, **S**, that represents each particle's spin is given by the right-hand (corkscrew) rule.

or process and its mirror image are indistinguishable. This principle is called *parity conservation.* We say that the laws of classical physics are invariant under the parity operation. Although the world is full of examples of left–right or object–mirror image asymmetries, ranging from the direction of the spirals of certain mollusc shells to the optical activity of many organic molecules, all such examples are thought to be accidental; they do not reflect any inherent asymmetry in the laws of nature. The predominance of the L-amino acids in living tissue is considered an evolutionary 'accident', not a requirement for life; in principle a 'mirror image man' constructed of D-amino acids should be a living being.[41]

The parity of the process by which certain unstable nuclei and particles decay was first questioned in 1956 by Lee and Yang.[42] That conservation of parity in electromagnetic interactions and strong interactions had already been established at the time. However, it had been found that a certain charged particle, the (positive or negative) *kaon,* $K^{\pm}$, could decay in two different ways, one that produced three pions, the other only two. In order to account for this anomaly, Lee and Yang postulated that parity is not conserved in processes of this type.

The interaction that underlies the decay of the kaon particle, K^{+}, is the same as that involved in β decay, namely, the *weak interaction.* Lee and Yang associated the non-conservation of parity with this interaction. On this view, β^{-} decay should also exhibit non-conservation of parity, i.e. the mirror image of β^{-} decay should be different from directly viewed β^{-} decay. Shortly after Lee and Yang put forward their remarkable hypothesis, this was confirmed in a landmark experiment.

Consider a nucleus whose spin is aligned in a certain direction; the mirror image of this nucleus has its spin aligned in the opposite direction. Under suitable circumstances, the spins of many β emitting nuclei such as Co-60, can be aligned by the action of a strong external magnetic field and a *polarised array* of the nuclei created. The effect is an indirect one: the field first acts on the atomic electrons and they, in turn, influence the nuclei. In the mirror image of such a *polarised array* of nuclei, the orientations of both the magnetic field and the nuclear spin vector are reversed (Fig. 5.25).

If parity is conserved in β^{-} decay, the emission of the electrons from the polarised array of Co-60 nuclei must look exactly the same in a mirror as it

[41]To answer Alice's question: Looking-glass milk should be good to drink; certainly for looking-glass cats.
[42]Tsung Dao Lee 1926–, Chen Ning Yang 1922–, Chinese/American physicists who were awarded the Nobel prize in 1957 for their penetrating investigation of the so-called parity laws which has led to important discoveries regarding the elementary particles.

does when viewed directly despite the opposing orientations of the nuclear spins. This requires that the emission of the β electrons be symmetrical; equal numbers should be emitted, on average, parallel and anti-parallel to the nuclear spins. If more are emitted parallel than anti-parallel, or *vice versa*, parity is not conserved.

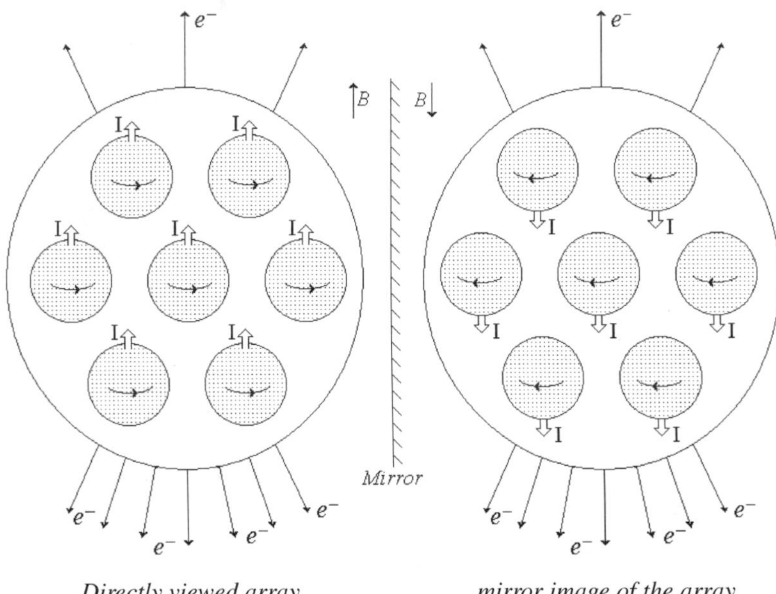

Fig. 5.25. A polarised array of Co–60 nuclei and its mirror image. The effect of the magnetic field, B, is to align the nuclear spin vectors, **I**, of the nuclei. Reversing the direction of the magnetic field reverses the direction of the axial spin vectors; this has the effect of transforming the array into its mirror image. In the directly viewed array, the beta electrons are emitted preferentially anti-parallel to the nuclear spins. Accordingly, if parity is conserved, in the mirror image array they should be preferably emitted parallel to the nuclear spins; parallel and anti-parallel emission should be equally probable. But this is not observed in nature, *ergo*, parity is not conserved in beta decay.

The emission of beta electrons from a polarised array of radioactive nuclei was first studied in an experiment conducted in 1957 at the US National Bureau of Standards by a group led by Chien Shiung Wu.[43] A sample of the β^- emitter Co–60 was placed in a strong magnetic field at a temperature close to absolute zero. The low temperature served to reduce the random motions of the atoms sufficiently for the magnetic field to align the spin of their nuclei. The

[43]Chien Shiung Wu 1912–1997, Chinese–American physicist, also known as 'Madam Wu: The First Lady of Physics.'

number of beta electrons emitted by the array in the direction of the applied magnetic field (the nuclear spins) and in the opposite direction was counted. It was found that significantly more beta electrons were emitted in the direction opposite to that of the field. The emission of the electrons was not symmetrical; parity was, therefore, not conserved. In a subsequent experiment with Co–58, which is a β^+ emitter, the opposite effect was observed: the positrons were preferentially emitted parallel to the nuclear spins.

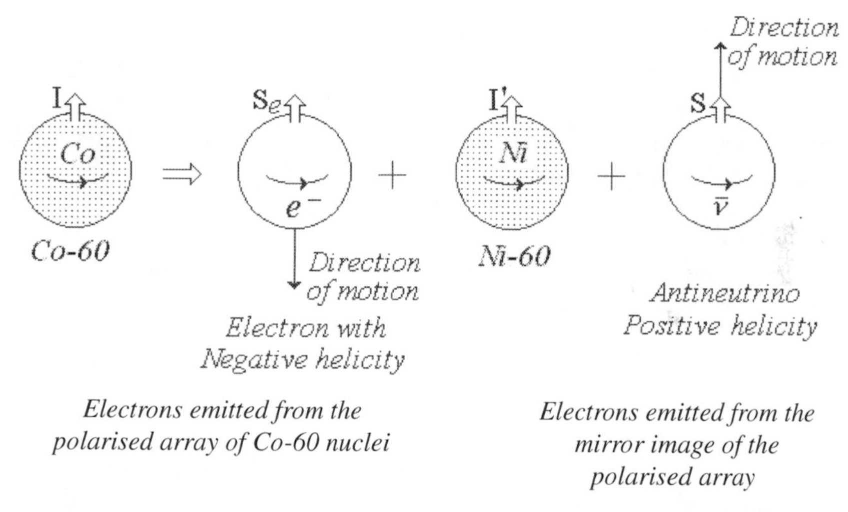

Fig. 5.26. The emission of beta electrons from the polarised array of Co–60 nuclei. The difference, $\mathbf{I} - \mathbf{I}'$, between the spin of the Co–60 nucleus, $\mathbf{I}$, and that of the Ni–60 nucleus, $\mathbf{I}'$, is carried away by the electron and the antineutrino. The spin of the electrons, $\mathbf{S}_e$, is parallel to that of the cobalt nuclei and so, as they preferentially emitted anti parallel to the nuclear spins, they have negative helicity.

In the beta decay of Co–60, the spin of the daughter Ni-60 nucleus is less than that of the parent nucleus by an amount equal to $\hbar$. This difference must be carried away by the beta electron and the accompanying antineutrino; their respective spins should therefore be parallel to that of the Co–60 nucleus. Since the beta electrons are emitted preferentially in the direction opposite to that of spin of the nucleus, they must have negative helicity. Evidently, β decay – the weak interaction – preferentially produces left-handed electrons and right-handed positrons (Fig. 5.26).

Helicity is an intrinsic property only of massless particles that move at the speed of light and cannot be overtaken. Particles that possess mass and whose velocities are a relative quantity, such as electrons and positrons, may have either positive or negative helicity, depending on the reference frame from which their direction of motion is viewed. The intrinsic 'mirror dependent property' that is unaffected by an observer's motion is called *chirality*: a figure is chiral if it is not identical to its mirror image. A hand or a shoe is chiral; it is different from its mirror image irrespective of the reference frame from which they are viewed. Such objects and their mirror images are said to be *enantiomorphs*, a term familiar to students of chemistry. The chirality of elementary particles is an inherent quantum mechanical property, such that particles with different chiralities are really different particles even though they may be identical in all other aspects (just like the asymmetric carbon atoms in optical isomers). If a particle exhibits left-handed helicity, there will also be a circumstance in which it exhibits right-handed helicity. But a particle with left-handed chirality will not necessarily have a right-handed partner. Helicity and chirality are the same only for massless particles, which appears to suggest a link between mass and chirality.

2. The weak interaction

All the fermions of the standard model participate in the weak interaction; it has three distinctive properties:

* It is the only route by which one type of quark (flavour) can change into another.
* It is the only interaction which violates parity symmetry.
* It is mediated by three massive carrier particles called *intermediate* or *gauge bosons*: the W^+, W^- and Z^0 bosons.

The physical source of the weak force has been designated the *weak charge* or *weak isospin* (T_3). This property, possessed by quarks and leptons alike,[44] is to the weak force what electric charge is to electromagnetism and colour is to the colour/strong force. Two kinds of weak isospin have been identified; they are designated by the quantum numbers $+\frac{1}{2}$ and $-\frac{1}{2}$. The type carried by u, c and t quarks is $T_3 = +\frac{1}{2}$; the type carried by d, s and b quarks is $T_3 = -\frac{1}{2}$. A quark never decays by the weak interaction into a quark of the same T_3. The 'up-type' quarks (u, c, t) always transform into a 'down-type' quark (d, s, b), or *vice versa*, as, for example, occurs in beta decay.

$$\begin{array}{ll} \beta^- \text{ emission:} & d \to u + e^- + \bar{\nu} \\ \beta^+ \text{ emission:} & u \to d + e^+ + \nu. \\ \text{electron capture:} & u + e^- \to d + \nu \end{array} \tag{5.30}$$

In general, fermions having $T_3 = +\frac{1}{2}$ always transform into fermions having $T_3 = -\frac{1}{2}$, and *vice versa.* A particle's weak charge, T_3, and its electric charge, q,

[44] A corollary of the parity violation observed in β decay is that weak isospin is only possessed by left-handed particles. Particles having right-handed chirality do not participate in the weak interaction.

are related through a quantity, $Y_W = 2(q - \mathrm{T}_3)$, called *weak hypercharge* which unifes weak interactions with electromagnetic ones.

Like other fundamental interactions, the weak interaction too can be understood as a quantum field theory in which photon-like particles, called *intermediate (gauge) bosons*, mediate (carry) the interaction. The mediating particles involved in β decay are the charged W^+ and/or W^- bosons; so too in the transmutation of a neutron into a proton (Fig. 5.27). Processes in which like weak charges interact have also been observed; these interactions are mediated by the uncharged Z^0 intermediate boson.

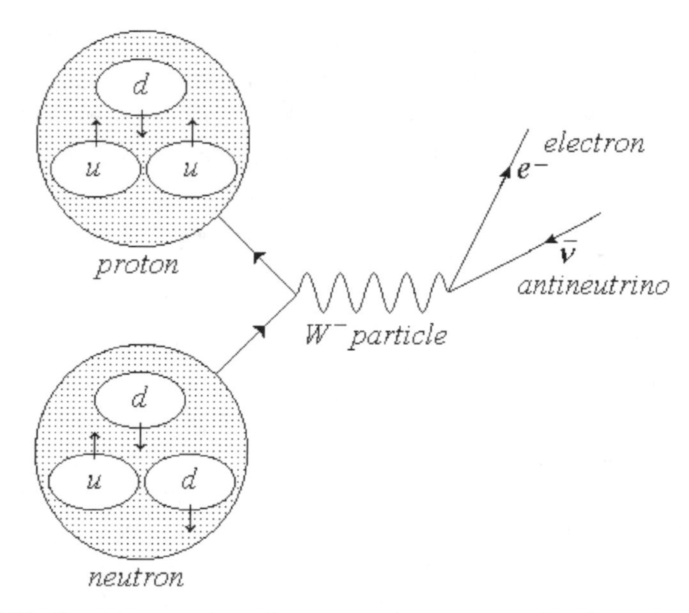

Fig. 5.27. The transmutation of a neutron into a proton, by the emission of a beta electron, viewed as the conversion of a *u* quark into a *d* quark. The process is mediated by the virtual W^- particle. By convention, in Feynman diagrams, particles are shown with forward arrows and anti-particles with backward arrows.

Because of the very short range of the weak interaction, up to distances of just 10^{-18} m, the intermediate bosons, W^+, W^- and Z^0, must all be massive particles.[45] Assuming that these particles are exchanged in weak interactions at close to the speed of light, they will have lifetimes of about of $10^{-18}/c \approx 10^{-26}$ s. Thus, their intrinsic energy will be of the order:

$$\Delta E = \frac{\hbar}{\Delta t} = \frac{6.58 \cdot 10^{-22}}{10^{-26}} \approx 60{,}000\mathrm{MeV}. \tag{5.31}$$

[45]In contrast to electromagnetism whose carrier particle, the photon, is massless and range is infinite.

β decay is a relatively slow process because of the low probability of generating such massive ($\sim$65 u) intermediate bosons, heavier even than iron atoms.

There is an important and fundamental difference between the bosons and gluons that mediate the weak and colour interactions, respectively, and the photons that mediate electromagnetic interactions. Whereas photons carry no electric charge, the W^+, W^- and Z^0 bosons that mediate the weak force all have non-zero weak isospin; the two charged species carry electric charge too. The gluons, likewise, that mediate the colour force themselves have colour. Thus, these latter 'carrier' particles can themselves emit still more carrier particles, complicating the interactions still further.

The existence of the W^+, W^- and Z^0 intermediate bosons was demonstrated in 1984 in experiments between colliding beams of protons and anti-protons carried out at CERN. They are amongst the heaviest elementary particles ever detected having a mass about 80 times that of the proton. Their discovery was a triumph for the *electroweak hypothesis* proposed by Weinberg,[46] Salam and Glashow in 1968, which showed how the weak and electromagnetic interactions could be united into a single *electroweak theory* in a way similar to that by which Maxwell had integrated electrical and magnetic phenomena into the single framework of electromagnetism 100 years earlier.

The rationale that underlies the unification of the weak and electromagnetic interactions into the single *electroweak interaction* hinges on the notion that apparently distinct particles, the regular photon and the $W^\pm$ and Z^0 bosons, can be regarded as differing forms of the same underlying quantum entity. However, their initial symmetric configuration is unstable and breaks down spontaneously, producing the asymmetry represented by the different properties they manifest. Restoring the symmetry requires the investment of energy. It is estimated that in excess of 1 TeV (10^{12} eV) is required for the symmetry of these particles to be made explicit. Such energies are unavailable in terrestrial laboratories but could have been available immediately after the big bang, before the initial symmetry was broken.[47]

3. The standard model

The standard model of elementary particles postulates the existence of three generations of leptons corresponding to the three generations of quarks. These are the electron, e, the muon, μ (the electron's heavy 'brother', $m_\mu \approx 200\, m_e$),

[46]Steven Weinberg 1933–, an American physicist who, together with the Pakistani physicist Abdus Salam 1926–1996 and the American physicist Sheldon Lee Glashow 1932–, was awarded the Nobel prize in 1979 for the development of the electroweak theory.

[47]Spontaneous symmetry breaking is exhibited when a system passes from a state of unstable equilibrium to one of stable equilibrium. A pencil standing on its end is symmetrical, but it falls over, losing its symmetry, under the slightest perturbation.

and the tauon, τ. Each of these leptons has a neutrino associated with it: the electron neutrino, υ_e; the μ-neutrino, υ_μ; and the τ-neutrino, υ_τ, respectively. The reality of all these particles has been confirmed in particle accelerator experiments (Table 5.16).

In addition to the leptons and quarks listed in Table 5.16 and the bosons that mediate the interactions between them, the standard model also predicts the existence of a massive particle of intrinsic energy $\sim$100 GeV/c^2, known as the *Higgs boson*,[48] whose existence would substantiate the simplest of the suggested hypothetical mechanisms by which elementary particles acquire mass. Thus, it would explain the fact that whereas electromagnetic photons are massless, the W and Z gauge bosons that mediate weak interactions are massive. Great expense and effort have been invested in the search for the Higgs boson. In July 2012, experimental teams at CERN reported the discovery of a previously unknown boson of mass $\sim$125 GeV/c^2 whose behaviour is "consistent with" a Higgs boson, adding a cautionary note that further data and analysis were needed.

Table 5.16. The weak and electric charges of the leptons and quarks.

	Leptons			Quarks		
	Particle	Weak Charge	Electric Charge (e)	Flavour	Weak Charge	Electric Charge (e)
First Generation	Neutrino, υ_e	$+\frac{1}{2}$	0	Up, u	$+\frac{1}{2}$	$+\frac{2}{3}$
	Electron, e	$-\frac{1}{2}$	-1	Down, d	$-\frac{1}{2}$	$-\frac{1}{3}$
Second Generation	μ−Neutrino, υ_μ	$+\frac{1}{2}$	0	Strange, s	$+\frac{1}{2}$	$-\frac{1}{3}$
	Muon, μ	$-\frac{1}{2}$	-1	Charm, c	$-\frac{1}{2}$	$+\frac{2}{3}$
Third Generation	τ−Neutrino, υ_τ	$+\frac{1}{2}$	0	Bottom, b	$+\frac{1}{2}$	$-\frac{1}{3}$
	Tauon, τ	$-\frac{1}{2}$	-1	Top, t	$-\frac{1}{2}$	$+\frac{2}{3}$

The successful unification of the weak and electromagnetic interactions into a single framework has stimulated the search for a theory – a *grand unified theory* or *GUT* – that would unite the four fundamental interactions: gravity and the weak, strong and electromagnetic interactions (Table 5.17). This search, unsuccessful to date, has centred on the symmetries that are thought to underlie all four interactions.

4. The solar neutrino problem; neutrino oscillations.
Neutrinos only interact through the weak interaction, which, in turn, only acts at distances below 10^{-18} m. Hence, the probability of a neutrino interacting with

[48]Named after Peter Higgs who was one of the early proponents of its existence.

Table 5.17. The four fundamental interactions – the standard model.

Interaction	Source	Particles Affected	Range	Relative Strength	Mediating Bosons	Role in Nature
Gravity	Mass	All	up to ∞	$\sim 10^{-38}$	Gravitons	Sustains the large-scale structure of the universe
Weak Force	Weak charge	Quarks and Leptons	$\sim 10^{-18}$ m	$\sim 10^{-13}$	W^+, W^- and Z^0	Stipulates and impels the transformations of quarks and leptons
Electro-magnetism	Electric charge: e	Charged Particles	up to ∞	$\sim 10^{-2}$	Photons	Sustains atomic and molecular structures and stipulates the allowed changes in their configurations
Colour Force	Colour: *red*, *blue* or *green*	Quarks	$\sim 10^{-15}$ m	1	Gluons	Sustains hadronic and nuclear structures

any other particle and undergoing a change as a result is close to zero; they are all but immortal.[49] Consequently, neutrinos are by far the most numerous of all the elementary particles in the cosmos. It is estimated that some 10^{16} neutrinos pass through our bodies unnoticed every second, leaving no trace. Many of these neutrinos may have been created in the big bang and survive today as part of the cosmic background. Yet more have been created since and are being produced still, by the nuclear fusion that takes place in the sun and other stars, by the fission processes in nuclear reactors and by the beta decay of radioactive nuclei; cosmic rays striking the atmosphere also contribute to the neutrino flux on earth.

The largest generator of neutrinos in our part of the cosmos is the sun. According to the generally accepted solar model, the neutrinos created by the nuclear fusion that takes place in the sun's core are exclusively electron neutrinos and their rate of production can be estimated from the power of the sun's radiation. However, in experiments designed to count the flux of solar neutrinos reaching the earth in which detectors sensitive to electron neutrinos were employed, only about a third of the number predicted by the solar model were observed. This anomaly became known as the *solar neutrino problem.*

As originally formulated, the standard model had postulated that neutrinos have zero mass. However, in order to account for the missing solar neutrinos, theorists hypothesised that this is incorrect and that they must actually possess

[49]Electron–neutrino cross-sections are of the order 10^{-21} barn ($10^{-49}\,\text{m}^2$).

a minute quantity of mass. Underlying this hypothesis was the notion that elementary particles – the quarks, leptons and gauge bosons – exhibit the same quantum elusiveness as minute matter particles and that, likewise, their parameters and behaviour can only be formulated in terms of wave functions and probabilities. The neutrinos were seen as a case in point; on this view, even their identity may be subject to a quantum superposition and it is this concept that has offered a solution to the anomaly of the missing solar neutrinos.

Neutrinos are always created with a particular *flavour*: electron, muon or tau. Similarly, they are also always detected as one or other of these types. However, it is conjectured, each *flavour eigenstate* (characteristic quantum state) can be equivalently described as a 'mixing' of three *mass eigenstates*, each of which is associated with one or other of the three flavours, electron, muon or tau. It is further conjectured that neutrinos propagate through space as a quantum superposition of these mass eigenstates, each of a slightly different mass and frequency. As the neutrino propagates, the phases of the three superposed mass states advance at slightly different rates due to their difference in mass. Interference between them produces a periodic series of flavour maxima and minima along the line of propagation; this, in turn, determines the probability of observing a particular flavour at points along it. After travelling some distance, a neutrino, born as an electron neutrino, will be some mixture of electron, muon and tau neutrino flavours. The effect is called *neutrino oscillation.* In this way the deficit observed in the number of solar electron neutrinos detected on earth can be accounted for.

Clearly, were neutrinos all of the same mass or massless, there would be no mass eigenstates to mix and, hence, no periodic interference pattern. Calculations show that for a hypothetical system of two flavours, the oscillation length in vacuum, L_ν, for neutrinos of energy, E_ν, is given by

$$L_\nu = \frac{4\pi E_\nu}{\Delta m^2},$$

where Δm^2 is the difference between the squares of the neutrino masses. Whereas solar neutrinos have energies of several MeV, the conjectured difference between the squares of the neutrino masses is of the order $10^{-5}(\text{eV})^2$. Consequently, neutrino oscillations will only be detected at some distance from their source.

Evidence for the phenomenon of neutrino mixing was first obtained in 1998 at the Super Kamiokande in Japan in experiments on the atmospheric neutrinos produced by the impact of cosmic rays on the upper atmosphere. The definitive solution to the solar neutrino problem came in 2010 from parallel measurements of the solar neutrino flux conducted at the Japanese Super-K detector and at the Solar Neutrino Observatory (SNO) in Canada. Whereas the SNO detector, which is sensitive only to electron neutrinos, detected the same one third of the expected flux as had previous measurements, the Super-K detector, which has some sensitivity to muon and tau neutrinos, detected one half of the expected value. Extrapolating the data obtained from the two detectors gave a number for the total solar neutrino flux that matched the value expected from the accepted model of solar fusion. All the neutrinos had been produced in the sun as electron neutrinos but had oscillated into other flavours during their propagation as mass-states.

5. The anthropic principle
The fortuitous properties of the elementary particles and the four fundamental forces are a cause for some wonder. Were the ranges and strengths of the forces otherwise, the cosmos would be strikingly different and might well not exist as we know it. The formation of stable nuclei depends on the ratio of the strong and electromagnetic forces: a small change in their relative strengths might allow the electromagnetic force between the nuclear protons to overcome the cohesive strong (nuclear) force and atoms could not exist. The strength of gravity is also critical: if it was any stronger, stellar matter would bind more strongly and stars would consume their nuclear fuel faster, leaving insufficient *deep time* for the evolution of life. On the other hand, if gravity was any weaker, elemental matter might not have 'clumped together' to form larger structures and stars might not have formed in the first place. The fact that the laws of nature and the parameters of the universe appear to take on values that are consistent with conditions for life, as we know it, has led some thinkers to formulate the *anthropic principle*, which asserts that this is a necessity because living observers would not be able to exist and observe the universe, were these laws and constants not constituted in this way.

Chapter 5.4

Nuclear Reactions and Nuclear Energy

In his *New System of Chemical Philosophy*, published in 1803, Dalton had proposed that the chemical elements comprise indestructible immutable atoms. This postulate gave a simple explanation for the failure of the alchemists during the previous thousand years to transmute base metals into gold: the base metals and gold were taken to be elements. According to Dalton, this meant that each comprises its own species of indestructible immutable atoms. Thus, no amount of heating, cooling, refracting or distilling of this or that mixture of materials could ever convert lead or any other base metal into gold. Dalton's model implied that any reactions induced by such processes could only change the way the atoms are ordered and grouped within the materials; they could not affect the essential attributes of the atoms themselves.

The discovery of the natural radioactivity of the heavy elements marked the first breach in Dalton's model. The atoms of these elements are not immutable; they undergo spontaneous changes that alter their elemental identity. The daughter nuclei that result from radioactive decay are invariably those of a different element than the parent nuclei.

The alchemists' dream of inducing the transmutation of one element into another was first realised by Rutherford, fortuitously, in 1919. It occurred in the course of experiments he was conducting into the 'stopping power' of various gases. In these investigations, a scintillation counter was used to count the number of α particles that penetrated a given distance through each gas. Gases such as oxygen and carbon dioxide gave the expected results, *viz.* the number of scintillations observed at a given distance from the radioactive source diminished when these gases were admitted into the apparatus. But with the admission of air into the apparatus quite unexpected and singular effects were observed. As Rutherford himself reported:

> '*A surprising effect was noticed, however, when dried air was introduced. Instead of diminishing, the number of scintillations was increased, and for an absorption corresponding to about 19cm of air the number was about twice that observed when the air was exhausted [removed].*'[50]

[50] (1919) 'Collision of α particles with light atoms, IV. An anomalous effect in Nitrogen,' *Phil. Mag.*, **37**, p. 581.

The additional scintillations were evidently caused by a different particle, not by the α particles, but by one whose penetrating power was greater. This suggested that they had a smaller mass. The 'brightness' of the scintillations produced by these 'long-range' particles was comparable to that produced in the same apparatus by high-energy hydrogen atoms. Further experiments showed that it was the nitrogen in the air that was responsible for the production of these additional 'long-range' scintillations.

Rutherford wrote:

> '*From the results so far obtained it is difficult to avoid the conclusion that the long-range atoms [particles] arising from the collision of α particles with nitrogen are probably atoms of hydrogen ... If this be the case, we must conclude that the nitrogen atom is disintegrated under the intense forces developed in a close collision with a swift α particle, and that the hydrogen atom which is liberated formed a constituent part of the nitrogen atom ... The results as a whole suggest that, if α particles – or similar projectiles – of still greater energy were available for experiment we might expect to break down the nucleus structure of many of the lighter atoms.*'[51]

This remarkable insight ushered in the age of nuclear and high-energy physics.

The reaction between the α particles and nitrogen atoms discovered by Rutherford can be formulated as

$$\alpha\left({}_{2}^{4}\mathrm{He}^{++}\right) + {}_{7}^{14}\mathrm{N} \rightarrow {}_{8}^{17}\mathrm{O} + p. \quad (5.32)$$

In subsequent experiments, all the other light elements were bombarded with α particles from natural radioactive sources, and in about ten cases protons were detected. Protons were clearly a constituent of the nuclei of all these elements.

The ensuing development of particle accelerators, in which the particle projectiles could be accelerated from energies of just a few million electron-volts up to hundreds of millions of electron-volts, enabled nuclear reaction experiments to be carried out at much higher energies. With the use of these particle accelerators, experiments could also be carried out with other particle projectiles, such as protons and gaseous ions.

5.4.1 Nuclear Reactions

For a nuclear reaction to occur, two nuclei, or a nucleus and a particle, must approach to within the range of the strong force they each exert. Under these circumstances, changes can take place in the way the nucleons are ordered

[51] (1919) 'Collision of α particles with light atoms, IV. An anomalous effect in Nitrogen,' *Phil. Mag.*, **37**, p. 587.

and grouped within the nuclei. The process is analogous to the changes in the order and grouping of the atoms within a molecule that accompanies a chemical reaction.

The close approach of the two nuclei, or of the nucleus and the particle projectile, is essentially a collision event. In general, and whether or not a reaction results from their encounter, energy, linear and angular momentum, mass number and electric charge are all conserved in the process.

These close encounters are usually achieved by bombarding target nuclei with particle projectiles. The process can be formulated as

$$\begin{array}{ccccccc} X_i & + & x_i & \rightarrow & x_f & + & X_f \\ \textit{Target} & & \textit{Incident} & & \textit{Scattered} & & \textit{Recoiling,} \\ \textit{nucleus} & & \textit{particle} & & \textit{particle} & & \textit{nucleus} \end{array} \tag{5.33}$$

where the symbols X_i and X_f designate the nuclei and x_i and x_f the particles. A short-hand designation often used to express nuclear reactions is:

$$X_i(x_i, x_f)X_f.$$

If the reaction leaves the identity of the particle projectile unchanged, i.e. $x_i \equiv x_f$, it is called *scattering.* If kinetic energy is conserved in the reaction the scattering is termed *elastic.* An example of elastic nuclear scattering is the classic Rutherford experiment in which α particles were scattered on passing through a gold foil. If kinetic energy is not conserved, as when the nucleus is afterwards in a changed state, the scattering is termed *inelastic.*

Nuclear reactions generally occur in two stages. In the first, a *compound excited nucleus* is formed that comprises all the nucleons and energy of the colliding species. Typically, this unstable intermediate nucleus is very short-lived, $\sim 10^{-16}$ s, and so it goes unobserved. In the second stage, the compound nucleus decays into the final products of the reaction. For example, in the first nuclear reaction discovered by Rutherford, $^{14}N(\alpha, p)^{17}O$, the intermediate compound nucleus is $^{18}F^*$:

$$\begin{aligned} \alpha + {}^{14}_{7}N &\rightarrow {}^{18}_{9}F^* \\ {}^{18}_{9}F^* &\rightarrow {}^{17}_{8}O + p. \end{aligned} \tag{5.34}$$

A particular compound nucleus may be produced in more than one way and may also decay by more than one mode. For example, bombarding ^{63}Cu nuclei with accelerated protons initially produces the compound nucleus $^{64}Zn^*$:

$$p({}^{1}_{1}H^+) + {}^{63}_{29}Cu \rightarrow {}^{64}_{30}Zn^*. \tag{5.35}$$

Bombarding ^{60}Ni nuclei with α particles also produces this compound nucleus:

$$\alpha(^4_2\text{He}^{++}) + ^{60}_{28}\text{Ni} \rightarrow ^{64}_{30}\text{Zn}^*. \quad (5.36)$$

This compound nucleus, ^{64}Zn*, exhibits three different decay modes.

$$\begin{aligned} ^{64}_{30}\text{Zn}^* &\rightarrow ^{63}_{30}\text{Zn} + n \\ &\rightarrow ^{62}_{30}\text{Zn} + n + n \\ &\rightarrow ^{62}_{29}\text{Cu} + p + n. \end{aligned} \quad (5.37)$$

The probability of each mode is a function of energy but is independent of the route by which the compound nucleus was produced.

Unstable (radioactive) nuclei that do not occur naturally can be manufactured by nuclear reactions. This artificial radioactivity was first discovered in 1934 by Irene and Frederic Joliot-Curie[52] during their study of the nuclear reactions between α particles and light atoms. They found that the product of the reaction between the stable isotope of boron, ^{10}B, and α particles was radioactive; it decayed by β^+ emission. The source of this radiation was found to be a hitherto unknown isotope of nitrogen, ^{13}N:

$$\begin{aligned} \alpha\left(^4_2\text{He}^{++}\right) + ^{10}_5\text{B} &\rightarrow ^{14}_7\text{N}^* \\ ^{14}_7\text{N}^* &\rightarrow ^{13}_7\text{N} + n. \end{aligned} \quad (5.38)$$

Artificial radioactive isotopes of all the elements have been manufactured. Many are produced commercially in substantial quantities for use in research and technology.

A particularly interesting group of nuclear reactions are those by which artificial elements possessing an atomic number greater than 92, *transuranium elements*, are produced. For example,

$$\begin{aligned} \alpha + ^{238}_{92}\text{U} &\rightarrow ^{241}_{94}\text{Pu} + n \\ ^{241}_{94}\text{Pu} &\rightarrow ^{241}_{95}\text{Am} + e^- + \bar{\nu} \\ ^{241}_{95}\text{Am} &\rightarrow ^{237}_{93}\text{Np} + \alpha. \end{aligned} \quad (5.39)$$

The isotope of plutonium, Pu$-$241, initially produced by the reaction between the α particles and the uranium nuclei, is a β^- emitter. The product of its decay is an isotope, Am$-$241, of another transuranium element, americium.

[52]Irene Joliot-Curie 1897–1956, daughter of Marie Curie, and her husband Frederic Joliot-Curie 1900–1958, French physicists who were awarded the Nobel prize in 1935 for their discovery of radioactive artificial isotopes.

This isotope is an α emitter and decays in turn into the relatively long-lived ($T_{\frac{1}{2}} = 2.14 \cdot 10^6$ years) isotope of neptunium, Np–237. To date, transuranium elements with atomic numbers up to 112 have been produced.

5.4.2 The Discovery of the Neutron

A year after discovering the nuclear reaction between α particles and nitrogen atoms, Rutherford put forward the hypothesis that nuclei comprise just protons and neutrons.[53] However, a further 12 years were to pass before the neutron's existence was confirmed experimentally. Neutrons were particularly hard to observe because they carry no net charge and so are undeflected by either electrical or magnetic fields. They were also initially confused with high-energy photons.

The first step in the discovery of the neutron was made by Bothe in 1930. He discovered that when the metal beryllium is bombarded with α particles, a radiation of great penetrating power is produced. This radiation was undeflected by either electrical or magnetic fields and so it was presumed to comprise very high-energy photons. Two years later, Irene and Frederic Joliot-Curie discovered that when the radiation released from the beryllium target was passed through substances rich in hydrogen atoms, like paraffin wax, protons with energies as high as 5.7 MeV were released from the material. Presuming that the protons had been released from the wax by the impact of photons in a kind of proton Compton effect, the Joliot-Curies calculated that the energy of the photons in the mysterious radiation was about 50 MeV.[54] However, 'it is difficult to account for the production of a quantum of 50 MeV from the interaction of a beryllium nucleus with an α particle of kinetic energy of 5 MeV'.

The mystery was solved by James Chadwick[55] in 1932. Applying elementary mechanics to the problem he suggested that:

> '*If we suppose that the radiation is not a quantum radiation, but consists of particles of mass very nearly equal to that of the proton, all the difficulties connected with the collisions disappear.*'[56]

---

[53] Previous theories had suggested that the basic components of nuclei were helium nuclei (α particles) and protons. Rutherford's hypothesis simplified the picture. However, though his neutron hypothesis ultimately proved correct, at the time he regarded the neutron as 'a proton and electron in close combination', a concept we now know to be wrong.

[54] Such *photonuclear* reactions in which a high-energy photon emits a nucleon from a nucleus are known. For example, $^{25}Mg(\gamma,p)^{24}Na$.

[55] Sir James Chadwick 1891–1974, English physicist who was awarded the Nobel prize in 1935 for the discovery of the neutron.

[56] (1932) *Proc. Roy. Soc. Lond.*, **A136**, p. 692.

A corollary of the classic theory of elastic collisions is that when a stationary particle is struck by a moving particle of the same mass, the moving particle stops dead and the stationary one moves off at the velocity with which it was struck. In the collision, all the kinetic energy of the incident particle is transferred to the target particle. Accordingly, if the particles striking the protons in the paraffin wax have nearly the same mass as the protons, their kinetic energy need be no more than 5.7 MeV (and not some ten times as much, as the Joliot-Curies had supposed) in order that a proton acquire a kinetic energy of 5.7 MeV as a result of a collision. Chadwick continued:

> '*In order to explain the great penetrating power of the radiation we must further assume that the particle has no net charge. We may suppose it (to be) the "neutron" discussed by Rutherford in his Bakerian Lecture of 1920.*
> *It is concluded that the radiation consists, not of quanta as hitherto supposed, but of neutrons, particles of mass 1, and charge 0.*'

Thus, the reaction between the α particles and the beryllium can be formulated as:

$$\alpha\,({}^4_2\mathrm{He}^{++}) + {}^9_4\mathrm{Be} \rightarrow {}^{13}_{6}\mathrm{C}^*$$

$${}^{13}_{6}\mathrm{C}^* \rightarrow {}^{12}_{6}\mathrm{C} + n. \tag{5.40}$$

Other methods of producing neutrons were soon discovered. For instance, it was found that the α particles could be replaced by protons or positive ions that had been accelerated in cyclotrons or by the high potentials of a van de Graff generator. Moreover, light elements other than beryllium could be used as the target for these projectiles. By means of these alternative techniques, neutrons in a whole range of different energies have been produced.

Having no electrostatic repulsion to overcome, neutrons can enter the range of the strong force exerted by a nucleus with ease. In a close encounter with a nucleus, a neutron may be scattered, elastically or inelastically, or it may be captured by the nucleus. The capture of a neutron does not alter the chemical identity of the nucleus; it just converts it into a heavier isotope of the same element.

For example, irradiating stable Al–27 nuclei with neutrons produces unstable (radioactive) Al–28 nuclei:

$$n + {}^{27}_{13}\mathrm{Al} \rightarrow {}^{28}_{13}\mathrm{Al} + \gamma. \tag{5.41}$$

In a similar way, almost every stable isotope can be converted into an artificial radioactive isotope. Such reactions are called *radiative capture.* They are the basis of the commercial production of the short-lived artificial isotopes such as I–131 and Tc–99 used for medical diagnoses. Nuclear (atomic) reactors, that release an abundance of neutrons, are the neutron source usually employed.

Neutrons do not themselves ionise the atoms and molecules with whose nuclei they collide. However, the products of their reactions with the nuclei may include ionising radiations. For example, α particles are produced in the reaction $^{10}B(n,\alpha)^7Li$ and high energy photons (γ-rays) are released in radiative capture reactions. The various devices used to detect neutrons operate by registering the appearance of these ionising radiations.

5.4.3 Nuclear Cross-Sections

In practice, inducing nuclear reactions is very much a hit or miss affair. In the hope that some of the nuclei and particle projectiles will collide and react, a target containing a multitude of the nuclei is placed in the path of a beam of the particles or is irradiated by a shower of them. Very few reactions are actually observed. For instance, in 1925 an attempt was made to photograph the tracks left in a cloud-chamber by the particles involved in the first nuclear reaction discovered by Rutherford. Some 20,000 photographs were taken. On average, each photograph revealed the tracks of 20 α particles making a total of 400,000 α particle tracks overall. However, clear signs of a reaction between an α particle and a nitrogen atom in the atmosphere inside the cloud-chamber were detected in just eight photographs.

The probability that a nuclear reaction will occur is expressed in terms of a parameter called the *nuclear cross-section*; the larger the nuclear cross-section the more probable the reaction. This factor can be viewed as the target area effectively presented to the incident particles by each nucleus (Fig. 5.28). Nuclear cross-sections are measured in a unit called the *barn*: $1b = 10_{-28}\ m^2$. They may be larger or smallar than the geometric cross-section of the target nucleus, though in most cases they are of the same order of magnitude. In general, the reduction, ΔI, in the intensity of a particle beam of intensity I on passing through a slice of homogeneous material of thickness, Δx, is given by

$$\Delta I = -\mu I \cdot \Delta x. \tag{5.42}$$

μ is the *macroscopic cross-section* or *linear attenuation coefficient.*

If each unit volume of the material contains N nuclei that can interact with the particles in the beam and if each such nucleus presents a target of area σ to the incident particles, then a slice of frontal area A effectively presents a 'wall' of interacting nuclei of area $AN\sigma\Delta x$ to the beam (Fig. 5.29). The relative attenuation, $\Delta I/I$, of the beam resulting from its passage through the slice

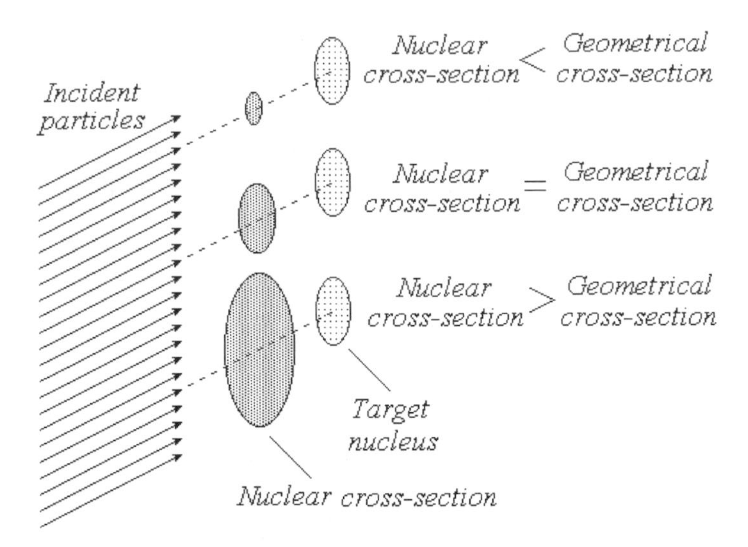

Fig. 5.28. Nuclear cross-sections viewed as the area effectively presented to the incident beam by each target nucleus. The larger the nuclear cross-section the greater the probability that an incident particle will react with a nucleus.

will be

$$\frac{\Delta I}{I} = -\frac{AN\sigma\Delta x}{A} = -N\sigma\Delta x. \tag{5.43}$$

From Eqs (5.42) and (5.43), the relationship between the nuclear cross-section, σ, and the macroscopic cross-section, μ, is simply

$$\mu = N\sigma. \tag{5.44}$$

On average, a particle entering the slice travels a distance of $1/N\sigma$ before reacting with a nucleus.

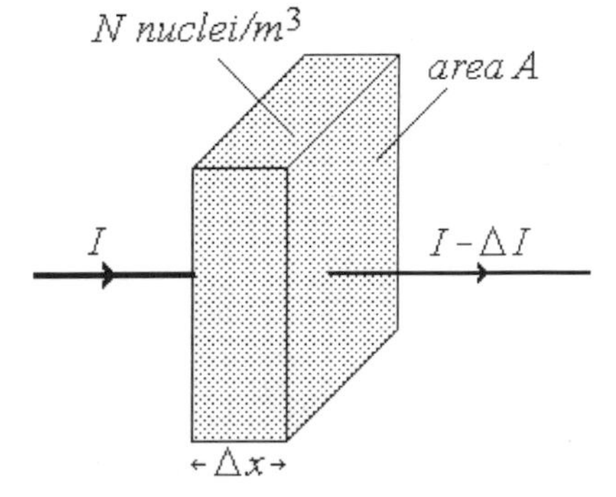

Fig. 5.29. The attenuation of a beam of particles by a slice of a material containing N reacting nuclei per cubic meter. The slice effectively presents an impervious 'wall' of area $AN\sigma\Delta x$ to the beam.

In general, nuclear cross-sections depend on two factors:

1. the energy of the particle projectile;
2. the nature of the nuclear reaction.

The dependence of cross-sections on the energy of the particle projectile is striking. Whereas, in reactions involving positively charged projectiles such as protons and and α particles, nuclear cross-sections generally increase as the projectiles' energy rises, the opposite is true in the case of neutrons. This key discovery was made almost inadvertently by Fermi in 1934. With 'no conscious prior reasoning', he decided one day to place a piece of paraffin wax between the high-energy neutron source he was employing and the target he was irradiating. The effect was dramatic. The activity of the irradiated sample increased dramatically, indicating that the neutrons 'filtered' through the wax had induced many more radiative capture reactions of the type shown in Eq. (5.41) than had the unfiltered neutrons.

Returning from lunch, Fermi explained the phenomenon: the high-energy neutrons emitted by the source were being slowed down as they passed through the paraffin wax. As Chadwick had previously pointed out, when high-energy neutrons collide with the hydrogen nuclei in the paraffin wax they are slowed almost to a standstill. After repeated collisions, they enter into thermal equilibrium, settling down to a kinetic energy of about 0.025 eV, this being the average energy of a gas particle at room temperature. These slow neutrons, called *thermal neutrons*, remain in the vicinity of the target nucleus for much longer than fast neutrons and so the probability of their capture is correspondingly greater. An alternative equivalent explanation is that the slower the neutron, the less is its linear momentum and the longer its De Broglie wavelength. However, a particle's De Broglie wavelength is a measure of its size, and so slow neutrons are effectively larger than fast ones and are, thus, more easily captured by the target nuclei.

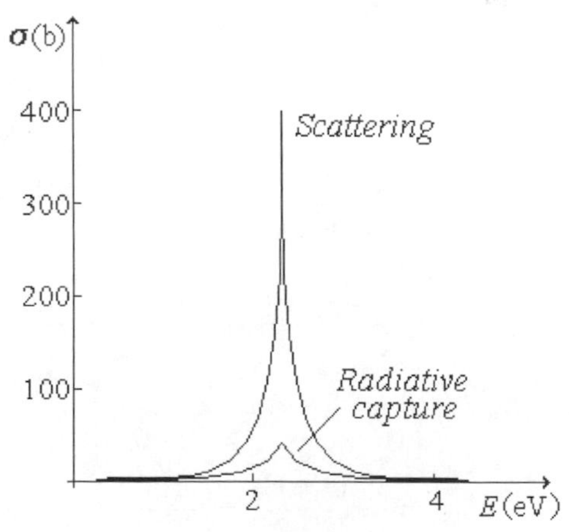

Fig. 5.30. The observed nuclear cross-sections, σ, as a function of the neutron energy, E, for the radiative capture, $^{123}\mathrm{Te}(n,\gamma)^{124}\mathrm{Te}$, and scattering, $^{123}\mathrm{Te}(n,n)^{123}\mathrm{Te}$, reactions. The latter reaction proceeds via the formation of a compound nucleus, $^{124}\mathrm{Te}^*$, which subsequently decays with the emission of a neutron.

Much can be learnt about nuclear structure and nuclear energy levels from the dependence of nuclear cross-sections on the energy of the neutrons. At certain *resonance energies*, relatively large nuclear cross-sections are observed indicating a high probability that neutrons with these energies will react with the nucleus. These resonance energies usually correspond to the energy required to raise the target nucleus to one of its excited states. Thus, many nuclear scattering reactions exhibit resonances analogous to those displayed by the mercury atoms in the Franck–Hertz experiment. For example, both the radiative capture, ${}^{123}\mathrm{Te}(n,\gamma){}^{124}\mathrm{Te}$, and scattering, ${}^{123}\mathrm{Te}(n,n){}^{123}\mathrm{Te}$, reactions of neutrons with the tellurium isotope Te−123 exhibit striking resonances with neutrons of energy 2.33 eV. The nuclear cross-section for the former is 40 b and that for the latter is 400 b. This is attributed to the presence of a 2.33 eV excited energy state in the tellurium nucleus (Fig. 5.30).

Example 5.9: Nuclear Cross-Sections

A cadmium foil of thickness 0.1 mm absorbs about two thirds of the thermal neutrons incident upon it through a ${}^{113}\mathrm{Cd}(n,\gamma){}^{114}\mathrm{Cd}$ reaction. Given that the isotope ${}^{113}\mathrm{Cd}$ constitutes 12% of natural cadmium and that the density of cadmium metal is 8,600 kg/m^3, calculate the nuclear cross-section of ${}^{113}\mathrm{Cd}$ for slow neutrons.

Calculation: By analogy with Eq. (5.9), the ratio, I_0/I, between the initial intensity, I_0, of the neutron beam and its intensity, I, after passing through a thickness x of the cadmium is given by

$$\ln\left(\frac{I}{I_0}\right) = \mu x = N\sigma x. \tag{5.45}$$

Substituting $I = 0.33 I_0$ and $x = 10^{-4}$m gives

$$\ln 3.03 = 10^{-4} N\sigma.$$

Since the isotope ${}^{113}\mathrm{Cd}$ constitutes just 12% of natural cadmium, the number, N, of ${}^{113}\mathrm{Cd}$ atoms per cubic metre is

$$N = 0.12 \cdot \frac{8640}{113\mathrm{u}} = 0.12 \cdot \frac{8640}{113 \cdot 1.66 \cdot 10^{-27}} = 5.5 \cdot 10^{27}\mathrm{m}^{-3}.$$

Substitution gives

$$\sigma = \frac{\ln 3.03}{10^{-4}N} = \frac{1.109}{5.5 \cdot 10^{23}} = 2.01 \cdot 10^{-24}\mathrm{m}^{-2} \approx 20,000\mathrm{b}.$$

Contrast this value with the geometric cross-section of ${}^{113}\mathrm{Cd}$, that is about 1.06b.

5.4.4 Nuclear Energy — Fusion and Fission

The energy changes involved in nuclear reactions such as

X_i	+	x_i	→	x_f	+	X_f
Target nucleus		*Incident particle*		*Scattered particle*		*Recoiling nucleus*

can generally be studied by the familiar tools of classical mechanics.[57] Applying, respectively, linear momentum and energy conservation to this reaction gives:

$$\sum \mathbf{p}_i = \sum \mathbf{p}_f \tag{5.46}$$

and

$$Q + \sum K_i = \sum K_f, \tag{5.47}$$

where Q is the difference between the final, ΣK_f, and initial, ΣK_i, kinetic energies of the system.

The value of Q is independent of the reference frame from which the linear momenta and kinetic energies are measured and is given by the formula:

$$Q = [(M_i + m_i) - (M_f + m_f)]c^2, \tag{5.48}$$

where M_i and M_f represent, respectively, the masses of the target and recoiling nuclei and m_i and m_f the masses of the incident and scattered particles. In an elastic collision $Q = 0$; in this case no change occurs in the internal energy of the system. When $Q > 0$, the reaction releases kinetic energy at the expense of the system's internal energy; the reaction is *exoergic*. When $Q < 0$, the reaction requires an investment of energy in order to raise the system's internal energy; the reaction is *endoergic*.

The value of Q can also be calculated from the average binding energy per nucleon, E_B/A, of the species reacting and produced in the reaction (Fig. 5.31). In general, if the average binding energy per nucleon is higher in the products of a nuclear reaction than in its reactants, Q will be positive and the reaction will be exoergic. If it is lower, Q will be negative, and the reaction will be endoergic.

An examination of the graph in Fig. 5.31 suggests two possible nuclear processes in which the products would have a greater average binding energy per nucleon than the reactants. One is the fusion of light nuclei, the other is the fission of heavy nuclei.

[57] In most nuclear reactions a non-relativistic treatment is adequate.

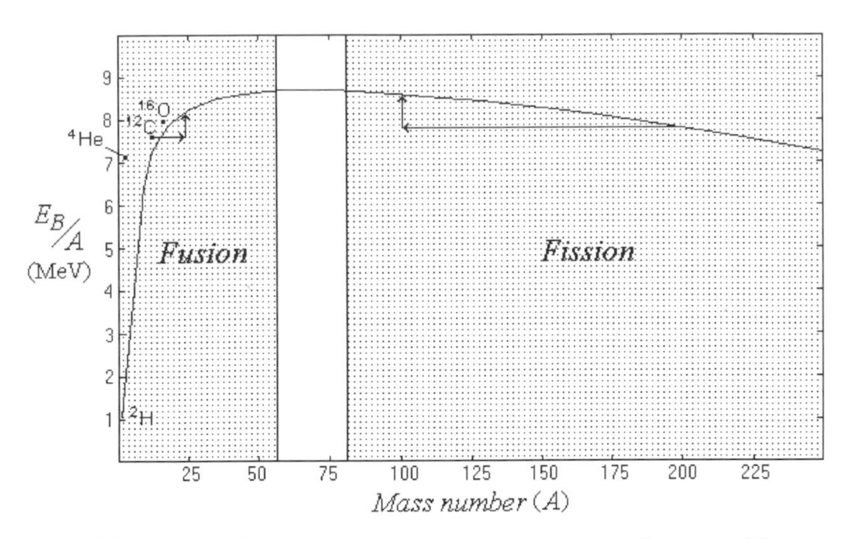

Fig. 5.31 The average binding energy per nucleon, E_B/A, as a function of the mass number, A. New nuclei in which the binding energy per nucleon is greater than it was in the original nuclei are produced by both the fusion of two light nuclei $(A < 50)$ and the fission of a heavy nucleus $(A > 80)$.

For instance, the fusion of two $^{12}_{6}C$ nuclei into a single $^{24}_{12}Mg$ nucleus would increase the average binding energy per nucleon from 7.7 MeV to 8.3 MeV, i.e. by 0.6 MeV per nucleon. Since a total of 24 nucleons would be involved in this fusion reaction, the fusion would lead to the release of about 15 MeV of energy.

The average binding energy per nucleon would also be increased if a heavy nucleus was to split into two medium-sized nuclei. For instance, the symmetrical fission of a nucleus containing 200 nucleons into two medium-sized nuclei each of mass number 100 would increase the average binding energy per nucleon from 7.9 MeV to 8.7 MeV, leading to the release of some 160 MeV of energy.[58] Furthermore, whereas the ratio, N/Z, between the number of its neutrons and protons in nuclei containing 200 nucleons is about 1.5, its value in stable medium-sized nuclei is less than 1.4. Thus, in addition to the energy released, the fission of heavy nuclei should be accompanied by a shower of free neutrons.

The demarcation line between those nuclei that can in theory fuse exoergically and those that can fission exoergically is marked by the band of medium-sized nuclei whose mass numbers are between 55 and 85, in

[58] 160 MeV is more than ten million times the heat of reaction per atom in ordinary combustion processes. It is enough energy to make a grain of fine sand visibly jump!

which the average binding energy per nucleon rises to its maximum value, ~8.8 MeV.[59]

Example 5.10: Endoergic Nuclear Reactions

Calculate the minimum (threshold) energy required to initiate endoergic nuclear reaction $^{14}N(\alpha,p)^{17}O$.

m_{N-14}:	14.00307u	m_{O-17} :	16.99913u
m_α :	4.00260u	m_p:	1.00783u

Calculation: The mass difference, Δm, between the reactants, ^{14}N and α, and the products, p and ^{17}O, is:

$$\Delta m = (14.00307 + 4.00260) - (1.00783 + 16.99913) = -0.00129\text{u}.$$

Thus, the energy, Q, that must be made available to the system for the reaction to occur is

$$Q = (931 \cdot 0.00129\text{u}) = 1.2\,\text{MeV}.$$

This must be supplied by the kinetic energy of the reactants, the α particles and the ^{14}N nuclei.

Relative to the laboratory from which the reaction is observed, the total kinetic energy of the reactants, $(\sum K_i)_{\text{Lab}}$, is equal to the sum of their kinetic energy, $(\sum K_i)_{\text{CM}}$, relative to their centre of mass plus the kinetic energy, $(\sum K_{\text{CM}})_{\text{Lab}}$, of their centre of mass relative to the laboratory:

$$\left(\sum K_i\right)_{\text{Lab}} = \left(\sum K_{\text{CM}}\right)_{\text{Lab}} + \left(\sum K_i\right)_{\text{CM}}.$$

The energy, $(\sum K_{\text{CM}})_{\text{Lab}}$, associated with the motion of the system's centre of mass cannot be converted into internal energy; the amount, Q, must be supplied by the reactants' kinetic energy relative to the centre of mass, i.e. $(\sum K_i)_{\text{CM}} \geq 1.2\,\text{MeV}$. But energy must also be provided for the motion of the products of the reaction; they cannot be at rest as this would violate the conservation of linear momentum. Detailed calculations show that only α particles with observed energies relative to the laboratory greater than 1.54 MeV can initiate the reaction.

5.4.5 Nuclear Fusion

The fusion of light nuclei is the primary energy-producing process in the observed universe; it is the power source that lights the stars and provides the sun with its energy. On earth, the uncontrolled (explosive) fusion of light

[59]The nucleus with the greatest binding energy per nucleon is Fe−56.

nuclei was first achieved in 1952 with the detonation of the first hydrogen bomb. Controlled non-destructive fusion has yet to be accomplished.

In order for fusion to occur, the nuclei must first approach to within the range of the strong force each exerts, *viz.* to an internuclear distance less than 10^{-14} m. They must then remain within this range long enough for the strong interaction between them to fuse them into a new and larger nucleus.

Bringing a proton to within 10^{-14} m of another proton requires the investment of a considerable amount of energy in order to overcome the long-range electrostatic repulsive forces that act between them. For example, at a separation of 10^{-14} m, the electrostatic potential energy, U_E, of a pair of protons is:

$$U_E = \frac{ke^2}{r} = \frac{9 \cdot 10^9 (1.6 \cdot 10^{-19})^2}{10^{-14}(1.6 \cdot 10^{-13})} = 0.144\,\text{MeV}. \tag{5.49}$$

One possible way of supplying this energy to the neuclei is to raise their temperature. Assuming, further, that the classical equation, $\overline{E} = kT$, for the average energy of an oscillator remains valid at very high temperatures, the average kinetic energy of a gas particle reaches 0.144 MeV at a temperature of $\sim 10^9$ K. Assuming that the kinetic energy has a Maxwellian distribution, some particles will possess this energy momentarily at a temperature of $\sim 10^8$ K. Quantum mechanical considerations suggest that even at a temperature of $\sim 10^7$ K some particles could 'tunnel' through the electrostatic energy barrier and fuse (Fig. 5.32).

At these extreme temperatures, the gas is completely ionised and comprises charged particles, ions (nuclei) and electrons; in this state it is called

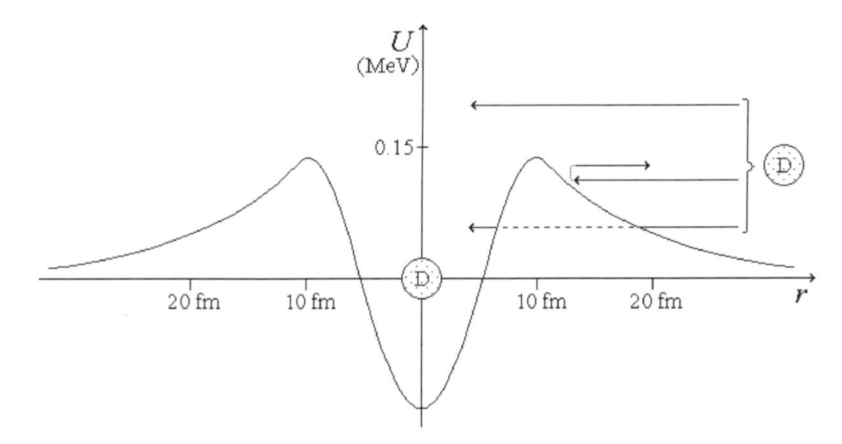

Fig. 5.32. Potential energy diagram of two deuterons. In order to clear the potential barrier and fuse with the target deuteron, an oncoming deuteron needs an energy greater than 0.144 MeV. Most deuterons possessing less energy will be reflected from the barrier though some may tunnel through.

a *plasma.* Attaining these temperatures is particularly difficult because of the fundamental principle, expressed in the Stefan–Boltzmann Law, $E \propto T^4$, that the energy density of a system varies as the fourth power of its absolute temperature. Hence, doubling the temperature requires 16 times the energy density.

Uncontrolled (explosive) fusion is the process by which the visible stars produce energy; it is also the process employed in thermonuclear[60] weapons. In thermonuclear weapons, the extreme initial conditions required for the fusion of the deuterium and tritium nuclei that comprise their 'fuel' are achieved by using a fission bomb to initiate the reaction.[61] At the great densities and pressures found at the core of stars such as the sun, hydrogen is 'burnt' at a temperature of about 10^7 K in nuclear reactions that convert it into helium.

$$\begin{array}{llll} {}^1_1\mathrm{H} + {}^1_1\mathrm{H} & \rightarrow {}^2_1\mathrm{H} + e^+ + \nu & Q = 1.19\,\mathrm{MeV} & \\ {}^1_1\mathrm{H} + {}^2_1\mathrm{H} & \rightarrow {}^3_2\mathrm{He} + \gamma & Q = 5.49\,\mathrm{MeV} & (5.50) \\ {}^3_2\mathrm{He} + {}^3_2\mathrm{He} & \rightarrow {}^4_2\mathrm{He} + {}^1_1\mathrm{H} + {}^1_1\mathrm{H} & Q = 12.85\,\mathrm{MeV}. & \end{array}$$

The process is called the *proton–proton cycle.* The net effect of these reactions is the conversion of four hydrogen nuclei (protons) into a helium nucleus (an α particle), two positrons and two neutrinos. The positrons collide almost at once with electrons in the plasma, mutually annihilating and releasing two γ photons. The neutrinos fly off unhindered, radiating away from the sun together with the solar energy; a flux of about 10^{15} solar neutrinos per square metre per second has been detected on earth. Note that no neutrons are produced in the process; such fusion reactions are termed *aneutronic.* Life on earth would be impossible if the sun emitted a flux of high-energy neutrons.

An alternative aneutronic process by which the fusion of four hydrogen nuclei into a helium nucleus can occur in stars was proposed by Bethe.[62] In this process, called the *carbon cycle*, the C–12 nucleus fulfils the role of a catalyst, being consumed in the first stage and regenerated in the final stage of the series

[60] So called because of the extreme temperatures, more than ten million degrees, required for their detonation.
[61] See Section 5.4.7 below.
[62] Hans Albrecht Bethe 1906–2005, German–American physicist who was awarded the Nobel prize in 1967 for his investigations of stellar processes.

of reactions.

$$
\begin{aligned}
{}^{1}_{1}\mathrm{H} + {}^{12}_{6}\mathrm{C} &\rightarrow {}^{13}_{7}\mathrm{N} + \gamma & Q &= 1.95\,\mathrm{MeV} \\
{}^{13}_{7}\mathrm{N} &\rightarrow {}^{13}_{6}\mathrm{C} + e^{+} + \nu & Q &= 2.22\,\mathrm{MeV} \\
{}^{1}_{1}\mathrm{H} + {}^{13}_{6}\mathrm{C} &\rightarrow {}^{14}_{7}\mathrm{N} + \gamma & Q &= 7.54\,\mathrm{MeV} \\
{}^{1}_{1}\mathrm{H} + {}^{14}_{7}\mathrm{N} &\rightarrow {}^{15}_{8}\mathrm{O} + \gamma & Q &= 7.35\,\mathrm{MeV} \\
{}^{15}_{8}\mathrm{O} &\rightarrow {}^{15}_{7}\mathrm{N} + e^{+} + \nu & Q &= 2.71\,\mathrm{MeV} \\
{}^{1}_{1}\mathrm{H} + {}^{15}_{7}\mathrm{N} &\rightarrow {}^{12}_{6}\mathrm{C} + {}^{4}_{2}\mathrm{He} & Q &= 4.96\,\mathrm{MeV}
\end{aligned}
\qquad (5.51)
$$

When all its hydrogen has been converted into helium, the star contracts, heating up until it reaches the temperature at which three helium nuclei can fuse into a carbon nucleus. After all its helium has been thus exhausted, the star contracts further, heating up until the carbon and heavier nuclei can fuse. The cycles are repeated successively until the Fe–56 nucleus is reached, at which stage fusion ceases to be exoergic.

It is hypothesised that the heavier elements, $A > 56$ and $Z > 26$, are then synthesised from the medium-sized nuclei by a series of repeated neutron captures and β^- decays, the former increasing the mass number of the nuclei and the latter their atomic number. The requisite neutrons are initially produced in the mature stars by reactions such as

$$
\begin{aligned}
&{}^{4}_{2}\mathrm{He} + {}^{13}_{6}\mathrm{C} \rightarrow {}^{16}_{8}\mathrm{O} + n \\
&{}^{4}_{2}\mathrm{He} + {}^{17}_{8}\mathrm{O} \rightarrow {}^{20}_{10}\mathrm{Ne} + n,
\end{aligned}
\qquad (5.52)
$$

and ultimately, after it has exhausted all its fusion reserves, by the star's final cataclysmic implosion (nova or supernova). The remaining debris blasts off into space to become *inter alia* the raw material for the planets, moons and asteroids.[63]

Reactions considered possible for the controlled fusion of light nuclei include:

$$
\begin{aligned}
{}^{2}_{1}\mathrm{H} + {}^{2}_{1}\mathrm{H} &\rightarrow {}^{3}_{2}\mathrm{He} + n & Q &= 3.3\,\mathrm{MeV} \\
{}^{2}_{1}\mathrm{H} + {}^{2}_{1}\mathrm{H} &\rightarrow {}^{3}_{1}\mathrm{H} + {}^{1}_{1}\mathrm{H} & Q &= 4\,\mathrm{MeV} \\
{}^{3}_{1}\mathrm{H} + {}^{2}_{1}\mathrm{H} &\rightarrow {}^{4}_{2}\mathrm{He} + n & Q &= 17.6\,\mathrm{MeV}.
\end{aligned}
\qquad (5.53)
$$

The basic fuel for these reactions is the deuterium nucleus, ${}^{2}\mathrm{H}$, the isotope that constitutes about 0.015% of the naturally occurring hydrogen on the earth's surface. The abundance of this isotope in the water of the oceans represents a

[63]Summing up the import of this hypothesis, romantic cosmologists have remarked that we are all the stuff of stardust.

virtually inexhaustible supply of fuel. Thus, induced controlled fusion reactions could supply all of man's future energy needs.

Confining the hot plasma at a sufficiently high density long enough for the fusion reaction to occur is the major unsolved problem in the search for controlled fusion. Obviously, at such extreme temperatures, material containers would be useless. One solution being tried is to confine the plasma along the axis of a toroidal (doughnut shaped) vessel by means of magnetic fields. However, sustaining the stability of such 'magnetic bottles' has proved very difficult. A second technique being tried is the sudden compression of solid fusible material, deuterium–tritium pellets. It is hoped that by 'zapping' the pellets simultaneously from all sides with the high-energy particles in intense laser or particle beams, the material's density will be raised abruptly by a factor of one thousand and its temperature to over ten million degrees. Under these conditions it is anticipated that fusion will occur. However, at the date of writing, neither technique has produced controlled fusion, despite the huge investment made into their research.

A further problem is that the deuterium–deuterium and deuterium–tritium fusion reactions that have been investigated so far produce a flux of high-energy neutrons. Regulating this flux is virtually impossible because it is unaffected by electric or magnetic fields. Extracting the energy of these neutrons in order to use it for generating electricity, without producing substantial quantities of unwanted and contaminating radioactive isotopes, also appears to be a major problem. In the light of this, it has been suggested that though they require a higher temperature for their initiation, other aneutronic reactions should be investigated, such as

$$ {}^{1}_{1}\mathrm{H} + {}^{11}_{5}\mathrm{B} \rightarrow {}^{4}_{2}\mathrm{He} + {}^{4}_{2}\mathrm{He} + {}^{4}_{2}\mathrm{He}. \tag{5.54} $$

5.4.6 Nuclear Fission

The fission of heavy nuclei was first observed by Hahn[64] and Strassman[65] in 1938, when they detected barium ions in a solution of uranium salts they were irradiating with slow neutrons. The appearance of the barium, whose atomic number 56 is about half that of uranium, was a complete mystery. Its presence in the irradiated solution was inexplicable in terms of the known modes of radioactive decay.

---

[64] Otto Hahn 1879–1968, German chemist who was awarded the Nobel prize in 1944 for the discovery of nuclear fission.

[65] Fritz Strassman 1902–1980, German chemist who was recognized by *Yad Vashem Holocaust Memorial* as Righteous Among the Nations.

The riddle was solved, just a few weeks after the discovery, by Meitner[66] and Frisch.[67] Their explanation was based on the liquid-drop model of the nucleus.

> '*They pictured the uranium nucleus as a liquid drop gone wobbly [Fig. 5.33] ... and imagined it hit by even a barely energetic slow neutron. The neutron would add its energy to the whole. The nucleus would oscillate. In one of its many random modes of oscillation it might elongate. Since the strong force operates only over extremely short distances, the electric force repelling the two bulbs of an elongated drop would gain advantage. The two bulbs would push farther apart. A waist would form between them. The strong force would begin to regain the advantage within each of the two bulbs. It would work like surface tension to pull them into spheres. The electric repulsion would work at the same time to push the two separating spheres even farther apart.*'[68]

Each of the two spheres would be a medium-sized nucleus; in the case of the fission of a uranium nucleus, $Z_U = 92$, they might well be a barium nucleus, $Z_{Ba} = 56$, and a krypton nucleus, $Z_{Kr} = 36$. The energy released in the process arises from the intense electrostatic repulsion that acts between the middle-sized nuclei at the moment of their formation. Most of this energy subsequently appears as the kinetic energy of these nuclei.

In order to undergo fission, a nucleus must first surmount (or tunnel through) the substantial energy barrier that corresponds, in the liquid-drop model of nuclear fission, to the shaping of the waist or constriction in the parent nucleus (Fig. 5.33). Thus, the spontaneous fission of heavy nuclei is a rare

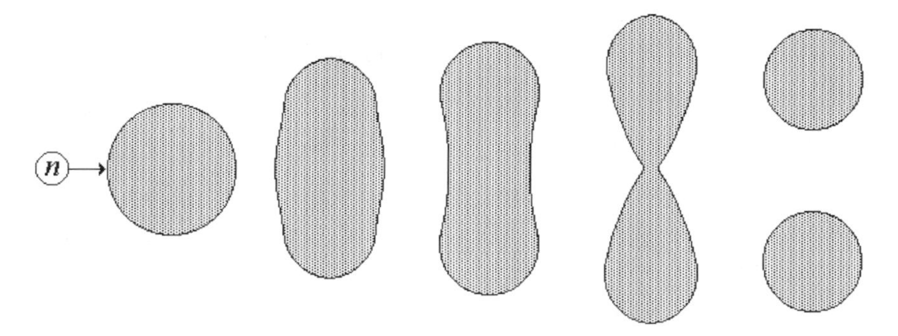

Fig. 5.33. The stages in the fission of a drop of nuclear 'fluid' into two drops after the capture of a neutron.

[66]Lise Meitner 1878–1968, Austrian–Swedish physicist.
[67]Otto Frisch 1904–1979, Austrian–British physicist.
[68]Richard Rhodes (1986) *The Making of the Atomic Bomb*, New York: Simon & Schuster, p. 259.

event. For example, just one out of every two million U–238 nuclei decays by spontaneous fission; all the rest decay by α emission.[69]

The nucleus can obtain the *fission threshold energy* it requires in two ways: by the absorption of a high-energy photon or the capture of a neutron. In the latter process, the neutron initially adds cohesion to the nucleus without adding any electrostatic repulsion. The extra cohesion adds to the binding energy of the nucleus. The surplus energy thereby released provides the threshold energy required for fission to occur (Fig. 5.34).

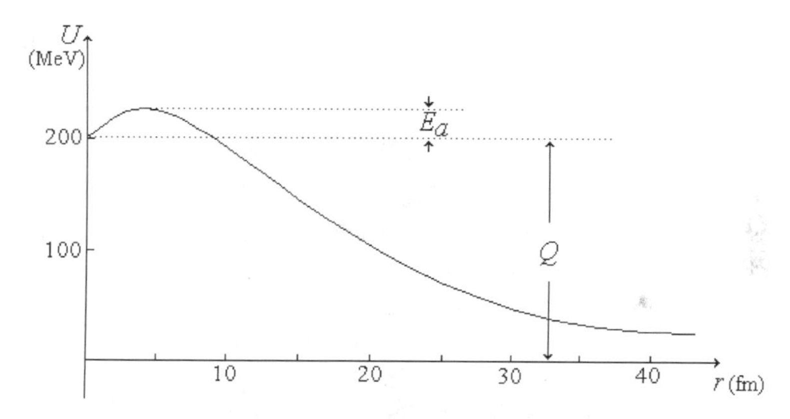

Fig. 5.34. The potential energy, U, of a fissioning nucleus as a function of a distortion parameter, r, that measures its deviation from a spherical shape. The minimum energy required to initiate a reaction, E_a, is its *threshold energy*.

Since the pairing of nucleons reduces their energy, the total amount of energy provided by neutron capture depends, *inter alia*, on the number of neutrons in the nucleus. The capture of a neutron by a nucleus containing an odd number of neutrons provides more nuclear energy than its capture by a nucleus containing an even number of neutrons. In the former case, an unpaired neutron becomes paired; in the latter case, a lone unpaired neutron is added to a previously fully paired nucleus. Thus, neutron capture induced fission should occur more readily in nuclei containing an odd number of neutrons.

This is borne out by the radiative capture and fission patterns of many heavy nuclei. For example, the capture of a neutron by Th−232 and U−238, both of which contain an even number of neutrons, produces the compound nuclei Th−233 and U−239, respectively. The incorporation of the neutron

[69] It has been suggested that the emission of an α particle from a nucleus should be regarded as a very asymmetrical form of fission.

into their structure makes about 5 MeV of energy available to these compound nuclei. However, the threshold energy for their fission is about 6 MeV. Consequently, without additional energy, the probability of these nuclei undergoing fission is negligible. Thus, irradiating Th$-$232 and U$-$238 with slow (thermal) neutrons results in radiative capture reactions, $^{232}\mathrm{Th}(n,\gamma)^{233}\mathrm{Th}$ and $^{238}\mathrm{U}(n,\gamma)^{239}\mathrm{U}$, respectively, and not fission. The products of these reactions undergo β^- decay. In the case of U$-$239, this produces the relatively stable transuranic isotope Pu$-$239 ($T_{\frac{1}{2}} = 24,000$ years).

$$\begin{aligned} n + {}^{238}_{92}\mathrm{U} &\rightarrow {}^{239}_{92}\mathrm{U} + \gamma \\ {}^{239}_{92}\mathrm{U} &\rightarrow {}^{239}_{93}\mathrm{Np} + \beta^- + \bar{\nu} \quad T_{\frac{1}{2}} = 24\,\mathrm{m} \\ {}^{239}_{93}\mathrm{Np} &\rightarrow {}^{239}_{94}\mathrm{Pu} + \beta^- + \bar{\nu} \quad T_{\frac{1}{2}} = 2.3\,\mathrm{d} \end{aligned} \tag{5.55}$$

Compound nuclei that do undergo fission can be obtained from Th$-$232 and U$-$238 by the capture of fast neutrons (> 1 MeV); the kinetic energy of the fast neutrons provides the additional energy required for their fission. For example, unless it is pre-empted by radiative capture, i.e. by the emission of a γ-ray from the excited compound nucleus, the incorporation of a 1.3 MeV neutron into a U$-$238 nucleus generally results in its fission.

By contrast, the nuclei U$-$233, U$-$235 and Pu$-$239 all contain an odd number of neutrons. The capture of a neutron by these nuclei produces the compound nuclei, U$-$234, U$-$236 and Pu$-$240, respectively. Incorporating the neutron into their structure makes about 6 MeV of energy available to these compound nuclei. This is more than the threshold energy, $\sim$5 MeV, for their fission. Thus, any neutron – slow (thermal) neutrons as well fast (high-energy) neutrons – can induce the fission of nuclei such as U$-$233, U$-$235 and Pu$-$239. Unless pre-empted by rapid radiative capture, the intermediate compound nuclei undergo fission within an instant ($\sim 10^{-12}$ s) of their formation without requiring any additional energy.

The values of the radiative capture and neutron induced fission cross-sections, $\sigma_{(n,\gamma)}$ and $\sigma_{(n,f)}$, listed in Table 5.18, reflect the probabilities of the two reactions occurring when the nuclei U$-$233, U$-$235, U$-$238 and Pu$-$239 are irradiated with fast or slow (thermal) neutrons. Except in the case of the U$-$238 nuclei, the probability of fission is over one hundred fold greater when slow rather than fast neutrons are employed.

The fission of a compound nucleus is generally asymmetric, producing two different medium-sized nuclei. It is also accompanied, as we would expect, by the release of free neutrons. A typical example of such a fission reaction is

$$n + {}^{235}_{92}\mathrm{U} \rightarrow {}^{236}_{92}\mathrm{U}^* \rightarrow {}^{144}_{54}\mathrm{Xe} + {}^{90}_{38}\mathrm{Sr} + 2n. \tag{5.56}$$

Table 5.18. The radiative capture and neutron-induced fission cross-sections of some fissionable (fissile) nuclei (in barns) for thermal and fast neutrons.

	Thermal neutrons (~0.025 eV)		Fast neutrons (>1 MeV)
Nucleus	Radiative Capture, $\sigma_{(n,\gamma)}$	Fission, $\sigma_{(n,f)}$	Fission, $\sigma_{(n,f)}$
U–233	47	530	–
U–235	95	586	1.22
U–238	2.7[70]	negligible	0.70
Pu–239	270	752	1.73

This reaction is shown schematically in Fig. 5.35. The *fission products* in this example, namely, the medium-sized isotopes Xe–144 and Sr–90, represent just one possible outcome. Any pair of isotopes whose combined atomic number equals that of the fissioning nucleus can in principle be produced, though some pairs are more often found than others. The isotopes are initially in an excited nuclear state and are usually radioactive; they generally stabilise through a series of β^- and γ decays.

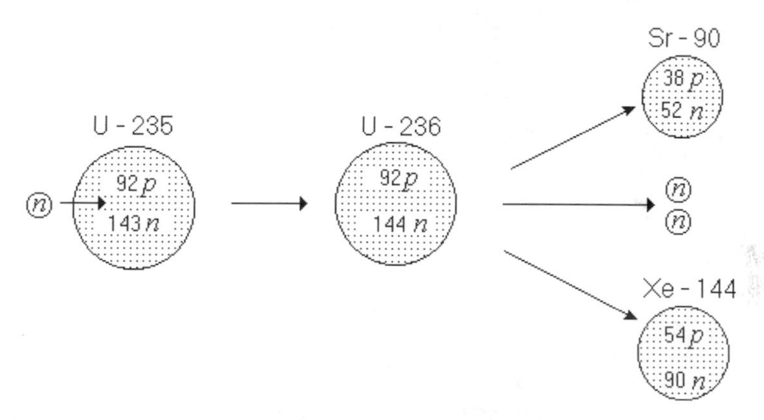

Fig. 5.35. One of the possible modes of the neutron capture induced fission of U–235. The radioactive isotopes Sr–90 and Xe–144 decay by β^- emission that converts their excess neutrons into protons until stability is reached.

Typically, the average binding energy per nucleon in the fission products is about 0.9 MeV greater than that in the original nucleus. Consequently, the energy released from the fission of a nucleus possessing 235 nucleons is about $235 \cdot 0.9 \approx 200$ Mev.[71] This energy is divided between (1) the kinetic energy of the fission products, the isotopes and fast neutrons; (2) the instantaneous

[70]The U–238 isotope has a relatively large cross-section for the radiative capture of 1 to 100 eV neutrons.
[71]The maximum amount of energy released in a chemical reaction is about 5 eV per atom.

radiation released by the process; (3) the radioactivity of the isotopes, as shown in Table 5.19.

Table 5.19. The average disposition of the energy released from the fission of U−235 in MeV.

Form of the Energy	Disposition	Amount
Kinetic energy:	Isotopes	165
	Fast neutrons	5
Instantaneous radiation:	γ-rays	7
Radioactivity of the isotopes:	β decay	23
Total		200

The fission of a U−235 nucleus produces on average 2.5 neutrons. These are fast neutrons, with a mean kinetic energy of about 2 MeV and a mean velocity of $\sim 10^7$ m/s.

5.4.7 Nuclear Chain Reactions

The notion that the considerable energy released by neutron induced nuclear fission could be usefully exploited was first contemplated by Szilard[72] in 1934; he even filed a patent application for the process.[73] Years later, he related that the idea occurred to him as he was waiting to cross the road in London:

> '*As the light changed to green and I crossed the street, it suddenly occurred to me that if we could find an element which is split by neutrons and which would emit two neutrons when it absorbs one neutron, such an element, if assembled in sufficiently large mass, could sustain a nuclear chain reaction.*
> *...In certain circumstances it might be possible to set up a nuclear chain reaction, liberate energy on an industrial scale and construct atomic bombs.*'[74]

Shortly after the discovery of the neutron induced fission of uranium by Hahn and Strassman, the surplus of fast neutrons released in the process was also detected. The element that emits two (or more) neutrons when split by one neutron had been found; it was uranium. Szilard's notion of a nuclear chain reaction[75] had become a reality. The two and a half neutrons released,

[72] Leo Szilard 1898–1964, Hungarian–American physicist.

[73] UK Patent No. 630726. A year earlier, Rutherford had asserted that 'anyone who looked for a source of power in the transformation of the atoms was talking moonshine'.

[74] W. Spencer and G. Weiss-Szilard (eds) (1978) *Leo Szilard: His Version of the Facts*, Cambridge, MA: MIT Press, p. 17

[75] Chemical combustion is a chain reaction. It requires that a molecule be provided with a certain activation energy for its initiation (ignition), but once the first molecules start burning the energy they thereby release provides the activation energy for the propagation of the reaction through the rest of the molecules. The process terminates when all the molecules have reacted.

on average, from the fission of a uranium nucleus could in turn induce the fission of other nuclei, and those released from these other nuclei could go on to induce the fission of still further nuclei. Thus, the chain reaction would be propagated with each extra fission adding its ~200 MeV of liberated energy to the total.

A nuclear chain reaction will only be self-sustaining if, on average, at least one of the neutrons released from each nuclear fission induces the fission of another nucleus. Or, equivalently, if at the beginning of each fission generation, at least the same number of neutrons are present in the system as were there at the start of the previous generation. This provision is expressed mathematically in terms of a parameter, k, called the *multiplication factor.* This is defined as the number of the neutrons released from the fission of each nucleus that, on average, induce the fission of another nucleus.

If $k < 1$, on average, too few neutrons induce further fissions and so the chain reaction stops. This is called the *sub-critical* state.
If $k = 1$, on average, at least one of the neutrons released from the fission of each nucleus induces the fission of another nucleus. The chain reaction will be self-sustaining, generating energy at a steady rate. This is called the *critical* state.
If $k > 1$, on average, more neutrons induce further fissions than are absorbed or lost. The rate at which the process generates energy will grow exponentially, terminating only when the system has exhausted all its fissile resources. This is called the *supercritical* state (Fig. 5.36).

The multiplication factor of an infinitely large mass of fissile material, one from which effectively no neutrons can escape, is designated k_∞. For pure U–235, the multiplication factor $k_\infty \approx 2$; for a mixture comprising 50% U–235 nuclei and 50% inert non-fissile nuclei, $k_\infty \approx 1$, and for a mixture containing less than 50% U–235, $k_\infty < 1$.

The multiplication factor of a finite body of fissile material will always be less than k_∞. How much less depends on the ratio between the number of neutrons that collide with the nuclei in the body of fissile material and the number that escape from its surface. This ratio, in turn, depends on the body's shape and dimensions. In general, the smaller the body the higher the proportion of the neutrons that escape and the lower is the value of k.[76]

[76]Consider a spherical lump of material. Whereas the leakage of neutrons depends on its surface area, $4\pi R^2$, the probability that a neutron will collide with a nucleus depends on its volume, $4\pi R^3/3$. Thus, the ratio between the number of neutrons colliding to the number escaping depends on the radius of the lump. Consequently, the larger the lump the lower the neutron loss.

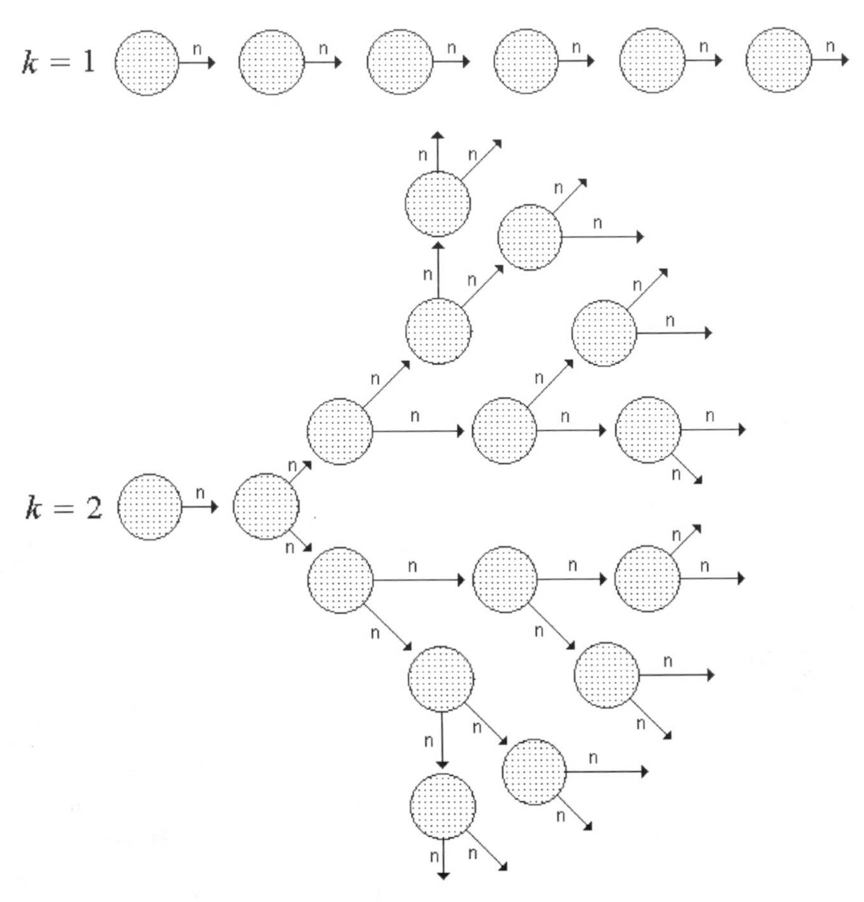

Fig. 5.36. Nuclear fission chain reactions. In the critical state, $k = 1$, just one neutron from each fission induces another neutron. In the supercritical state shown, $k = 2$, two neutrons from each fission induce further fissions.

The mass of a body or formation of fissile material for which $k = 1$ is called its *critical mass.* A body or formation of fissile material larger than the critical mass will be supercritical; one smaller than the critical mass will be sub-critical.

Whether the initial fission in the chain reaction is induced by a fast or slow neutron, all the neutrons released in subsequent fissions are fast neutrons. Thus, unless special arrangements are incorporated to slow the neutrons down, a chain reaction in a supercritical chunk of fissile material will be overwhelmingly a fast neutron reaction.

Once initiated, the chain reaction will propagate rapidly, terminating only when all the fissile resources have been exhausted or after the chunk has blown apart, whichever comes first, at which point the system becomes sub-critical.

Since there are always some free neutrons around to initiate the reaction, for instance from cosmic rays, any supercritical body of fissile material is potentially explosive. In order to prevent such eventualities, fissile material must be kept in small pieces each of whose mass is less than the critical mass.

Given that σ_t, is the cross-section for a particular chain of neutron induced fission reactions, the radius of a spherical critical mass of highly fissile material containing N fissile nuclei per unit volume is approximately equal to the probable distance,

$$d = \frac{1}{N\sigma_t}, \tag{5.57}$$

that a neutron must travel in the material, during a chain reaction, before inducing fission. With a density, $\rho_U = 19,000\,\mathrm{kg/m^3}$, the number of atoms, N, in a unit volume of solid U$-$235 is

$$N = \frac{N_{Avogadro} \cdot \rho_{\mathrm{U}}}{0.235} = \frac{6.0 \cdot 10^{23} \cdot 19,000}{0.235} \approx 4.85 \cdot 10^{28}\mathrm{m}^{-3}. \tag{5.58}$$

The cross-section for a chain of neutron induced fission reactions is equal to twice the fission cross-section: $\sigma_t \approx 2\sigma_{n,f}$.[77] The fast neutron fission cross-section of U$-$235 is $\sigma_{n,f}, = 1.22\mathrm{b}$. Thus, the radius of a critical sphere of pure U–235 is

$$d = \frac{1}{N\sigma_t} = \frac{1}{4.85 \cdot 10^{28} \cdot 2 \cdot 1.22 \cdot 10^{-28}} = 0.085\,\mathrm{m}. \tag{5.59}$$

The mass of a sphere of pure U$-$235 of radius 0.085 m is about 50 kg. This is the critical mass of a sphere of pure U–235; a larger sphere would detonate the moment it was assembled. The critical mass of a sphere of Pu$-$239 is about a quarter as large, ~10 kg.

Naturally occurring uranium, which predominantly comprises the two isotopes, U$-$238 (99.3%) and U$-$235 (0.7%), can be safely kept in ingots of any size; any quantity of it, no matter how large, is sub-critical. This is because of the high threshold energy, ~6 MeV, for neutron induced U$-$238 fission. Of this total, about 1 MeV must be supplied by the neutron's kinetic energy. However, although their average kinetic energy is ~2 MeV, only three-quarters of the fast neutrons released by fissioning uranium nuclei actually have energies greater than 1 MeV and only a quarter of these induce

[77] The neutrons are preferentially emitted from the fissioning nucleus parallel to the direction of the incident neutron. One of these neutrons can be regarded as being the original incident neutron and it has not been significantly scattered. This effectively increases the cross-section for the reaction.

a fission before being slowed to energies below 1 MeV. Thus, the multiplication factor, k_∞, for an infinite mass of pure U–238 has a maximum value of

$$k_\infty = 2.5 \cdot 0.75 \cdot 0.25 = 0.47. \tag{5.60}$$

With a concentration of just 0.7%, the more fissile U–235 isotope cannot make up the difference required for the material to sustain a fission chain reaction and so naturally occurring uranium is safe against fast neutron fission. Only if the U–235 concentration was increased more than tenfold, would an infinite mass of the mixture become critical. Making such 'enriched' uranium requires separating the isotopes in the naturally occurring material. This can be done by mass spectography, or, more efficiently, by first converting the element into the gaseous compound uranium hexafluoride, $^{235}UF_6$ and $^{238}UF_6$, and then segregating the molecules comprising the respective isotopes by membrane diffusion or centrifuging.

In order to make nuclear fission both sustainable and controllable, the neutrons emitted from the fissioning nuclei must first be slowed down. This requires the incorporation of a neutron-slowing substance such as paraffin wax into the body of fissile material. The incorporation of such a *moderator* into the fission process is a key feature of the devices used for the controlled release of nuclear energy known as *atomic piles* or *atomic reactors.*

5.4.8 Fission by Fast Neutrons — Bombs

Within a few months of the discovery of nuclear fission, the governments of all the major world powers had been advised by their scientists of its possible military applications:

> '*Work began first in Germany, where the Reich Ministry of Education convened a secret conference on April 29th, 1939, that led to a research program and a ban on uranium exports. Independently, during the same month, a Japanese army general ordered military applications explored. In the United States, Hungarian emigré physicists Leo Szilard, Edward Teller*[78] *and Eugene Wigner*[79] *communicated their concerns about German developments to President Franklin D. Roosevelt through a letter from Albert Einstein that Roosevelt reviewed on October 11, 1939. The British took a first look in 1939, stalled, and then, catalysed by a memorandum from the physicists Otto Frisch*

[78]Edward Teller 1908–2003, Hungarian–American physicist, the 'father' of American thermonuclear weapons.

[79]Eugene P. Wigner 1902–1995, Hungarian–American physicist who was awarded the Nobel prize in 1963 for his contributions to the theory of the nucleus.

and Rudolph Peierls,[80] *emigrés from Nazi Germany, began in earnest early in 1940. The Soviet physicist Igor Kurchatov*[81] *alerted his government to the possible military significance of nuclear fission in 1939.'*[82]

All of the governments were informed of the possibility of constructing a 'super-bomb' the blast from which would destroy life in a wide area: 'The size of this area is difficult to estimate, but it will probably cover the centre of a big city ... As a weapon, the super-bomb would be practically irresistible'.

Explosive power, i.e. the intensity of the shock (pressure) wave produced by an explosion, depends not only on the amount of energy released but also on the rate at which it is released.[83] In an efficient explosive device, as much energy as possible must be released in as short a period of time as possible.

The heat released by the exploding material raises its temperature and this, in turn, raises the pressure inside the device. It is the sudden release of this pressure as the device bursts that generates the destructive shock-wave of an explosion. The longer it is delayed, the stronger the explosion. For this reason, conventional explosives are often encased in heavy materials – *tampers* – whose inertia confines the blast for a few more microseconds, delaying the release of the pressure. Taking the typical muzzle velocity of a rifle, $\sim 1000\,\text{m/s}$, as indicative of the average velocity with which the fragments separate, the distance between them will be $\sim 1\,\text{cm}$ just $10\,\mu\text{s}$ after detonation.

The detonation of a fission bomb requires the almost instantaneous assembly of a supercritical body of fissile material; the presence of just a single free neutron in this body will then initiate a nuclear fission chain reaction. Once initiated, the chain reaction in the assembly must proceed both fast enough and long enough to produce a powerful explosive blast. Expansion of the fissile material by just a few centimetres will halt the chain reaction; the assembly will no longer be supercritical. Thus, the chain reaction must be completed within one or two microseconds.

At a typical velocity of $10^7\,\text{m/s}$, a fast neutron will cover the probable distance it must travel between fissions in pure U–235, *viz.* 0.085 m, in about

---

[80] Sir Rudolph Peierls 1907–1995, German–British physicist. Ironically, because he and Frisch were technically enemy aliens, they were excluded from the critical work that led to the invention of radar being done at the time in Birmingham.

[81] Igor Kurchatov 1903–1960, Soviet physicist who headed the Soviet Union's nuclear weapons programme.

[82] Robert Serber (1992) *The Los Alamos Primer – The First Lectures on How to Build an Atomic Bomb*, Berkeley: University of California Press, p. xiii. The odyssey of the heroes of modern physics in their search for nuclear fission is recounted in Richard Rhodes (1986) *The Making of the Atomic Bomb*, New York: Simon & Schuster.

[83] More energy is released from the combustion of one kilogram of coal than from the explosion of one kilogram of dynamite.

$0.01\,\mu$s. This, then, is the typical lifetime of a fast fission generation. Slowing down the neutrons requires about a microsecond for each generation, i.e. 100,000 times the lifetime of each fast fission generation. This is far too slow for an explosion to develop. Thus, even though the fission cross-sections of U–235 and Pu–239 for slow neutrons are some 300 times those for fast neutrons, atomic bombs employ fast neutron fission.

A critical mass of fissile material, *viz.* about 50 kg of pure untampered U–235 or 10 kg of pure untampered Pu–239, is the minimum quantity required for a fission bomb. Since the mass begins to disintegrate and to fly apart once the reaction starts, the fission of the entire critical mass is unrealisable. Typically, just one kilogram, ~4 moles, of the fissile material undergoes fission; an efficiency of just 2% for a 50 kg mass of U–235.

Under the conditions prevailing in an explosion, the multiplication factor $k \approx 1.5$. Hence, the number of generations, ν, required for the fission of 4 moles of fissile material is given by the sum:

$$\begin{aligned} 4 \cdot 6 \cdot 10^{23} &= 1 + k + k^2 + \cdots + k^{v-1} \\ &= \frac{(1 - 1.5^{\nu})}{1 - 1.5}, \end{aligned} \tag{5.61}$$

from which $\nu \approx 130$. Given that the lifetime of a fast fission generation is typically just $0.01\,\mu$s, the fission of the one kilogram of U–235 is completed in under $1.5\,\mu$s. Thus, the reaction is all over before the device has had time to expand to any significant degree.

Assuming that the fission of each uranium nucleus releases 165 MeV of mechanical (kinetic) energy, the total blast energy released by the fission of one kilogram (~4 moles) of U–235 will be about $4 \cdot 6 \cdot 10^{23} \cdot 165 \approx 4 \cdot 10^{26}$ MeV. The energy released in the explosion of a ton of conventional chemical explosive, TNT or dynamite, is about $4 \cdot 10^9$ J. Thus, the fast fission of just one kilogram of U–235 is equivalent to the detonation of

$$\frac{4 \cdot 10^{26} \cdot 1.6 \cdot 10^{-13}}{4 \cdot 10^9} \approx 16,000 \text{ tons} \tag{5.62}$$

(16 kilotons) of conventional chemical explosive.

The efficiency of the device can be enhanced by encasing the fissile material in a heavy neutron-reflecting tamper. This has the double effect of slowing the expansion of the fissioning material and reducing the neutron loss from its surface. The tampers are typically made of beryllium that reflects neutrons back into the fissioning mass and natural uranium metal which is one of the densest of metals. Their overall effect is to reduce the critical mass.[84]

Two different methods have been employed to assemble the supercritical body of fissile material required for the efficient detonation of a fission bomb. The first is the rapid fusion of several sub-critical masses of fissile material into a single supercritical mass; the second is the rapid compression of a single sub-critical mass of fissile material into a supercritical body of denser material. Compressing the fissile material increases its density, which, in turn, reduces the probable distance a neutron must travel in the material before it induces fission. Thus, in the compressed state, the fissile material has a smaller critical mass; doubling the density of the fissile material results in a four-fold reduction in the critical mass.

The first method was employed in the bomb nicknamed 'Little Boy' that destroyed Hiroshima on 6 August, 1945. In this device, a sub-critical bullet of almost pure U–235 was fired into sub-critical target forming a tampered supercritical assembly. In order to ensure the efficient detonation of the bomb the moment the supercritical assembly was created, an initiating neutron source, a small amount of an α emitter such as polonium coated with a light element like lithium or beryllium, was incorporated into the bomb (Fig. 5.37).

The assembly of a supercritical mass of Pu–239 from a number of subcritical pieces will not work; plutonium is so reactive that the device will predetonate, i.e. the fission reaction will start too soon, 'melting the bullet and target before the parts had time to join to produce a high-yield explosion'. Thus, the second method of detonating a fission bomb, compressing the fissile material to increase its density, must be used in plutonium based bombs. The bomb, nicknamed 'Fat Man', that devastated Nagasaki on 9 August, 1945 was a plutonium bomb.

Although a solid 5 kg tampered sphere of Pu–239 would explode immediately on assembly, a hollow thick-walled spherical shell of the same mass is safely sub-critical. The compression and inward collapse of the hollow shell would raise its density and at a sufficiently high density it would become

[84] Published values for the critical mass of a sphere of weapons-grade (enriched) materials are:

Mode	U–235	Pu–239
Untampered	56 kg	11 kg
Thick uranium tamper	15 kg	5 kg

Source: J.K. King (ed.) (1979) *International Political Effects of the Spread of Nuclear Weapons*, U.S. Government Printing Office.

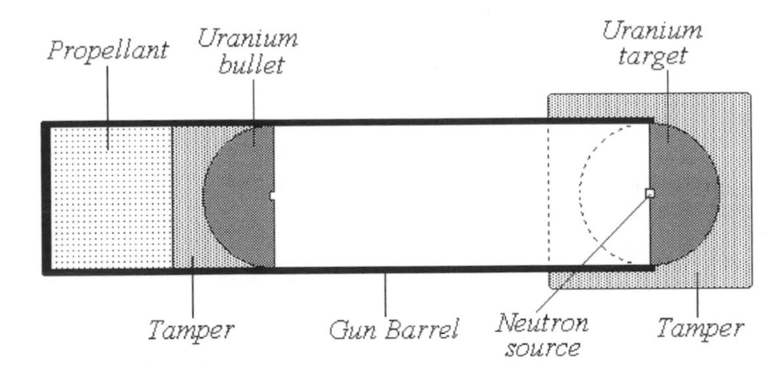

Fig. 5.37. Schematic diagram of the U−235 bomb 'Little Boy'. Igniting the propellant fires the sub-critical bullet at the sub-critical target. Together they form a supercritical assembly. The neutron source ensures the bomb's efficient detonation.

supercritical. In bombs of the 'Fat Man' type, a plutonium shell is enveloped in an intricate assembly of different types of conventional explosives.[85] The timed detonation of these explosives produces an implosive shock-wave that is focused on the centre of the shell and crushes the plutonium into a dense supercritical mass (Fig. 5.38).

The plutonium isotope, Pu−239, used in the 'Fat Man' bomb was manufactured by irradiating natural uranium with neutrons. The radiative capture reaction, $^{238}U(n,\gamma)^{239}U$, between the uranium's preponderant isotope, U−238, and the neutrons is followed by two β^- decays. These raise the atomic number by two, thus producing the plutonium – Eq. (5.55). The plutonium was extracted chemically from the irradiated material. The isotope Pu−239 that accumulates in the irradiated material during its exposure also reacts with the neutrons. This radiative capture reaction, $^{239}Pu(n,\gamma)^{240}Pu$, produces the isotope Pu−240. The longer the material is irradiated, the more Pu−240 accumulates in it.

Although the isotope Pu−240 does not undergo neutron induced fission, it has a small but not insignificant probability of undergoing spontaneous fission, with the release of a flood of neutrons. Thus, the presence of this isotope in the plutonium used in a bomb could cause it to go off too soon. Imploding at a speed of $\sim 1000\,\text{m/s}$, the plutonium shell takes about $4\,\mu\text{s}$ to achieve the desired density. The release of a neutron during this time from the spontaneous fission of a Pu−240 nucleus would cause the core to pre-detonate and 'fizzle' instead

[85]The assembly is often referred to as an *explosive lens*. It was the brain-child of John von Neumann 1903–1957, a brilliant Hungarian–American mathematician who first demonstrated the mathematical equivalence of Schrödinger's wave mechanics and Heisenberg's matrix mechanics, laid the foundations of game theory and developed the construction of large computers.

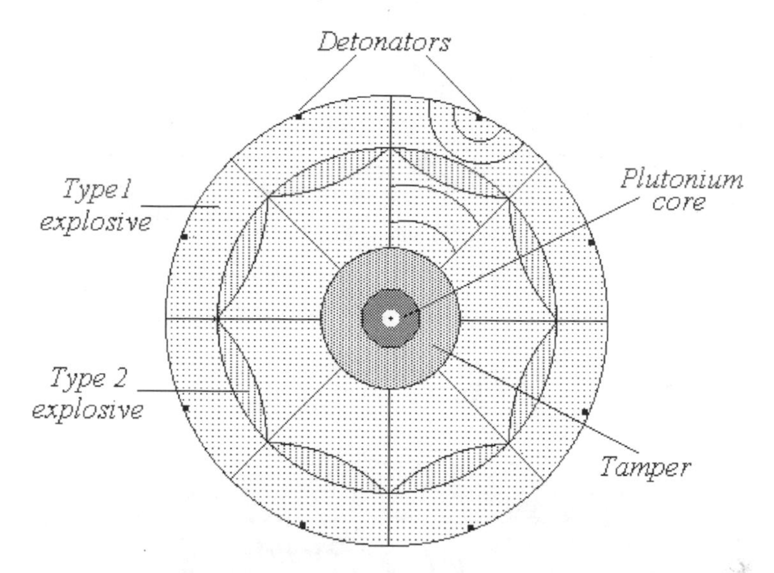

Fig. 5.38. The convex shock-wave generated by the detonation of the outer layer of the fast burning type 1 explosive is focused by the lenses of the slower burning type 2 explosive. The intensity of the concave wave that emerges from the lenses is reinforced by the inner layer of type 1 explosive. It is smoothed out by the heavy tamper and then crushes the plutonium core into a supercritical state.

of exploding. Thus, the plutonium extracted from uranium that has been irradiated by neutrons for more than a few weeks is generally too contaminated with Pu−240 to be considered weapons-grade material.

For a few seconds after detonation, the temperature of the fireball produced from a nuclear explosion exceeds that of the surface of the sun; any object exposed to the fireball receives an almost instantaneous intense dose of thermal radiation. A 20 kiloton explosion delivers enough thermal radiation to ignite combustible material at a distance of 2,000 m. About one third of the fatalities at Hiroshima and Nagasaki were due to flash burns. Since the bombs were detonated at an altitude of about 2,000 feet, blast damage was maximised. However, although radioactive material with an activity of $\sim 10^6$Ci was generated in the blast, a month afterwards virtually no trace of radioactive fall-out was found in Hiroshima or Nagasaki. It appears, that after the initial burst of nuclear radiation, an air-blast leaves little radioactive debris in the vicinity of the explosion. Both cities were subsequently rebuilt on the same sites.[86]

[86]The site of the first nuclear test explosion in New Mexico (code named *Trinity*) is now a picnic area. The need for the explosive testing of nuclear devices has declined of late with the introduction of super-computer simulations.

More modern nuclear weapons employ both fission and fusion. In a typical example, the instantaneous burst of γ-rays released by the detonation of a primary fission device initiates more powerful fission and fusion reactions in a secondary nuclear assembly. Because the γ-rays travel at the speed of light, they arrive at the secondary assembly well ahead of the shock-wave generated by the primary explosion; this provides time for the secondary assembly to react before being blown apart. The secondary assembly comprises layers of thermonuclear material, lithium–6 deuteride ($^6\text{Li}^2\text{H}$), and fissile material, Pu–239, that are encased in a U–238 tamper and surrounded by configured layers of dense plastic foam (Fig. 5.39).

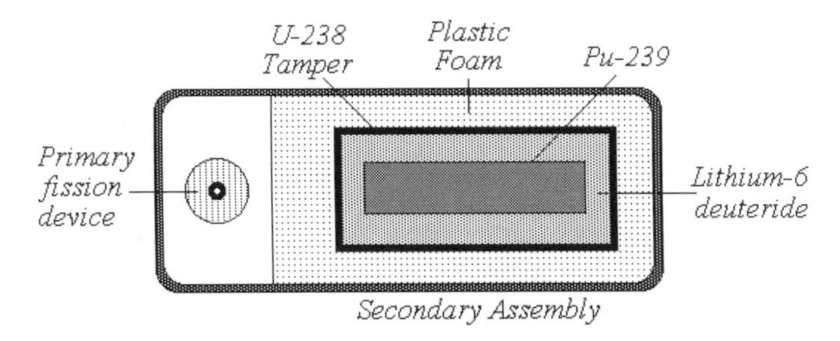

Fig. 5.39. Schematic diagram of a fission-fusion-fission nuclear bomb. The γ-rays released from the primary fission device vaporise the plastic foam. This compresses the core of fissile and thermonuclear materials initiating both fusion and fission reactions. These rapid reactions are sustained by the heat and neutrons they release until the bomb blows itself apart.

The first reaction that occurs in the secondary assembly is the absorption of the γ-ray energy from the primary explosion by the shaped pieces of foam. This causes their instantaneous vaporisation and conversion into a rapidly expanding plasma. The pressure-wave generated by the plasma compresses and implodes the fissile and thermonuclear materials. The now supercritical layers of fissile Pu–239 react, releasing a flood of neutrons and still more energy. Neutrons escaping into the thermonuclear material from the chain reaction in the Pu–239 react with the Li–6, releasing tritium, ^{3}H, and helium, ^{4}He, nuclei and yet more energy:

$$^6_3\text{Li} + n \rightarrow {}^4_2\text{He} + {}^3_1\text{H} \quad Q = 4.9\,\text{MeV}. \tag{5.63}$$

At the tremendous temperatures and pressures by now prevailing in the secondary assembly, the tritium produced from this reaction fuses with the deuteron in the thermonuclear material to produce still more neutrons and even more energy:

$$ {}^{3}_{1}\mathrm{Li} + {}^{2}_{1}\mathrm{H} \rightarrow {}^{4}_{2}\mathrm{He} + n \quad Q = 17.6\,\mathrm{MeV} \tag{5.64} $$

Any surplus neutrons remaining in the device at this stage will possess energies in excess of the threshold energy required to induce the fission of the U−238 in the tamper and so this too will start to react, adding its contribution to the aggregation of energy and neutrons.

All these reactions take place in parallel and at great speed until the bomb finally bursts. The total amount of energy released depends on the amount of thermonuclear and fissile material packed into the secondary assembly. Thermonuclear weapons have TNT equivalencies between 100,000 tons (100 kilotons) and 10,000,000 (ten megatons) depending on the target for which they are intended.

At a distance r from the point of detonation, the maximum value, p_{max}, of the pressure in the shock-wave from a powerful explosion of energy E is approximately given by[87]

$$ p_{\mathrm{max}} = \frac{E}{r^3}. \tag{5.65} $$

Given that the destructive radius of a bomb containing 0.25 tons of TNT is about 50 m, that of a one megaton bomb is about $\sqrt[3]{4 \cdot 1{,}000{,}000} \cdot 50 \approx 8000\,\mathrm{m}$. This is wide enough to encompass even the largest cities in the world.

It is a sobering thought that at the time of writing, some 25 years after the end of the Cold War, there are still over 10,000 nuclear weapons of various types and capacities in the arsenals of the world's armies ready for use at a moment's notice.[88]

5.4.9 Fission by Slow Neutrons — Nuclear Reactors

The first atomic reactor was constructed by a team led by Fermi in 1942. Its successful operation established two important points:

1. Nuclear fission could be controlled and so could be used as a useful energy source;

[87] The maximum pressure varies as $1/r^3$ rather than at the usual $1/r^2$ because of the particular form of the wave that accompanies a very strong compression.

[88] Placing other so-called 'tools of mass destruction' such as chemical and biological weapons under the same heading as nuclear weapons exaggerates the threat of the former and masks that of the latter. Chemical and biological weapons may be cheaper and easier to make but whereas defence against them is possible, there is no way of surviving a direct nuclear strike.

2. The fissile artificial isotope Pu–239, previously prepared only in microgram quantities by irradiating uranium with the relatively weak neutron flux produced from a cyclotron, could now be manufactured in industrial quantities.

Fermi's reactor was fuelled with natural uranium. As the following reasoning shows, although natural uranium, whose composition is 99.3% U–238 and 0.7% U–235, is not susceptible to fast fission, because of the very large slow neutron fission cross-section of U–235, a sufficiently large mass of natural uranium can sustain a slow (thermal) neutron fission chain reaction.

Whether the initial fission occurring in the fissile material is induced by a fast or slow neutron, all the neutrons released in subsequent fissions will be fast neutrons. Thus, sustaining a chain reaction in natural uranium requires slowing down – *thermalising* – the neutrons emitted from each fissioning uranium nucleus before they collide with another nucleus. This is achieved by embedding the natural uranium in a 'moderating' material like paraffin wax that absorbs the kinetic energy of the neutrons.

The choice of 'moderator' involves striking a balance between two factors: the material's efficiency in slowing neutrons versus its nuclear cross-section for neutron capture. Whereas the former should be as high as possible, the latter should be as small as possible. The presence in the assembly of a material that readily captures neutrons is tantamount to 'poisoning' the chain reaction; neutrons that should be available for propagating the chain reaction would instead be captured by the material. In addition, the whole assembly must be made very large in order to prevent the loss of neutrons as a result of their leakage from its surface.

In elastic collisions between neutrons and stationary nuclei of mass number A, the energy of the incident neutrons is reduced on average by about $A/(1+A)$ in each collision. Thus, from an initial energy E_0, the energy, E, of a neutron will have been reduced to about

$$E = E_0\left(\frac{A}{(1+A)}\right)^n \tag{5.66}$$

after n collisions. Clearly, the smaller the mass number of the stationary nucleus, the less collisions will be required to thermalise fast neutrons, i.e. to reduce their average energy from 2 MeV to 0.025 eV. This suggests that the best moderating material to use would be hydrogen nuclei, for which $A = 1$. According to Eq. (5.66) just 23 collisions between the neutrons and the hydrogen nuclei would be required to achieve the required reduction in neutron energy. However, hydrogen nuclei have an appreciable cross-section,

$\sigma = 0.31$b, for absorbing slow neutrons. Thus, the presence of hydrogen in the moderating material would cause a significant loss in the number of neutrons available for fission.

The second best choice is deuteron, ^{2}H. Thermalising the 2 MeV neutrons would require more collisions between the neutrons and the deuterons than with hydrogen. However, deuteron's cross-section for slow neutron capture is only 0.00065b. The common isotope of oxygen, ^{16}O, also has a negligible cross-section for neutron capture. Consequently, heavy water, $^2H_2^{16}O$, is a particularly good neutron moderator. Whereas, on average, a slow neutron will be absorbed in ordinary water after just 150 collisions, in heavy water it will only be absorbed after 10,000 collisions. Ordinary water can only be used as a moderator in reactors fuelled by enriched uranium, i.e. uranium in which the concentration of the U–235 is 2.5% or more.[89]

Since, at the time, heavy water was unavailable in sufficient quantities, Fermi had to use the next easily available small nucleus with a negligible cross-section for neutron capture; this was the carbon–12 nucleus. Over 100 collisions would be needed to thermalise each neutron but few would be lost as they could collide on average 1,000 times before being absorbed by a ^{12}C nucleus.[90] In the reactor Fermi constructed, alternate layers of solid blocks of very pure graphite and of blocks whose cores had been drilled out and loaded with natural uranium were piled up on one another; the assembly was called an *atomic pile*. Calculations showed that a sufficiently large pile would be supercritical.

In order to prevent the pile from becoming critical, or even supercritical, before its assembly was completed, removable cadmium rods were incorporated into its design. Natural cadmium has a remarkably high cross-section, $\sigma \approx 2000$b, for the radiative capture of thermal neutrons.[91] Thus, so long as the cadmium rods remained inside the pile, it could never reach criticality, whatever its size. Any neutrons released inside the pile would be absorbed by the cadmium before they could induce fission. Removing the cadmium rods from an inherently supercritical reactor would start off the fission chain reaction at once (Fig. 5.40).

Controlling or halting the chain reaction in the pile after it had become supercritical appeared, at first, to present an insurmountable problem. Since the

[89] Weapons-grade uranium is much more enriched than this, up to 90%.

[90] In Nazi Germany, Heisenberg was leading the attempt to construct a heavy water reactor. The destruction, by British commandos, of the Norsk Hydro plant that produced the heavy water effectively put an end to this programme.

[91] See Example 5.7.

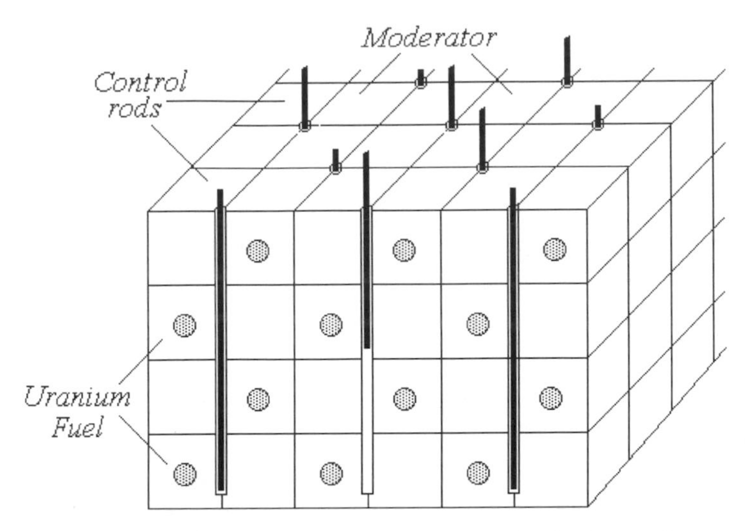

Fig. 5.40. Diagram of a typical reactor core. Horizontal rods of natural or enriched uranium are packed into the moderating material. The flux of neutrons in the reactor and, hence, the rate of the fission reaction is controlled by raising or lowering the vertical neutron absorbing control rods.

pieces of uranium are some 10–20 cm apart in the reactor and the mean velocity of thermal neutrons is $\sim$1000 m/s, the average time spent by the neutrons in travelling between fissions is about 100 μs. Thus, 1,000 generations can pass in just one-tenth of a second. Suppose that a reactor in a critical state, $k = 1$, becomes just barely supercritical, $k = 1.001$. Whereas in the critical state, the rate at which the reactor produces energy is constant, in the supercritical state its energy output grows exponentially. Every tenth of a second, its energy output grows by $1.001^{1000} = 2.7$, and after just one second it will be almost 22,000 times greater than it was in the critical state. No mechanical control can react as quickly as the nuclear chain reaction propagates; the cadmium rods could not be re-inserted into the reactor in time. Thus, once it becomes supercritical, it would appear that a reactor will rapidly and unavoidably run out of control and 'melt down'.

Fortuitously, there is a factor that makes reactor control possible; it is the delay in the release of a small number of the neutrons. On average, 0.6% of the neutrons are released only after a delay that can be as long as 30 seconds after the actual fission. These 'delayed' neutrons are released by the unstable fission fragments, the radioactive medium-sized nuclei into which the fissile nucleus splits. In the estimate we made of the average time interval between fissions, we assumed, wrongly, that all the neutrons are released promptly; we only took into account the movement of the slowed neutrons from one piece of uranium in the reactor to another. Allowing for the delayed neutrons, the average time interval between fissions in a reactor is about one second, 1,000 times longer than we had estimated. Consequently, changes in a reactor's power output take place only very slowly; starting up a reactor from scratch can take hours.

Although their release is delayed, these neutrons still contribute to the overall neutron flux in the reactor; without them it would not become critical. Consequently, so long as a reactor is inherently sub-critical without them, i.e. $k_{\text{reactor}} < 1.006$, its power production is controllable. This sets an upper-limit on the rate at which nuclear reactors may be safely operated.

The reactor assembled by Fermi produced energy at a rate of just 200 W; modern commercial reactors used for the production of electricity typically operate at power ratings of thousands of megawatts. Nuclear reactors that employ slow neutron fission are called *thermal reactors.* The simplest form of commercial thermal reactor is the *pressurised-water reactor.* These reactors are fuelled by slightly enriched uranium, 3% U–235; the water serves both as the coolant and moderator. The reactor core is situated inside a pressurised vessel, which is itself enclosed inside a strong containment building. Superheated water, at a pressure of about 150 atmospheres, enters the pressurised vessel at a temperature of about 280 °C, heats up further as it flows through the reactor core and leaves at a temperature of about 310 °C. From here it passes into a heat exchanger, where it heats the water in a secondary circuit producing steam. This, in turn, spins turbines which drive electric generators (Fig. 5.41). A typical commercial reactor operates at an overall power of 3,000 MW and generates electricity at a rate of 900 MW.

The slow-neutron fission of its U–235 is not the only nuclear reaction that takes place in the reactor. Surplus neutrons released from the fission of the U–235 nuclei react with the more plentiful U–238 nuclei also present in the fuel elements, resulting in its conversion into Pu–239 by the series of reactions listed in Eq. (5.53). The artificial transuranic isotope Pu–239 is even more susceptible to slow-neutron fission than is U–235; its nuclear cross-section for slow-neutron fission is 752b. Thus, as the percentage of U–235 in the fuel elements depletes, its role in the reactor is assumed by the fissile Pu–239. After a few months more energy is being produced by the fission of the plutonium than the uranium. Thus, a freshly fuelled reactor can operate continuously for a year or more before having to be shut down for maintenance and re-fuelling.

Reactors that manufacture more fissile material than they consume are called *breeder reactors.* By using highly enriched uranium or plutonium as the

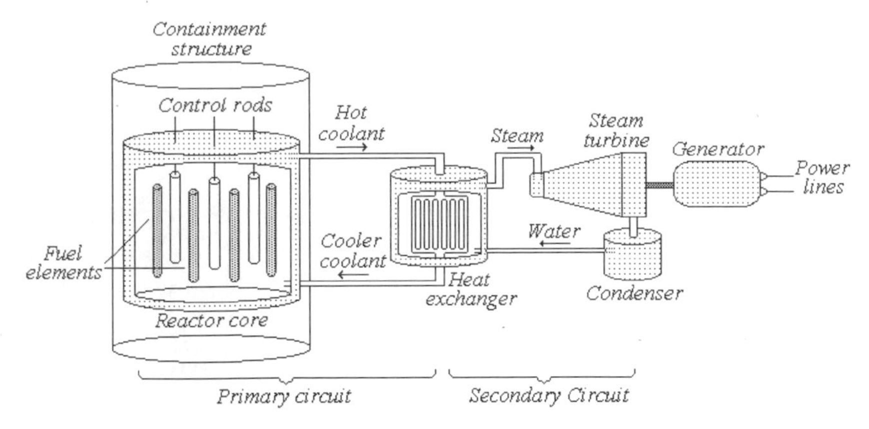

Fig. 5.41. Schematic diagram of a typical nuclear power station. If the reactor is of the pressurised-water type, the fuel is enriched uranium and the coolant is superheated water; the water also acts as the moderator. In other reactor designs, natural uranium is the fuel, graphite is used as the moderator and the coolant may be steam, helium or carbon dioxide gas or liquid sodium.

fuel, reactors can be constructed that dispense with the need for a moderator. These reactors work by fast neutron fission and are called *fast reactors* to distinguish them from the *thermal reactors* that operate with slow (thermal) neutrons. In other reactors called *fast-breeders*, the core of enriched nuclear fuel is enveloped in a blanket of U−238; under suitable conditions, radiative capture converts the U−238 to Pu−239, producing more of it than was required to enrich the fuel in the core.

Although the fissile isotope Pu−239 is a by-product of every uranium-fuelled reactor, unless it is removed from the reactor within days or weeks of its manufacture it is usually so contaminated with Pu−240 that it is unusable for bomb production. In almost all reactors designed primarily for electricity production, the removal of the plutonium requires closing down the reactor. Since, for reasons of cost and service, this is done only at the longest possible intervals, by which time there is considerable Pu−240 contamination, the plutonium produced by such reactors is not usually of weapons grade. Nevertheless, in their attempts to control the proliferation of fissile material and nuclear weapons, the nation states possessing the technology required for the construction of commercial reactors demand the return of the spent fuel elements containing the plutonium as a condition for supplying the fresh fuel elements without which the reactor will no longer work.

The isotope Pu−240 is not the only non-fissile substance that accumulates in a reactor's fuel elements during its operation; radioactive fission fragments, unstable medium-sized nuclei, also accumulate in its fuel elements. One of these nuclei is the iodine isotope I−135 which decays via Xe−135 into Cs−135:

$$^{135}_{53}\mathrm{I} \rightarrow {}^{135}_{54}\mathrm{Xe} + \beta^- + \bar{\nu} \quad T_{\frac{1}{2}} = 6.7\,\mathrm{h}$$

$$^{135}_{54}\mathrm{Xe} \rightarrow {}^{135}_{55}\mathrm{Cs} + \beta^- + \bar{\nu} \quad T_{\frac{1}{2}} = 9.2\,\mathrm{h}. \tag{5.67}$$

The isotope Xe−135 has an enormous radiative capture nuclear cross-section, $\sigma_{(n,\gamma)} = 2.6\cdot 10^6$b. Its production in the reactor slows the chain reaction; its accumulation in the reactor would 'poison' it completely. Because of its enormous appetite for neutrons, no Xe−135 will accumulate in a reactor as long as it is in operation and there is a substantial neutron flux through it; any Xe−135 produced from the decay of the I−135 will be immediately swept up and removed by the radiative capture reaction, $^{135}\mathrm{Xe}(n,\gamma)^{136}\mathrm{Xe}$. However, if the reactor is shut down, the Xe−135 does start to accumulate; there are no longer neutrons around with which it can react. The Xe−135 reaches its maximum concentration in the reactor about 10 hours after the shut down; it then starts to fall as the stock of I−135 declines.[92] Thus, after shutting down or drastically reducing the operation of a reactor, it cannot be immediately restarted. It must remain idle until the accumulated Xe−135 has decayed into Cs−135.

Example 5.11: The Slow Neutron Chain Reaction of Natural Uranium

By calculating the multiplication factor for an infinite mass of natural uranium show that a sufficiently large mass of the material can sustain a slow neutron chain reaction.

Calculation: Table 5.20 shows the weighted macroscopic cross-sections, $\mu_{(n,f)}$, and $\mu_{(n,\gamma)}$ for slow neutron fission and slow neutron radiative capture reactions of the principal isotopes, U−238 (99.3%) and U−235 (0.7%), present in natural uranium. These values are based on those given in Table 5.18, for the respective nuclear cross-sections.

Table 5.20. The weighted macroscopic cross-sections for slow neutron fission and slow neutron radiative capture of the principal isotopes in natural uranium. N is the number of uranium nuclei per unit volume of the material.

	Weighted macroscopic fission cross-section – $\mu_{(n,f)}$	Weighted macroscopic radiative capture cross-section – $\mu_{(n,\gamma)}$
U−235	$0.7N\cdot 586 = 410.2N$	$0.7N\cdot 95 = 66.5N$
U−238	negligible	$99.3N\cdot 2.7 = 268.1N$

Slow neutrons captured by U−238 nuclei will initiate the series of reactions – Eq. (5.33) – that produce the artificial transuranic isotope Pu−239. Of those captured

---

[92] See Appendix 5.5.2.

by the U−235 nuclei, approximately 85% will induce their fission; the rest undergo radiative capture. The ratio, η, between the number of slow neutrons that induce fission in natural uranium and the number undergoing radiative capture is

$$\eta = \frac{0.7N \cdot 586}{99.3N \cdot 2.7 + 0.7N \cdot 95} = 1.22. \tag{5.68}$$

Consequently, the proportion, $\frac{\eta}{1+\eta}$, of the slow neutrons absorbed by natural uranium that induce fission is $\frac{1.22}{1+1.22} = 0.55$. Thus, natural uranium behaves towards slow neutron fission as though it was a mixture of 55% U−235 and 45% U−238. The multiplication factor, k_∞, for an infinite mass of such a mixture is

$$k_\infty = 0.55 \cdot 2 = 1.1. \tag{5.69}$$

Hence, a sufficiently large mass of natural uranium will be supercritical for slow neutron fission.

5.4.10 Nuclear Engineering: The Chernobyl and Fukushima Catastrophes

The experience of the past generation has shown that well-designed commercial reactors, in the hands of competent operators, are safe. Over 1,200 atomic reactors of various types are presently in operation throughout the world. About one third are installed in ships or submarines providing the energy for their locomotion; another third are used to generate electricity; the rest are used for research and the production of plutonium. Over 15% of the world's electricity is currently produced by nuclear reactors; in certain countries this figure exceeds 50%.

The occupational and health hazards faced by workers in the nuclear generating industry are no greater than those met in other primary energy sectors, for instance, by underground coal miners. Moreover, unlike the fossil fuel industry, nuclear power produces no air pollution, acid-rain, oil-slicks, open-cast coal mines, leaking oil pipe-lines, rusty drilling rigs or abandoned refineries. On the other hand, disposing of the radioactive waste that is unavoidably produced during the operation of nuclear reactors and decommissioning the reactors after the end of their useful lives, presents great problems.[93] These are probably technically solvable but the cost appears to be prohibitive. The proliferation of the potential weapons-grade fissile material produced in the reactors is another factor predisposing the wisdom of their use.

[93] During the Cold War period, insufficient care was taken in the handling and disposal of the radioactive waste produced in the manufacture of nuclear weapons. The contamination resulting from this ineptitude will take years and billions of dollars to clean up.

Since the 1980s, world-wide construction of new reactors for power generation has virtually ceased. The reasons for this are many, doubts about the true cost of the electricity they produce and their safety being chief amongst them. Notwithstanding these reservations, faced with (1) a growing world population and the demand for improved living standards; (2) the inevitable future exhaustion of fossil fuels; (3) the limits placed on the supply of agricultural energy by the availability of cultivable land and fresh water; and (4) the practical and aesthetic restrictions on the wide-scale development of solar power and other renewables, sometime in the near future there will probably be no choice but to return to nuclear processes for the required extra energy production. The alternatives may be either the perpetuation of the present inequalities in world living standards or a degree of population control and social engineering that would undermine basic human freedoms.

The greatest threat to the safe operation of any power station is a sudden or undetected breakdown in the mechanism that extracts the heat from the reactor or furnace. In a typical power station, nuclear or conventional, heat is being continuously generated at a rate of ~3,000 MW. This is sufficient heat to convert more than 1,000 kg of water into steam each second. Under normal operating conditions this heat is extracted from the reactor or furnace by the coolant and transferred to the turbine-generators. However, if the heat generated during just a few seconds was to accumulate in the reactor or furnace, it would either melt or explode. Thus, an undetected or sudden breakdown in the system which circulates the coolant round the reactor or burners, or, alternatively, a loss by leakage or evaporation of the coolant itself, is the most dangerous event that can occur in a power station.[94]

In reactors fuelled by enriched uranium, such as the so-called *light water reactors* (*LWR*) of which the pressurised water reactor is an example, the coolant water is also the moderator. Consequently, such reactors are inherently safe against a loss of coolant. No action is required to close them down if coolant is lost. This is because losing the coolant means losing the moderator, and without the moderator to slow the neutrons the only fission possible is fast-neutron fission, whose probability (nuclear cross-section) is some 500 times less than that of slow neutron fission. Consequently, if it loses its coolant, a LWR closes

[94] Similar to today's concerns about the safety of nuclear power plants, were the questions raised in the late 1800s concerning the safety of tubular steam boilers, many of which had exploded causing numerous casualties and considerable loss of life. In most instances, poorly riveted boiler sections were blasted hundreds of feet into the air, though the dials on the smashed safety valves recovered from the debris indicated that their respective pressures were 'well within safe limits' at the moment they exploded. It seemed just a matter of time before Jules Verne's prophecy that the world would come to a violent end as a result of a huge boiler explosion would be realised.

down of its own accord. The reactor immediately becomes sub-critical and the chain reaction automatically ceases.

In the accident in the LWR at the Three Mile Island installation in 1979, a loss of coolant caused the almost instantaneous close-down of the nuclear chain-reaction. Thus, there was never any danger of a run-away nuclear fission reaction. However, a small proportion of the heat generated by a nuclear reactor (1 or 2%) comes from the radioactive disintegration of the relatively long-lived fission fragments, the unstable medium-sized nuclei, that accumulate in its fuel elements. These heat generating processes continue even after the reactor has closed down the fission chain reaction. Unless measures are taken to cool the reactor, even after the chain reaction has ceased, there is a danger that the assemblage will gradually heat up and ultimately even melt. However, the probability of the melt self-organising into a nuclear bomb is essentially zero. When the cooling system in the reactor at Three Mile Island broke down, the containment structure enclosing the reactor prevented the spread of any significant amounts of radiation to the surroundings.

The reactor that exploded catastrophically at Chernobyl in April 1986 was designed for the concurrent production of electricity and weapons-grade plutonium. It was constructed in such a way that its fuel elements could be removed and replaced after as short an interval as was required without closing down the reactor. In order to achieve this capability, it was necessary to dispense with the containment structure. Like all the early reactors designed to produce weapons-grade plutonium, the moderator used in the Chernobyl reactor was pure graphite. The coolant was water, which boiled as it flowed through pipes in the reactor. The steam thereby produced drove the turbine-generators.

Unlike a light water reactor, a Chernobyl-type reactor does not automatically close down in the event of a loss of coolant. Just the opposite. Hydrogen nuclei have an appreciable nuclear cross-section, $\sigma = 0.31\text{b}$, for absorbing slow neutrons. Consequently, the water coolant, H_2O, absorbs some of the neutrons in the reactor; water is a mild 'reactor poison'. Hence, a loss of the water coolant results in an increase in the net neutron flux inside the reactor core and so enhances the chain reaction. Thus, there is an imminent danger of an explosion or melt down in the event of a sudden or undetected loss of the water coolant.

Its designers knew that reactors of the Chernobyl type are inherently unsafe against a loss of coolant. Because of this, they incorporated an array of alarm systems as well as several automatic secondary (back-up) cooling and pumping systems into the plant. Furthermore, in the event of the failure of these systems, there was a mechanism by which neutron absorbing cadmium rods would be automatically released into the reactor, shutting it down. In the search for still greater safety, emergency diesel generators had been installed in order to keep the coolant pumps working in the event of a loss of main power. It had been further suggested that, in an extreme emergency, the angular kinetic energy of the power station's own rotating turbines could be used to generate the electricity needed to keep the coolant pumps working for the few minutes until the emergency generators came on line. The catastrophe occurred during an experiment that was being performed to investigate this possibility.[95]

It was originally intended to carry out the experiment on the spring day the reactor was to be shut down for its annual maintenance. The experiment was to have been conducted with the reactor operating at about 30% (700–1,000 MW) of its full capacity (~3,000 MW), a level at which it was still controllable. However, because of a sudden bout of cold weather, at the last moment, the shut down was put off for a few days. It was reauthorised when the weather improved, but because of the pressure of time, the reduction from full capacity was then carried out too quickly. As a result, a quantity of 'poisonous' Xe–135 accumulated in the reactor. At this stage the reactor was operating at just 1% (~30 MW) of its capacity. Instead of immediately aborting the experiment, the operators attempted to 'rekindle' the reactor by withdrawing some of its cadmium control rods. After about two hours, they had managed to coax it back up to about 7% (200 MW) of its maximum capacity. Because of the Xe–135 that had accumulated in the reactor, this was the highest they could achieve.

At this point the experiment was started. All the alarm systems and the many automatic safety devices were first neutralised. Leaving them on would have closed down the reactor automatically the moment the main power supply was cut off. In accordance with the original plan, the flow of water coolant into the reactor was then increased. Since water is a reactor poison, this had the effect of slowing the chain reaction. To compensate for this and in an attempt to maintain power, most of the remaining cadmium control rods were now withdrawn. This left the neutron flux dependent on the liquid water coolant in the reactor alone. Any loss of coolant would release so many more neutrons for the chain reaction, that the reactor's multiplication factor would almost instantly rise above the permitted level, putting it beyond control.

[95] For a full account of the events at Chernobyl, including a detailed explanation of the physics involved, the reader is referred to Bernard L. Cohen (1987) 'The Nuclear Reactor Accident at Chernobyl, USSR', *Am. J. Phys.*, **55**, 1076.

At 01:22 local time on April 26, the flow of coolant into the reactor was reduced but no control rods were reinserted. Ignoring the alerts to shut down the reactor at once that were now flashing on their computer terminals, the operators continued with the experiment. At 01:23:04, the supply of steam was cut off from the turbines, leaving them to rotate freely. It was in this state that their kinetic energy was to have been utilised to operate the coolant pumps. But by now, the chain reaction in the reactor was going out of control. Reducing the coolant flow without reinserting control rods was catastrophic. With less 'poison' flowing through the reactor, the chain reaction could speed up. This, in turn, boiled off more of the water, increasing the steam pressure inside the reactor. But with its flow reduced, there was less water to be boiled and to cool the reactor. But less water meant less 'poison' and so the chain reaction could proceed even faster. This vicious circle of cause and effect was driving the reactor out of control.

The control rods were now reinserted, but to no avail; the heat production in the reactor continued to grow. At 01:23:40 the order was given to release the emergency rods into the reactor, but they jammed. The shafts down which they were supposed to drop had been bent out of shape by the heat generated in the reactor.[96] At about 01:24 two explosions were heard and fires broke out throughout the reactor core. It is estimated that at the moment of these explosions, parts of the reactor were operating at a capacity 20 to 100 times its normal maximum rating.

Within two hours, the fires in the 1,500 tons of graphite that had been the reactor's moderator were extinguished. During the following days, some 5,000 tons of boron compounds (to absorb the neutrons and ensure that there was no danger of the mass becoming supercritical), sand, stone and lead were dropped from helicopters onto the smouldering debris. Most of the pilots who carried out these sorties, as well as a number of the firemen, received fatal doses of radiation and have since died. By 6 May, 1986 the level of radiation from the site had been reduced to reasonable levels. Since then, the reactor site, together with the debris from the explosion, has been entombed in a concrete sarcophagus.

Until Friday, 11 March, 2011, no catastrophic incident on a similar scale had been reported at any nuclear installation during the 25 years following the events at Chernobyl. However, on that day at 14:46 local time, Japan was struck by a magnitude 9.0 earthquake – the largest recorded for Japan – followed,

[96]In Chernobyl-type reactors the emergency rods do not fall freely. The chains holding them are wound onto capstans. As the rods fall the chains unwind, rotating the capstans. Thus, the weight of the rods must accelerate the capstans as well as themselves and so they fall with an acceleration of only $g/4$.

about an hour later, by a tsunami of terrifying proportions; some 20,000 people were swept away by the wave which reached a height of tens of metres in places and penetrated kilometres inland.

Japan has few primary energy resources of its own and so, despite being the only country to have suffered the devastating effects of nuclear weapons in wartime, it has embraced nuclear technology to generate a significant proportion of its electricity requirements; in 2011, some 30% of its electricity was supplied by its 50 light water reactors. Six of these reactors, known collectively as the Fukushima I plant, were situated close to the shore on the east coast about 200 kms north of Tokyo. The installation had been designed and built over a long period from 1960 to 1979; all six reactors were of the boiling water type (BWR).

In a BWR, steam is produced inside a pressurized vessel (~75 atmospheres) when the water coolant, which also acts as the moderator, boils as it rises through the fuel elements in the reactor core. The hot vapour that emerges is dried and the resultant pure steam is directed into turbine generators which produce the electricity. On leaving the turbines the steam is condensed by heat exchangers cooled with sea-water and the pure water obtained is recycled to the reactor. On the fateful day, just reactors 1, 2 and 3 were operating at power; units 4, 5 and 6 had been offline for some time due to maintenance work. All the fuel elements from reactor 4 had been offloaded into its adjacent fuel pond where they were kept under water; reactors 5 and 6 contained a full complement of fuel but both were in shutdown mode.[97]

The reactors at Fukushima I were fitted with a shutdown system linked to seismic sensors which automatically inserted the control rods to halt the fission reaction in the event of significant levels of seismic activity. At the onset of the earthquake on 11 March, 2011, shutdown levels of ground motion were encountered and the three working reactors successfully tripped. This was not the first time the plant had been hit by a seismic event. In 1978 it had shut down automatically following a magnitude 7.4 quake and, having suffered minimal damage, was fully operational again in a matter of days. But on that occasion there was no tsunami.

The Fukushima I plant was built on a raised embankment of a height of 10 m above the normal sea level. In addition, it was protected by a sea-wall about 5 m high. However, the tsunami wave on the afternoon of 11 March, 2011 is estimated to have been 14–15 m high at that point along the coast. The incoming

[97]This account of the events at Fukushima 1 is based on the interim report by the UK Chief Inspector of Nuclear Installations: *Japanese Earthquake and Tsunami; Implications for the UK Nuclear Industry*, Office for Nuclear Regulation (May 2011).

wave completely inundated the plant and entered the buildings through ground level access doors. Portions of the reactor and turbine buildings, including the hall which housed the emergency generators, were below ground level and remained flooded for days after.

Off-site power from the grid had been lost almost immediately after the onset of the earthquake and the on-site emergency diesel generators had come online, as they were intended to do in an emergency. This kept the pumps that circulated the coolant to the reactors working. However, all on-site power was lost when less than an hour later the tsunami flooded the entire plant. This resulted in an immediate loss of coolant to the reactors which began over-heating; a melt down was imminent. There was also the matter of the fuel ponds in which the spent fuel elements were held. They too needed cooling to prevent the stored fuel elements from over-heating, breaking up, even catching fire and releasing volatile radioisotopes into the atmosphere.

Without cooling or sufficient water injection, the fuel elements inside the reactors would be exposed; they would over-heat, lose their geometry and release gases previously contained within their zirconium-based cladding. In addition, the exposed zirconium cladding would react with any hot steam inside the containment vessel to produce explosive hydrogen gas. Fire trucks were brought to the site to pump sea-water into the reactors (freshwater was unavailable), but it would appear that the fuel elements were never more than two thirds immersed in water. The reactor vessels were vented during the days immediately following the tsunami to reduce pressure. This released both radioisotopes into the atmosphere and hydrogen gas which evidently collected in upper levels of the reactor buildings and finally exploded, causing extensive damage to the structures.

A secondary, though no less severe problem, was the inability to cool the spent fuel ponds, particularly the one in the building that housed reactor 4 which contained over two full reactor complements of fuel elements. The proximity of the ponds just above the reactors in the Fukushima design, increased the risk of pond degradation and system loss. It is probable that water leaked from the pools in all the units as a result of structural damage suffered in the initial earthquake and from the hydrogen explosions.

The operators struggled to introduce cooling water into the reactors and the fuel ponds, by previously untried and unplanned means, for over a week. Heavily contaminated water collected in open areas at the site and leaked out to sea. The damage to the buildings caused by the explosions provided a direct route for any activity released from the spent fuel to escape into the atmosphere; it is also likely that debris from the structural damage fell into the ponds and may have further damaged the fuel rods. On the other hand, the same structural damage made it possible to spray water directly onto the pools from outside water cannons, helicopters and concrete pumping trucks. Eventually, measures to curtail the various toxic discharges were successful.

The initial 3 km exclusion/evacuation zone declared by the authorities was first extended to 10 km and then to 20 km. Significant levels of the isotopes I–131 ($T_{\frac{1}{2}} \approx 8$ days) and Cs–137 ($T_{\frac{1}{2}} \approx 30$ years) have been detected at sampling points away from the Fukushima 1 site. The relatively high amounts of the short-lived iodine isotope found would suggest that these radioisotopes had been released from the reactors and not the fuel ponds; any I–131 originally present in the stored spent fuel elements would have long decayed. Notwithstanding, the ponds still represent a threat as there is no containment to prevent release.

The country-wide devastation and chaos caused by the earthquake and tsunami made bringing in emergency generators to replace those lost and reconnecting the off-site power particularly difficult; the mobile diesel generators had to be brought by helicopter because of the severe disruption to the road network. Off-site power was only slowly returned to the various plant components over the next three weeks; not until 3 April, 2011 were all the reactor and fuel pond cooling systems back online. Up to 12 April, 2011, 419 aftershocks had been recorded of a magnitude greater than 5, two of magnitude 7.1. Some equipment was moved to high ground in case there was another tsunami.

Whilst it would appear that the Fukushima 1 plant passed the magnitude 9 earthquake with minimal damage, as did all the other nuclear power stations in Japan, it was totally unprepared for the tsunami that followed. It had been determined that the plant needed to be protected against a wave of height 6 m, far less than that actually encountered; it is unclear why this was the case. Over the last 100 years Japan's east coast has suffered several tsunami waves greater than 12 m in height following earthquakes, some even over 20 m.

At the time of writing, the reactors appear to be in a steady state with a constant supply of fresh water being injected into their cores. However, the *closed loop* cooling systems have not been fully restored and the reactors have not reached the steady state usually described as 'cold shutdown.' The status of the containment structures in units 1, 2 and 3 is still not entirely clear; they appear to be largely intact though to have suffered damage in the explosions. The integrity of the fuel ponds is likewise unclear.

What happened at Chernobyl was not an accident. It was the result of poor design and an inflexible management philosophy. Instead of relying on the natural imperatives of physics to ensure the plant's safety, its designers put their trust in planning. This was the underlying philosophy of their political system at the time. But whereas nature, though unforgiving, can never go wrong, planning is always subject to Murphy's Law: 'If it can go wrong, it will!' Within five years of the events at Chernobyl, the citizens of that unfortunate town, together with their fellow country-men, had overturned the system.

Fukushima, on the other hand, was an 'act of God', but one that might well have been foreseen. In retrospect, the decision to erect the plant so close to the shore-line and at such a low elevation relative to the sea level on a coast known to be subject to tsunamis, seems incomprehensible. Nature has an unpredictable streak; a substantial margin of error should always be allowed for when making risk assessments. The planners and designers of the Fukushima 1 installation were at fault in this regard.

Appendices to Part Five

5.5.1 The Mean Lifetime of a Radioactive Nucleus

At a given time, t, all the nuclei in a radioactive sample that have not yet decayed have lifetimes longer than t. The number, dN, of nuclei that will decay on average in the infinitesimal time interval, dt, between t and $(t + dt)$ is

$$dN = N\lambda dt, \tag{5.70}$$

where λ is the decay constant of the radioactive isotope.

In a sample, that at time $t = 0$ contains N_0 radioactive nuclei, the probability, $P(t)dt$, that one of these nuclei will decay during the time interval dt between t and $(t + dt)$ is given by

$$P(t)dt = \frac{|dN|}{N_0} = \frac{N}{N_0} \cdot \lambda dt. \tag{5.71}$$

From Eq. (5.19), $N = N_0 e^{-\lambda t}$, and so

$$P(t)dt = e^{-\lambda t} \cdot \lambda dt. \tag{5.72}$$

The mean value, τ, or equivalently, the expectation value, $\langle t \rangle$, of the lifetime of a nucleus is given by

$$\begin{aligned}\langle t \rangle &= \int_0^\infty tP(t)dt \\ &= \lambda \int_0^\infty te^{-\lambda t}dt \\ &= \lambda \left[-\frac{1}{\lambda} te^{-\lambda t} - \frac{1}{\lambda^2} e^{-\lambda t} \right]_0^\infty = \frac{1}{\lambda}.\end{aligned} \tag{5.73}$$

5.5.2 Radioactive Decays of the Type A → B → C

The isotope I−135 is one of the possible fission products produced as the reactor operates. It decays by two β^- emissions into Cs−135:

$$\begin{matrix} {}^{135}_{53}\mathrm{I} & \rightarrow & {}^{135}_{54}\mathrm{Xe} & \rightarrow & {}^{135}_{55}\mathrm{Cs} \\ \mathrm{A} & \rightarrow & \mathrm{B} & \rightarrow & \mathrm{C} \end{matrix}$$

In such a series of decays, how many nuclei of each type, A, B and C, are present in a sample at any time?

Let us assume that at $t = 0$, only nuclei of type A (I–135) are present in the sample, i.e. $N_{B_0} = N_{C_0} = 0$. At a later time t, the number, N_A, of A nuclei remaining in the sample will be

$$N_A = N_{A_0} e^{-\lambda_A t}, \tag{5.74}$$

where N_{A_0} is the number of A nuclei that were present in the sample at time $t = 0$.

The rate at which nuclei of type B (Xe–135) are produced at any moment equals the activity, $\lambda_A N_A$, of the A nuclei at that moment. Similarly, the rate at which B nuclei decay equals their activity, $\lambda_B N_B$, where N_B is the number of B nuclei present in the sample at the time. Thus, B nuclei are produced at the net rate of

$$\begin{aligned}\frac{dN_B}{dt} &= \lambda_A N_A - \lambda_B N_B \\ &= \lambda_A N_{A_0} e^{-\lambda_A t} - \lambda_B N_B.\end{aligned} \tag{5.75}$$

Given that $N_{B_0} = 0$, the solution of Eq. (5.75) is

$$N_B = \frac{\lambda_A N_{A_0}(e^{-\lambda_A t} - e^{-\lambda_B t})}{\lambda_B - \lambda_A}, \tag{5.76}$$

from which it follows that the number of B nuclei in the sample will reach its maximum value at the time given by

$$t_{B\max} = \frac{\ln\left(\frac{\lambda_A}{\lambda_B}\right)}{\lambda_A - \lambda_B}, \tag{5.77}$$

decreasing to zero at $t = \infty$.

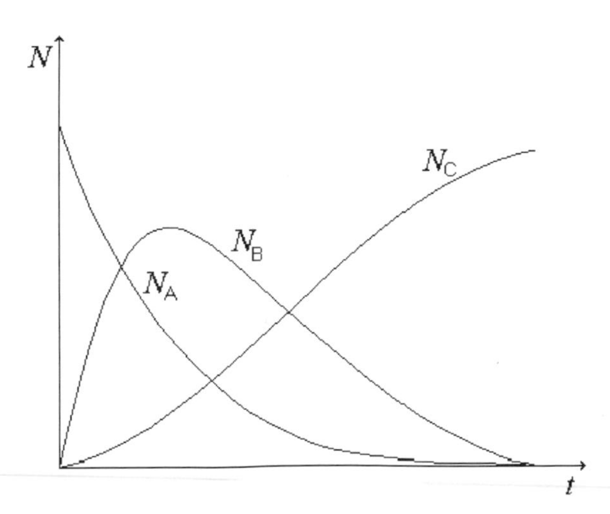

Fig. 5.42. The number of nuclei of each type present in a sample in which the decay series A → B → C occurs.

The number, N_C, of nuclei of type C present in the sample at any time is given by

$$N_C = N_{A_0} - N_A - N_B. \tag{5.78}$$

Substituting Eqs (5.74) and (5.76) in Eq. (5.78) gives

$$N_C = N_{A_0} \left(1 - \frac{\lambda_B}{\lambda_B - \lambda_A} e^{-\lambda_A t} - \frac{\lambda_A}{\lambda_B - \lambda_A} e^{-\lambda_B t} \right). \tag{5.79}$$

These results are summarised in the graphs in Fig. 5.42.

Part Five: Questions, Exercises and Problems

Questions

1. a. Explain why isotopes of the same element have the same chemical properties.
   b. Which physical properties do isotopes of the same element share and in which do they differ?
2. How many protons, neutrons and electrons are there in the following neutral atoms: ^{4}He, ^{14}C, ^{17}O and ^{206}Pb?
3. Why do heavy nuclei have a greater neutron excess than lighter ones?
4. Explain the difference between the half-life of an isotope and the mean lifetime of a radioactive nucleus.
5. The inherent uncertainty in the rest-energy of an unstable nucleus is $\hbar/\tau$ where τ is its mean lifetime. Based on the data in Table 5.12, with what precision can the masses of typical radioisotopes be correctly expressed?
6. Show that the rest-mass of the intermediate boson of an interaction of infinite range, such as the electromagnetic interaction, will be zero.
7. Is the assumption that the decay constant of a radioactive isotope is a constant independent of time a requirement of a theory, an axiom or an empirical conclusion?
8. Could carbon−14 dating be used to determine the age of fibres estimated to be one million years old?
9. Show that whereas the β^- decay of a neutron into a proton and electron is possible, the β^+ decay of a proton into a neutron and positron is impossible.
10. Whereas the average kinetic energy of the electron emitted in β^- decay is $0.3K_{max}$, that of the positron emitted in β^+ decay is $0.4K_{max}$. Explain.
11. Whereas the nuclear cross-section of reactions with neutron projectiles decreases as the projectile's energy increases, it increases if the projectiles are protons. Explain.
12. The emission of X-rays always accompanies electron capture decay. Explain.
13. Although the daughter nucleus, ^{15}N, of the β^+ decay of ^{15}O is produced in its ground state, a sample containing ^{15}O emits photons with an energy of 0.51 MeV. Explain.
14. The pools of water in which small research reactors are kept often glow with a blue Çerenkov radiation. Which particles could be the source of this radiation?
15. 'Neutron leakage is a surface effect; neutron production is a volume effect'. Explain this statement.
16. What is the difference between a delayed neutron and a slow neutron? Between a fast neutron and a prompt neutron?
17. It is estimated that a γ-ray photon produced at the sun's core takes about one million years to diffuse to the surface because of its particle interactions on the way. What might these interactions be?
18. Whereas photons take a long time to rise from the sun's core to its surface, neutrinos make the journey in about one second. Explain.

19. What is the electric charge of the particle composed of a down quark, an up quark and a strange quark?
20. Why must the constituent quarks in a hadron each have a different colour?
21. Mesons and baryons are both hadrons. What are the differences between them and how does the quark theory account for them?
22. What is the difference between a boson and a fermion? Between a hadron and a lepton?
23. Baryons and leptons are both fermions. What is the difference between them?
24. Is a neutrino or an antineutrino emitted when an antineutron decays into an antiproton?
25. If the stable decay products of a particle include a proton, the original unstable particle was a baryon; if there is no proton amongst them, it was a meson. Explain this rule.
26. Although the gravitational interaction is by far the weakest of the four fundamental interactions, it alone governs planetary and stellar motions. Explain.

Exercises

1. Calculate the radius of a neutron star whose mass equals that of the sun, $2 \cdot 10^{30}$ kg.
2. Find the average binding energy per nucleon in an α particle. ($m_{\text{He}-4} = 4.002603\,\text{u}$).[98]
3. Calculate the energy required to break up a tritium nucleus, ^{3}H, whose mass is 3.016054 u into its component nucleons.
4. Calculate the binding energy per nucleon in the nuclei ^{16}N, ^{16}O and ^{16}F ($m_{\text{N}-16} = 16.00610\,\text{u}$; $m_{\text{O}-16} = 15.99491\,\text{u}$; $m_{\text{F}-16} = 16.01146\,\text{u}$).
5. Calculate the maximum energy of the electrons emitted from the β^- decay of ^{12}B ($m_{\text{B}-12} = 12.014353\,\text{u}$; $m_{\text{C}-12} = 12.000000\,\text{u}$).
6. Show that the α decay of ^{16}O is impossible ($m_{\text{O}-16} = 15.994915\,\text{u}$; $m_{\text{C}-12} = 12.000000\,\text{u}$; $m_\alpha = 4.002603\,\text{u}$).
7. After passing through 17 cm of water the intensity of a beam of 200 keV photons is one tenth of its original value. What depth of water is required to reduce the beam's intensity by half?
8. A photomultiplier tube contains ten electrodes.

   a. If the multiplying factor is six, how many electrons reach the final electrode for each one that reaches the second electrode?
   b. What is the electrical current leaving the tube if 20 electrons are released from the first electrode each second?

9. The half-life of ^{226}Ra is 1,600 years. After how long will just one gram of the isotope remain in a sample that originally contained 32 grams?
10. After 60 days, the activity of a radioactive material is one sixteenth that of its original value. What is the half-life of the material?

[98] In this and other questions the masses quoted are those of the atomic masses of the neutral isotopes.

11. ^{218}Po is an α emitter. How many α particles are emitted each second from a milligram of pure ^{218}Po whose half-life is 3.05 minutes?
12. The half-life of a radioactive substance that decays into a stable nucleus is 36 hours. A sample of the substance emits 900 α particles per minute. What will the sample's activity be 48 hours later?
13. What mass of the radioisotope ^{60}Co, often used as a γ-ray source for medical and industrial purposes, has an activity of one million curies? The half-life of ^{60}Co is 5.27 years.
14. Whereas the half-life of ^{235}U is $7 \cdot 10^8$ years, that of ^{131}I is just eight days. What is the ratio between the activity of a micromole of ^{235}U and a micromole of ^{131}I.
15. The effective half-life of ^{131}I in the human body is just four days because apart from radioactive decay some is removed from the body with the urine. How long after receiving a dose of ^{131}I will the body's activity be reduced by a factor of 1,000?
16. The half-life of ^{40}K is $1.3 \cdot 10^9$ years. The abundance of this isotope in naturally occurring potassium is 1% that of the stable isotope ^{39}K. How many years ago were they equally abundant?
17. If the half-lives of ^{235}U and ^{238}U are, respectively, $7.1 \cdot 10^8$ and $4.5 \cdot 10^9$ years, how long ago were they equally abundant in the earth's crust?
18. The activity of a radioactive substance fell by a factor of four during 48 hours. What is the probability that a nucleus in the substance will decay in the next 30 minutes, 6 hours, 12 hours, 24 hours, 48 hours, week?
19. On average, each US resident consumes $3 \cdot 10^{11}$ J of energy each year. How much reactor fuel must fission to supply this?
20. When a 6.0 μA beam of protons is aimed at an iron foil of thickness 1 μm, neutrons are emitted at a rate of $1.9 \cdot 10^8$ per second as a result of the reaction $^{56}\mathrm{Fe}(p, n)^{56}\mathrm{Co}$. Find the nuclear cross-section for the reaction ($\rho_{\mathrm{Fe}} = 7900\,\mathrm{kg/m^3}$).

Problems

1. a. Calculate the electrostatic potential energy of a pair of protons at a distance of 1.7 fm from one another.
   b. Calculate the difference between the binding energy per nucleon in ^{3}H and ^{3}He ($m_{\mathrm{H-3}} = 3.016049\,\mathrm{u}$; $m_{\mathrm{He-3}} = 3.016029\,\mathrm{u}$).
   c. Comment on your answers to a. and b. with regard to the charge dependency of the nuclear (strong) force.
2. Show that if the abundance of the daughter nuclei in the radioactive decay series $\mathrm{A} \to \mathrm{B} \to \mathrm{C} \to \cdots$ is constant then $N_A\lambda_A = N_B\lambda_B = N_C\lambda_C = \ldots$.
3. Show that if the activities of a radioactive sample are, respectively, a_1 and a_2 at times t_1 and t_2 then $N_1 - N_2 = \tau(a_1 - a_2)$.
4. The activity of a sample of a radioactive isotope that decays into a stable nucleus fell from 137 Bq to 125 Bq during one hour. Calculate:

   a. the half-life of the isotope;
   b. the number of nuclei of the radioactive isotope in the sample at the beginning and end of the hour;

c. the mean lifetime of a nucleus of the radioactive isotope;
d. how long must elapse until just one radioactive nucleus is left in the sample.

5. 99.96% of the disintegrations of the radioisotope ^{218}Po are α decays and 0.04% are β decays. The half-life of ^{218}Po is 3.05 minutes.
   a. Write down the formulae of the two decay modes.
   b. Calculate the partial half-lives of each decay mode.
   c. What is the probability that a nucleus will emit an α particle or a β particle during any second?
6. A radioactive source contains two radio-isotopes: one an α emitter with a half-life of six days, the other a β emitter with a half-life of ten days. The source has an activity of 1,000 Bq. However, if a piece of thick cardboard is placed between the source and the counter the measured activity drops to 600 Bq. What will be the activity of the source 30 days later?
7. Use the following data to calculate the energy made available to ^{235}U, ^{238}U and ^{239}Pu nuclei by the incorporation of a neutron and show how it accounts for the possibility of inducing their fission with thermal or fast neutrons.

$$m_{\mathrm{U-235}} = 235.043925\,\mathrm{u} \quad m_{\mathrm{U-236}} = 236.045563\,\mathrm{u} \quad m_{\mathrm{U-238}} = 238.050786\,\mathrm{u}$$
$$m_{\mathrm{U-239}} = 239.054291\,\mathrm{u} \quad m_{\mathrm{Pu-239}} = 239.052158\,\mathrm{u} \quad m_{\mathrm{Pu-240}} = 240.053809\,\mathrm{u}$$

8. Assume that immediately after the fission the medium-sized fragment nuclei in Fig. 5.35 are just touching. What is their electrostatic potential energy at this moment?

Part Six Selected Topics

The potential benefits or dangers of a scientific discovery are seldom apparent at the time it is made. Like a new-born baby, there is no way of knowing how it will turn out: a serial murderer, a great artist, a poet or just another honest hard-working citizen? The same is true of scientific discovery. Will it be used in a holocaust, to enrich human culture or to save lives? Or will it just add to the general store of human knowledge without having had any apparent impact?

Knowledge has an imperative of its own. The notion that Becquerel's musings about the source of the radiation emitted from uranium salts would lead to Hiroshima, could never have occurred to anyone at the time. But in retrospect, we can see that it was inevitable. Try as we may, we cannot unthink our thoughts nor can we uninvent the things we have invented. Whether we want it or not, nuclear weapons will now be with us for ever more.

But there is more to modern physics than nuclear weapons. In addition to expanding our understanding of the physical world around us, it has provided countless inventions, technologies and concepts without which our lives would all be poorer.

In this part of the book, we will examine five topics, each in its own way representative of the contribution modern physics has made to our thinking and way of life as we advance into the 21st century. They are, in order:

* The Laser.
* The Mössbauer Effect.
* Nuclear Magnetic Resonance.
* The Conduction of Electricity Through Solids.
* Invariance, Symmetry and Conservation Laws.

In each case, the underlying physics is explained in detail along with the practical applications of the technique, where appropriate.

The choice of the topics is not unbiased. It reflects an optimistic outlook; the hope that the benefits and wonders of science will continue to outweigh its calamities.

Chapter 6.1

The Laser

The word 'laser' is an acronym of light amplification by the stimulated emission of radiation. Since their invention in 1960, lasers of various types have found numerous applications in research, enterprise and technology. The key features of the light produced by a laser are: its almost pure monochromacity; its coherence; its high intensity and, fourthly, its low beam divergence. For example, the beam from a typical 10 mW helium-neon laser has a monochromacity of 632.81 ± 0.02 nm, a beam diameter at source of only 0.5 mm and a beam divergence of just 0.001 radians; at a distance of 100 m from its source the beam's diameter is just 10 cm.

To gain an understanding of laser action, we must first review the mechanisms by which atoms and molecules absorb and emit radiation.

6.1.1 The Spontaneous and Stimulated Emission of Radiation

Excited atoms and molecules can emit radiation (photons) in two different ways: spontaneously or as a result of being induced to do so.

The *spontaneous emission* of radiation from an excited atom or molecule is a pure quantum effect that has no classical analogue. Ideal classical oscillators do not spontaneously lose energy; an undamped pendulum would go on swinging for ever.[1]

$$\text{Spontaneous emission:} \quad \text{Atom}^* \rightarrow \text{Atom} + \text{photon}$$

Spontaneous emission is a chance phenomenon; the emissions occur haphazardly, at unrelated times. Consequently, spontaneously emitted radiation is incoherent; spontaneously emitted photons have random phases.

Excited atoms and molecules can also be induced to emit radiation. When radiation strikes an excited atom or molecule, it may trigger the ejection of a photon from it. This second mechanism is called *stimulated*

[1] For a comprehensive treatment of spontaneous emission see Section 3.3.2.

emission.

$$\text{Stimulated emission:} \quad \text{Atom}^* + \text{photon} \rightarrow \text{Atom} + 2 \text{ photons}$$

However, stimulated emission only occurs if the energy of the incident photon is equal to one of the transition energies in the excited atom or molecule, i.e. if the frequency of the incident photon matches that given by Bohr's equation, $f_{ij} = (E_i - E_j)/h$, where E_i and E_j are the energy levels of two stationary states in the particular atom or molecule.

The new photon emitted from the stimulated emission of an excited atom or molecule is always identical to the triggering photon; it moves in the same direction and has the same energy, polarisation and phase as the incident photon (Fig. 6.1).

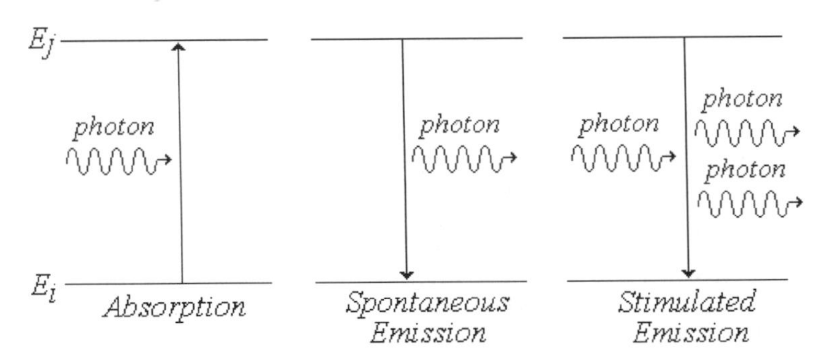

Fig. 6.1. Energy level diagrams for the processes of absorption, spontaneous emission and stimulated emission.

Essentially, the stimulated emission of radiation and the absorption of radiation involve the same occurance, *viz.* the collision of a photon with an atom or molecule. They also share a classical analogue: the phenomenon of resonance. In both cases, we say that there is resonance between the radiation and the atoms or molecules.

In classical mechanics, when a periodic force and an ideal oscillator both possess the same frequency, they are said to be in resonance. If the periodic force is coupled to the oscillator, they will exchange energy. Whether the oscillator gains or loses energy as a result depends on the phase difference between the periodic force and the oscillator's vibrations. If the two are in phase, the oscillator gains (absorbs) energy; if the phase difference between them is equal to π, the oscillator loses (emits) energy. Thus, absorption and stimulated emission differ essentially only in terms of phase.

A gain or loss in the energy of an atomic or molecular oscillator is manifested, respectively, as the absorption or emission of radiation. The former corresponds to a 'jump' and the latter to a 'fall' or 'drop' between the energy levels.

That excited atoms and molecules emit radiation by stimulated as well as spontaneous emission can be shown by considering the dynamic equilibrium that exists in cavity radiation. Suppose, for simplicity's sake, that the system comprises just two energy levels, E_1 and E_2, that are occupied by N_1 and N_2 particles, respectively. Assuming that the distribution of the particles between the two levels follows the Maxwell–Boltzmann rule, the ratio between their populations will be

$$\frac{N_1}{N_2} = e^{(E_2 - E_1)/kT} = e^{hf/kT}, \tag{6.1}$$

where f is the frequency of the photons absorbed or emitted in transitions between the two levels. At equilibrium, the ratio N_1/N_2 must on average be constant, i.e. the rates of the upward and downward transitions must be equal.

The rate at which upward transitions occur clearly depends on the intensity of the radiation in the cavity: the greater the energy density of the radiation, the more collisions there will be between photons and atoms in the cavity's walls and the greater the number of upward transitions that occur. On the other hand, spontaneous emission is unaffected by changes in the energy density of the radiation in the cavity; there is no way it can adjust to changes in this density. Hence, spontaneous emission cannot of itself balance these upward transitions. It follows that in order for the upward and downward transition rates to balance, an emission mechanism that depends on the energy density must also exist. This mechanism is the stimulated emission that occurs when photons strike excited atoms in the cavity's walls.

It can be shown[2] that, at equilibrium, the transition probabilities, P_{St} and P_{Ab}, of stimulated emission and absorption, respectively, are equal, i.e. $P_{St} = P_{Ab}$. This result was first obtained by Einstein in 1917. He also showed that the ratio, P_{Sp}/P_{St}, between the transition probability for spontaneous emission, P_{Sp}, and that for stimulated emission, P_{St}, is given by

$$\frac{P_{Sp}}{P_{St}} = \kappa_f(e^{hf/kT} - 1), \tag{6.2}$$

where f is the frequency of the emitted radiation and κ_f is the average photon density in the system. Thus, at high frequencies, where $hf \gg kT$, spontaneous emission is the more probable of the two emission mechanisms; this is the

[2]See Appendix 6.5.1.

case with electronic transitions in atoms and molecules as well as for nuclear radiative transitions. However, at low frequencies, where $hf \ll kT$, stimulated emission becomes more important; this is the case in microwave transitions and nuclear spin transitions.[3]

6.1.2 Laser Action

The two coherent photons that emerge from a stimulated emission can go on to induce further emissions in other excited atoms in the system. Thus, in principle, the conditions exist for propagating a chain reaction that can amplify the intensity of the radiation. However, as we have seen, the probability that a photon will induce the emission of a second photon from an excited atom is equal to the probability that it will itself be absorbed by an atom in the lower state. Thus, a chain reaction of stimulated emission can only develop if the number of atoms in the excited state is greater than the number in the lower state. Achieving such a *population inversion* is the first requirement in the construction of a laser device.

Because of the equal probabilities of absorption and stimulated emission, a population inversion cannot be achieved in systems that comprise only two energy levels; the system must have at least three energy levels.[4] The simplest configuration that can produce laser action is shown schematically in Fig. 6.2. Atoms, initially in the ground state, E_1, are excited into the E_3 state from which they quickly ($\sim 10^{-9}$ s) fall by non-radiative transitions into the metastable state, E_2. The E_2 state is typically a triplet state ($S = 1$) that has an average lifetime of about one millisecond; the probability of spontaneous emission from this state is low. Thus, atoms rapidly accumulate in the metastable E_2 state and within a short time its population overtakes that of the ground state, E_1.

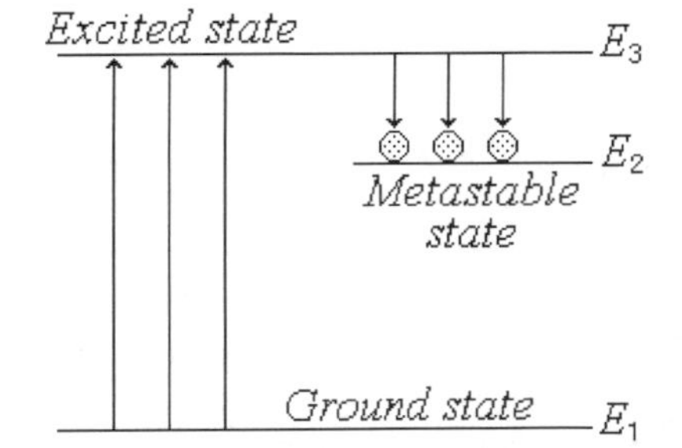

Fig. 6.2. Population inversion in a three level system. As a result of the accumulation of atoms in the metastable state its population exceeds that of the ground state.

[3] Nuclear spin transitions are the topic of Chapter 6.3.

[4] The semiconductor diode laser is an exception to this rule; it is in some sense a two level laser system. See Section 6.4.9.

Having achieved a population inversion, the stimulated emission chain reaction will only propagate if each of the photons emitted by the atoms in the metastable state triggers the emission of at least one more photon before escaping from the assembly. This is the second requirement for the construction of a laser device. It can be satisfied:

1. By making the laser component in which the population inversion is achieved long enough to ensure that each photon strikes an atom in the metastable state before escaping (most X-ray lasers work this way).
2. By installing a pair of parallel mirrors at opposite ends of the component and reflecting the photons emitted parallel to the axis of the device to and fro along its length (Fig. 6.3). In essence the component is repeatedly folded up between the mirrors. The chain reaction propagates with each passage of the photons between the mirrors as more and more atoms are triggered into emitting a photon and the photon flux in the component increases.

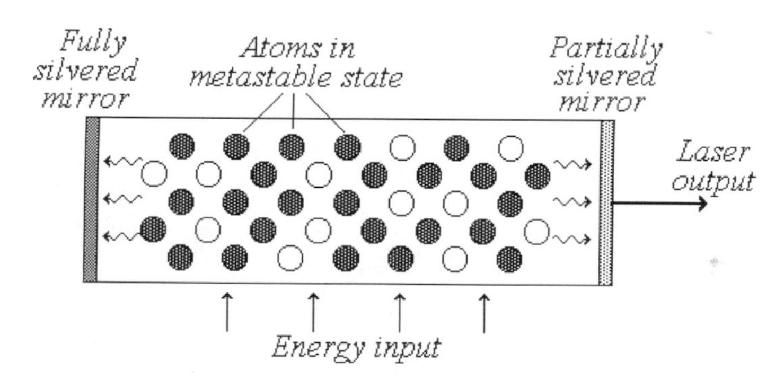

Fig. 6.3. A schema of laser design. The atoms are 'pumped' up to the metastable state by the energy input. The parallel end mirrors reflect the stimulating photons to and fro through the inversed population of atoms. The laser output emerges from the partially silvered end as a narrow beam.

If the distance between the mirrors is equal to an integer number of half wavelengths of the light emitted by the atoms, a highly monochromatic standing wave, whose amplitude grows as more and more photons accumulate, is generated along the component's length. The component may be regarded as an optical resonator. If the mirror at one end is only partially silvered, light ultimately leaks out from it as a narrow highly monochromatic coherent[5] beam.

[5] The emerging beam is a single plane wave all of whose component waves are in phase. A perfectly coherent light source must be monochromatic.

6.1.3 The Ruby Laser

Laser action with visible light[6] was first achieved by Maiman[7] in 1960 using a cylindrical ruby whose ends had been polished and silvered. Ruby is a crystal of aluminium oxide (Al_2O_3) doped with a small amount of chromium. The chromium atoms in the crystal, present in a concentration of about 0.1%, possess a three level system of the type shown in Fig. 6.2. The energy difference, $E_3 - E_1$, of the chromium system corresponds to a photon wavelength of 550 nm (Fig. 6.4).

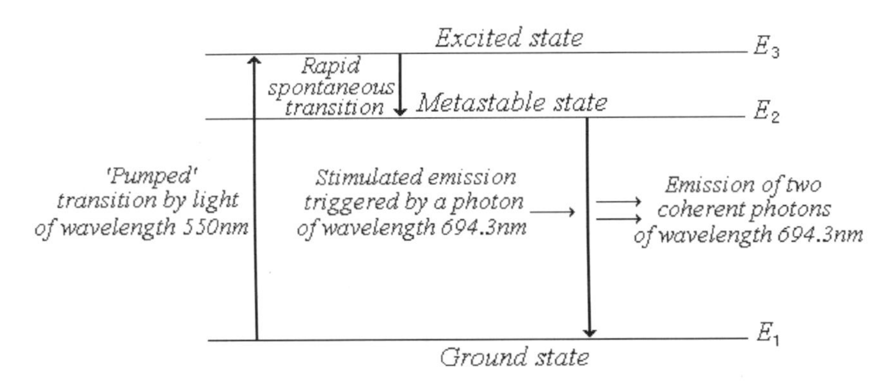

Fig. 6.4. Diagram of the transitions that occur in a ruby laser. The chromium atoms are raised to the higher excited state by absorbing a 550 nm photon. They rapidly and spontaneously fall into the metastable state from which they return to the ground state by stimulated emission.

Maiman obtained the required population inversion of the chromium atoms by 'pumping' the ruby crystal with flashes of green/yellow light of wavelength 550 nm. The metastable E_2 state has a lifetime of about 3 ms after which the excited atom returns to its ground state with the emission of a 694.3 nm (red) photon.

The stimulated emission chain reaction in the ruby crystal is initiated by the few photons emitted spontaneously from chromium atoms in the metastable state. In order to propagate the chain reaction, these 694.3 nm (1.79 eV) photons are reflected to and fro between the polished and silvered ends of the ruby crystal. The chain reaction propagates with each passage of the photons as more and more chromium atoms are triggered into emitting a photon. The amplitude of the standing wave set up in the crystal grows as the photons accumulate. Ultimately, light leaks out from the partially silvered end as a

[6]The amplification of microwaves by stimulated emission had been achieved previously.
[7]Theodore Maiman 1927–2005, American physicist.

narrow highly monochromatic coherent[8] beam. Typically, about 1% of the photons incident on the partially silvered surface emerge from it (Fig. 6.5).

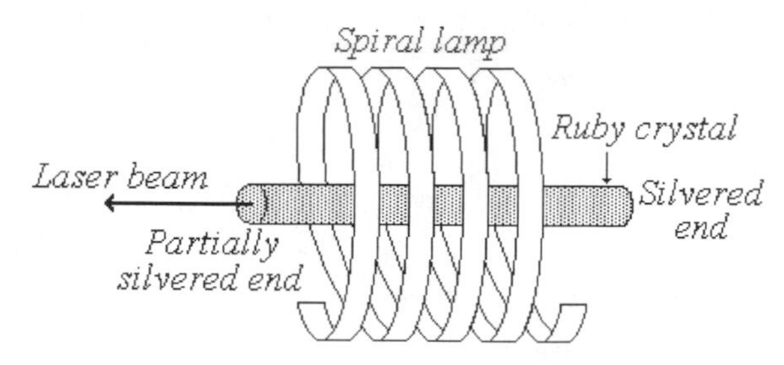

Fig. 6.5. Ruby laser. Intense flashes of light from the spiral lamp excite the chromium atoms in the cylindrical crystal into a metastable state; a population inversion is achieved. The repeated reflections of the emitted light from the silvered ends of the crystal amplify the light intensity in the crystal until it emerges as a pulse of monochromatic coherent light from the partially silvered end of the crystal.

The ruby laser is an example of a solid state laser. Each pumping flash of light lasts for a few milliseconds during which the device emits a number of rapid pulses of laser light each of about 0.2–0.3 ms duration. The cross-sectional area of its beam is about $0.1\,mm^2$ and so at a mean operating power of 2 kW the intensity of each pulse is about $2 \cdot 10^{10}\,W/m^2$. Even greater intensities can be obtained by suppressing the stimulated emission until most of the 'pumping' energy of the light flash has been absorbed by the system. This can be achieved by replacing one of the fixed mirrors at the crystal's ends by a vibrating or rotating mirror that is phased with the pumping lamp to produce the resonant conditions in the ruby only when a highly inverted population has been established. In this way single pulses lasting just 100 ns in which all the energy absorbed during a single pumping flash is emitted can be produced. Such ultra-short pulses have a power of $\sim$10 MW.

The ruby laser has a number of drawbacks:

* Because of the heat generated in the crystal during its operation, it cannot be operated continuously. It must be pumped by spaced flashes of light.
* For laser action to occur, at least half of the chromium atoms in the crystal must be in the metastable state. This is the reason why the ruby laser emits light in rapid pulses during excitation and not continuously. Once the chain reaction has started, the metastable state depopulates faster than it is

[8] The emerging beam is a single plane wave all of whose component waves are in phase.

repopulated. As a result, after each pulse of laser action several tenths of a millisecond are needed to restore the population of the metastable state.

Both of these disadvantages are avoided in the four level helium-neon gas laser.

6.1.4 The Helium-Neon Laser

In the helium-neon laser, the crystal is replaced by a discharge tube containing a mixture of helium (85%) and neon (15%) at low pressure. The tube has parallel mirrors at its ends just like the crystal in the ruby laser. The laser is pumped by the electrical discharge across the tube.

The population inversion in a helium-neon laser does not involve the ground state and so its operation does not require that at least 50% of the atoms involved in the laser action be in the metastable state. The inversion is achieved by stages. Initially, the more abundant helium atoms in the discharge tube are electrically excited from their ground state up to various higher states from which they spontaneously and rapidly fall into the metastable $1s^1 2s^1$ state of energy 20.61 eV. Neon atoms possess an excited state, $2p^5 5s^1$, of almost the same energy, *viz.* 20.66 eV. In collisions between helium atoms in the excited metastable state and neon atoms in the ground state, the helium atoms transfer their energy to the neon atoms. In the process, the neon atoms are raised to the $2p^5 5s^1$ state and the helium atoms return to their ground state

$$\mathrm{He}^* + \mathrm{Ne} \rightarrow \mathrm{He} + \mathrm{Ne}^*.$$

Neon atoms in the $2p^5 5s^1$ state can fall by stimulated emission to the short-lived $2p^5 3p^1$ state of energy 18.7 eV, from which they spontaneously and rapidly drop to the $2p^5 3s^1$ state. The neon atoms finally return to their ground state from the $2p^5 3s^1$ level.

Because of the continuous transfer of energy from the excited helium atoms to the neon atoms and because of the short life of the $2p^5 3p^1$ state, the $2p^5 5s^1$ state will always be more highly populated by the neon atoms than the lower $2p^5 3p^1$ state. This is the population inversion that produces the laser action (Fig. 6.6).

The visible light emitted by a helium-neon laser has a wavelength of 632.8 nm corresponding to a photon energy of $20.66 - 18.7 = 1.96$ eV. The output of a typical helium-neon laser is about $5 - 10$ mW; they are low-power lasers. Other gas lasers can produce continuous beams of very high intensities; powerful carbon dioxide lasers (~200 W) can easily cut through sheet metal. Short pulses (~50 ns) with a peak power of 50 MW can be produced by

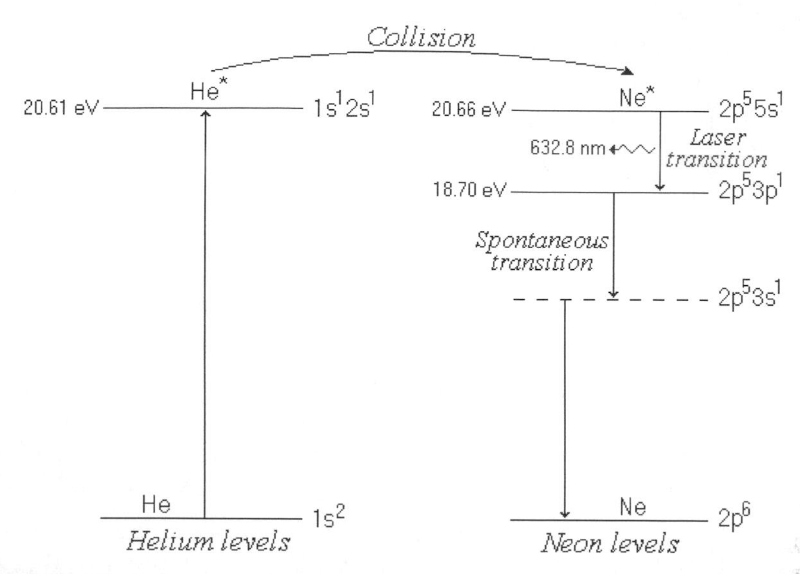

Fig. 6.6. The helium-neon laser. Energy is continuously supplied to the neon atoms by the more abundant excited helium atoms. This maintains the population of the $2p^5 5s^1$ state. The population of the $2p^5 3p^1$ state is always negligible due to its short life.

chemical lasers in which the population inversion is achieved by exciting the gas molecules participating in an energetic chemical reaction into the metastable state.

6.1.5 Laser Applications

The different uses to which laser light is put can be grouped under the headings of its key characteristics: monochromacity, coherence, high intensity and low beam divergence.

1. Monochromacity

The lines in atomic emission spectra are not absolutely monochromatic; if they were, the lines would be infinitely narrow and invisible.[9] Typical spectral lines have a variance of about $\pm 0.001\%$. A red spectral line might comprise wavelengths in the range 700 ± 0.007 nm; its linewidth is of the order 10^{-2} nm. However, laser light can be produced with a linewidth as fine as 10^{-13} nm. Stabilised lasers that maintain a constant wavelength to within 2 parts in 10^{12} have enabled the measurement of the speed of light and other physical constants with unprecedented accuracy.

[9]The factors affecting the width of spectral lines are considered in Section 6.2.1.

The availability of almost pure monochromatic laser light has had a great impact on spectroscopy, particularly since the invention of the *tuneable dye laser.* In this device the length of the optical resonator in which the population inversion is achieved can be finely adjusted. The active medium contained in the resonator is an organic fluorescent pigment. The molecules of organic pigments have large numbers of closely spaced energy levels and so, in a given range, they can emit light over an almost continuous spectrum of wavelengths. Excitation of the fluorescent pigment into its metastable states and adjustment of the length of the optical resonator, enables selection of monochromatic laser light of any wavelength within the given range.

Scanning a substance with the range of frequencies available from a tuneable laser, reveals the precise frequencies at which it absorbs; at these frequencies the attenuation of the laser beam passing through the sample is greatest. Tuneable lasers have also been used for isotope separation by means of selective ionisation. Notwithstanding the very small, almost negligible, differences between their energy levels, the frequency of the laser beam can be so finely tuned that it will excite one isotope in a sample but not the others.

2. Coherence

Ordinary light sources, such as white hot objects (the filaments in electric bulbs) or discharge tubes, emit light in short wavetrains, each about 1 m long or $\sim 10^{-8}$ s. Lasers, however, can emit wavetrains hundreds of kilometres in length ($\sim 10^{-3}$ s). For this reason and because of the great intensity of their light, lasers are an excellent source for experiments on the interference and diffraction of light. The wavetrains of lasers are so long that it has even been possible to observe the phenomenon of beats between two independent laser sources, a phenomenon well known in sound but unobservable in light before the advent of the laser.

The use of the holographic technique of imaging, invented by Gabor[10] in 1948, has been greatly extended by the availability of coherent high-intensity laser light. Traditional photographic imaging has the disadvantage that the camera lens must be focused on a particular plane in the field of view. Only this plane is in focus in the photograph taken; all other planes are out of focus to a greater or lesser extent. Holography overcomes this limitation.

The essential ideas underlying holography are:

a. That an interference or diffraction pattern contains all the information needed to construct an image of the physical system in which it was formed;

[10]Dennis Gabor 1900–1979, English physicist who was awarded the Nobel prize in 1971 for discovering the principles of holography.

b. That this information can be stored for future retrieval as a photographic etching of the interference or diffraction pattern called a *hologram.*

To prepare a hologram, a laser beam is first split into two beams by means of a suitable optical arrangement of half-silvered mirrors. One of the beams, called the 'reference beam' is then aimed directly at a high resolution photographic plate, whereas the second beam, called the 'object beam', is aimed at the object whose image is to be recorded (Fig. 6.7).

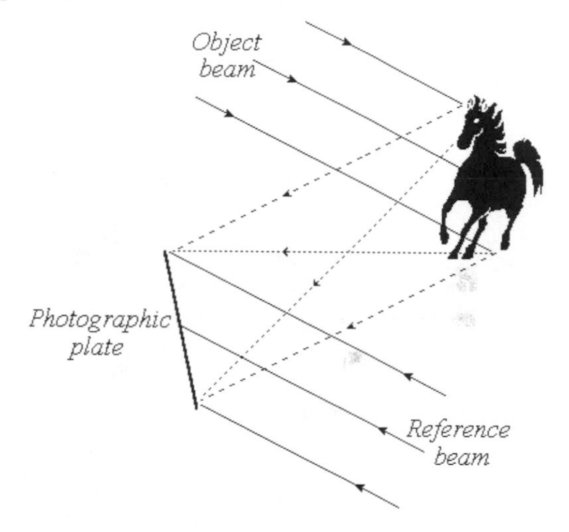

Fig. 6.7. The laser light scattered by the object (the horse) interferes on the photographic plate with the light in the reference beam.

The light scattered by the object strikes the photographic plate where it combines with the light from the reference beam to produce an interference pattern (rather than the usual areas of light and dark intensity that appear on a conventional photographic negative). In the absence of the object the two beams would combine on the plate to give the typical two-slit interference pattern associated with Young's experiment. The presence of the object modulates this basic pattern just as a signal modulates the carrier wave in radio or telecommunications. In order that just one interference pattern be formed on the plate, the highly coherent light of a laser must be used in preparing the hologram. Using non-coherent light would result in 'noise' instead of a clear signal.

After developing, the photographic plate acts as a diffraction grating. An image of the original object can be recreated by passing a beam of coherent light, generally but not necessarily laser light from the original laser, through the grating. Two strongly diffracted beams are obtained: one can be projected onto a screen to give a real image of the object; the other appears as a virtual stereoscopic image when looking through the transparent photographic plate (Fig. 6.8).

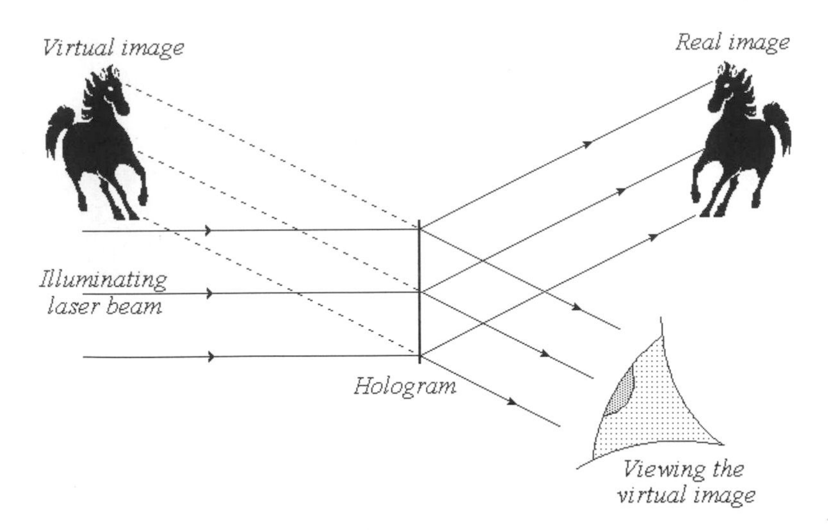

Fig. 6.8. The hologram transparency is illuminated by a beam of laser light. The first order diffracted beam provides a real image of the object on one side of the zero and a high quality virtual image on the other.

Because there is no specific correspondence between points on the object and points on the hologram, all parts of the image can be seen by viewing through any part of the hologram. Thus, all parts of the image can be obtained from just a fragment of the hologram.

Full colour holograms, which can be viewed in ordinary incoherent white light, can also be produced. For example, those embossed on credit cards to prevent their forgery, it being virtually impossible to exactly copy them. The full colour holograms are etched on a thick photographic emulsion using three laser beams, each in one of the primary colours – red, green and blue.

3. High intensity and low beam divergence

The unfocused beam from a low intensity 1 mW helium-neon laser has a brightness equal to that of sunlight ($\sim 1\,\text{kW/m}^2$). However, laser beams can in principle be focused to a width no greater than their wavelength with a corresponding increase in intensity. Thus, the beams from 1 W lasers can be focused to an intensity of $10^9\,\text{kW/m}^2$ and pulsed ruby lasers have produced beam intensities as high as $10^{17}\,\text{kW/m}^2$; beams of this intensity can ionise the air and puncture metal sheets. These high intensities have found many applications in industry, medicine and research.

In industry, focused laser beams have been used to cut metals and to weld metal parts together, for the accurate computer controlled cutting of cloth and for the spot hardening of sensitive mechanical components.

The use of a focused laser beam instead of a scalpel in surgery has the advantages of total sterility and the instantaneous cauterising of severed blood vessels, thus reducing bleeding. Detached retinas are routinely 'welded' back into place using lasers. Using bundles of *optical fibres* (*endoscopes*), laser light can be introduced through the natural orifices of the body, conducted around its inner organs and aimed at specific locations within the body without the need for conventional surgery. The optical fibres comprise long thin cylindrical crystals of a transparent material such as glass. Light introduced at one end of the fibre is repeatedly reflected off its inside surface as it propagates along its length. Thus, the light is kept trapped inside the fibre until it emerges at the other end with little attenuation.

Because of their low beam divergence, laser beams can be used to determine the location or movement of distant objects. For example, lasers have been used to monitor the movement of dams as the level of the water in their reservoir rises and falls as well as the geological movements associated with plate tectonics. By accurately measuring the time taken for a pulse of intense laser light reflected from mirrors positioned on the moon by the Apollo astronauts to return to earth, the distance from the earth to the moon has been measured to an accuracy of ± 10 cm.

The most widespread use of laser light is in the field of information processing. In all these applications, the information is encoded as a modulation of the laser light. A familiar application is the optical scanner, of the type seen at supermarket check-outs, that reads bar codes. The laser beam scans the bar code and is reflected from it. The intensity of the light reflected from the bar code varies in time with the pattern of its white and black lines. These variations are detected and converted into an electrical signal, which is then fed into a computer. Low power helium-neon lasers are typically used in these scanners.

Information is stored on the compact optical discs used as read-only memories for computers and to provide digital audio and video signals, as a series of small pits or holes in the disc's reflecting surface. As the disc spins, the light from a small semiconductor laser is focused onto its surface. The pattern of pits is read by detecting the modulations in the light reflected from the disc's surface and converting them electronically into a digital signal. Since the laser beam can be made as narrow as a single wavelength, the pits on the disc need be no more that about one micron in size. In this way, vast amounts of information can be stored on a single disc.

Chapter 6.2

The Mössbauer Effect

One of the most exciting aspects of scientific research is the way that technical advances made in one field can impact on other seemingly unrelated fields. The life sciences were twice revolutionised by advances in microscopy. First, by the invention of the optical microscope which revealed the existence of previously unseen forms of life; secondly, by the invention of the electron microscope which revealed the existence of the viruses and of cellular architecture.

Experimental evidence for Einstein's theory of general relativity and for the existence of the exotic bodies such as 'black holes' whose existence it posits, has been hard to come by. The extreme accelerations and strong gravitational fields under which its effects are clearly manifested only exist in deep space or in deep time. The detection of its predicted effects on earth requires the ability to make measurements of gravitational effects at an accuracy greater than one part in 10^{14}. This only became possible with the discovery by Mössbauer[11] in 1958 of a technique that enabled such measurements to be made. The technique, known as the *Mössbauer effect*, is associated with the emission and absorption of γ-ray photons.

As part of their decay process, many radioactive nuclei emit high energy electromagnetic radiation – γ-rays. These γ-rays are emitted as the nuclei fall from an excited nuclear state to a lower one. However, under ordinary conditions, nuclei already in the lower state do not absorb the γ-rays emitted by identical excited nuclei as they drop into that state, i.e. though the nuclei are identical, there is no resonance between them. This contrasts with the situation in atoms and molecules. Generally, atoms and molecules in the ground state do absorb the radiation emitted by identical excited atoms and molecules as they drop into that state, i.e. there is resonance between identical atoms and molecules.

[11]Rudolf Mössbauer 1929–2011, German physicist who was awarded the Nobel prize in 1961 for the discovery of the effect named after him.

Elucidating the conditions under which resonance does or does not exist between identical nuclei, atoms and molecules, requires a prior understanding of both the nature of spectral lines and the mechanics of photon emission and absorption.

6.2.1 The Width of Spectral Lines

Spectral lines are not absolutely monochromatic; they are not composed of just the single photon frequency, $f_{ij} = (E_i - E_j)/h$, given by Bohr's equation. If they were they would be infinitely thin and, therefore, invisible. What then are the reasons for the finite width (broadening) of spectral lines?

A primary cause of the broadening of spectral lines is the thermal motion of the particles emitting or absorbing the electromagnetic radiation. For example, as a result of this random motion, some of the photons are emitted from particles moving towards the observer and others from particles moving away from him. This, in turn, affects the frequencies detected by the observer due to a Doppler effect. Some of the photons appear to have a frequency slightly greater than f_{ij} and some a frequency slightly less, depending on the direction and magnitude of the velocity of the emitting particles. The result is a broadening of each spectral line. The effect is particularly pronounced in gas spectra.

However, even after allowance is made for such effects, spectral lines are still found to have an irreducible finite width. They always extend over a range of frequencies greater than the single frequency predicted by Bohr's equation for the particular transition. This inherent broadening of spectral lines is explained by Heisenberg's uncertainty principle, $\Delta E \cdot \Delta t \approx \hbar$, according to which there is no way of knowing exactly how long a particle will remain in an excited state before it emits the photon. All that can be determined is the average lifetime of the state, within the limits of an uncertainty, Δt. For the same reason, the energies of stationary states (their energy levels) are not precisely known; there is always an inherent uncertainty, $\Delta E \approx \hbar/\Delta t$, called the *energy width*, associated with them – see Fig. 6.9.

It can be reasonably assumed that the uncertainty, Δt, in the average lifetime of a stationary state is of the order of magnitude of the state's average lifetime. It follows that the energy width of a state will be inversely proportional to its average lifetime, i.e. the less stable a state, the broader its energy width. Hence, whereas unstable excited states have finite energy widths, the energy width of the ground state (the most stable state) can be taken as zero.

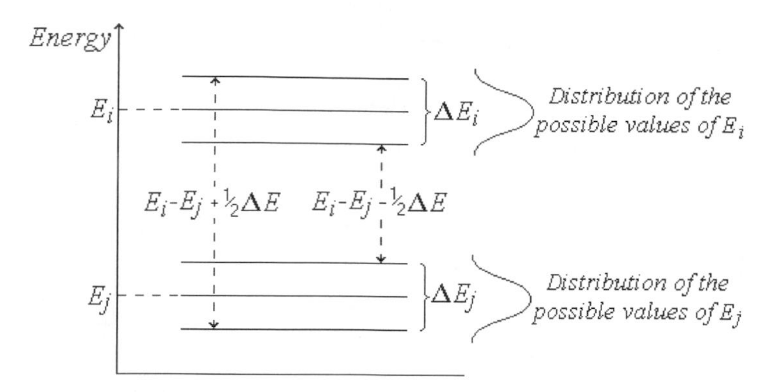

Fig. 6.9. The variance and energy widths, ΔE_i and ΔE_j, of the stationary states i and j, respectively. ΔE_i is the uncertainty in the value of E_i and ΔE_j is the uncertainty in the value of E_j: $\Delta E = \Delta E_0 + \Delta E_1$.

Consider two stationary states, i and j. Because of the variance (uncertainty) in their energy levels, the difference between their energies does not have a single precise value. On the contrary, it can have any value in the range

$$(E_i - E_j) \pm \frac{\Delta E}{2}, \tag{6.3}$$

where $\Delta E = \Delta E_i + \Delta E_j$ is the sum of the energy widths. The photons emitted or absorbed in transitions between these states will exhibit a similar range of energies. Hence, even in the absence of any Doppler broadening, resonance is not limited to the unique frequency, f_{ij}, given by Bohr's equation. All the photons whose frequencies fall in the range

$$f_{ij} \pm \frac{\Delta E}{2h}, \tag{6.4}$$

where f_{ij} is the transition frequency given by Bohr's formula, may be absorbed or emitted. This results in an inherent or 'natural' broadening of spectral lines. From the fundamental uncertainty expression, $\Delta E \cdot \Delta t \approx \hbar$, and the relationship $\Delta E = h\Delta f$, the *natural width*, Δf, of a spectral line is given by

$$\Delta f = \frac{1}{(2\pi \Delta t)}. \tag{6.5}$$

Example 6.1: The Width of Spectral Lines

Calculate

1. the natural width of the spectral line associated with the emission of a photon from a stationary state of average lifetime 10^{-8} s;

2. the fractional natural (or minimal) broadening of the spectral line if its wavelength is 500 nm;
3. the fractional Doppler broadening of a spectral line of wavelength 500 nm, emitted from a stationary state of average lifetime 10^{-8} s, for hydrogen atoms at a temperature of 1,000 K.

Solutions:

1. From Eq. (6.5) we have

$$\Delta f \approx \frac{1}{(2\pi \Delta t)} = \frac{1}{2\pi 10^{-8}} \approx 1.6 \cdot 10^{7}\text{ Hz}.$$

2. The central frequency, f_0, of the spectral line is given by the equation

$$f_0 = \frac{c}{\lambda} = \frac{3 \cdot 10^8}{500 \cdot 10^{-9}} = 6 \cdot 10^{14}\text{ Hz}.$$

The fractional inherent or natural broadening of the spectral line is given by

$$\frac{\Delta f}{f_0} = \frac{1.6 \cdot 10^7}{6 \cdot 10^{14}} = 2.65 \cdot 10^{-8}$$

3. According to the kinetic theory of gases, the average translational kinetic energy of a gas molecule at an absolute temperature T is given by the formula

$$\overline{K} = \frac{3kT}{2} = \frac{mv^2}{2},$$

where k is Boltzman's constant, m the mass of the molecule and v its average velocity. For a molecule of mass $m = 1.67 \cdot 10^{-27}$ kg (a free hydrogen atom) at a temperature of 1,000 K:

$$\begin{aligned} v^2 &= \frac{3kT}{m} \\ &= \frac{3 \cdot 1.381 \cdot 10^{-23} \cdot 1000}{1.67 \cdot 10^{-27}} \\ &= 2.48 \cdot 10^7 \\ v &= 4.98 \cdot 10^3\text{ m/s}. \end{aligned}$$

The fractional change in the observed frequency due to the Doppler effect is given by the formula $\Delta f/f_0 = v/c$, where Δf is the difference between the frequency, f_0, when the source is at rest and the frequency, f, when the source is moving with velocity v; c is the wave velocity. This gives the following value for the fractional Doppler line broadening:

$$\frac{\Delta f}{f_0} = \frac{4.98 \cdot 10^3}{3 \cdot 10^8} = 1.66 \cdot 10^{-5}.$$

Thus, the fractional broadening of the spectral line due to the Doppler effect is two orders of magnitude greater than the natural fractional broadening due to uncertainty. In the case of a spectral line of wavelength 500 nm emitted by a hydrogen atom at a temperature of 1,000 K, the Doppler broadening is ± 0.0083 nm.

6.2.2 The Mechanics of Photon Emission and Absorption

We now turn to the mechanics of a radiative transition between the two stationary states, i and j. Consider a particle of mass m, initially at rest in state i, that emits a photon. The particle recoils backwards as it drops to the lower j stationary state. Since the particle was initially at rest, its total linear momentum before the emission of the photon equalled zero. The conservation of linear momentum requires that after the emission of the photon the sum of its linear momentum, $p_{particle}$, and that of the photon, p_{photon}, must also be zero:

$$p_{particle} + p_{photon} = 0. \tag{6.6}$$

Thus, the particle and the photon emitted from it move apart in opposite directions, each possessing an equal amount of linear momentum.

The linear momentum of a photon is given by $p_{photon} = E_{photon}/c$. Since, after the emission, the photon and particle possess equal amounts of linear momentum, this must also be the magnitude of the linear momentum of the particle. Hence we have

$$|p_{particle}| = |p_{photon}| = \frac{E_{photon}}{c}. \tag{6.7}$$

The total energy of the system after the emission of the photon is the sum of three quantities:

1. the particle's energy in its new stationary state, E_j;
2. the kinetic energy of recoil of the particle, K;
3. the energy of the emitted photon,E_{photon}.

Energy conservation requires that the sum of these three energies equal the energy, E_i, possessed by the particle in its original stationary state, i.e.

$$E_i = E_j + K + E_{photon}. \tag{6.8}$$

Thus, the energy of the photon emitted when the particle falls from the i state to the j state is

$$E_{photon} = (E_i - E_j) - K. \tag{6.9}$$

It can be shown, by similar reasoning, that when a photon strikes the particle and is absorbed, and as a result the particle rises from the j state to the i state, the energy of the absorbed photon is

$$E_{photon} = (E_i - E_j) + K. \tag{6.10}$$

Hence, the energy of the photons absorbed when the particle rises from state j to i is greater by an amount equal to $2K$ than that of the photons emitted when it falls from stationary state i to stationary state j.

The question immediately posed by this result is that if transitions between the same two stationary states involve photons of different energies, depending on whether the photon is emitted or absorbed, how can the observed resonance between identical atoms and molecules be explained?

As we have seen, the difference between the energy levels of a particle is neither an exact nor an absolute quantity; there is always an inherent uncertainty in its value. Consequently, resonance between identical particles does not require the emission and absorption of identical photons. So long as the difference between the energies of the emitted and absorbed photons, $2K$, is less than the sum of the energy widths of the two states, $\Delta E = \Delta E_0 + \Delta E_1$, resonance can occur, i.e. on condition that

$$\Delta E \geq 2K. \tag{6.11}$$

The kinetic (recoil) energy, K, of the particle is related to its linear momentum, $p_{particle}$, by the equation[12]

$$K = \frac{(p_{particle})^2}{2m}. \tag{6.12}$$

Substituting Eq. (6.7) in this expression gives

$$K = \frac{E^2_{photon}}{2mc^2}. \tag{6.13}$$

It follows that there will be resonance between the particles so long as

$$\Delta E \geq \frac{E^2_{photon}}{mc^2}. \tag{6.14}$$

6.2.3 Recoilless Emission and Absorption

The term mc^2 in Eq. (6.13) is equal to the rest-mass of the particle; for atoms and nuclei its value is of the order 10^9 to 10^{11} eV. The energy of the photons emitted or absorbed by atoms in visible and ultra-violet transitions is between 1 and 10 eV, and so the recoil energy, $K = E^2_{photon} 2mc^2$, is of the order 10^{-11} to 10^{-7} eV. The sum of the energy widths, ΔE, in such radiative transitions is usually substantially greater than this, and so the range of photon frequencies absorbed by an atom largely overlaps the range of frequencies emitted by identical atoms – see Fig. 6.10a. In this case the recoil energy can be ignored, resonance conditions exist, and, within experimental limits, Bohr's equation is valid for transitions in both directions.

[12] See Appendix 2.5.1.

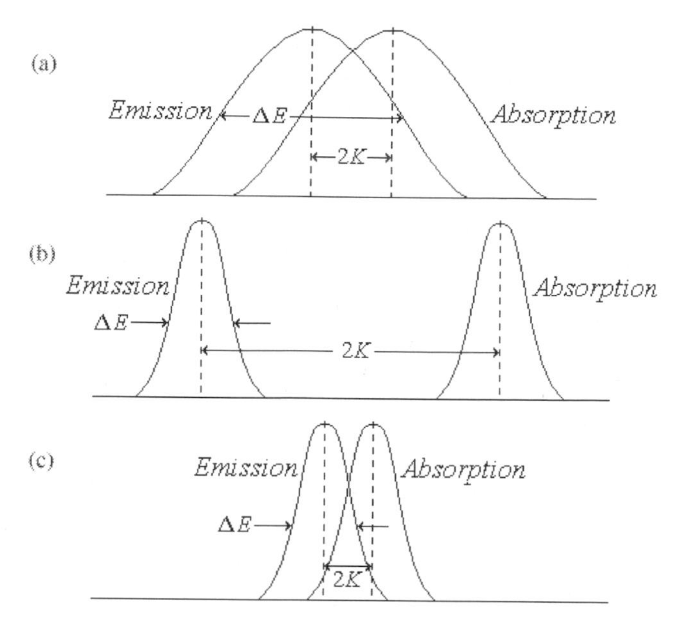

Fig. 6.10. The inherent variance of the energy widths of two states in (a) a typical atomic or molecular system: the absorption and emission peaks overlap and there is resonance between the states; (b) a free nuclear system: the peaks do not overlap and there is no resonance; (c) a nuclear system in which the nuclei are embedded in a crystal: the peaks overlap and there is resonance.

However, the energy, E, of the γ-ray photons emitted by radioactive nuclei is of the order 10^4 to 10^6 eV. This gives a recoil energy, K, of the order 10^{-3} to 10^3 eV. This amount of energy cannot be ignored; the sum of the energy widths, ΔE, is much less than this. Typical values for the lifetimes of excited nuclear states are 10^{-12} to 10^{-8} s, and so the natural width of the spectral lines is of the order 10^{-7} to 10^{-3} eV. Consequently, the photons emitted by a radioactive nucleus are not absorbed by other identical nuclei; resonance conditions do not exist between the nuclei – see Fig. 6.10b.

The conditions needed for γ-ray resonance would be obtained, were it possible to reduce the recoil energy of the nucleus. It is clear from Eq. (6.13) that the larger the particle mass m, the smaller the recoil energy K. Radioactive nuclei that are embedded in a crystal structure do not behave, under certain conditions, as free particles. The particle mass, m, in this case is effectively equal to the mass of the crystal, which is about 10^{23} times that of a single nucleus. Consequently, the sum of the energy widths, ΔE, will be much larger than the recoil energy. The recoil energy of the crystal will be negligible, whatever the energy of the emitted or absorbed photon. Thus, photons emitted by a radioactive nucleus embedded in a crystal will be absorbed by identical nuclei embedded in another crystal – see Fig. 6.10c. This is the phenomenon discovered by Mössbauer in 1958.

Because the nuclei are embedded in a crystal, their movement is restricted and there is no Doppler broadening of the spectral lines. As a result, the width of the lines in the spectrum of Mössbauer radiation is just their natural width, that determined by the uncertainty principle. The lines are very thin, and so resonance conditions are lost if there is even the slightest change in the radiation's frequency. For example, should the emitting crystal be moving relative to the absorbing one, a Doppler change in the frequency of the radiation occurs. Mössbauer found that a relative velocity of just 4 cm/s between the emitting and absorbing crystals was enough to cause the loss of resonance between them (Fig. 6.11). Both the energy width of the excited state in the radioactive nucleus and its average lifetime can be calculated from this result. Nuclear lifetimes shorter than 10^{-11} s have been measured using the Mössbauer effect.

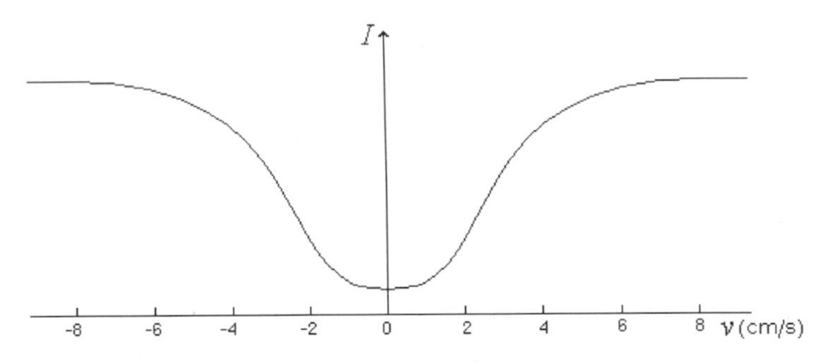

Fig. 6.11. Typical results of an experiment using the Mössbauer effect. I is the intensity of the radiation transmitted by the absorber and v is the relative velocity between the emitter and the absorber. When the relative velocity is zero, almost all the radiation is absorbed. A relative velocity of just 4 cm/s is sufficient to destroy resonance conditions and allow transmission of the radiation through the absorber.

Since any change in the energy levels of a nucleus affects the frequencies of the photons it emits and absorbs, the Mössbauer effect can be used to examine how changes in the immediate environment of a nucleus affect its energy levels. For example, when a nuclear quantum state possesses a magnetic or electric moment that can interact with the internal electrical and magnetic fields of the crystal, the energy levels in the nucleus will be affected by its electrical and magnetic environment. Consequently, identical nuclei, embedded in different crystals, will not necessarily be in resonance even when both crystals are at rest if there is a difference in their immediate environments. The associated changes in the nuclear energy levels can be revealed and calibrated by determining the relative velocity between the crystals required to obtain resonance. Alternatively, if the nuclear energy levels are known, the internal electrical and magnetic fields of the crystal can be studied, yielding information on its structure.

One of the principal nuclei employed for crystal research with the Mössbauer effect is the isotope of iron, Fe–57. The energy of this isotope's first excited state is $14.4 \cdot 10^3$ eV, and it has the relatively long average lifetime of 10^{-7} s. Its nuclear energy levels are affected by:

1. its valence state – whether the iron is bivalent or trivalent;
2. its neighbours in the crystal structure and by the crystal structure itself – whether, for instance, it is tetrahedral or octahedral;
3. whether the compound is ferromagnetic or paramagnetic.

Because the element iron is so widespread in nature and has an isotope that is particularly suited for Mössbauer effect experiments, it is widely used in the study of the solid state, and in chemistry, mineralogy and biology.

6.2.4 The Gravitational Shift — Black Holes

In 1960, Einstein's theory of general relativity was tested experimentally with the help of the Mössbauer effect. We will describe the experiment's theoretical background and enumerate some of its repercussions.

Figure 6.12 illustrates a sequence of events which suggest, at first sight, that an infringement of mass-energy conservation might be a practical possibility.

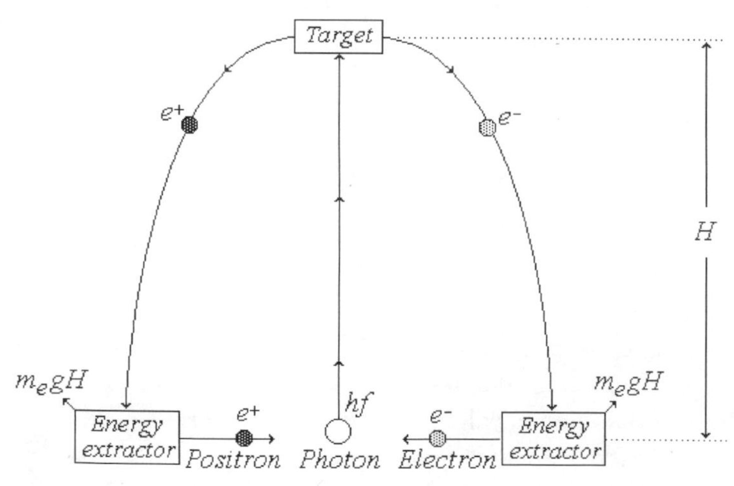

Fig. 6.12. A photon of energy 1.02 MeV strikes the target producing an electron–positron pair. These particles fall, meet and annihilate producing a new photon of energy 1.02 MeV. In each cycle an amount $2m_e gH$ of energy is usefully extracted.

A photon of energy 1.02 MeV (twice the rest-mass energy equivalence, $2m_ec^2$, of an electron and a positron) is aimed upwards from the earth's surface. At a height H, the photon strikes a target nucleus and an electron–positron pair is produced. The electron and positron fall back to earth and in so doing acquire an amount of energy $2m_egH$ from the gravitational field. This energy is extracted and used to perform useful work on the earth's surface. After releasing this energy, the electron and positron collide, mutually annihilate and produce a new photon of energy 1.02 MeV that is aimed upwards from the earth's surface restarting the cycle.

Thus, it would appear, that in each cycle an amount of energy, $\Delta E = 2m_egH$, can be usefully extracted from the process without any investment being made. Clearly, if mass-energy is conserved, this is impossible. The photon must lose an equal amount of energy during its rise from the earth's surface as the electron–positron pair gain during their fall if mass-energy is to be conserved. This energy loss would be manifested as a fractional reduction, $\Delta f/f$, in the frequency of the radiation associated with the photon.

In terms of the energy, E, of the photon produced by their mutual annihilation, the mass, m_e, of the electron and of the positron is

$$m_e = \frac{E}{2c^2} \tag{6.15}$$

and the amount of energy, ΔE, that must be added during each cycle for mass-energy conservation to be maintained is

$$\Delta E = 2m_egH = \frac{EgH}{c^2}. \tag{6.16}$$

As a fraction of the photon's energy, E, this can be expressed as

$$\frac{\Delta E}{E} = \frac{gH}{c^2}. \tag{6.17}$$

Substituting $E = hf$ gives for the required fractional reduction in the photon's frequency,

$$\frac{\Delta f}{f} = \frac{gH}{c^2}. \tag{6.18}$$

The frequency of a photon that 'falls' a distance H should increase by the same extent. The phenomenon is called the *gravitational shift.*

In an experiment that was carried out in 1960, the Mössbauer effect was employed to determine whether the frequency of a 'free-falling' photon does indeed change as predicted by Eq. (6.18). In the experiment, γ-ray photons emitted from radioactive Fe–57 nuclei embedded in a stationary crystal 'fell' a distance of 22.5 m. The predicted fractional frequency change for this height was just $2.46 \cdot 10^{-15}$. However, the Mössbauer effect is so sensitive that even such small changes can be detected by it.

If, indeed, their frequency changed as predicted, the γ-ray photons would not be absorbed by an identical stationary crystal placed directly beneath the emitter; there would be no resonance between them. However, if the absorbing crystal was itself receding from the emitter it would ascribe a lower frequency to the γ-ray photons due to a Doppler effect. At the right speed, resonance would occur and the radiation would be absorbed.

In the experiment, the relative velocity between the emitting and absorbing crystals was adjusted until resonance was obtained, i.e. until the reduction in the photons' frequency due to the Doppler effect exactly cancelled out the increase in their frequency due to the gravitational field. The Doppler change in the frequency of the photons was calculated from the velocity; it exactly matched that predicted by Eq. (6.18).

The shift observed in the frequency of the radiation can be explained in two different ways:

1. In terms of gravitation: by ascribing a gravitational mass to the photons.
2. In terms of an accelerated motion: the consequence of a Doppler effect in an accelerated system.

1. Assuming that the photons have a gravitational mass hf/c^2, the system's gravitational potential energy, U, will change when a photon rises or falls a short distance H by an amount

$$\Delta U = m_{photon} gH$$
$$= \left(\frac{hf}{c^2}\right) gH. \tag{6.19}$$

This change in gravitational potential energy is matched by a corresponding change in the photon's energy, $\Delta E_{photon} = h(f - f')$, when its frequency changes from f to f'. Equating the energy changes gives

$$h(f - f') = \left(\frac{hf}{c^2}\right) gH, \tag{6.20}$$

from which we obtain the expression

$$\frac{f - f'}{f} = \frac{\Delta f}{f} = \frac{gh}{c^2}, \tag{6.21}$$

which is identical to Eq. (6.17).

2. Alternatively, suppose that the experiment illustrated in Fig. 6.12 is conducted in a free-falling laboratory close to the earth's surface; a gravityless laboratory. Moving with a velocity c, the photon strikes the target at a time $\Delta t \approx H/c$ after its upward release. During this time the velocity, v, of the laboratory increases by an amount

$$\Delta v = g\Delta t = \frac{gH}{c}. \tag{6.22}$$

As a result of this change in velocity, there will be a Doppler change in the frequency of the photon:

$$\frac{\Delta f}{f} = \frac{\Delta v}{c} = \frac{gH}{c^2}. \tag{6.23}$$

This relationship is also identical to Eq. (6.17).

Hence, the frequency shift can be explained as being the result either of a gravitational force or of an acceleration, as required by the principle of equivalence.[13] Accordingly, this experiment is regarded as giving confirmation to the theory of general relativity.

The frequency of a photon that 'rises' in a gravitational field decreases; this effect is called a *red-shift.*[14] The *gravitational red-shift* of the radiation emitted by galaxies and stars is usually masked by the greater Doppler shift in its frequency attributed to their recessional motion. Nevertheless, we can imagine a body whose gravitational field is so strong that a photon has insufficient energy to escape from it; such a body is called a *black hole.*

It has been conjectured that black holes could be formed from the complete gravitational collapse of the supernova core that remains after the explosive demise of a massive star. The boundary of a black hole, known as its *event horizon*, is the envelope formed by the space–time paths of the light rays that just fail to escape from it. Hawking[15] has suggested that despite the tremendous gravitational force exerted by a black hole, mass-energy in the form of radiation and particles should nevertheless be able to leak from it. His hypothesis,

[13] See Chapter 1.5.
[14] Red light has a lower frequency than blue light.
[15] Stephen W. Hawking 1942–, English mathematician and physicist.

which is based on general relativity and quantum theory, can be understood as follows:

One of the corollaries of the uncertainty principle is that empty space cannot be a complete void. If it were, the value of the electromagnetic and gravitational fields in the empty space would be precisely zero, in contravention of the uncertainty principle. Empty space must be regarded as a fluctuating field of particle pairs, that are continuously appearing and disappearing; particles being constantly created and annihilated. A particle and its antiparticle form, move apart, come together again a short time later and, finally, mutually annihilate, the process being repeated unceasingly throughout the expanse of 'empty' space–time.

Suppose, that as a result of these *vacuum fluctuations*, a particle pair forms just beyond the event horizon of a black hole and that, before they manage to converge and mutually annihilate, one of them 'falls' into the black hole.[16] Bereft of its partner, the particle remaining outside the event horizon has little prospect of annihilation; it becomes destined to exist. Alone, it moves away from the event horizon and to an observer at some distance from the black hole it appears as though the particle was emitted from it. The particle possesses mass-energy that can only be accounted for as having been acquired at the expense of the black hole. Thus, black holes appear gradually to lose mass-energy, ever shrinking as they radiate more and more particles until they finally disappear in one last cataclysmic explosion.

[16]Possibly as a result of the tremendous tidal forces exerted on the pair by the black hole's powerful but non-uniform gravitational field.

Chapter 6.3

Nuclear Magnetic Resonance

The concept of nuclear magnetism dates back almost to the time of Rutherford's original discovery of the atomic nucleus. The components of the nucleus were initially thought to be protons, electrons and perhaps α particles, and it was presumed that these charged particles would be involved in some sort of rotational motion and would, therefore, give rise to a magnetic moment. In 1924, Pauli suggested that the hyperfine structure of atomic spectra resulted from an interaction between the orbital motion of the electrons and the angular momentum of the nucleus. Later researchers suggested that the interaction might be magnetic and the first values of nuclear magnetic moments were inferred from spectral measurements.

Initially, precise measurements of nuclear magnetic moments proved hard to achieve; they are some three orders of magnitude smaller that that of the electron. The breakthrough in accurate measurements came when techniques were discovered for inducing the transition of the magnets associated with nuclear spin from one energy level to another. Under the influence of an oscillating magnetic field, the nuclear spins resonantly absorb energy from the magnetic field or release energy to the field by stimulated emission. This can be detected by suitable instrumentation.

At first, these nuclear magnetic phenomena, termed *nuclear magnetic resonance* (*NMR*), were of interest mainly to physicists. But when, in 1951, it was reported that the magnetic moments associated with the protons in ethanol (CH_3CH_2OH) absorb/emit energy at three different frequencies and at intensities in the ratio of 3:2:1, nuclear magnetic resonance had arrived for the organic chemist. Since then, it has established itself as one of the most valuable tools for studying chemical compounds. Few analytic or diagnostic techniques provide such an abundance of accurate data.

In many cases the nucleus under investigation is that of a hydrogen atom, i.e. a proton; the effect is then often referred to as *proton magnetic resonance* (*PMR*). A similar mechanism is involved in *electron spin resonance* (*ESR*) where the magnetic dipole concerned is that associated with an electron's angular motion.

For an understanding of how nuclear magnetic resonance works, we must start by examining the sources of the magnetic properties of nuclei.

6.3.1 Magnetism and Angular Momentum

Despite the repeated attempts that have been made to isolate a natural unit of magnetism similar to the natural unit of electricity – a magnetic monopole or charge – one has never been discovered or identified. Magnetic poles always seem to occur in pairs; a north pole facing one way and a south pole facing the other. Thus, the fundamental 'particle' of magnetism appears to be the *magnetic moment* or *magnetic dipole* whose existence is attributed to the angular motion of charged elementary particles. Examples of this angular motion are:

* the motion of an electron around a conducting loop;
* the orbital motion of an atomic electron relative to the nucleus;
* the intrinsic spin of an electron or of any other charge carrying elementary particle.

In general, the magnetic dipole moment, $\boldsymbol{\mu}$, associated with a particle's angular momentum, **J**, is given by

$$\boldsymbol{\mu} = \gamma \mathbf{J}\left(\frac{e}{2m}\right), \tag{6.24}$$

where m is the particle's mass, e is the natural unit of charge and γ a parameter of the system called the *gyromagnetic ratio.* For example, the magnetic moment associated with the orbital motion of a particle is given by

$$\boldsymbol{\mu} = \gamma_L \mathbf{L}\left(\frac{e}{2m}\right), \tag{6.25}$$

where **L** is the particle's orbital angular momentum. In the case of a single electron, the orbital gyromagnetic ratio $(\gamma_L)_{electron} = -1$. For an orbiting proton, the gyromagnetic ratio $(\gamma_L)_{proton} = +1$.

The magnetic moment associated with a particle's spin, **S**, is given by

$$\boldsymbol{\mu}_S = \gamma_S \mathbf{S}\left(\frac{e}{2m}\right), \tag{6.26}$$

where γ_S is the particle's spin gyromagnetic ratio. The spin gyromagnetic ratios of the electron, proton and neutron are respectively

$$(\gamma_S)_{electron} = -2.0023; \quad (\gamma_S)_{proton} = +5.5851; \quad (\gamma_S)_{neutron} = -3.8256.$$

In general, the energy possessed by a magnetic dipole situated in a magnetic field is given by the scalar multiplication

$$E = -\boldsymbol{\mu} \cdot \mathbf{B} = -\mu \mathrm{B} \cos\theta, \tag{6.27}$$

where θ is the angle between the direction of the magnetic dipole vector, $\boldsymbol{\mu}$, and that of the magnetic field vector, $\mathbf{B}$. The dipole has its minimum energy, $E_{\min} = -\mu B$, when it is aligned with the direction of the field, i.e. when $\theta = 0°$, and its maximum energy, $E_{\max} = +\mu B$, when its direction is opposite to that of the field, i.e. when $\theta = 180°$.

6.3.2 Nuclear Magnetic Moments

Atomic nuclei possess an intrinsic angular momentum, $\mathbf{I}$, called *nuclear spin*, that arises from the combination of the intrinsic angular momentum (spin) of their constituent nucleons with their motion inside the nucleus.

Nuclear spin is quantised. The eigenvalues of the square, I^2, of the magnitude of the nuclear spin are given by

$$I^2 = l(l+1)\hbar^2, \tag{6.28}$$

where l is the *nuclear spin quantum number.*[17] Relative to any arbitrarily chosen axis, the nuclear spin can take up $(2l+1)$ orientations. The permitted values of the component, I_z, of nuclear spin parallel to an arbitrarily chosen Z-axis are

$$I_z = m_l\hbar \tag{6.29}$$

where $m_l = -l, (-l+1), (-l+2)\ldots, +l$.

It has been found, empirically, that the values of the nuclear spin quantum number are subject to the following rules:

* In nuclei containing an even number of nucleons, the nuclear spin quantum number, l, is an integer.
* In nuclei containing an odd number of nucleons the nuclear spin quantum number, l, is an integer multiple of a half.
* In almost all nuclei containing an even number of both protons and neutrons, the nuclear spin quantum number $l = 0$.

These rules suggest that the protons and neutrons in a nucleus each tend to pair their angular momenta in opposite directions.

By convention, the nuclear magnetic moments associated with nuclear spin are expressed in terms of the mass of the proton, m_p. Accordingly, the magnetic moment of a nucleus is given by

$$\boldsymbol{\mu}_{Nucleus} = \gamma_N \mathbf{I}\left(\frac{e}{2m_p}\right), \tag{6.30}$$

[17] To avoid confusion, a different font is used for the nuclear spin quantum number, l, than for the magnitude, I, of the nuclear spin.

where γ_N is the *nuclear gyromagnetic ratio.* The allowed components, $(\mu_{Nucleus})_z$, of the nuclear magnetic moment parallel to an arbitrarily chosen Z-axis are given by

$$(\mu_{Nucleus})_z = \gamma_N I_z \left(\frac{e}{2m_p}\right)$$
$$= \gamma_N m_l \hbar \left(\frac{e}{2m_p}\right). \quad (6.31)$$

This equation can be written in the concise form

$$(\mu_{Nucleus})_z = \gamma_N \mu_N m_l \quad (6.32)$$

where $m_l = -l, (-l+1), (-l+2)\ldots, +l$.

The quantity μ_N is called the *nuclear magneton*:

$$\mu_N = \frac{e\hbar}{2m_p} = 5.051 \cdot 10^{-27}\mathrm{J} \cdot \mathrm{T}^{-1}; \quad (6.33)$$

its value is 1,836 times smaller than the Bohr magneton.

In a magnetic field, **B**, the energy levels, E_{m_l}, associated with the z-component of the nuclear magnetic moment are given by

$$E_{m_l} = -(\mu_{Nucleus})_z B_z$$
$$= -\gamma_N \mu_N m_l B_z \quad (6.34)$$

where $m_l = -l, (-l+1), (-l+2)\ldots, +l$.

In the case of a hydrogen nucleus (a proton), whose nuclear spin quantum number is $l = \frac{1}{2}$, just two energy levels are possible. The lower level, $E_{+\frac{1}{2}}$, corresponds to the $m_l = +\frac{1}{2}$ state and the higher, $E_{-\frac{1}{2}}$, to the $m_l = -\frac{1}{2}$ state. The difference between the two energy levels is

$$\Delta E = -\gamma_N \mu_N B_z \left[-\frac{1}{2} - \left(+\frac{1}{2}\right)\right]$$
$$= 5.5851 \cdot 5.051 \cdot 10^{-27} B_z$$
$$= 28.2 \cdot 10^{-27} B_z. \quad (6.35)$$

Note that the difference between the energy levels is a linear function of B_z, the magnitude of the z-component of the magnetic field (Fig. 6.13).

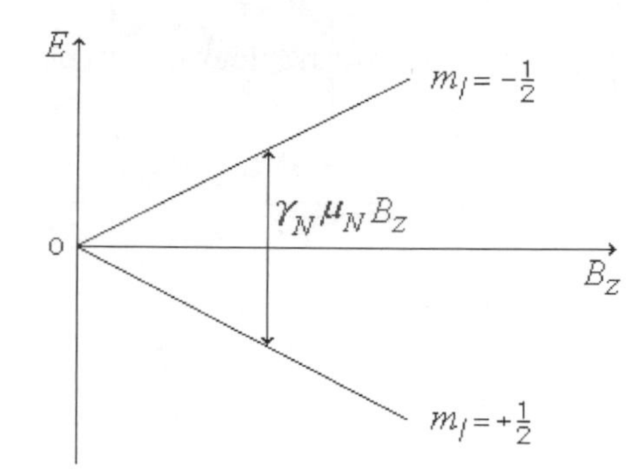

Fig. 6.13. Diagram of the energy levels of the z-component of the magnetic moment of a proton in an external magnetic field. The stronger the magnetic field, the greater the difference between the levels.

Because the nuclear magneton, μ_N, is some three orders of magnitude smaller than the Bohr magneton, μ_B, the differences between the energy levels associated with nuclear magnetic moments are substantially lower than those associated with electronic magnetic moments. For a proton immersed in a magnetic field of magnitude $B_z = 1\text{T}$, the difference, ΔE, between its two spin energy levels is

$$\Delta E = \frac{28.2 \cdot 10^{-27}}{1.6 \cdot 10^{-19}} \cdot 1 = 17.63 \cdot 10^{-8}\,\text{eV}. \tag{6.36}$$

This is a small amount compared to the thermal energy of the atoms. Hence, given the Maxwell–Boltzmann distribution, the ratio between the populations of the two spin states at room temperature ($\sim$300 K) will be

$$\begin{aligned} N_{Lower}\Big/N_{Higher} &= e^{\Delta E/kT} \\ &= e^{\left(17.63 \cdot 10^{-8}/8.62 \cdot 10^{-5} \cdot 300\right)} \\ &= 1.000007, \end{aligned} \tag{6.37}$$

i.e. the population of the lower state is only very slightly greater than that of the higher state. The ratio can be increased by cooling the system or by increasing the magnitude of the magnetic field.

The smallness of the nuclear magneton also affects the transition frequencies between the nuclear spin energy levels. In radiative transitions between the proton spin energy levels in a magnetic field of magnitude $B_z = 1\text{T}$, a photon of frequency

$$f = \frac{\Delta E}{h} = \frac{17.63 \cdot 10^{-8}}{4.14 \cdot 10^{-15}} = 4.26 \cdot 10^{7}\,\text{Hz} \tag{6.38}$$

will be emitted or absorbed. This frequency is in the radio region of the electromagnetic spectrum. By contrast, radiative transitions associated with electronic spin energy levels are in the microwave region.

Typical values for nuclear magnetic data are given in Table 6.1.

Table 6.1. Magnetic data of selected nuclides.

Nuclide	Nuclear spin quantum number, l	Transition frequency in a magnetic field of magnitude 1.4094 T[18]	Nuclear gyromagnetic ratio - γ_N
^{1}H	$\frac{1}{2}$	60.000 MHz	5.585
^{2}D	1	9.210	0.857
^{7}Li	$1\frac{1}{2}$	23.317	2.171
^{14}N	1	4.334	0.403
^{15}N	$\frac{1}{2}$	6.080	−0.567
^{17}O	$2\frac{1}{2}$	8.134	−0.757
^{19}F	$\frac{1}{2}$	56.446	5.257
^{23}Na	$1\frac{1}{2}$	15.871	1.478

6.3.3 Nuclear Magnetic Resonance

In semi-classical terms, the behaviour of a nuclear magnetic moment in a magnetic field is analogous to that of the orbiting electron in the Bohr–Sommerfeld model of the hydrogen atom.[19] It can, likewise, be visualised as precessing around the direction of the field (Fig. 6.14). On this view, the transition frequency of the nuclear magnetic dipole is the Larmor frequency of its precessional motion, f_L, and the photons emitted or absorbed in these transitions will be of energy hf_L. Since the Larmor frequency is itself a function of the applied magnetic field, B, the frequency of the photons emitted or absorbed will also depend on the field strength.

At the radio frequencies involved in magnetic dipole transitions, the probability of the spontaneous emission of a photon is negligible. Thus, both the absorption and emission of photons from nuclear magnetic dipoles requires the stimulus of an external source. On the semi-classical model, the absorption or emission of a photon by a nuclear magnetic dipole involves a change in its orientation relative to the applied field, B. In the case of a proton, where $l = \frac{1}{2}$,

[18] Conventionally, the transition frequencies of nuclear magnetic moments are given in relation to a magnetic field of magnitude 1.4094 T, the magnetic field in which the transition frequency of a proton is 60 MHz.
[19] See Section 3.2.2.

this requires flipping the magnetic dipole through 180°, i.e. from $m_l = +\frac{1}{2}$ to $m_l = -\frac{1}{2}$, or *vice versa.*

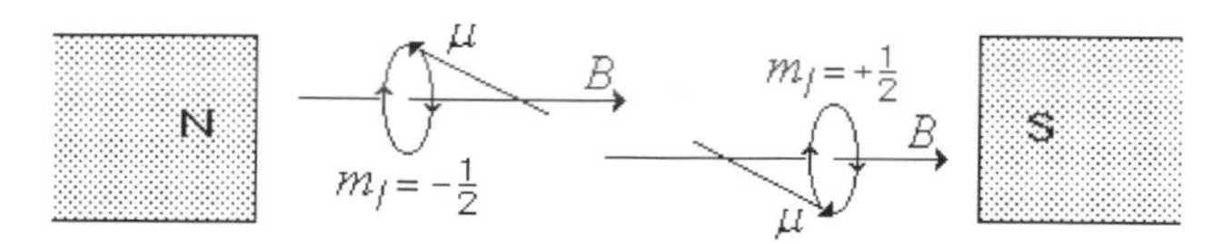

Fig. 6.14. Semi-classical visualisation of the precession of a proton's magnetic dipole moment, μ, around the direction of a uniform magnetic field, B. The diagram shows the two possible orientations of the dipole, one corresponding to $m_l = +\frac{1}{2}$, the other to $m_l = -\frac{1}{2}$.

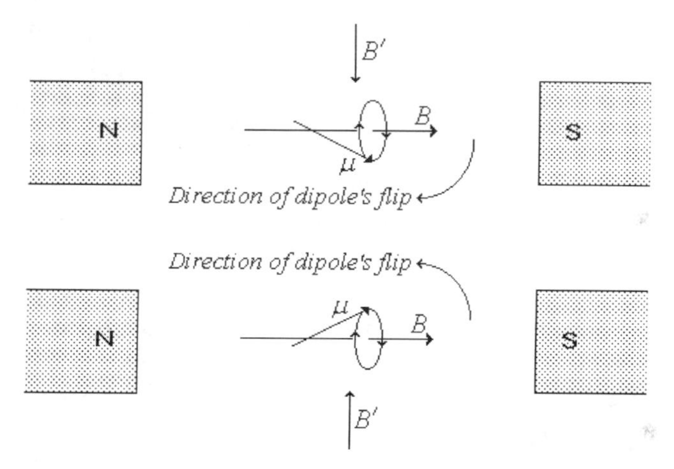

Fig. 6.15. Nuclear magnetic resonance. Only when the frequency of the oscillating secondary magnetic field, B', equals that of the dipole's precessional motion, does a turning moment that tends to flip the axis of its precession through 180° act continuously on the dipole. Under these conditions, the dipole can flip from the $m_l = +\frac{1}{2}$ state to the $m_l = -\frac{1}{2}$ state.

In order to flip the precessing magnetic dipoles through 180°, a force must be applied to them at right angles to the axis around which they are precessing. This can be achieved by the action of a secondary oscillating magnetic field, B', whose frequency equals the Larmor frequency of the dipoles' precessional motion. Only when the oscillations of the secondary 'tipping' field are exactly synchronised with the precessional motion of the dipole is the transition induced and is energy exchanged between the field and the dipole (Fig. 6.15).

6.3.4 Observing Nuclear Magnetic Resonance

Nuclear magnetic resonance can be detected by several techniques. We will describe three of them: molecular-beam resonance, absorption resonance and induction resonance. In all of these, the material whose nuclear magnetic dipole moments are under examination is immersed in a uniform magnetic field, B, that is crossed by an oscillating magnetic field, B' (Fig. 6.16).

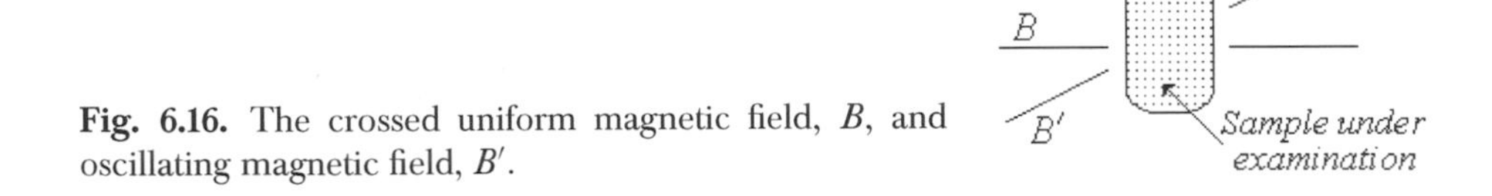

Fig. 6.16. The crossed uniform magnetic field, B, and oscillating magnetic field, B'.

Although, in principle, resonance can be located either by adjusting the frequency of the oscillating field, B', to that of the dipoles' precessional (Larmor) frequency or by varying the magnitude of the uniform field, B, until the frequency of the dipoles' precession equals that of the oscillating magnetic field, in practice, the latter option is usually technically easier. The three methods we describe differ in the way they detect that resonance has been achieved between the precessing nuclear dipole and the oscillating magnetic field.

1. Molecular-beam resonance

This technique, which is closely related to the design of the Stern–Gerlach experiment, was first developed by Rabi.[20] It likewise exploits the effect of the non-uniform magnetic field produced in the space between the differently shaped poles of an electromagnet on spin magnetic moments.

Gaseous molecules emitted by a heated source are introduced into one end of the space between the differently shaped poles of an electromagnet (Fig. 6.17). The molecules emerging from the slit at the far end of this space enter a narrow region in which a uniform magnetic field, B, is crossed by an oscillating magnetic field, B'. The molecules enter the region at right angles to both fields.

If the orientation of their nuclear magnetic dipole moments is unchanged during the passage of the molecules through the crossed fields, any deflections they had sustained in their passage through the non-uniform magnetic field can be annulled by passing them through a second identical but reversed non-uniform field. However, should there be resonance between the nuclear magnetic dipoles and the oscillating magnetic field, they will flip into new orientations. In these new orientations, the deflections they had sustained in the first non-uniform magnetic field will not be annulled by passing them through an

[20] Isidor Rabi 1898–1988, American physicist who was awarded the Nobel prize in 1944 for his work on molecular-beam resonance.

identical but reversed second non-uniform magnetic field. Resonance between the dipoles and the oscillating field, B', will be indicated by a reduction in the beam intensity registered by a detector positioned at the exit from the second non-uniform field.

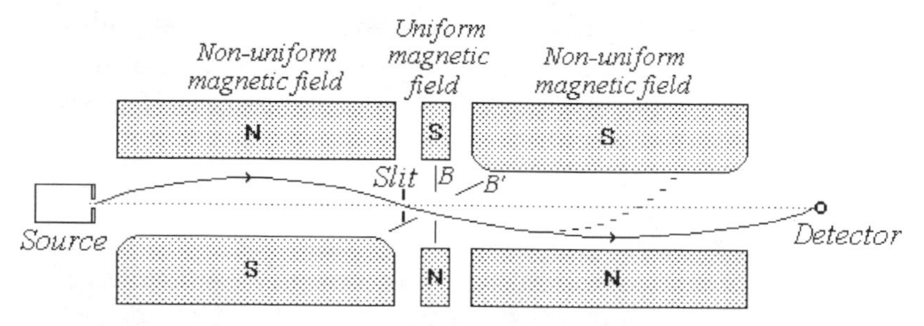

Fig. 6.17. Molecular-beam resonance. The oscillating magnetic field B' is generated by passing a radio frequency alternating current through a coil. If the orientation of the nuclear magnetic dipole moments is unaffected by the crossed magnetic fields, B and B', the molecular-beam reaches the detector. However, if a change occurs in their orientation the path taken by the molecules through the second non-uniform magnetic field (the broken line) will not match the one they took through the first field and so they will not reach the detector.

In the experiments carried out by Rabi and his coworkers, the frequency of the oscillating magnetic field, B', was held constant as the magnitude of the uniform field B was varied. Typical results they obtained are shown in Fig. 6.18.

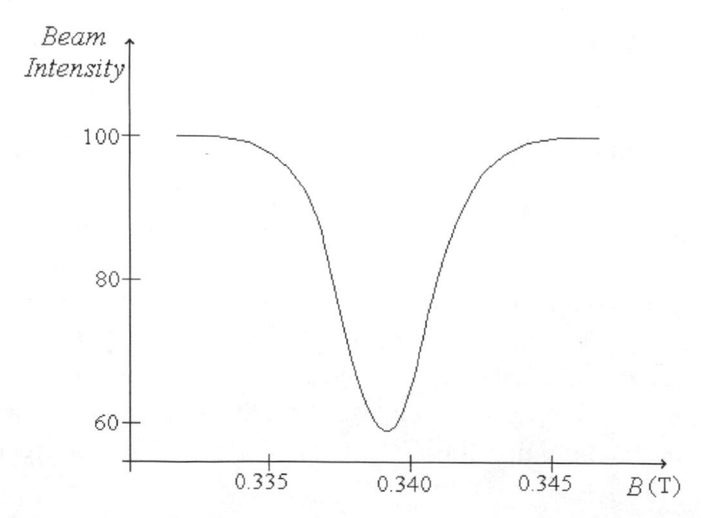

Fig. 6.18. The beam intensity as a function of the magnitude of the uniform field, B, in an experiment with LiCl molecules. The frequency of the oscillating field was 5.585 MHz. Resonance due to the ^{7}Li isotope occurred at $B = 0.3384$ T.

The molecular-beam technique can only be employed with gases. Nuclear magnetic resonance in liquids and solids cannot be investigated by this method. The following two methods were subsequently developed to answer this need.

2. Nuclear absorption resonance

This technique for observing nuclear magnetic resonance in liquids and solids was developed by Purcell.[21] It involves measuring the nuclear absorption of the energy carried by radio waves whose oscillating magnetic field supplies the 'tipping' B' field (Fig. 6.19).

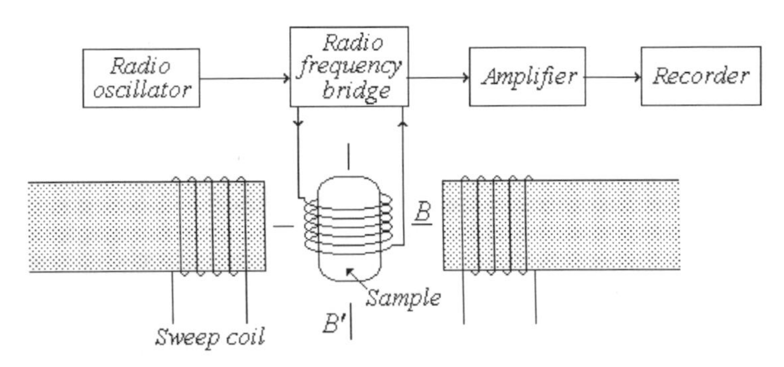

Fig. 6.19. Schematic diagram of nuclear absorption NMR spectrometer. The frequency of the secondary field, B', is held constant. The primary field, B, is varied by means of the sweep coil. Changes in the intensity of the radio waves are recorded as a function of the primary field B.

The material under examination is irradiated with radio waves of a fixed frequency. At resonance, nuclei in the sample absorb energy from the oscillating field. This is indicated by a sharp attenuation of the intensity of the radio waves. The recorded spectrum shows the intensity of the radio waves as a function of the magnitude of the primary field (Fig. 6.19). Absorption peaks correspond to resonance between the field and the nuclei.

Only those nuclei that are in the lower spin state can absorb energy from the field. Thus, the strength of the radio frequency signal picked up by the spectrometer depends on the number of nuclei rising from the lower to the higher state. However, as Eq. (6.37) shows, the population of the ground state is only marginally greater than that of the excited state. The situation can be improved by cooling or by using very strong magnets to produce the B-field – up to 10 T. However, maintaining the uniformity of such fields is a great technical challenge.

[21] Edward Purcell 1912–1997, American physicist who was awarded the Nobel prize in 1952 for his work on nuclear magnetic resonance.

A further problem is that if the sample is irradiated by a continuous strong oscillating field, the populations of the two states rapidly equalise. With equal populations in each state, the number of photons absorbed by the sample will equal the number emitted and no signal will be received. This can be avoided either by employing weak oscillating fields or, alternatively, by using strong fields in short pulses, between which the populations of the two states can return to their equilibrium numbers.

3. Nuclear induction resonance

This technique for observing nuclear magnetic resonance in liquids and solids was developed by Bloch.[22] It involves detecting the radiation emitted by the excited nuclei as they fall from a higher spin state to a lower one (Fig. 6.20).

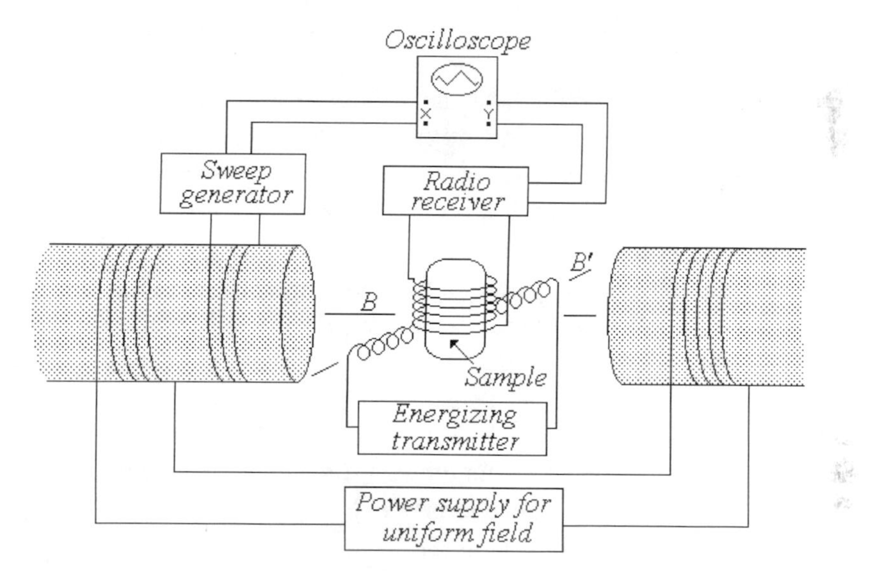

Fig. 6.20. Schematic diagram of nuclear induction NMR spectrometer. The frequency of the secondary field, B', induced by the energising transmitter is held constant. The magnitude of the primary field, B, is varied by means of the sweep coil. Changes in the intensity of the radio waves emitted are recorded as a function of the primary field B. A computer can replace the oscilloscope.

The radio frequency radiation is picked up by a coil (antenna) whose axis is at right angles to both the uniform filed, B, and the oscillating field, B'. After amplification, the signal from the coil is typically fed into the Y-plates of an oscilloscope whose X-plates are connected to the sweep coil that varies the magnitude of the uniform field, B.

[22]Felix Bloch 1905–1983, American physicist who was awarded the Nobel prize in 1952 for his work on nuclear magnetic resonance.

At resonance, nuclei are excited into a higher spin state by the oscillating field, B', generated by the energising transmitter and the intensity of the radiation they subsequently emit increases correspondingly. The spectrum appears on the oscilloscope screen as a trace showing the intensity of the emitted radiation as a function of the magnitude of the uniform field, B (Fig. 6.21).

In an adaptation of the induction spectrometer, the sample is irradiated with a range of radio frequencies – *white radiation.* In order to detect the dominant frequencies in the ensuing radiation emitted by the sample, it is subjected to a Fourier transform. This technique is extremely sensitive and can be used to study the NMR spectrum of nuclei present in low concentrations, such as ^{13}C.

6.3.5 Chemical Shift

The effective magnetic field, $\mathbf{B}_{Eff}$, that actually determines the energy levels of the nuclear magnetic moments in the sample under examination in NMR absorption or induction spectroscopy differs from the applied B-field. Under the influence of the B-field, the atomic electrons in the material are set into motion. This motion in turn induces a secondary magnetic field, $\mathbf{B}_{Ind}$. The effective field experienced by the nuclear magnetic moments is the resultant of the applied field, $\mathbf{B}$, and the induced field, $\mathbf{B}_{Ind}$:

$$\mathbf{B}_{Eff} = \mathbf{B} + \mathbf{B}_{Ind}. \tag{6.39}$$

The induced field generally opposes the applied B-field. This is an instance of the familiar phenomenon of diamagnetism discovered by Faraday.[23]

The nature of the induced secondary field, $\mathbf{B}_{Ind}$, depends on the electronic configuration of the system. In general,

$$\mathbf{B}_{Ind} = \sigma\mathbf{B}, \tag{6.40a}$$

where σ is a constant characteristic of the nuclear environment called the *shielding constant.* It follows that the effective field experienced by the nuclear magnetic moment is

$$\mathbf{B}_{Eff} = \mathbf{B}(1 - \sigma). \tag{6.40b}$$

The frequency of the photons absorbed and emitted by a nuclear magnetic moment depends on the actual magnetic field it experiences, which in turn depends on its electronic environment. Thus, any change in the immediate electronic environment of the nuclear magnetic moments will result in a shift

[23]The complementary paramagnetic effect of field enhancement also occurs, though much more rarely.

in the photon frequency. This is the basis for the widespread use of NMR spectroscopy in the study of molecular structure.

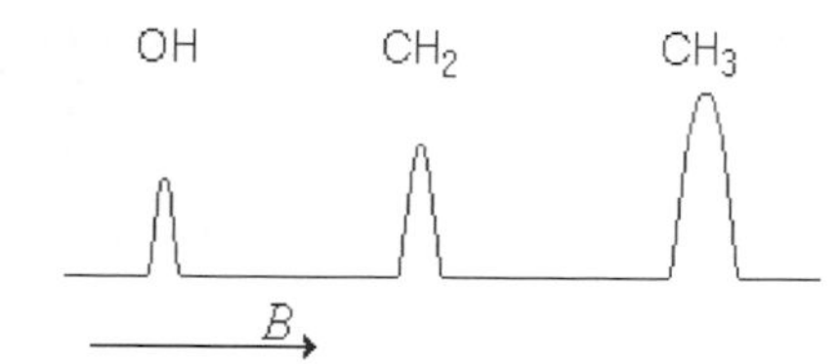

Fig. 6.21. The low resolution NMR spectrum of ethanol.

For a given nuclear magnetic dipole moment, say that of a proton, the greater the diamagnetic shielding it experiences, the stronger the external B-field required for resonance to be observed. The low resolution NMR spectrum of ethanol, CH_3CH_2OH shown in Fig. 6.21 is an example of this effect. Three distinct bands appear in the spectrum corresponding to the CH_3, CH_2 and OH groups. The area of each band is proportional to the number of hydrogen atoms each comprises. It is apparent that a stronger field is required for the resonance of the CH_3 protons than for the CH_2 protons, indicating the former's greater diamagnetic shielding; the proton in the OH group is the least shielded.

Because the B-field magnitude at which resonance occurs depends on the frequency of the oscillating B'-field, it is meaningless to express NMR data in absolute units; a value of the B-field magnitude that gives rise to resonance in one spectrometer will not necessarily produce resonance in another spectrometer that operates at a different frequency. For this reason, NMR data are always expressed with reference to a standard substance.

A standard commonly employed in proton magnetic resonance is Tetramethylsilane or TMS, $Si(CH_3)_4$, which contains 12 magnetically identical protons. The shielding constant of TMS is relatively large and its NMR spectrum is a single sharp line. Moreover, TMS is chemically inert and miscible with many solvents and so can be added as an internal standard to the samples under investigation without affecting them chemically or magnetically.

Suppose that in a particular NMR spectrometer, the protons in TMS resonate at a field magnitude of B_{TMS} and those in the CH_3 group in ethanol at a field magnitude of B_{CH_3}. This will be reported in the form

$$\delta_{CH_3} = \frac{B_{TMS} - B_{CH_3}}{B_{TMS}} \cdot 10^6 \text{ parts per million,} \tag{6.41}$$

where δ_{CH_3} is called the *chemical shift* of the protons comprising the CH_3 group. The NMR spectrum of ethanol exhibits three chemical shifts: that corresponding to the three protons in the CH_3 group, that of the two protons in the CH_2 group and that of the single proton in the OH group. All the protons in each group share the same chemical environment. The range of chemical shifts for most hydrogen nuclei (protons) is about ten parts per million.

Chemical shifts can also be expressed in terms of frequencies:

$$\delta_{Nucleus} = \frac{f_{Nucleus} - f_{\mathrm{TMS}}}{f_{Spectrometer}} \cdot 10^6 \text{ parts per million,} \tag{6.42}$$

where $f_{Nucleus}$ is the resonance frequency of the magnetic moment of the nucleus under investigation, f_{TMS} is the resonance frequency of the protons in the TMS molecule and $f_{Spectrometer}$ is the operating frequency of the NMR spectrometer.

The secondary magnetic fields responsible for the chemical shifts are very weak, typically less than 0.001% of the applied B-field. In a B-field of magnitude 1.4094 T, the maximum displacement in the resonance frequencies of hydrogen nuclei (protons) due to their chemical environment is only 600 Hz. This is a very small shift compared to the resonance frequency of free protons in such a field, *viz.* 60 MHz. Thus, the detection of chemical shifts requires great experimental precision. In order to achieve a resolution of ± 1 Hz, the uniformity of the B-field and the frequency of the B'-field must vary by no more than one part in 60 million. Higher resolutions require still greater precision.

The ultimate limit on the resolution that can be obtained by any spectroscopic technique is set by the inherent natural width of the spectral lines. From Heisenberg's uncertainty principle, $\Delta E \cdot \Delta t \approx \hbar$, and the relationship $\Delta E = h\Delta f$, the natural width, Δf, of a spectral line is given by

$$\Delta f \approx \frac{1}{(2\pi \Delta t)}, \tag{6.43}$$

where Δt is the lifetime of the excited state. The lifetimes of nuclear spin states are often called their *relaxation times.*

Nuclear relaxation times are affected by the coupling that occurs between the nuclear magnetic moments. Relaxation times in solids are short, $\sim 1\,\mu$s, because of the direct spin–spin dipole coupling that occurs between their magnetic moments. As a result, the NMR spectral lines of solids are broad, $\Delta f \approx 100,000$ Hz, and can provide no information about chemical shifts. In certain cases, the lines can be narrowed significantly by spinning the solid sample at high speed at so-called *magic angles* relative to the direction of the B-field.

On the other hand, because of the rapid random motion of their molecules, the coupling of the nuclear magnetic moments in liquids is negligible. Thus, relaxation times in liquids are relatively long, ~ 1 s, and the NMR spectral lines are correspondingly narrow, ~ 0.15 Hz. For example, the inherent uncertainty of the standard 60 MHz/1.4904 T proton line is less than one part in

100 million. To take advantage of this inherent accuracy and the opportunity it presents for gaining information about the proton and its close magnetic environment, the magnetic field, B, in both the absorption and induction NMR spectrometers must be uniform to a degree of one part in 100 million. To average out any remaining field inhomogeneities, the sample tube is usually spun at about 60 revolutions per minute about an axis perpendicular to the B-field. The variation in the frequency of the B'-field must also be less than one part per 100 million.

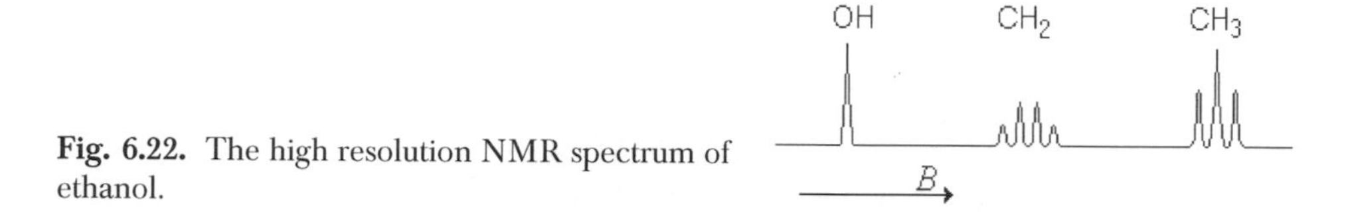

Fig. 6.22. The high resolution NMR spectrum of ethanol.

The high resolution NMR spectrum of ethanol shows that the CH_2 and CH_3 bands are split into a number of narrower bands (Fig. 6.22). The phenomenon, called *spin–spin coupling*, results from the electron mediated interactions of the various nuclear spins in the molecule.

Because of its mobility, the proton in the hydroxyl group does not couple with the other protons in the molecule. However, because the nuclear spins of the three protons in the CH_3 group can be arranged in four distinct ways,

$$\begin{array}{cccc} & \uparrow\uparrow\downarrow & \downarrow\downarrow\uparrow & \\ \uparrow\uparrow\uparrow & \uparrow\downarrow\uparrow & \downarrow\uparrow\downarrow & \downarrow\downarrow\downarrow \\ & \downarrow\uparrow\uparrow & \uparrow\downarrow\downarrow & \end{array}$$

each of the two protons in the CH_2 group can couple with this neighbour in four ways; the result is the splitting of its spectral line into a quartet. On the other hand, because the nuclear spins of the two protons in the CH_2 group can only be arranged in three distinct ways,

$$\begin{array}{ccc} & \uparrow\downarrow & \\ \downarrow\downarrow & & \uparrow\uparrow \\ & \downarrow\uparrow & \end{array}$$

each of the three protons of the CH_3 group can couple with their neighbour in just three ways; the result is the splitting of its spectral line into a triplet. The extent of the band splitting is expressed in terms of a coupling constant, J, that, unlike the chemical shift, is independent of the magnitude of the external B-field. Typically, J is of the order of a few Hertz.

6.3.6 Applications of Nuclear Magnetic Resonance

Since its initial development, nuclear magnetic resonance has become a standard research and diagnostic tool. Initially, its major application was in the study of molecular structures. By measuring chemical shifts and examining the breadth and fine structure of the spectral lines, a vast quantity of detailed and accurate information could be obtained about the shape and environment of molecules.

Important information concerning the structure of biological membranes and the proportions of phospholipid vesicles and detergent and bile-salt micelles has been obtained from the NMR spectrum of the ^{1}H, ^{2}D, ^{13}C and ^{31}P nuclei incorporated in these species. Well-defined narrow spectral lines were found in the spectra of small vesicles and micelles (radii up to 20 nm). However, the lines widened and even disappeared in large multi-layered vesicles (radii of $\sim 1\,\mu m$). Using the NMR spectrum of ^{31}P, phase transitions from bilayers to other configurations have been identified in membranes. Similarly, increases in the viscosity of phospholipid membranes caused by cations such as Ca^{++}, Mg^{++} and Na^{+} have been detected.

More recently, non-invasive techniques have been developed by which images of the structure of individual cells, tissues and even whole organisms can be obtained using nuclear magnetic resonance. The technique is called *NMR imaging* or *magnetic resonance imaging* (*MRI*). Unlike X-ray images, where the signal strength (whether the image is black or white) is determined by one factor alone – attenuation – MRI signals are affected by many factors, chief among them proton density and tissue type.

The process hinges on three factors:

1. The dependence of the resonant (Larmor) frequency on the strength of the magnetic field.
2. The dependence of the relaxation time on the environment of the nucleus.
3. The power of electronic computers to collate and display the mass of data obtained in a useful format.

If a specimen containing nuclear magnetic moments is placed in a non-uniform magnetic field (a magnetic field with a spatial gradient), the resonant frequency exhibited by each nucleus will depend on its position in the field. If the contours (the map) of the magnetic field are known, the observed frequency of given nuclei in the specimen, say the protons, will specify their spatial coordinates (Fig. 6.23). In this way, nuclei in one part of the specimen can be distinguished from those in another. By changing the contours or the orientation of the field, other sets of coordinates can be obtained. Combining the various sets of coordinates produces a matrix of the overall distribution of the nuclei in the specimen. For diagnostic purposes, this can be displayed as a series of two-dimensional slices through the specimen in which regions of high proton density show up brighter. Such cross-sectional images are called *tomographs* and the technique is known as *tomography*.

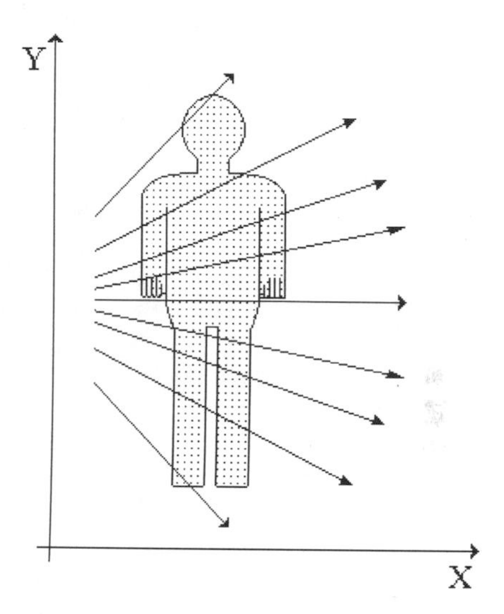

Fig. 6.23. A non-uniform magnetic field. The field at each point is a function of the point's coordinates.

Contrast can be added to the images by exploiting the different relaxation times of the nuclei in the specimen. Protons in tissues such as subcutaneous fat and bone marrow have shorter relaxation times than those in fluids such as the cerebrospinal and ocular fluids. These differences are revealed by applying the oscillating magnetic field, B', in carefully timed pulses of a fairly broad band of frequencies. If the time period, τ, between the pulses is less than the relaxation time, T, of a particular nucleus in the sample, the nucleus will not respond to subsequent pulses. The first pulse elevates the nucleus into the relatively stable higher state and as long as it remains in this elevated state, it will not respond to the later pulses. Only those nuclei whose environment does not hinder their relaxation into the ground state, i.e. those nuclei whose relaxation time, T, is less than τ, will respond to the later pulses. These differences between the nuclei are detected electronically and add contrast and detail to the two-dimensional images of the nuclei in the specimen. By taking a sequence of images as the sample traverses the magnetic field, 'whole-body' scans can be obtained. In order to obtain the strong ($>1\,T$) and spatially large magnetic fields required for human whole-body imaging, superconducting magnets are used.

The use of NMR imaging in medicine has two significant advantages over X-ray imaging, what is commonly known as CT (*computed tomography*):

* There is no evidence that magnetic fields have a detrimental effect on the human body, whereas the damage caused by ionising radiation such as X-rays is well documented.
* X-ray images are essentially just pictures of the electron density in the specimen; NMR images are sensitive to a wide range of different chemical and physical conditions in the tissue and so can provide more information.

A third non-invasive tomographic diagnostic tool is *positron emission tomography* (PET). In contrast to CTs and MRIs which only provide information about the patient's condition at a given moment, PET scans are dynamic and can be used to follow processes taking place within the subject in real time (see Section 3.3.4).

Chapter 6.4

The Conduction of Electricity Through Solids

No single development of modern physics has touched the lives of people everywhere more than the invention of the transistor. Requiring less energy to process a single unit of information – a *bit* – than do brain neurons, transistor-based devices have reduced the cost of computing and of information to an unprecedented level. By revolutionising electronics, they have put the whole world in the lap of every human being on earth. How this was done is the central theme of this chapter.

6.4.1 The Electrical Conductivity of Solids

The electrical conductivity of a solid material is generally specified in terms of the following three parameters:

1. the material's *resistivity*, ρ;
2. the *temperature coefficient*, α, of its resistance;
3. the *number of charge carriers*, N, in each unit volume of the material.

A material's resistivity indicates the opposition it presents to the passage of an electric current. At room temperature, certain materials, such as diamond and PVC, do not conduct electricity to any measurable degree. Such materials are called *insulators*; their resistivity is exceedingly high: $\rho_{diamond} \approx 10^{16}\Omega{\cdot}\mathrm{m}$. On the other hand, metals like copper and silver present very little opposition to the passage of an electric current. Such materials are called *conductors*. Copper and silver are excellent conductors of electricity; their resistivity is very low: $\rho_{copper} \approx 10^{-8}\Omega{\cdot}\mathrm{m}$.

The resistivity of a material depends *inter alia* on its temperature. Typically, the resistivity of metals increases with temperature, i.e. the temperature coefficient of their resistance is positive: $\alpha_{copper} \approx +4{\cdot}10^{-3}\,\mathrm{K}^{-1}$. Metals also contain large numbers of charge carriers: $N_{copper} \approx 8.5{\cdot}10^{28}\,\mathrm{m}^{-3}$. These are the particles whose flow through the material constitutes the electric current. The charge carriers in metals are generally electrons.

Table 6.2. The electrical conductivity parameters of diamond, copper and silicon.

	Diamond	Copper	Silicon
ρ – Resistivity ($\Omega \cdot$m)	10^{16}	$1.6 \cdot 10^{-8}$	$3 \cdot 10^{3}$
α – Temperature Coefficient of Resistance (K^{-1})	N/A	$+4 \cdot 10^{-3}$	$-70 \cdot 10^{-3}$
N – Number of Charge Carriers per Unit Volume (m^{-3})	Negligible	$8.5 \cdot 10^{28}$	10^{16}

A third group of materials, called the *semiconductors*, have properties between those of the insulators and metal conductors; silicon is a typical semiconductor. As the data in Table 6.2 show, the resistivity of the semiconductor silicon, lies between that of the insulator diamond and that of the conductor copper and it contains fewer charge carriers per unit volume than copper.

Like so much in modern physics, the study of semiconductors originated with Faraday. As long ago as 1833, Faraday discovered that in contrast to regular metallic conductors, the resistance of certain poor electrical conductors (he was working with silver sulphide) falls with rising temperature.

That the conductivity of semiconductors was essentially different from that of regular metallic conductors became evident from measurements of the *Hall effect*. Hall[24] had discovered that if a strip of conducting material is placed in a magnetic field whose direction is perpendicular to its length and an electric current flows along the strip, a potential difference is set up between the edges of the strip, i.e. perpendicular to the direction of the current. The direction of this *Hall potential* depends on the sign of the charged particles that carry the current flowing through it.

Measurements of the Hall potential showed that whereas the *charge carriers* in regular metallic conductors are invariably negatively charged, those in semiconductors may be negatively or positively charged. Moreover, whereas the number of these charge carriers is virtually independent of the temperature in metallic conductors, their number increases greatly with temperature in semiconductors.

By the end of the 19th century, three other important properties of these materials had been discovered, all of which have since found numerous practical applications:

1. The *photovoltaic effect*: the production of an electromotive force (an e.m.f.) when a junction of dissimilar materials, one of which is a semiconductor, is illuminated by visible or near ultraviolet light. The junction may be an

[24] Edwin Hall 1855–1929, American physicist.

electrolytic cell or simply a thin mesh of the semiconductor material overlaid on a metal substrate. *Solar cells* are a development of this effect.

2. The rectifying property of a metal-semiconductor junction: the fact that such junctions do not obey Ohm's law. Their resistance – the ratio between the voltage across them and the current passing through them – is not a constant but depends on the sign as well as the magnitude of the voltage.
3. *Photoconductivity*: the increased conductivity of a semiconductor when it is illuminated by visible or near ultraviolet light. It is the basis of the dry copying process known as *xerography*, by which powdered ink – 'toner' – is transferred from an optically induced image on a negatively charged transfer device to a piece of positively charged paper. The transfer device is a rotating drum coated with selenium, which retains electrostatic charge almost indefinitely in the absence of light, but loses it at once wherever light falls upon it. The image is 'etched' onto the charged drum by projecting light onto its surface. Wherever the light strikes, charge is lost; elsewhere, it is retained. Toner is next sprayed onto the drum and sticks to the charged areas. The drum finally passes over a positively charged sheet of paper onto which the toner is transferred and fixed by heat treatment.

6.4.2 The Electron Gas

A short time after the discovery of the electron, Lorentz and Drude[25] proposed a simple classical model to explain the electrical conductivity of metals. The outstanding feature of this model was its hypothesis that a 'gas' of free electrons pervades the interstices of metallic crystals. It was postulated that the lattice energy released as a metal crystallises, frees the *valence electrons* from its atoms, leaving the atoms positively charged. Thus, the model postulated that a metal is composed of a lattice of positive metal ions whose interstices are filled by the gas of free valence electrons. The excellent electrical conductivity of metals was credited to the relative ease with which this gas of *conduction electrons*, as they were subsequently called, could move through the lattice under the influence of an electric field.

The electron gas was regarded at the time as being essentially no different from any other gas such as hydrogen or oxygen. Thus, the average velocity, v, of the electrons comprising the gas due to their random thermal motion at an absolute temperature T, was calculated from the classical

[25] Paul Drude (1863–1906), German physicist.

relationship

$$\frac{m_e v^2}{2} = \frac{3}{2}kT, \tag{6.44}$$

which gives a value at room temperature of $v \approx 10^5$ m/s.

Placing a metal in an electric field causes the conduction electrons to flow. Because of their negative charge, the electrons are accelerated in the opposite direction to that of the field.[26] As their velocity increases, they acquire kinetic energy and linear momentum in the direction of their flow. In the classical model, the progress of the conduction electrons through the lattice is slowed by their collisions with the metal ions, in which they lose both energy and momentum to the lattice. When the average rate at which the conduction electrons acquire energy and momentum from the electric field is balanced by the average rate at which they lose energy and momentum to the lattice ions, a steady electric current flows and heat is generated in the metal at a constant rate.

It can be shown that the classical model predicts both Ohm's law, i.e. the direct proportionality of electric current to potential difference, and the positive sign of the temperature coefficient of a metal conductor's resistance. However, its quantitative predictions of the resistivity of metals and of its temperature dependence are not borne out by experiment. For instance, whereas the model predicts that a metal's resistivity should be a function of the square root of its absolute temperature, i.e. $\rho \propto \sqrt{T}$, at low temperatures $(30 < \mathrm{K} < 120)$ resistivity is actually found to vary approximately linearly with temperature, i.e., $\rho \propto T$.

The classical model takes no account of the wave nature or spin of the electron; this is a fundamental flaw. Because electrons are fermions and are subject to the exclusion principle, the electron gas cannot be compared to an ordinary gas; electrons obey a different statistics from normal gas molecules.[27] Furthermore, the intuitive idea that the conduction electrons must collide with the metal ions as they flow through the lattice is untenable in wave mechanics. Matter waves propagate unhindered through regular structures and so a regular lattice of positive ions would not oppose the flow of the conduction electrons at all; they would only be scattered from any deformations there may be in the lattice. And thirdly, electrons can only accept energy from the applied field in discrete packets, rising each time from one energy level to a higher one. If there

[26]By convention, the direction of an electric current is always given relative to the motion of the positive charges.

[27]See Section 3.4.5.

are no higher levels both vacant and accessible, the electron cannot accept the energy and a current cannot flow.

The use of quantum mechanics in the study of the electrical conductivity of solids has been one of its most fruitful applications. A full account of this achievement is beyond the scope of this book. The following descriptive account is intended just to outline its underlying principles and some of its more outstanding results.

Example 6.2: The Relaxation Time of Conduction Electrons

Assuming they are repeatedly accelerated by the electric field and halted by collisions with the lattice, find the average time, τ, (the *relaxation time*) the conduction electrons in copper are actually in motion between collisions.

Calculation: Consider a metal conductor of length L and uniform cross-section S, between whose ends a uniform electric field, E, is applied (Fig. 6.24). Ohm's law relates the potential difference, $V = EL$, between the ends of the conductor to the current, I, flowing along it:

$$V = EL = IR. \tag{6.45}$$

R is the electrical resistance of the conductor.

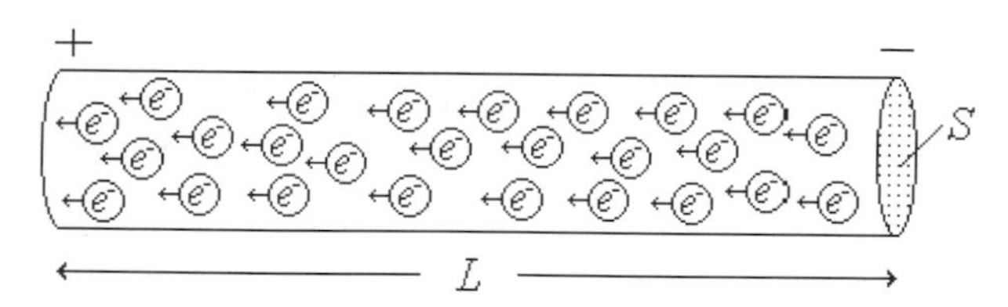

Fig. 6.24. The drift of the conduction electrons along a uniform metal conductor between whose ends an electrical potential is applied.

In terms of the *drift velocity*, v_d, of its conduction electrons along the conductor, the electric current flowing through a metal conductor of cross-section S is given by

$$I = NSev_d, \tag{6.46}$$

where N is the number of conduction electrons in each unit volume of the conductor.

In terms of the resistivity, ρ; of the conducting material, the resistance of a conductor of length L and cross-section S is

$$R = \frac{L\rho}{S}. \tag{6.47}$$

Combining Eqs (6.45), (6.46) and (6.47) gives

$$\rho = \frac{E}{Nev_d}. \tag{6.48}$$

The electric field exerts a force of magnitude $F = eE$ on the electrons in the conductor. Applying Newton's second law of motion, $a = F/m_e$, to the conduction electrons gives

$$a = \frac{eE}{m_e}. \tag{6.49}$$

Hence, the drift velocity, $v_d = a\tau$ of the electrons is given by

$$v_d = \frac{eE}{2m_e}\cdot\tau. \tag{6.50}$$

Substitution in Eq. (6.48) gives

$$\tau = \frac{2m_e}{Ne^2\rho}. \tag{6.51}$$

The relaxation time of copper obtained from the values of the parameters in Table 6.2 is:

$$\tau = \frac{2\cdot 9.11\cdot 10^{-31}}{8.5\cdot 10^{28}\cdot(1.6\cdot 10^{-19})^2\cdot 1.6\cdot 10^{-8}} = 5\cdot 10^{-14}\,\text{s}. \tag{6.52}$$

6.4.3 Energy Levels in Solids — Band Theory

Free atoms, such as those in sodium vapor, can be regarded as isolated systems whose energy levels are determined by a single set of quantum numbers. However, in an aggregate of atoms such as a liquid or solid, where the atoms are close to one another, the energy levels of each atom are affected by the presence of the other atoms. In terms of quantum mechanics, the wave functions of the atoms overlap. The wave function describing the aggregate is obtained from the appropriate superposition of the wave functions of the individual atoms.

The effect that the approach of another atom has on a system's wave function is most simply shown by considering the convergence of two ground state hydrogen atoms. So long as the atoms are widely separated, their wave functions are autonomous and their electrons each have the same energy; both occupy the $1s$ state. However, as the atoms converge, their orbitals (spatial wave functions) interfere, either constructively to give the symmetric orbital $\phi_S = \phi_1 + \phi_2$ (covalent bond formation that produces the H_2 molecule) or destructively to give the anti-symmetric orbital $\phi_A = \phi_1 - \phi_2$ (the anti-bonding orbital). The atomic electrons can now occupy either the anti-symmetric or the symmetric orbital. The symmetric orbital corresponds to a lower energy than the anti-symmetric orbital (Section 3.4.8).

When two hydrogen atoms converge the number of electronic energy levels is doubled. It is tripled by the convergence of three atoms, quadrupled by the convergence of four atoms, and so on. The cluster of energy levels that corresponds to each atomic energy level is called an *energy band* (Fig. 6.25).

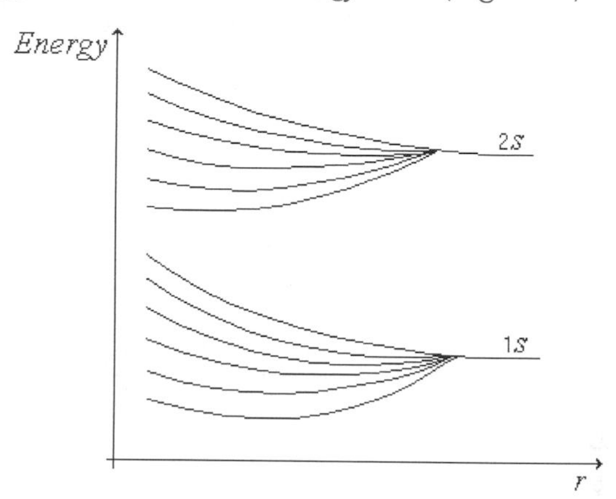

Fig. 6.25. The energy levels of six hydrogen atoms arranged in a straight line as a function of the distance, r, between them. Each atomic level is split into a band comprising six sub-levels.

In general, the number of sub-levels in each band is equal to the number of atoms in the system. Accordingly, the number of electrons required to fill a band in a system comprising N atoms is $2N(2l + 1)$, where l is the orbital quantum number of the corresponding atomic energy level. The exclusion principle permits the deployment of no more than two electrons in each sub-level. Thus, $2N$ electrons are needed to fill a band corresponding to an atomic s state and $6N$ will be needed for one corresponding to a p state.

The atoms in almost every crystalline solid, whether metallic or non-metallic, are so close to one another that their valence electrons constitute a single distinct system that pertains to the entire crystal. As a result, each atomic energy level is replaced by an energy band comprising a myriad of discrete sub-levels.

The typical width of an energy band is a few electron-volts. Since the average separation between the sub-levels comprising a band is so very small, $\sim 10^{-23}$ eV, each may in practice be regarded as being a continuum of energy values. The width of the energy bands depends on the extent to which the orbitals overlap. In aggregates of multi-electron atoms, it is the orbitals of the outermost electrons – the valence electrons – that overlap the most. Consequently, the broadest energy level bands are those associated with these electrons. Commonly, these outermost bands are so broad that they overlap. For example, whereas the inner $2p$ band in metallic sodium is unmoved under usual conditions, the bands corresponding to the $3s$ and $3p$ atomic levels overlap one another.

The bands corresponding to the closed inner atomic shells are narrower and are usually full. The electrons in these inner bands are unable to change their state; they are 'frozen' in them.

Figure 6.26 shows typical band configurations in systems in which the bands do not overlap. The *conduction bands* are so called because it is the presence or otherwise of electrons in this band that determines the electrical conductivity of the solid. The gaps between the bands indicate forbidden energy values, like the gaps between atomic energy levels; they are often referred to as *forbidden bands.*

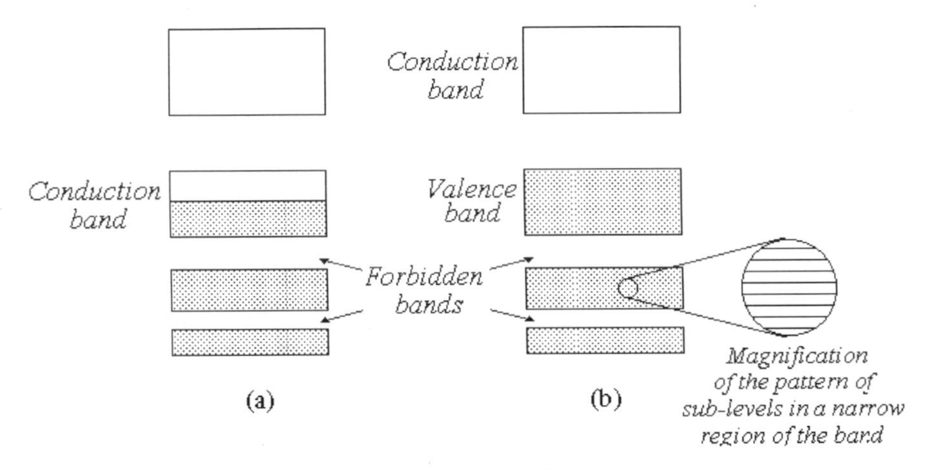

Fig. 6.26. Typical energy band diagrams of two different solids. Each band comprises a myriad of very closely spaced sub-levels. The lower energy bands are narrower since the orbitals of these inner electrons overlap less. When, as in (a), the highest populated band is only partially filled, it is called the *conduction band.* When the highest populated band is full, as in (b), it is called the *valence band* and the empty band above it is the conduction band.

6.4.4 Insulators

Diamond is a pure transparent crystalline form of carbon; it is a typical solid insulator. The electronic configuration of the carbon atom is $1s^2 2s^2 2p^2$. At the inter-atomic distances found in diamond, the $2s$ and $2p$ energy levels form two bands each of which can accommodate up to $4N$ electrons. The four outermost electrons of each atom are normally found in the lower valence band; this leaves the higher conduction band empty. There is a significant energy difference of $\sim 6\,\mathrm{eV}$ between the two bands (Fig. 6.27).

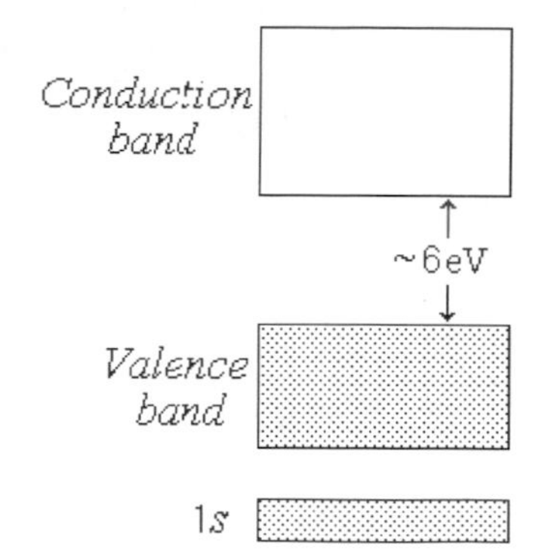

Fig. 6.27. Energy band diagram of diamond. There are no vacant states in the valence band. The excitation of an electron from the valence band to the conduction band requires at least 6 eV.

There are no vacant states in the diamond's valence band. On the other hand, there are normally no electrons in the conduction band. This situation sets a minimum on the amount of energy an electron in diamond can absorb. The only transition open to an electron is to rise from the valence band to the conduction band, but this requires at least 6 eV of energy.

At ordinary temperatures, the probability that an electron will accumulate this amount of energy from the thermal energy of the lattice is in effect nil. The average thermal energy of a lattice particle is equal to kT, which at 300 K is only $8.62 \cdot 10^{-5} \cdot 300 = 0.025\,\text{eV}$.

Nor, in practice, can the electrons be physically displaced. Assuming an electron might move a distance of about 10^{-8} m through a diamond before being halted by a collision with a lattice particle, the electric field required for it to acquire 6 eV during the move would be $6 \cdot 10^{8}$ V/m. Such a field is 10^{10} times stronger than that needed to establish an electric current in copper. This explains why diamond is such an excellent insulator.

The large energy gap between its valence and conduction bands also explains the transparency of diamond. The photons in visible light have energies between 1.8 and 3.1 eV. However, the minimum amount of energy an electron in diamond can accept is 6 eV. Consequently, the photons of visible light pass through diamond unabsorbed.

6.4.5 Metallic Conductors

For the purposes of the descriptive account given here, we shall assume that the electrons moving through a metal conductor have zero potential energy, i.e. they possess only kinetic energy. In fact, electrons experience a periodic potential as they move through a metal lattice; the form of this potential being determined by the lattice structure. However, for the purposes of the elementary treatment given here, the periodic nature of the electric potential in metal conductors will be ignored.

Sodium is a typical metal conductor. The electronic configuration of the sodium atom is $1s^2 2s^2 2p^6 3s^1$. In the metal solid the bands corresponding to the $1s$, $2s$ and $2p$ energy levels are fully occupied, but the band corresponding to the $3s$ level is only half full. Thus, the $3s$ band is the conduction band (Fig. 6.28).

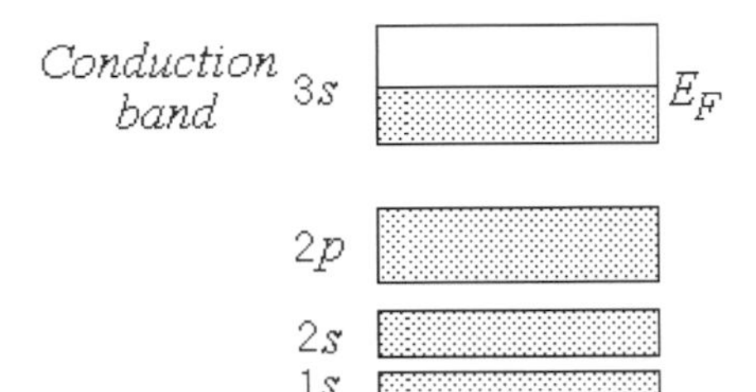

Fig. 6.28. Energy band diagram of sodium metal at 0 K. The electrons in the conduction band have kinetic energies from zero up to the Fermi energy, E_F.

The appropriate function for the probability, $f(E_i)_{Electron}$, of an electron occupying the E_i energy level in any system at an absolute temperature T is that given by the Fermi–Dirac statistics, namely,

$$f(E_i)_{electron} = \left(Ce^{E_i/kT} + 1\right)^{-1}. \tag{6.53}$$

In the case of the electron gas in a metal conductor the parameter C is the exponential function

$$C = e^{-E_F/k_T}. \tag{6.54}$$

E_F is called the *Fermi energy* of the metal.[28] Thus, the probability, $f(E_i)_{Electron}$, of an electron occupying the E_i energy level in the conduction band in a metal conductor at an absolute temperature T is given by

$$f(E_i)_{electron} = \left(e^{(E_i - E_F)/kT} + 1\right)^{-1}. \tag{6.55}$$

It follows, that at a temperature of 0 K (absolute zero), the probability of finding a conduction electron in an energy level $E_i < E_F$ is

$$\begin{aligned} f(E_i)_{electron} &= \left(e^{(E_i - E_F)/kT} + 1\right)^{-1} \\ &= (e^{-\infty} + 1)^{-1} \\ &= 1, \end{aligned} \tag{6.56}$$

i.e. all such energy levels will be occupied.

[28]The Fermi energy also constitutes the material's electrochemical free energy and must, therefore, have the same value on both sides of a phase boundary at equilibrium. The contact potential set up at equilibrium between two touching conductors is equal to the difference in their Fermi energies.

On the other hand, the probability of finding a conduction electron in an energy level $E_i > E_F$ is

$$f(E_i)_{electron} = \left(e^{(E_i - E_F)/kT} + 1\right)^{-1} \tag{6.57}$$
$$= (e^{\infty} + 1)^{-1}$$
$$= 0,$$

i.e. all such energy levels will be empty. Thus, at absolute zero, all the energy levels up to the Fermi energy are occupied by electrons, whereas all those above it are empty.

From the values of the Fermi energy given in Table 6.3, it is clear that even at absolute zero, many of the electrons in a metal conductor possess a kinetic energy of several electron volts. Classical gas molecules must be heated to temperatures of tens of thousands of degrees for them to possess kinetic energies of this magnitude.

Table 6.3. Typical Fermi energies.

Metal		Fermi Energy (eV)
Sodium	Na	3.12
Caesium	Cs	1.53
Copper	Cu	7.04
Zinc	Zn	11.0
Silver	Ag	5.51

The additional thermal energy possessed by the electrons at higher temperatures – ~0.025 eV at room temperature – is almost insignificant compared to the Fermi energies. Consequently, the electronic configuration of a metal conductor changes only slightly as its temperature rises (Fig. 6.29). Little of the heat energy absorbed by a metal as it warms is taken up by the electron gas; almost all of it goes to increase the thermal vibrations of the lattice.

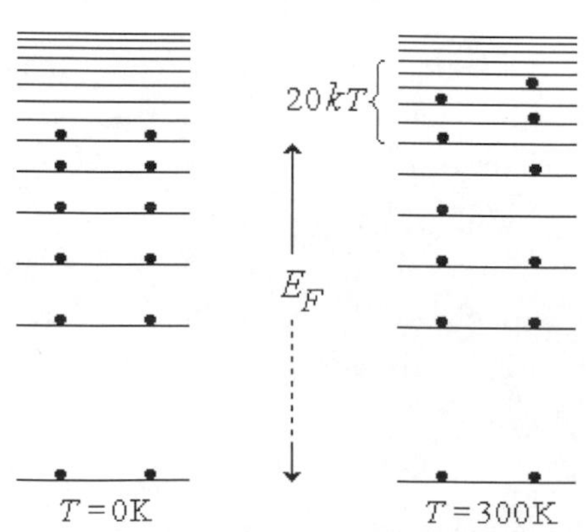

Fig. 6.29. The energy levels of the conduction electrons in a metal at absolute zero, $T = 0$ K, and room temperature, $T = 300$ K. The excited levels occupied by electrons at room temperature cover a range of about 20 kT above the Fermi energy.

The key feature of the electronic structure of metal conductors is the abundance of empty electronic states in the conduction band just above the Fermi energy. As a result of these, the conduction electrons occupying energy levels close to the Fermi level can absorb a whole range of energies, from just a fraction of an electron volt to several electron volts, thereupon rising to one of the many vacant levels. This accounts for (1) the opacity of metals and (2) their excellent electrical conductivity.

1. The photons in visible light have energies between 1.8 and 3.1 eV. These are readily absorbed by the conduction electrons occupying energy levels close to the Fermi level and so metals are opaque.
2. In the presence of an electric field, even a very weak field, conduction electrons occupying energy levels close to the Fermi level can easily 'jump' into the vacant higher levels, thereby absorbing energy from the field. Under the influence of the field, the electrons flow through the metal lattice producing an electric current.

The mechanism by which a steady electric current arises and heat is continuously generated in the metal is similar in the quantum-mechanical model to that sketched by the classical model. Both models posit collisions between the electrons comprising the electric current and the particles constituting the lattice, such that the rate at which the conduction electrons absorb energy from the applied field is balanced by that at which they lose energy to the lattice. However, the lattice particles involved in the transfer of the energy differ in each model. Whereas in the classical model the particles with which the conduction electrons collide are the metal ions that constitute the lattice, in the quantum-mechanical model they are the defects in the lattice. The metal ions themselves do not impede the propagation of the electron waves; matter waves pass unhindered through a regular lattice.

Two different types of lattice defects impede the passage of the electrons:

1. Defects in the structure and arrangement of the lattice itself. For example, atoms that have been displaced from their position in the structure or impurity atoms that have been incorporated into the lattice. These defects are independent of temperature.
2. Thermal vibrations of the lattice that result in random variations in the spacing of the lattice ions.[29] These defects are temperature dependent; the higher the temperature, the more vibrational defects there are.

[29]The atoms do not vibrate independently. The oscillations propagate through the solid as extremely high frequency acoustic (sound) waves. The energy transmitted by these mechanical waves is quantised. The quantum of energy is called a *phonon* and like the photon its value is hf, where h is Planck's constant.

The positive temperature coefficient of resistance typical of metal conductors is accounted for by the temperature dependent lattice defects. Raising the conductor's temperature increases the number of lattice defects. As a result, the conducting electrons are scattered to a greater degree and a weaker current is produced in the metal conductor by a given electric field, i.e. the conductor's resistance increases.

Example 6.3: The Velocity of Conducting Electrons

Calculate the *Fermi velocity*, v_F, of electrons whose kinetic energy, K, equals the Fermi energy, E_F, in copper metal, 7.04 eV.

Calculation: Assuming that the kinetic energy of the electrons is given by the classical expression $K = mv^2/2$, their Fermi velocity, v_F, is given by

$$v_F = \sqrt{\frac{2E_F}{m_e}} = \sqrt{\frac{2 \cdot 7.04 \cdot 1.6 \cdot 10^{-19}}{9.11 \cdot 10^{-31}}} = 1.57 \cdot 10^6 \text{ m/s}.$$

Even at the absolute zero, conduction electrons are still in random motion through the copper metal lattice with this considerable velocity. In contrast, only at a temperature of about 60,000 K do gas molecules possess this velocity.

The Fermi velocity is the average velocity of the conduction electrons between their random collisions with the defects in the lattice. It must not be confused with the drift velocity, v_d, at which the conduction electrons advance through the lattice when an electric current flows through the conductor. The drift velocity is superposed on the Fermi velocity when an electric field is applied to the conductor. Drift velocities are of the order 10^{-5} m/s.

Example 6.4: The Mean Path-length of the Conduction Electrons

The classical Lorentz–Drude model and the quantum mechanical model of conduction differ as to the value each ascribes to:

1. the average velocity, v, of the random motion of the conduction electrons;
2. the electrons' mean path-length, $\lambda = v\tau$.

According to the classical model, the random velocity of the conduction electrons is a function of the square root of their absolute temperature, *viz.* $v \propto \sqrt{T}$, and the mean path-length of the conduction electrons is the average distance between the ions in the metal lattice. At a temperature of 300 K, $v \approx 10^5$ m/s. Given a typical relaxation time $\tau \approx 10^{-14}$ s, this gives a mean path-length of about

$$\lambda = v\tau = 10^5 \cdot 10^{-14} = 1\text{nm},$$

which is consistent with the typical nearest neighbor distances in metal lattices, ~0.25 nm. This appears to substantiate the classical model.

However, if the random velocity of the conduction electrons is a function of the square root of their absolute temperature, the relaxation time, $\tau = \lambda/v$, should be an inverse function of the square root of the absolute temperature, $\tau \propto 1/\sqrt{T}$. Given that a conductor's resistivity is given by $\rho = \frac{2m_e}{Ne^2\tau}$, its resistivity should be a function of the square root of the absolute temperature, *viz.* $\rho \propto \sqrt{T}$. This prediction of the classical model is not borne out by experiment; resistivity is in fact a function of the absolute temperature, $\rho \propto T$, and not of its square root.

The correct relationship between resistivity and temperature is given by the quantum mechanical model according to which the random velocity of the conduction electrons is the Fermi velocity, $v_F \approx 10^6\,\mathrm{m/s}$, and is almost totally independent of the conductor's temperature. Given a typical relaxation time $\tau \approx 10^{-14}\,\mathrm{s}$, this gives a mean path-length of about

$$\lambda = v\tau = 10^6 \cdot 10^{-14} = 10\,\mathrm{nm},$$

i.e. conduction electrons pass by tens of lattice ions before undergoing a collision. Hence, electrical resistance cannot be the result of repeated collisions between the conduction electrons and the lattice ions. According to the quantum mechanical model, it is the collisions with the lattice defects – the structural and vibrational defects – that account for electrical resistance.

It follows that the temperature dependence of a material's resistivity is determined by the vibrational lattice defects. The greater the macroscopic cross-section, σ, of these vibrational defects the greater the material's resistance. The macroscopic cross-section of the vibrational defects depends on the square, A^2, of their amplitude, i.e. on their average energy; this, in turn, depends on the system's temperature, T. Thus, the macroscopic cross-section is proportional to the absolute temperature, i.e. $\sigma \propto T$, and since the material's resistivity ρ depends on the macroscopic cross-section, $\rho \propto \sigma$, it is, in turn, a function of the conductor's absolute temperature.

6.4.6 Superconductivity

In 1911, Kammerlingh Onnes[30] discovered that at temperatures below 4.15 K the metal mercury exhibits zero resistivity. Other superconducting metals and the critical temperatures, T_C, below which they exhibit this remarkable property are listed in Table 6.4.

Table 6.4. The critical temperature of some metal conductors.

Metal		T_C (K)
Aluminum	Al	1.19
Tin	Sn	3.7
Lead	Pb	7.18
Zinc	Zn	0.55
Niobium	Nb	9.46

[30]Heike Kammerlingh Onnes 1853–1926, Danish physicist who was awarded the Nobel prize in 1913 for investigations into the behaviour of materials at very low temperatures. He was the first to liquefy helium.

Note that the best normal conductors such as silver and copper do not exhibit superconductivity, suggesting that normal conductivity and superconductivity are fundamentally different phenomena. The discovery by Bednorz[31] and Muller[32] in 1986, that certain ceramic materials, which are insulators at room temperature, exhibit superconductivity when cooled to temperatures around 100 K, reinforces this impression. These materials are remarkable as they exhibit superconductivity at much higher temperatures, above the boiling point of nitrogen (77 K), than do metal superconductors.

Maintaining a steady current in normal conductors requires an electric field. However, electric currents induced in rings of superconducting materials will persist for years without diminution, without the need for a potential source such as a battery. Very strong magnetic fields can be produced by these persistent currents without the need for large amounts of electrical power. However, superconductivity is also destroyed by intense magnetic fields, whether these are external fields or those produced by the current in the superconductor. This sets a limit on the strength of the magnetic fields that can be produced using superconductors.

Superconductors are not only perfect conductors, they are also perfect diamagnets, i.e. the magnetic field within a superconductor is always zero. The applied magnetic field is neutralised inside the superconductor by eddy currents induced in its surface.[33] The expulsion of the magnetic field from a superconductor situated in a magnetic field is called the *Meissner effect*.[34] As a result of the exclusion of magnetic flux from the superconductor, a bar magnet placed over the superconductor levitates above its surface (Fig. 6.30).

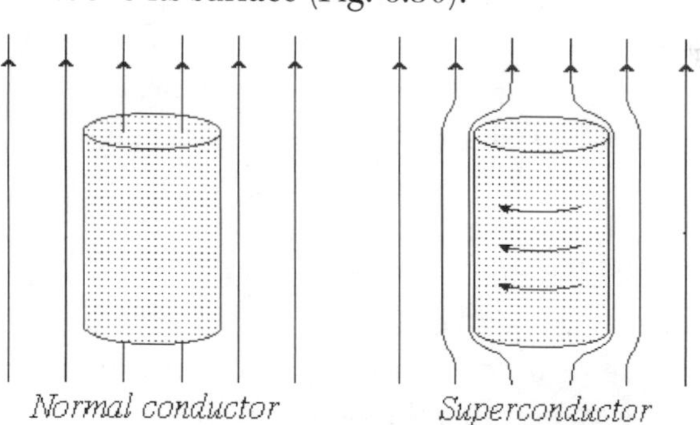

Fig. 6.30. The Meissner effect. As a metal situated in a magnetic field changes from a normal conductor into a superconductor eddy currents are set up in its surface which expel the magnetic flux that was passing through it.

[31] J. Georg Bednorz 1950–, German physicist who shared the Nobel prize in 1987 for the discovery of superconductivity in ceramic materials.

[32] Karl Alexander Muller 1927–, German physicist who shared the Nobel prize in 1987 for the discovery of superconductivity in ceramic materials.

[33] Strictly speaking, this is only true of the so-called *Type-I superconductors*. Tubes of magnetic flux can penetrate the *Type-II superconductors* that are typically made of metal alloys. These superconducting alloys can also sustain relatively intense magnetic fields.

[34] Named after the German/American physicist Karl Meissner 1891–1959.

A theory that elucidates the mechanism underlying superconductivity was only spelled out some 50 years after its initial discovery. The theory, known as the *Bardeen–Cooper–Schrieffer* (*BCS*) *theory*,[35] postulates that the charge carriers in superconductivity are not single electrons but pairs of electrons. At the very low temperatures at which superconductivity is exhibited, a lattice-mediated attractive interaction acts between the conduction electrons as a result of which they are paired. As they move through a conductor, the negatively charged electrons cause slight distortions in the lattice of positive ions, distortions that are so slight that they only assume significance when the thermal vibrations of the lattice become insignificant, i.e. at very low temperatures. At these low temperatures, the net effect of the local intensifications of positive charge associated with these lattice distortions is to couple the electrons into pairs called *Cooper pairs.* The involvement of the lattice in superconductivity has been demonstrated by the effect on the transition temperature, T_C, of the presence of different isotopes in the lattice.

If one of the electrons comprising a Cooper pair has spin $-½$ and linear momentum p, its partner will have spin $+½$ and linear momentum $-p$. In classical particle terms this is equivalent to saying that they are moving in opposite directions, a concept that defies intuition; how can particles moving away from each other maintain a coupling interaction? However, in quantum mechanical terms the particles can continue to interact as long as their matter waves overlap. The implication is that the wave function of a Cooper pair extends over a large volume, so large a volume that it overlaps the wave functions of as many as one million other Cooper pairs. Hence, the Cooper pairs in a superconductor can be regarded as occupying a collective state that extends over the whole volume of the superconductor; all the Cooper pairs occupying this state cooperate with one another. This configuration of the conduction electrons is reminiscent of the behaviour of bosons and has been interpreted as indicating that the superconducting state should be regarded as a distinct state of matter, essentially different from the familiar solid, liquid, gas and plasma states.

According to the BCS theory, the conduction electrons in a superconductor can be regarded as having condensed into a single 'macromolecule' that pervades the entire volume of the system and can flow through it as a whole. An energy gap of order 10^{-3} eV separates the electrons in the 'macromolecule' from the first excited state and thus ensures its stability. This gap represents the energy required to split a Cooper pair and it accounts for the essential properties of superconductors.

[35] Named after John Bardeen 1908–1991, Leon N. Cooper 1930– and John R. Schrieffer 1931–, American scientists who shared the Nobel prize in 1972 for developing the theory of superconductivity.

1. The 'macromolecule' of conduction electrons moves unhindered – without resistance – through the superconductor because the energy of its friction with the lattice irregularities is insufficient to enable electrons to separate from their Cooper partners and rise into the excited state.
2. Under the influence of the external magnetic field, the motion of the conduction electrons in the superconductor is accelerated such that eddy currents are set up whose secondary magnetic fields oppose the external field. This is the familiar phenomenon of diamagnetism. However, so long as the external field is not too strong, the acceleration of the electrons is not accompanied by a transfer to a higher energy level; the 'macromolecule' of Cooper pairs remains intact and so there is nothing to prevent eddy currents being induced that totally neutralise the magnetic field inside the superconductor.

6.4.7 Semiconductors

The configuration of the energy bands in pure (*intrinsic*) semiconductors is similar to that in insulators except that the gap between the valence and conduction band is smaller. For example, the energy gap, E_g, in silicon is just 1.1 eV (Fig. 6.31a). At low temperatures, silicon is as poor a conductor of electricity as diamond. However, as the temperature rises, the probability of a few electrons acquiring sufficient thermal energy from the lattice to cross the energy gap separating the valence from the conduction band increases (Fig. 6.31b). At a temperature of 450 K the average population of the conduction band is a

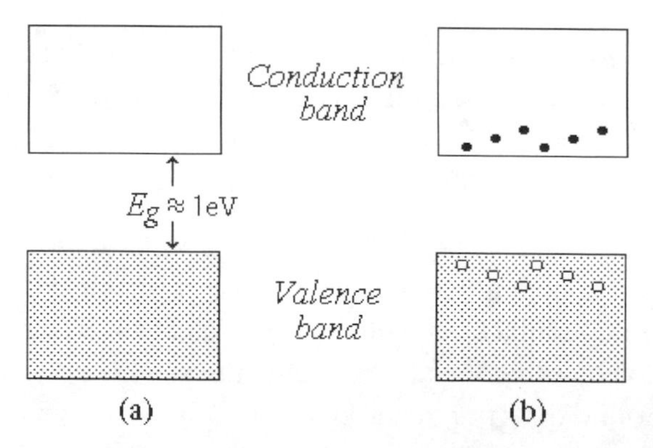

Fig. 6.31. (a) The energy band diagram of an intrinsic semiconductor at 0 K; the valence band is full and the conduction band is empty. (b) At higher temperatures a few electrons (represented by the black dots) acquire sufficient thermal energy to rise from the valence to the conduction band leaving an equal number of 'holes' (represented by the empty squares) in the valence band.

million times greater than at 250 K. Thus, the temperature coefficient of resistance, α, of semiconductors has a negative value, i.e. their resistance falls with rising temperature.

An electron that happens into the semiconductor's conduction band leaves a vacancy or positive 'hole' in the valence band. This hole can subsequently be filled by an electron that drops back into the valence band from the conduction band. Thus, positive holes are continuously and randomly created and annihilated in the semiconductor lattice. The holes are treated like electrons except that their charge is positive and their energy in energy level diagrams is measured downwards.

The charge carriers in intrinsic semiconductors are both the electrons in the conduction band and the holes in the valence band. As the conduction electrons drift through the lattice in one direction, the holes they left in the valence band drift through it in the opposite direction. The maximum current an intrinsic semiconductor can carry – its saturation current – is determined by the number of conduction electrons and positive holes it contains and their respective mobilities.[36] Until the saturation current is reached, intrinsic semiconductors generally obey Ohm's law.

The presence of even minute amounts (one part in one million) of certain impurities has a dramatic effect on the conductivity of intrinsic (pure) semiconductors such as silicon. Without causing a distortion to the lattice, these trace impurities can increase the number of charge carriers in the material by a factor of one million or even more. These doped materials are called *extrinsic semiconductors.* Two types of materials are usually used as the impurities, Group III elements and Group V elements.

* Semiconductors doped with traces of Group V pentavalent elements such as phosphorus, arsenic and antimony are called n-type semiconductors.
* Semiconductors doped with traces of Group III trivalent elements such as aluminium, indium and gallium are called p-type semiconductors.

Two-dimensional schematic representations of the lattices of intrinsic (pure), n-type and p-type semiconducting materials are shown in Fig. 6.32.

In n-type semiconductors, the fifth valence electron in the Group V atoms is not required for bonding the atoms in the lattice. This non-bonding electron is only loosely bound to its parent atom (Fig. 6.32b). The energy, E_D, it requires to break loose from its parent atom is typically $\sim$0.05 eV.[37] Consequently, at room temperature, the electron can easily escape from the atom and wander

[36]Electrons are more mobile than holes. Typically, electrons account for about 75% of the conductivity of intrinsic semiconductors.
[37]See Problem 5 at the end of Part 6.

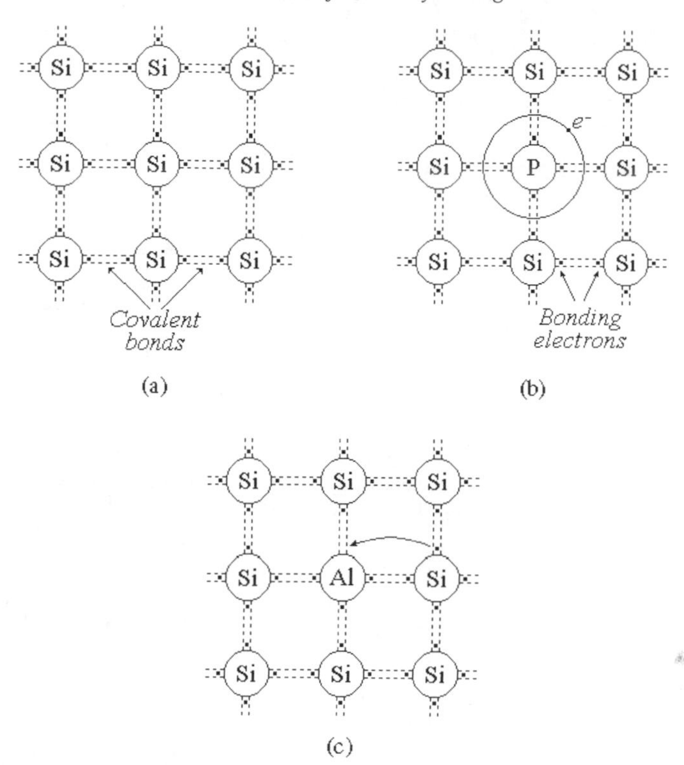

Fig. 6.32. Two-dimensional schematic representations of semiconductor lattices. (a) Pure silicon; each quadrivalent silicon atom is bonded covalently to its four closest neighbours. (b) An n-type semiconductor in which a pentavalent phosphorus atom replaces one of the silicon atoms; the surplus fifth valence electron is only loosely held to the phosphorus atom. (c) A p-type semiconductor in which a trivalent aluminium atom replaces one of the silicon atoms; the missing electron is equivalent to a positive hole which effectively moves randomly through the lattice as it is repeatedly filled by an electron only to reappear at the site previously occupied by the electron.

freely through the lattice; the necessary energy is obtained from the thermal vibrations of the lattice.

In terms of energy levels, the surplus electrons in n-type semiconductors occupy levels that are just below the semiconductor's conduction band, i.e. $E_D \ll E_g$. These electrons can easily rise from these *donor levels* into the semiconductor's conduction band, thereby enhancing its population and the conductivity of the material (Fig. 6.33). This, however, does not increase the number of labile positive holes in the valence band because the positive charge that balances the negative charge of the surplus electrons is 'locked' in the nuclei of the donor Group V atoms. Thus, it is the electrons in the conduction band that constitute the principal (majority) charge carriers in n-type semiconductors. The holes in the valence band do make a small contribution to the current passing through the semiconductor; however, theirs is a minority role.

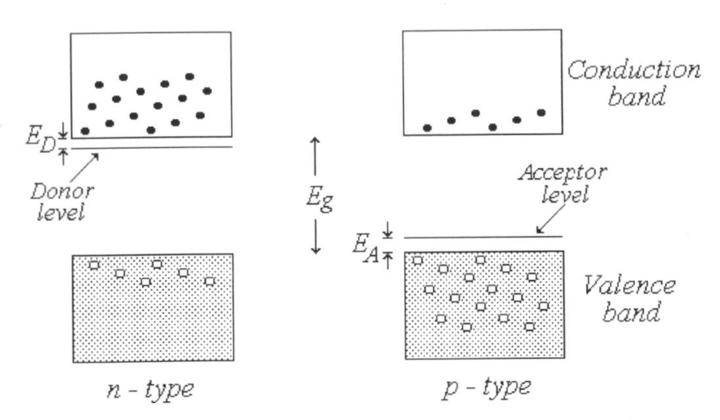

Fig. 6.33. The energy bands and energy levels in n-type and p-type semiconductors. In the n-type, most of the electrons in the conduction band come from the donor level and just a few from the valence level. In the p-type, the valence band contains many holes that are the result of electrons being promoted to the acceptor level and just a few that are the result of electrons promoted to the conduction level. The electrons remaining in the donor level of the n-type material as well as those that have risen to the acceptor level in the p-type are not shown in the diagram as they make no direct contribution to the materials' electrical conductivity.

In p-type semiconductors, the absence of a fourth valence electron in the Group III atom leaves one of the valence electrons in some of the silicon atoms unpaired (Fig. 6.32c). To offset this deficiency, an electron from a nearby covalent bond may jump across and fill the gap. Since, overall, this action produces no change in the lattice bonding, it requires just a small amount of energy, E_A, typically $\sim$0.05 eV. This amount of energy is readily obtained from the thermal vibrations of the lattice. However, the volunteer electron leaves a positive ion at the position it vacates; this is equivalent to the creation of a hole in the valance band. The hole can be filled by the transfer of another electron from a different covalent bond, but this in turn just creates a new positive hole somewhere else. Thus, the net result of introducing Group III impurity atoms into the silicon lattice is a random motion of positive holes from one lattice site to another. This random motion of the positive holes in p-type semiconductors is analogous to that of the conduction electrons in n-type semiconductors.

In terms of energy levels, the valence electrons flitting between the positive holes in the semiconductor material occupy an *acceptor energy level* that is just above the valence band, i.e. $E_A \ll E_g$. This increases the number of holes in the valence band and so enhances the material's conductivity. However, the number of electrons in the conduction band is unaffected (Fig. 6.33). It follows that the principal (majority) charge carriers in p-type semiconductors are the holes in the valence band; the electrons in the conduction band are the minority charge carriers.

6.4.8 The p-n Junction

The extent to which an extrinsic semiconductor's conductivity is enhanced closely depends on the concentration of the impurity. Techniques exist for manufacturing single semiconductor crystals with precisely determined impurity concentrations. It is even possible to produce single crystals one end of which is an n-type material and the other a p-type; such crystals are called *p-n junction diodes*.[38]

The moment a junction diode is formed (Fig. 6.34a), majority charge carriers start diffusing across the line between the two materials, i.e. electrons diffuse from the n-type material into the p-type material and holes from the p-type into the n-type. This exchange of majority charge carriers is called the *diffusion current*. It effectively transfers positive charge from the p-type side of the seam to the n-type side, thereby establishing a *contact potential difference*, V_0, of several tenths of a volt across the boundary between them (Fig. 6.34b).

Under the influence of the contact potential, minority charge carriers now start to move in the opposite direction across the narrow seam joining the two materials; holes pass from the n-type material in to the p-type and electrons from the p-type into the n-type (Fig. 6.34c). This exchange of minority charge carriers, called the *drift current*, effectively transfers positive charge from the n-type side of the seam to the p-type side.[39]

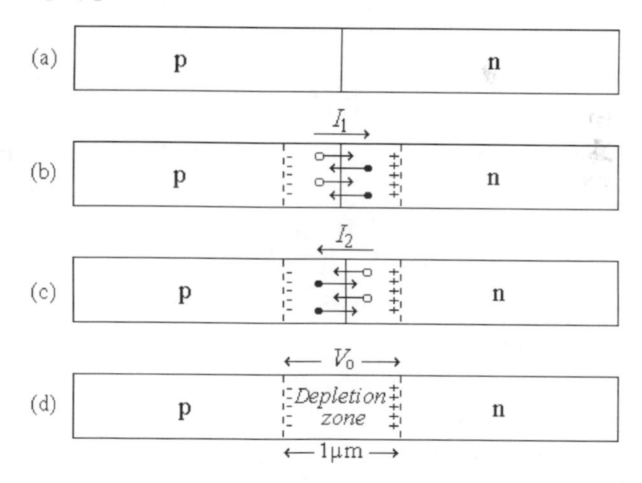

Fig. 6.34. (a) A p-n junction diode. (b) The diffusion current, I_1, of the majority charge carriers. (c) The drift current, I_2, of the minority charge carriers. (d) The contact potential, V_0, set up across the depletion zone.

[38] With one of its ends masked, the crystal is exposed to a gas of donor (Group V) atoms. These are absorbed and diffuse into the lattice at the unmasked end of the crystal. The exposed end is now masked and after removing the masking from the unexposed end, the crystal is immersed in a gas of acceptor (Group III) atoms. As a result one end is diffused with donor atoms and the other with acceptor atoms.

[39] In some texts the diffusion current is referred to as the *recombination current* and the drift current as the *thermal current*. The former term derives from the fate of the majority charge carriers, i.e. their recombination as when the electrons fill holes after descending into the valence band, whereas the latter refers to the origin of the minority charge carriers, *viz.* the thermal agitations of the lattice.

At equilibrium, the rate, I_1, at which positive charge is transferred from the p-type material to the n-type by the diffusion of the majority charge carriers is balanced by the rate, I_2, at which it is returned by the drift of the minority charge carriers from the n-type material to the p-type, i.e.

$$I_1 = I_2. \tag{6.58}$$

The exchange of charge carriers takes place in a narrow strip of material, typically of width $\sim 1\ \mu$m, that overlaps the boundary between the two materials on both sides. This strip, called the *depletion zone*, is effectively emptied of charge carriers. The only charged entities remaining in the depletion zone are the positive ions of the Group V atoms in the n-type material that have lost their fifth valence electron and the negative ions of the Group III atoms in the p-type material that have accepted these electrons; all the holes have been filled. These ions are locked in the lattice and so are immobile; it is they that maintain the contact potential at equilibrium.

In terms of energy levels, the formation of a junction diode involves the merging of the conduction and valence bands of the two materials across the boundary between them. This enables the relatively numerous electrons in the n-type material's conduction band and the relatively numerous positive holes in the p-type material's valence bond to migrate across the junction and neutralise one another (Fig. 6.35a). The potential difference thus set up across the junction lowers the electronic energy levels on the n-type side relative to the p-type side by an amount eV_0. This energy difference causes electrons in the conduction band to drift across the junction from the p-type material and holes in the valence band to drift in the opposite direction (Fig. 6.35b).

The potential difference across the boundary in a p-n junction can be modulated by means of an external potential source, V. By connecting the positive terminal of the external source to the n-type end of the junction and the negative terminal to the p-type end, the potential is raised to $(V_0 + V)$. This arrangement, called a *reverse bias connection*, has the effect of significantly reducing the diffusion current, I_1, the thermal vibrations of the lattice being too weak at normal temperatures to provide the majority charge carriers with the additional energy needed to overcome the increased potential. On the other hand, the additional potential facilitates the drift current, I_2, and widens the depletion zone (Fig. 6.36). The net current, $I = I_2 - I_1$, that flows through a reverse biased diode results from the drift of the minority charge carriers and is very weak, $\sim 1\ \mu$A. The diode effectively exhibits a high resistance ($> 10^5\,\Omega$) in the reverse bias connection that impedes the passage of any significant current.

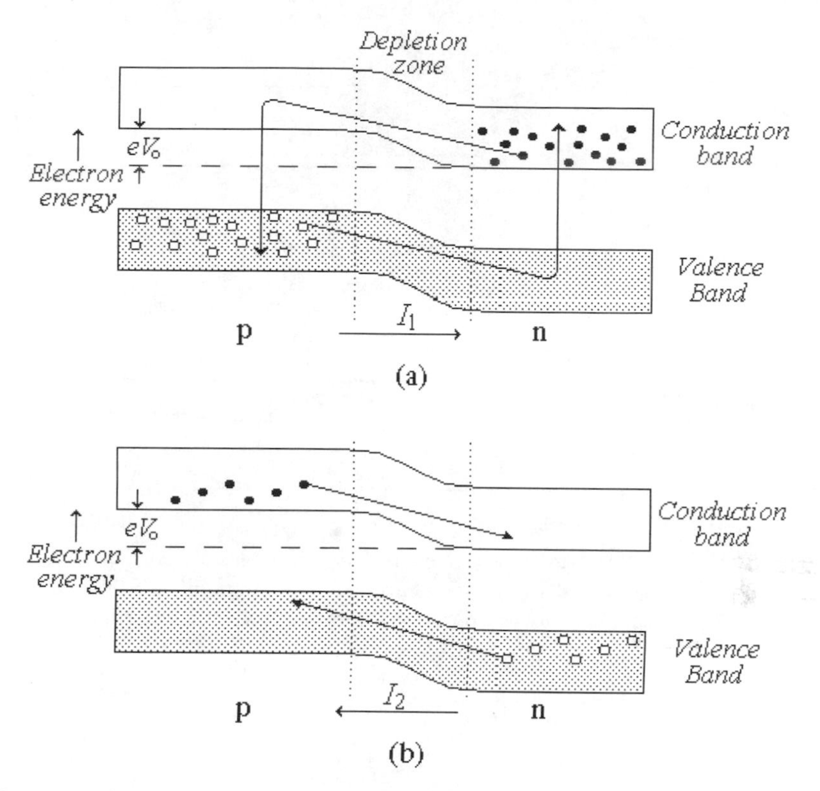

Fig. 6.35. The electronic energy level diagram of an unbiased p-n junction. (a) The diffusion current of the majority charge carriers. Electrons in the conduction band migrate from the n-type material into the p-type where they fall into the valence band and neutralise positive holes. Simultaneously, holes in the valence band migrate in the opposite direction and 'fall' into the conduction band where they are neutralised by electrons. (b) The drift current of the minority charge carriers.

When the positive terminal of the external source is connected to the p-type end of the junction and the negative terminal to the n-type end, the potential across the junction diode is reduced to $(V_0 - V)$. This arrangement, called a *forward bias connection*, has the effect of significantly enhancing the diffusion current, I_1, of the majority charge carriers. The reduced potential has little effect on the drift current, I_2, of the minority charge carriers though it does significantly narrow the depletion zone (Fig. 6.37).

The net current, $I = I_1 - I_2$, flowing through a forward biased diode results from the injection of majority charge carriers into and across the junction and is quite substantial, $\sim 1\,\text{mA}$. Thus, in the forward bias connection the diode effectively exhibits a relatively low resistance $(\sim 100 - 1{,}000\Omega)$.

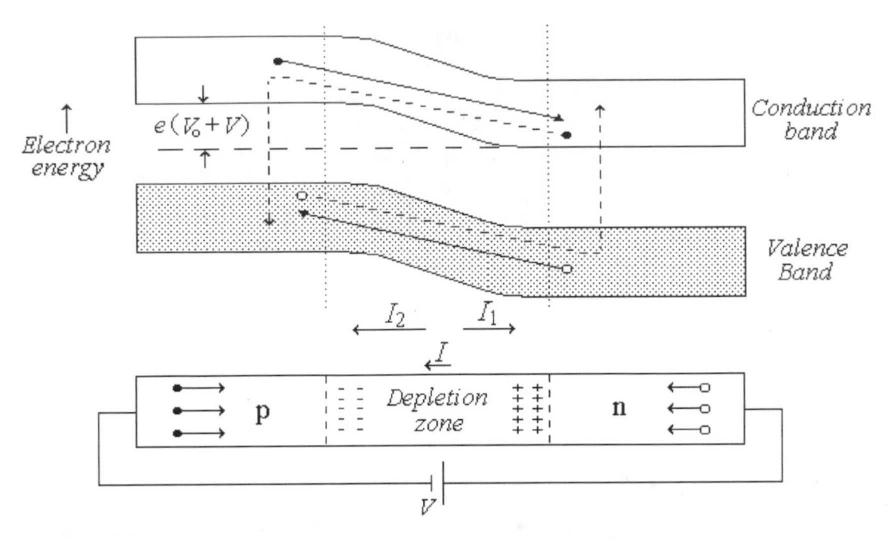

Fig. 6.36. A reverse biased p-n junction. The diffusion current, I_1, is greatly reduced and is swamped by the drift current, I_2. As a result, a feeble current, $I = I_2 - I_1$, of minority charge carriers flows from the n-type to the p-type material.

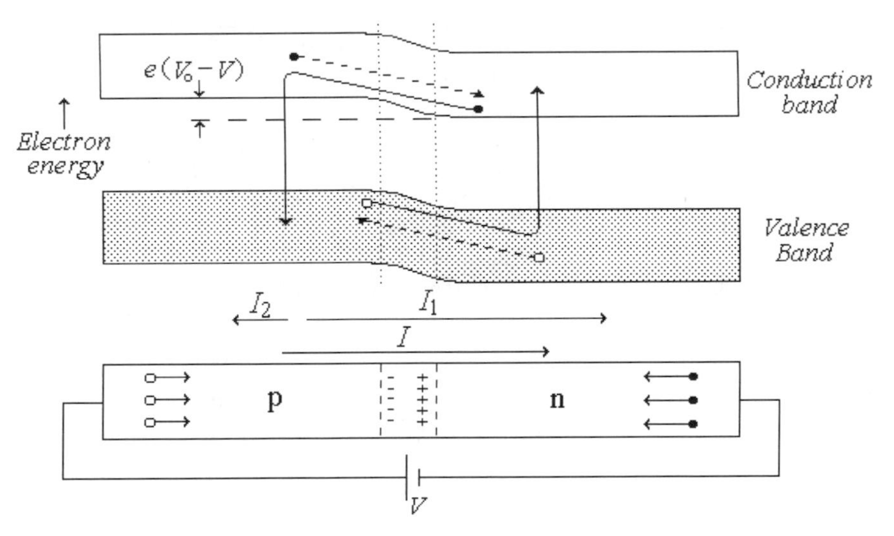

Fig. 6.37. A forward biased p-n junction. The drift current, I_2, is swamped by the greatly enhanced diffusion current, I_1 of the majority charge carriers. The result is a net current, $I = I_1 - I_2$, whose direction is from the p-type to the n-type material.

In general, the magnitude of the diffusion current, I_1, depends on the number of majority charge carriers that, at a given absolute temperature, T, possess sufficient energy to overcome the potential barrier, $(V_0 \pm V)$, between the two ends of the p-n junction. Assuming a Maxwell–Boltzmann distribution of the energies of the majority charge carriers, this gives

$$I_1 = Ae^{-(V_0 \pm V)/kT}, \tag{6.59}$$

where A is a constant of the system. At equilibrium, when $V = 0$, the diffusion current equals the drift current, i.e. $I_1 = I_2$. Thus,

$$I_2 = Ae^{-V_0/kT}. \tag{6.60}$$

Accordingly, the current, I, through a junction diode is given by the function

$$\begin{aligned} I &= I_1 - I_2 \\ &= Ae^{-(V_0 \pm V)/kT} - Ae^{-V_0/kT} \\ &= I_2(e^{\pm V/kT} - 1), \end{aligned} \tag{6.61}$$

which, to a good approximation, matches the experimental diode characteristic shown in Fig. 6.38.

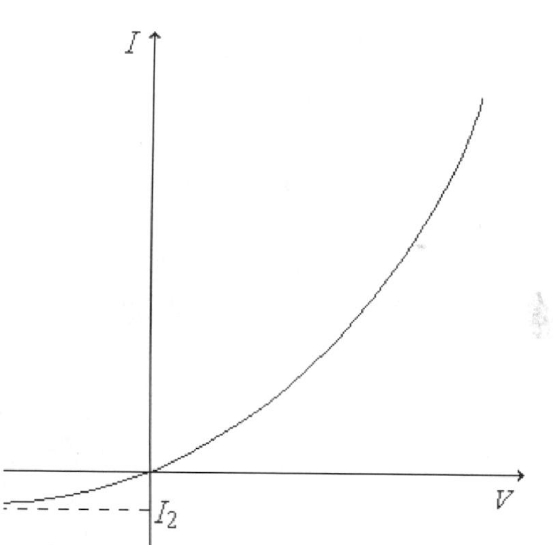

Fig. 6.38. The characteristic of a junction diode. The diode exhibits excellent conductivity in one direction but poor conductivity in the opposite direction.

6.4.9 Semiconductor Devices

The three most important functional properties of p-n junction diodes are:

1. they are excellent conductors of electricity in one direction but not in the other;
2. their interactions with radiation are similar to those of ordinary atoms;
3. they react almost instantaneously to electrical changes.

Moreover, by the techniques of miniaturisation employed in the production of integrated circuits,[40] p-n junctions can be made microscopically small. It is for these reasons that junction diodes are the principal constituents of most modern electronic devices. The electronic and computer revolution of the late 20th century could never have taken place without the development of p-n junction diodes and the semiconductor devices such as the transistors and photodiodes that are based upon it.

1. Photodiodes

Photons of sufficient energy that strike an unbiased p-n junction may be absorbed by electrons in the semiconductor's valence bond. The released electrons are excited into the conduction band leaving positive holes in the valence band. The net result of the process is the annihilation of photons and the production of electron-hole pairs. The contact potential across the depletion zone sweeps the electrons into the n-type material and the holes into the p-type material. If the ends of the crystal are connected by a closed circuit, a current will flow, and the irradiated junction acts as a source of EMF. This effect is the basis of the image sensors (CCD – charge coupled devices) used in digital cameras (Fig. 6.39).

The solar cells, such as those in the panels that provide the electrical energy for space probes, typically comprise a thin transparent layer of p-type material deposited on a crystal of n-type material. The EMF of such solar cells is ~0.6 V. However, they only produce weak currents; a 5 cm^2 silicon based solar cell produces just 0.1 A in full sunlight. Solar cells convert about 10% of the incident radiation into electrical energy.

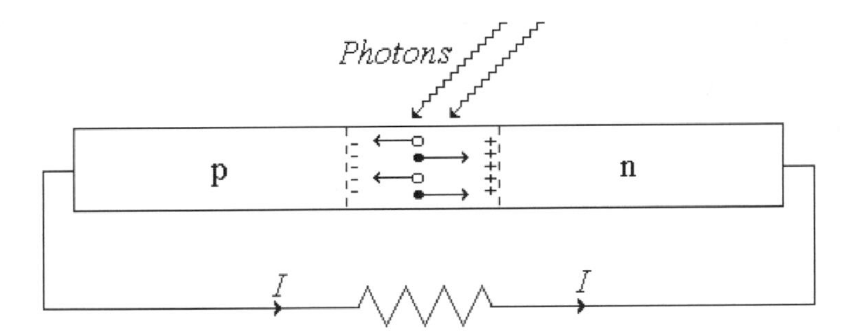

Fig. 6.39. A p-n junction operating as a source of EMF – a solar cell.

[40]Integrated circuits – 'chips' – are thin sheets of pure silicon on which resistors, capacitors, diodes and transistors are built or etched. Very large scale integration (VLSI) has enabled millions of such components to be incorporated and suitably connected on a chip no larger than 1 cm^2. These 'chips' are the computing and memory elements of modern computers and electronic devices.

Electron-hole pairs are also produced when an ionising particle passes through the depletion zone of a reverse biased p-n junction. Each pair results in a short pulse of current. By counting these pulses, the intensity of the radiation can be measured. This is the basis for the use of p-n junctions as radiation detectors.

Just as p-n junction diodes absorb radiation, they can also be made to emit radiation. Although the energy released when an electron falls from the conduction band into the valance band and combines with a positive hole usually appears as heat, in forward biased junction diodes made out of heavily doped semiconductor materials such as gallium arsenide, some of this energy may be released as radiation of frequency $f = E_g/h$. Typically, the energy difference, E_g, between the conduction band and the valence band in semiconductors is in the range 0.5 to 3.0 eV. Photons of infra-red and visible light possess energies in this range and so the diodes emit radiation in the infra-red and visible regions of the spectrum. These *light emitting diodes* have many uses and they are both reliable and cheap to produce. They are the small red lights typically used to indicate that a radio receiver or a component in a computer such as a disk drive is in operation.

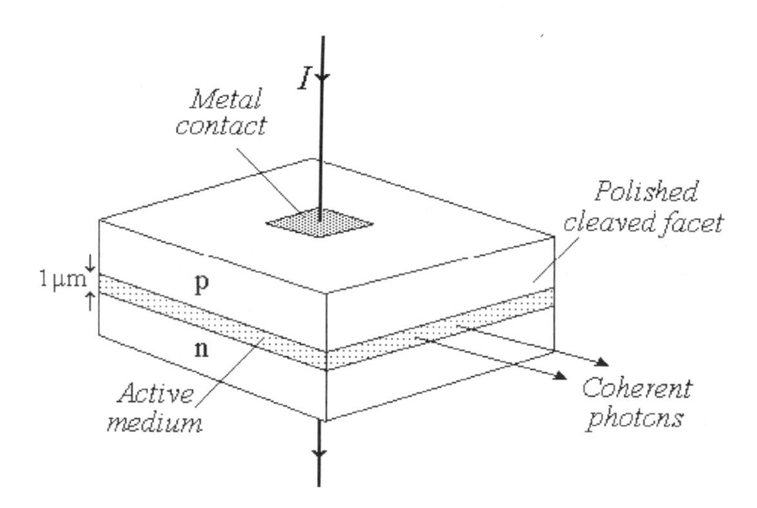

Fig. 6.40. Typical layered structure of a gallium arsenide laser diode. The laser is pumped by the current, I. The active medium acts as a resonator.

Light emitting diodes exhibit both spontaneous and stimulated emission. The injection of electrons into the conduction band on the p-side of a forward biased p-n junction and of holes into the valence band on the n-side establishes a population inversion across its plane; large numbers of electrons enter the p-region of the diode and holes enter the n-region. The greater the current flowing through the junction, the stronger the inversion. At low currents, i.e. modest population inversions, spontaneous emission dominates. At high currents, i.e. strong population inversions, stimulated emission dominates. Thus, a suitably proportioned and pumped light emitting diode can function as a laser (Fig. 6.40).

In their simplest form, laser diodes are small rectangular pieces of semiconducting material with two cleaved facets at opposite ends that act as mirrors. The active medium in which the lasing occurs is sandwiched between a layer of p-type material and a layer of n-type material. The beam emitted at the output facet is typically just a few micrometers wide. Beam wavelengths are generally between 780 nm and 905 nm and their intensities between 10 mW and 25 mW. Laser diodes are both robust and highly efficient (25%–30%). Laser diodes operate at modest currents, ~20 mA, and react almost instantaneously to modulations in the pumping current. Thus, laser diodes can be repeatedly and rapidly switched on and off; it is this property that underlies their widespread use, in conjunction with optical fibers, by the telecommunications industry.

2. Junction transistors

The junction (bipolar) transistor was invented by Bardeen, Brattain and Shockley[41] in 1948. The essential properties of a junction transistor are its capabilities to amplify an electrical signal and to switch a current on or off almost instantaneously. The junction transistor is a semiconductor crystal, composed of three layers of doped materials, either p-n-p or n-p-n. This arrangement effectively comprises two p-n junctions connected back to back (Fig. 6.41).

The three layers of material constituting the transistor are called the *emitter*, *base* and *collector*, respectively. The majority charge carriers in p-n-p transistors are holes; the majority charge carriers in n-p-n transistors are electrons. The base is very thin, ~1 μm, and is usually made of more lightly doped material that the emitter and collector.

Junction transistors are most usually connected in a circuit with the p-n junction between the emitter and the base forward biased and that between the base and the collector reverse biased (Fig. 6.42). As a result, a relatively large number of majority charge carriers flow across the junction from the emitter into the base. This flow of majority charge carriers constitutes the *emitter current*, I_E.

[41]John Bardeen 1908–1991, Walter Brattain 1902–1987 and William Shockley 1910–1989, American physicists who were awarded the Nobel prize in 1956 for the discovery of the transistor. It was the first of the two Nobel prizes in physics awarded to Bardeen.

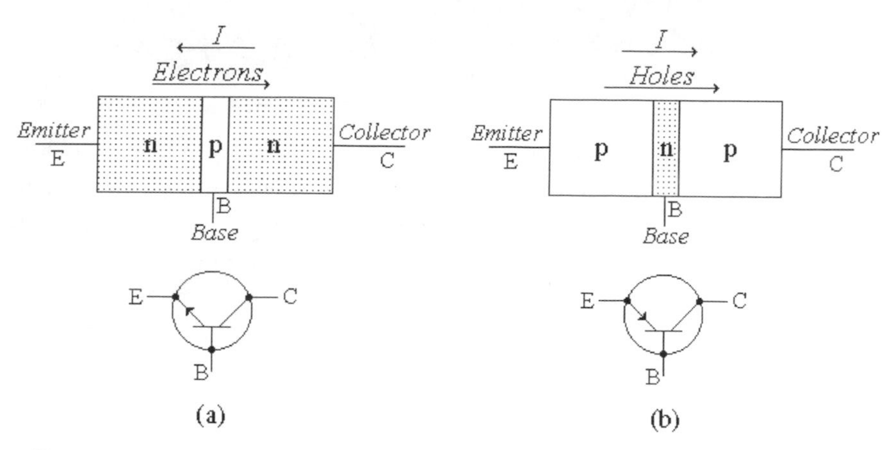

Fig. 6.41. The junction transistor. Transistors are three terminal devices. (a) An n-p-n transistor. The majority charge carriers are the electrons that flow from the emitter to the collector in the opposite direction to that of the current. (b) A p-n-p transistor. The majority charge carriers are the holes that flow from the emitter to the collector in the direction of the current.

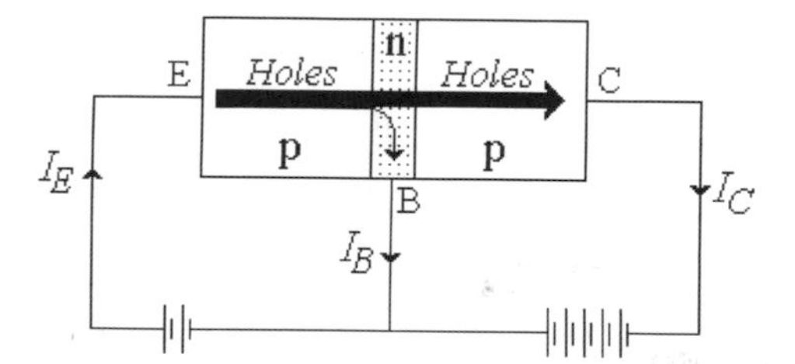

Fig. 6.42. The flow of the majority charge carriers – the holes – and the currents through a p-n-p transistor.

Because of the thinness of the base and the sparsity of its impurity atoms, most of the holes entering the base from a p-type emitter (or the electrons entering from an n-type emitter) do not encounter electrons (or holes) with which they can combine. Since both the emitter and the collector share the same type of majority charge carrier and the p-n junction between the base and the collector is reverse biased, i.e. it encourages the flow of the collector's majority charge carriers from the base into the collector, most of the emitter's majority charge carriers entering the base pass straight through it and are swept up into the collector. This transfer occurs despite the fact that reverse biased p-n junctions usually exhibit a high resistance (often $> 10^5\,\Omega$). The name *transistor* was derived from this 'transfer-resistor' effect. The flow of majority charge carriers into the collector constitutes the *collector current*, I_C.

The small number of charge carriers entering the base from the emitter that do encounter an electron (or hole, as the case may be), or that leak out of the base and back into the emitter constitute the *base current*, I_B

The relation between the three currents, I_E, I_C and I_B, is given by

$$I_C = I_E - I_B. \tag{6.62}$$

The ratio, $\beta = I_C/I_B$, between the collector current and the base current is a characteristic parameter of junction transistors. Its value depends on the geometry of the transistor and the composition of its component layers. Typical values for β vary between 20 and 200. Thus, even weak base currents can produce considerable collector currents, small fluctuations in the former appearing as large variations in the latter. It is this effect that underlies the use of transistors in amplifying circuits (Fig. 6.43).

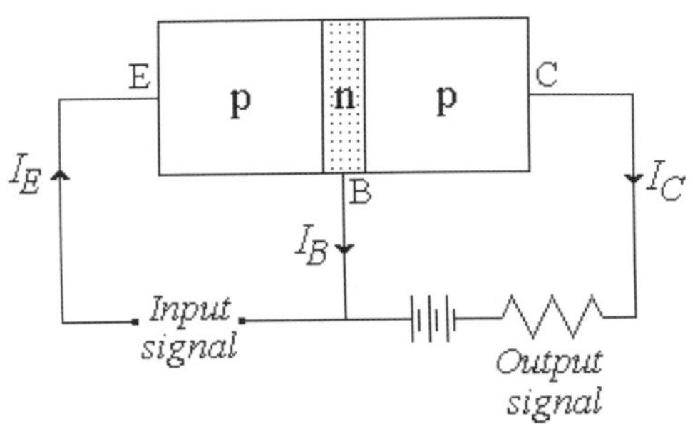

Fig. 6.43. A simple transistor amplifier circuit.

In the simplest arrangement, the signal to be amplified is applied across the p-n junction between the emitter and the base and constitutes its forward biasing potential. The collector current multiplies the resulting base current by the factor β. Any modulations in the signal are almost instantaneously reflected by variations in the collector current with very little distortion.

The power of the output signal, P_{Output}, depends not only on the collector current, I_C, but also on the voltage, V_{BC}, in the collector-base circuit,

$$P_{Output} = I_C V_{BC}. \tag{6.63}$$

Since the p-n junction in this circuit is reverse biased, substantial collector-base voltages can be used. For example, the collector-base voltage, V_{BC}, can typically be 200 times the emitter-base voltage, V_{BE}. This has the effect of further multiplying the power of the output:

$$\begin{aligned} P_{Output} &= I_C \cdot 200\, V_{BE} \\ &= \beta I_B \cdot 200\, V_{BE}, \\ &= 200 \beta P_{Input} \end{aligned} \tag{6.64}$$

where $P_{Input} = I_B V_{BE}$. In this way a current ratio of $\beta \approx 100$ can produce a power amplification > 20,000.

3. Field effect transistors

Whereas the magnitude of the current flowing through junction transistors is determined by the stock of majority charge carriers provided by the base current, in field effect transistors it is the size (cross-section) and shape of the semiconductor channel through which the charge carriers flow that modulates the current; typical channels are about 0.1 to $3\,\mu$m in length from 1 to $100\,\mu$m in width. These basic geometrical dimensions are, in turn, modulated during their operation by an applied electrical field and hence the name *field-effect transistors – FETs*.

Like the junction transistor, the field-effect transistor is also essentially a three terminal device. However, the terminals are denoted the *source*, *drain* and *gate* and they correspond, respectively, to the emitter, collector and base of the junction diode. The gate is so called because it controls the opening and closing of the channel through which the charge carriers pass between the source and the drain; however it is insulated from the channel by a thin layer (2 to 5 nm) of non-conducting material, typically silicon dioxide (glass). FETs do have a fourth terminal called the *base* or *substrate* (the bulk of the device's semiconductor material which may be a p-type or an n-type) but it is usually earthed and is not used as an input or output connection.

Of the many different configurations of FETs, we will describe the n-channel *metal oxide semiconductor field effect transistor* (*MOSFET*) in which the substrate is a lightly doped p-type semiconductor material and the source and drain terminals are connected to nuggets of heavily doped n-type material embedded at opposite ends of the substrate. The only difference in a p-channel MOSFET is that the two types of semiconductors are exchanged; the substrate is n-type and the nuggets p-type.

Both n-channel and p-channel MOSFETs can be made in two different modes *depletion-mode* and *enhancement-mode*, each with its particular conductive properties (Fig. 6.44). To make an n-channel depletion-mode device, a thin layer of n-type material is deposited onto the surface of the p-type semiconductor substrate between the nuggets of n-type material so as to link them into an unbroken conductive section. A thin coating of the oxide insulator is then laid over the sliver of n-type material and this, in turn, is plated above with a metallic film. A terminal is attached to this metal film and constitutes the transistor's gate terminal. In the alternative enhancement-mode, no n-type material is applied onto the substrate; the thin coating of oxide insulator is applied directly onto the surface of the p-type substrate and the metallic film and gate terminal above it. Note that in both modes, the gate and the semiconductor

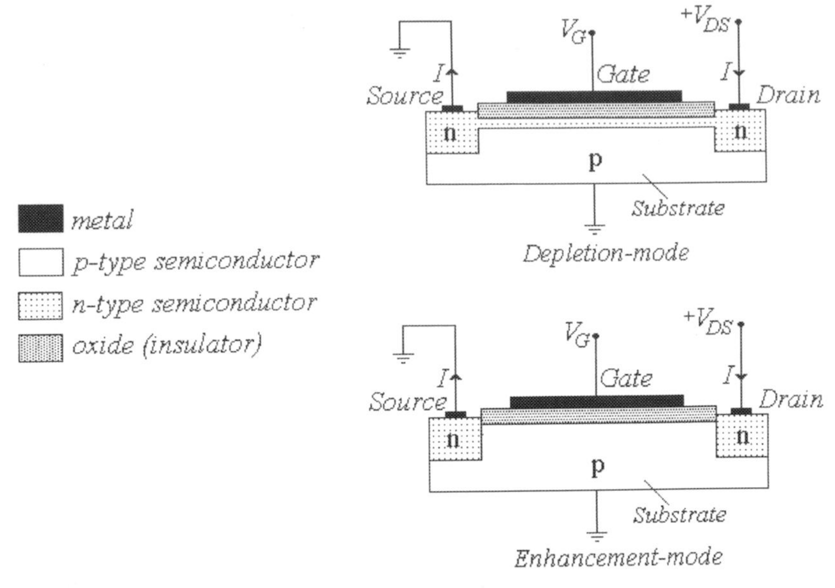

Fig. 6.44. The two types of n-channel MOSFETs. Whether or not a current, I, flows between the drain and the source, depends on the availability of charge carriers in the channel immediately below the gate. The depletion-mode is inherently conductive by reason of the charge carriers (electrons) present in the sliver of n-type material on the surface of the substrate and so its default state is 'ON'; this is equivalent to a normally closed switch. By contrast, no charge carriers are inherently available in the enhancement-mode and so its default state is 'OFF'; this is equivalent to a normally open switch.

materials are insulated from one another such that no current can flow between them.

In operation, two potentials are applied to the MOSFET. The first, V_{DS}, is applied between the drain and the source. In the depletion-mode, whose default state is 'ON', this causes a current, I, to flow down the thin channel of n-type material. The majority charge carriers are electrons and these flow from the source to the drain (against the conventional direction of the electric current). In order to modulate the current, a negative potential, $-V_G$, is applied between gate and the substrate of p-type material. Because of the layer of insulating oxide, no current flows between the gate and the substrate; the situation is comparable to that in a parallel-plate capacitor. An electric field is established between the gate and the earthed substrate which has the effect of *depleting* the number of charge carriers in the channel. Lowering the potential *depletes* their number still further. When the gate-substrate potential is sufficiently low, their number is reduced to zero; the current ceases and the device is now 'OFF'.

In the enhancement-mode MOSFET, whose default state is 'OFF', a conductive channel must first be established (induced) between the source and the drain since one does not exist naturally within the device. This is achieved by applying a positive gate-substrate potential, $+V_G$, which has the effect of driving away positive holes in the channel of p-type substrate adjacent to the oxide layer and attracting electrons into it. The electrons pile up on the surface of the substrate just beneath the oxide layer, creating an 'inversion layer' of free electrons which constitutes the n-channel of charge carriers connecting the two nuggets of n-type material. The potential required to create this inversion layer is known as the device's *threshold potential*. Raising the gate-substrate potential further *enhances* the number of electrons (charge-carriers) in the channel; the device is now 'ON'.

To summarise the two modes of the n-channel MOSFET:

Mode	Depletion		Enhancement	
Bias	ON	OFF	ON	OFF
V_G	zero	negative	positive	zero

Field-effect transistors are particularly suitable for use in electronic computers. The '0' of the binary arithmetic on which their logic is based can be made to correspond to the transistor being 'OFF' (gate closed) and the '1' to the transistor being 'ON' (gate open). MOSFETs can switch between these two modes almost instantaneously. Moreover, since the operation of the gate draws no current from the circuit, it does not corrupt or affect the information in the signals being exchanged between the computer's components.

Moreover, FETs are relatively simple to make and are readily incorporated into the VLSI circuits that are at the heart of modern computers.[42] Their power usage is minimal and they have proved to be both robust and reliable under a range of operating conditions. From their ubiquitous presence as the vital components in personal computers and electronic devices, in the central processing units of super-computers and, more recently, in biosensors used for DNA sequencing, they have become the indispensable core of the modern world.

MOSFETs can also be made into solid-state memory devices by the addition of a second gate, a *floating gate*, which is insulated from above and below by thin layers of oxide (Fig. 6.45). Although it is not electrically connected to the rest of the transistor, electrons can be placed on the floating gate, and removed

[42] Circuits involving millions of resistors, capacitors, diodes and transistors can be incorporated onto just 1 cm^2 of a VLSI chip.

from it, by means of quantum-mechanical tunnelling.[43] Moreover, because the floating gate is electrically insulated, any charge, once placed on it, will remain there for a long time, even years under normal circumstances; the device does not need to be connected to a power supply in order to retain this charge.

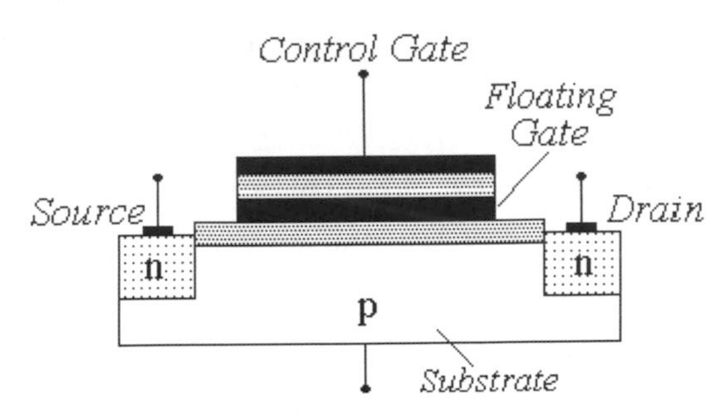

Fig. 6.45. A floating gate memory cell. When electrons are added to the floating gate, the device is said to be in a *programmed state*; when they are removed it is said to be in an *erased state*. It is this process of adding, removing and storing electric charge on the floating gate that makes it a memory device.

When the floating gate is charged with electrons, it screens the electric field set up by the applied gate-substrate potential. This changes the threshold potential of the device which, in turn, affects its conductivity. In general, the MOSFET will be less conductive when the floating gate is charged with electrons than in their absence.

When no electrons are on the floating gate, the device is in its default (raw) state and with the application of appropriate voltages to the control gate ($\sim$5 V) and drain ($\sim$1 V), current will flow through the induced source-drain channel. This state represents binary '1' (gate open). If, however, the floating gate is charged with electrons, no channel will be formed under these potentials and no current will flow. This state represents binary '0' (gate closed). The state of the memory cell can thus be determined (read) under these circumstances by sensing whether current is flowing through the channel or not.

The memory cell can be programmed from binary '1' to binary '0' and reset back to binary '1' as follows:

1. A relatively high voltage ($\sim$12 V) is applied to the control gate and to the drain ($\sim$7 V). Electrons will flow through the induced channel from the source to the drain. Some higher energy electrons in this stream may tunnel through the first thin oxide layer from the channel onto the floating gate;

[43]Tunnelling through thin oxide films was first reported by R.H. Fowler and L. Nordheim in 1928, and the phenomenon is usually referred to as *Fowler–Nordheim tunnelling*.

this process has been termed *hot-electron injection.* These electrons cannot rise further onto the control gate because of the second layer of oxide insulator nor can they return into the channel; they are trapped on the floating gate and will remain there even after the gate and drain potentials have been removed. The device is now in binary state '0'.

2. Erasing the binary '0' state requires removing the electrons from the floating gate. Since the gate is surrounded above and below by insulating oxide, this can only be achieved by quantum-mechanical tunnelling through the oxide layer back into the channel. Applying a negative voltage to the control gate (~ -9 V) and a moderate positive potential to the source (~ 6 V) achieves this reversal.

Assembled into suitable arrays, these memory cells are at the heart of the *flash memories* inside the ubiquitous electronic devices – mobile phones, pads and personal computers, GPSs, MP3 players, digital cameras and disks-on-key, to list just a few – without which 21st century life would be unthinkable.

Chapter 6.5

Invariance, Symmetry and Conservation Laws

Symmetry is a familiar concept. Examples of it abound in nature: snowflakes, crystals and vertebrate animals all exhibit certain symmetries. In general, the invariant features of a system, i.e. those that remain unchanged after some operation has been performed upon them, constitute its symmetry.

The simplest type of symmetry is the *bilateral symmetry* we observe when reflection about a line leaves the form of the object unaltered (Fig. 6.46a). Another type is the *rotational symmetry* exhibited by the pentagram; rotations around its centre through angles that are integral multiples of 360°/5 leave the figure unaltered (Fig. 6.46b).

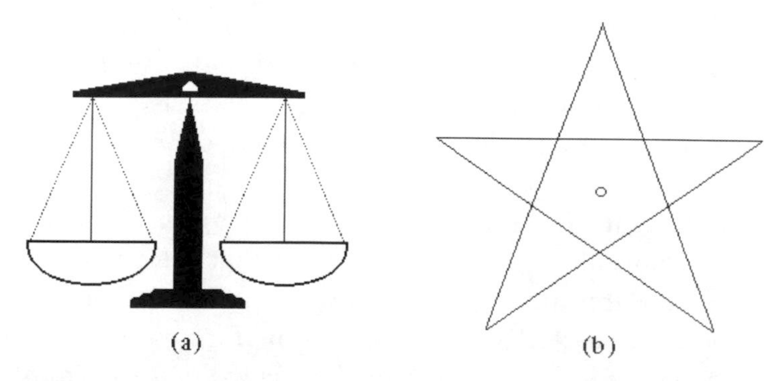

Fig. 6.46. Examples of symmetry. (a) Bilateral symmetry: reflection about a line drawn down the central column leaves the figure unaltered. (b) Rotational symmetry: the pentagram is carried over into itself by rotations around its centre through angles that are an integer multiple of 360°/5. The pentagram also exhibits bilateral symmetry.

Considerations of symmetry are often of great use in physics. The motion of a simple pendulum (or of any simple harmonic motion) is symmetrical. Quantities such as the system's velocity and acceleration have the same magnitude at symmetric positions (Fig. 6.47). Likewise, the system's potential energy is symmetric relative to the central equilibrium point. In a quantum system, the probability densities of two symmetrical points must be equal: $|\psi_A|^2 = |\psi_{A'}|^2$.

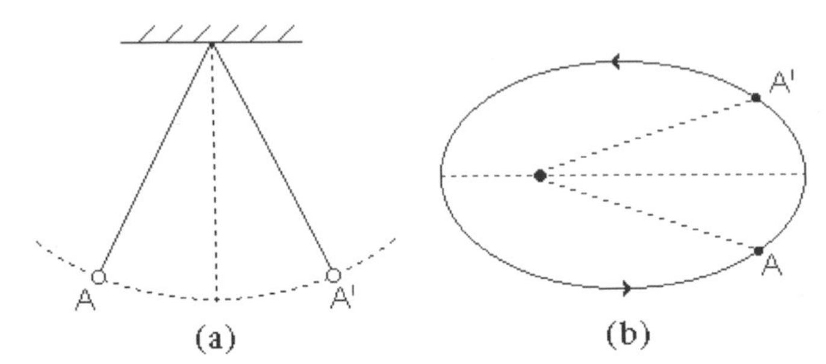

Fig. 6.47. Symmetrical motions. (a) A pendulum. (b) A planetary orbit. The points A and A′ are symmetric.

However, what concerns us here is not the symmetry of physical systems, objects or geometrical shapes but that of the fundamental laws which direct the workings of the physical world.

6.5.1 The Symmetry of the Laws of Physics

In modern times, the recognition of the importance of symmetry in understanding the physical world began largely with Einstein's theories of relativity. Underlying these theories was Einstein's conviction that the laws of physics must take the same form in all reference frames. The laws must be covariant and suitable transformations must be found and formulated to ensure this. This axiom incorporated the concept of observer equivalence that was implicit in the belief in a physical reality that is independent of any percipient entity.

In symmetry terms, the principle of covariance requires that the operation of moving from one reference frame to another must not alter the laws of physics. Like the appearance of the pentagram, the form of the laws must remain unaltered when certain operations are performed upon them. Those operations which do not change the form of the laws are their symmetry operations. They correspond to the transformations of the conventional approach to relativity.

Important symmetries other than those imputed by Einstein had already been incorporated into scientific thinking when he published his theories of relativity. The symmetry of nature under translations in time, however distant that time may be, was first expressed by Hutton[44] in the *Principle of Uniformitarianism*, which asserts that all geologic phenomena may be explained as the result of existing forces having operated uniformly from the origin of the earth

[44]James Hutton 1726–1797, Scottish geologist who is often called 'the father of geology'.

to the present time. It is the belief in this symmetry implicit in modern theories of evolution and cosmology that distinguishes them from creation myths.

The notion of identical elementary particles, whether they are atoms of a certain element or the electrons, protons and neutrons that comprise these atoms, expresses another kind of symmetry. It is a symmetry under exchange. Replacing an oxygen atom in a molecule by another oxygen atom makes no difference to the chemical properties of the molecule; all oxygen atoms are chemically identical.

Time reversal, the operation usually designated by the symbol T, is a perplexing operation. Whereas it can be easily shown that the laws of classical mechanics and electromagnetism are invariant under time reversal, i.e. if t is replaced by $-t$ wherever it appears in the equations, our intuition and daily experience point to strong evidence that time flows in just one direction, from the past to the future. In physical terms, the effect of time reversal is to exchange the final and initial states. Like running a motion picture backwards; starting with the fried egg and finishing with a freshly laid egg. As far as we can tell, this macroscopic irreversibility is a consequence of the very large number of particles involved. It underlies the thermodynamic concept that the entropy of the physical universe is always rising. But if we look at individual atoms we find that the laws that govern their behaviour are time reversible.

Not that the laws of nature are all invariably symmetrical. For example, their asymmetry under changes of scale was first noted by Galileo, when he recognised that:

> '*Similar solid objects do not exhibit a strength that is proportional to their size.... A large beam or column falling from a given height will go to pieces, when, under the same circumstances, a small scantling or small marble cylinder will not break.... Clearly then, if one wishes to maintain in a great giant the same proportion of limb as that found in an ordinary man, he must either find a harder and stronger material for making the bones or he must admit a diminution of strength.... Whereas, if the size of a body be diminished, the strength of that body is not diminished in the same proportion; indeed the smaller the body the greater its relative strength.*'[45]

By providing a simple rational explanation for so many familiar natural phenomena, this simple inference did perhaps even more than Galileo's astronomical observations to undermine the traditional and hallowed beliefs of his time.

[45] *Dialogues Concerning Two New Sciences.* Translated by Henry Crew & Alfonso de Salvio (1954) New York: Dover Publications, p. 108f.

6.5.2 Group Theory

The notion that the concept of symmetry can be rigorously applied to abstracts such as the laws of nature is quite new. It is based on the concept of a *group* introduced into mathematics by Galois.[46]

> **A group is any set for which a rule, symbolised by *, can be defined such that the product, *a*b*, of combining any two elements of the set is itself a member of the set.**

For instance, the set of integers

$$\ldots, -3, -2, -1, 0, +1, +2, +3, \ldots$$

forms a group with the combination rule 'addition'. As required, the product of combining any two integers by addition gives another integer:

$$(-3) + (+2) = (-1). \tag{6.65}$$

In this case, the symbol * is replaced by +. Such a group is called an *additive group*.

There are three other conditions that must also be satisfied by a group:

1. The product of combining a particular element of the set, called the *identity*, with any other element of the set must leave the latter unchanged, i.e. $a^*I = a$, where I symbolises the identity. In the additive group, the zero fulfils this role:
$$n + 0 = n. \tag{6.66}$$
2. Every element, a, has an inverse element, a^{-1}, such that their combination equals the identity, i.e. $a^*a^{-1} =$ *the identity*. In the additive group, for each integer, n, there is a negative integer, $-n$, such that their sum equals zero, and *vice versa*:
$$n + (-n) = 0. \tag{6.67}$$
3. The combination rule must be associative, i.e. $a^*(b^*c) = (a^*b)^*c$. This is satisfied by the additive group
$$(+2) + [(-3) + (-1)] = -2 = [(+2) + (-3)] + (-1). \tag{6.68}$$

[46] Evariste Galois 1811–1832, French mathematician who produced a method of determining when a general equation could be solved by radicals. This theory solved many long-standing unanswered questions including the impossibility of trisecting the angle and squaring the circle. He introduced the term 'group' when he considered the group of permutations of the roots of an equation. Having spent some time in prison for political offences, he was killed in a duel at the age of 21 shortly after his release.

Although the rule must be associative, it is not necessarily commutative. That is a^*b need not equal b^*a. A group, all of whose elements commute, is called an *Abelian group*; the additive integer group is Abelian

$$[(-3)+(-1)] = -4 = [(-1)+(-3)]. \quad (6.69)$$

Non-commutative groups are termed *non-Abelian.*[47]

Many different entities and operations can constitute a group. For example, vectors form an additive group similar to the additive integer group, though the operation performed when adding vectors is different (Fig. 6.48). But other than that, the rules for dealing with vectors are exactly the same as those for dealing with integers. The zero vector is a vector of zero magnitude; the inverse is a vector in the opposite direction.

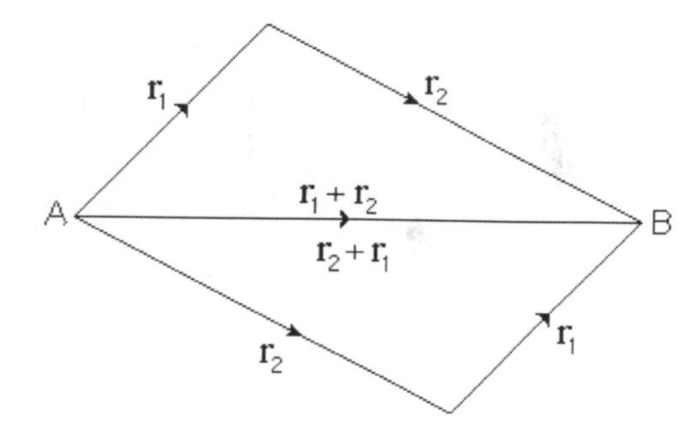

Fig. 6.48. The displacement from A to B is itself a vector which can be defined as the sums $\mathbf{r}_1 + \mathbf{r}_2$ or $\mathbf{r}_2 + \mathbf{r}_1$. From the diagram it is clear that $\mathbf{r}_1 + \mathbf{r}_2 = \mathrm{r}\mathbf{r}_2 + \mathbf{r}_1$.

Another example of an additive group is the rotations of a line in a given plane (Fig. 6.49). The zero of this system represents no rotation. The inverse is a rotation in the opposite direction.

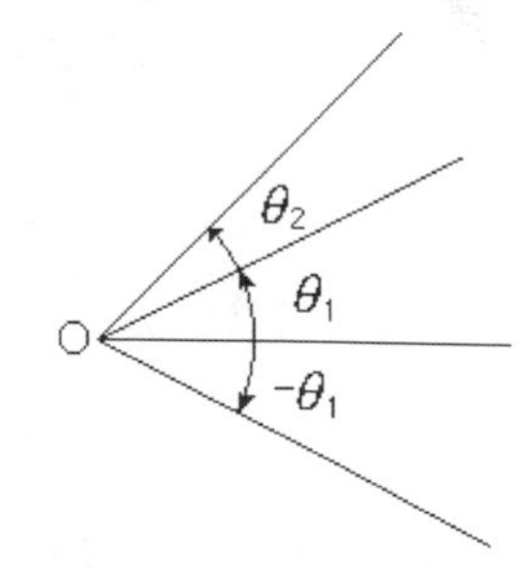

Fig. 6.49. Rotations in a plane about the point O. Rotations are positive if they are anti-clockwise and negative if clockwise. The rotations are commutative because evidently $\theta_1 + \theta_2 = \theta_2 + \theta_1$.

The integer, vector and rotation groups we have so far considered are all examples of *infinite groups*; they all contain an infinite number of elements. A simple example of a *finite group* is given by the symmetric rotations of the three pronged shape – *tripod* – shown in Fig. 6.50.

[47]Niels Henrik Abels 1802–1829, Norwegian mathematician.

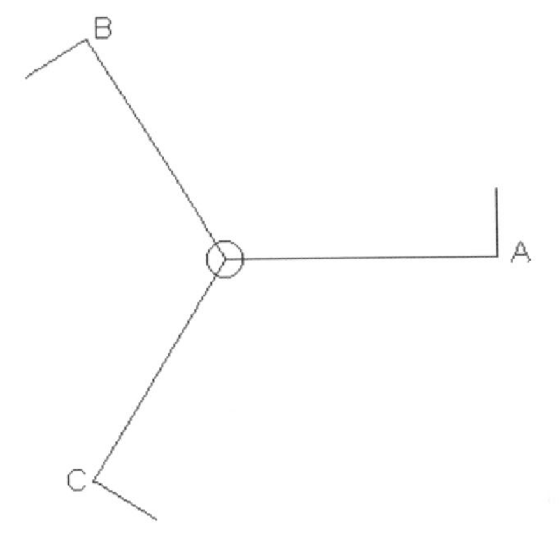

Fig. 6.50. Rotational symmetry. There are just three rotations around the tripod's centre which leave its appearance unchanged: an anti-clockwise rotation through 120°; an anti-clockwise rotation through 240°; an anti-clockwise rotation through 360°, which amounts to no rotation at all or to leaving things as they were.

There are just three rotations around its centre which leave its appearance unchanged: an anti-clockwise rotation through 120°, which we designate $\mathbf{Rot}_{+120°}$; an anti-clockwise rotation through 240°, which we designate $\mathbf{Rot}_{+240°}$; an anti-clockwise rotation through 360°, which we designate the identity, $\boldsymbol{I}$, as it leaves things as they were.

We can disregard clockwise rotations of the tripod as they are indistinguishable from the anti-clockwise rotations:

$$\mathbf{Rot}_{+120°} = \mathbf{Rot}_{-240°} \text{ and } \mathbf{Rot}_{+240°} = \mathbf{Rot}_{-120}. \tag{6.70}$$

The combination rule for this symmetry group is 'after applying the first operation, apply the second one'. Applying this rule we see that two successive rotations through 120° have the same effect as one rotation through 240°. We can express this symbolically as

$$\mathbf{Rot}_{+120°}{}^{*}\mathbf{Rot}_{+120°} = \mathbf{Rot}_{+240}. \tag{6.71}$$

Similarly, two 240° are equivalent to one 120° rotation:

$$\mathbf{Rot}_{+240°}{}^{*}\mathbf{Rot}_{+240°} = \mathbf{Rot}_{+120}. \tag{6.72}$$

The group of rotational symmetry operations that can be performed on the tripod can be summarised in a matrix termed a *multiplication table* in group theory.

*	$\boldsymbol{I}$	$\mathbf{Rot}_{+120}$	$\mathbf{Rot}_{+240°}$
$\boldsymbol{I}$	$\boldsymbol{I}$	$\mathbf{Rot}_{+120°}$	$\mathbf{Rot}_{+240°}$
$\mathbf{Rot}_{+120°}$	$\mathbf{Rot}_{+120°}$	$\mathbf{Rot}_{+240°}$	$\boldsymbol{I}$
$\mathbf{Rot}_{+240°}$	$\mathbf{Rot}_{+240°}$	$\boldsymbol{I}$	$\mathbf{Rot}_{+120°}$

An examination of the table shows that this group is commutative (Abelian).[48] For example,

$$\mathbf{Rot}_{+240^\circ}{}^{*}\mathbf{Rot}_{+240^\circ} = \boldsymbol{I} = \mathbf{Rot}_{+120}{}^{*}\mathbf{Rot}_{+240^\circ}. \tag{6.73}$$

Thus, we see that symmetry operations can be described in terms of an algebraic system. This is the key to the application of symmetry to the study of physical laws.

6.5.3 Noether's Theorem

A systematic study of the importance and significance of symmetry in the formulation of physical laws was only undertaken following the appearance of Einstein's theory of general relativity in 1915. The incentive for this was a vexing problem that became immediately apparent with the publication of the theory, namely, that energy is not conserved locally in general relativity.

In classical field theories and in the theory of special relativity, it can be shown that the net outflow of energy across the boundaries of any arbitrary volume element is equal to the rate at which the energy inside the element decreases. Physically, one has a localised conserved energy density. However, in general relativity, the *localisation of energy* has no meaning. A quantity can be defined that is analogous to the energy-momentum concept of special relativity but it is not invariant under general coordinate transformations; different results are obtained in different reference frames. In physical terms, there is a continuous transfer of energy to and from the all-pervading gravitational field which makes it meaningless to speak of the localisation of the energy of the gravitational field.[49]

Following a series of lectures on his new theory of general relativity given by Einstein at the University of Göttingen in 1915, this problem was taken up by Emmy Noether,[50] who had just been invited to join the world-famous team of

[48] For an example of a non-commutative group consider the operation of getting dressed. Putting on each garment may be considered as an element of a group whose inverse is taking it off. The combination rule is 'putting on one garment after another'. If a garment is taken off straight after being put on, the net result is the identity; effectively, nothing has been done. If **S** represents putting on a shirt and **J** putting on a jacket, then **S*J** represents putting on the shirt first and the jacket second. Accordingly, the operation **J*S** would represent putting on the jacket first and the shirt second, which gives a different result. Evidently, the group of 'getting dressed operations' is non-Abelian: $\mathbf{S}^*\mathbf{J} \neq \mathbf{J}^*\mathbf{S}$.

[49] In general relativity, the equivalent of the classical potential energy is the energy of the gravitational field. But this energy can be made to vanish by employing the principle of equivalence and replacing the gravitational field in the volume element by an acceleration.

[50] Emmy Noether 1882–1936, German mathematician. Like so many other Jewish scientists, she was forced to leave Germany following the successes of the Nazi Party in the national elections and their rise to power in 1933 with the choice of Hitler as Chancellor. She emigrated to the United States in 1934 and died shortly afterwards.

mathematicians led by David Hilbert.[51] In the landmark paper she published in 1918, Noether solved the problem, and in doing so, discovered general theorems concerning symmetry and the conservation of physical quantities that have profoundly affected modern physics.

In the first part of her paper, Noether revealed the general relationship that exists in field theories between symmetry and conservation:

> **To every symmetry there is a corresponding conservation law, and *vice versa*.**

She then went on to show, in the second part of the paper, that the symmetry underlying the coordinate transformation of general relativity is different from that which underlies coordinate transformations in classical field theories and the theory of special relativity.[52] Hence, a quantity that is conserved in classical field theories will not necessarily be conserved in the same way in general relativity.

Prior to Noether's theorem, the conservation of energy was something of a mystery. There were even those like Poincaré[53] who had contended that the conservation of energy is merely a definition of energy, since, whenever it appears to break down, we simply invent a new form of energy to restore the balance. Noether showed that energy conservation followed directly from symmetry under time translations; from the fact that the laws of nature are the same on every day of the week. Consider the law of gravity. Suppose the world was such that gravity was weaker on Mondays than on Tuesdays. We could then raise objects on Mondays at a low expenditure of energy and make an energy profit by letting them fall back again on Tuesdays. The process could be repeated week after week, thus increasing the store of energy in the universe as time went by.

The conservation of linear momentum follows from translational symmetry. Consider a system of interacting particles and suppose that its potential energy, U, is independent of its position. In this case, the derivatives of the potential energy with respect to the coordinates x, y and z would be zero, which implies

---

[51] David Hilbert 1862–1943, German mathematician whose position at the Mathematical Institute at the University of Göttingen was recognised as the most prestigious mathematical position in Germany, and possibly in the world. When the university Senate refused to grant Emmy Noether the status of a *Privatdozent* that would entitle her to lecture in her own right, Hilbert retorted: 'Gentlemen, I do not see that the sex of the candidate is an argument against her admission as a *Privatdozent*. After all, the Senate is not a bathhouse.' Notwithstanding, Hilbert arranged for her to remain at Göttingen by having her lectures announced under his name.

[52] Whereas the symmetry group of classical field theories, the Poincaré or Lorentz group, is finite, that of general relativity is infinite.

[53] Jules Henri Poincaré 1854–1912, French mathematician.

that no external force is acting on the system, e.g. $F_x = -\partial U/\partial x = 0$. Hence, the momentum of the system is unchanged: $\partial p_x/\partial t = F_x = 0$.

The conservation of angular momentum arises from the symmetry of the laws of physics under rotation of the coordinate system. For a free object to suddenly start spinning of its own accord would certainly violate the law of angular momentum conservation. It would also violate the symmetry of space for the spin axis would define a particular direction in space, distinguishable from all the others.

The symmetry of the classical laws of physics under translations in space and rotations of the coordinate axes is ensured by expressing them in terms of vectors. If the vector equation $\mathbf{F} = \mathbf{r}$ is true in one coordinate system, it is true in any coordinate system. The great usefulness of vector relationships is that they carry their transformations within them; they are inherently symmetrical. They incorporate the notion that empty space is the same in all directions.

6.5.4 The Conservation Laws of Particle Physics

The study of elementary particles has revealed new conservation laws and new symmetries. Certain of these laws are obeyed under all circumstances; others are not. Moreover, whereas some have arisen from theoretical considerations, others are just empirical relationships. Notwithstanding, any reaction not expressly forbidden by the conservation laws will generally occur, if perhaps at a slow rate. This expectation is often referred to as the *totalitarian principle* and accords with quantum mechanics. Unless the barrier between the initial and final states is infinitely high, there is always a non-zero probability that a system will make the transition between them.

In addition to the familiar conserved quantities listed above – energy, linear and angular momentum – those absolutely conserved in particle physics are (1) electric charge; (2) lepton number; (3) baryon number. Noether's theorem posits that there should be symmetries associated with each of these conserved quantities too. However, a corresponding symmetry has only been identified in the case of charge conservation.

1. The conservation of electric charge

The conservation of electric charge is associated with the symmetry called *gauge symmetry*.[54] A gauge transformation is one that has no observable consequence, like changing the arbitrary zero of the electric potential. This has no effect on the electric field or on observable potential differences. Like the height between

[54] The origin of the term 'gauge' is historical. A better name would be *phase symmetry*.

two levels, the difference in potential between two points is independent of the base chosen for the zero. Nor is the form of Maxwell's equations affected by the arbitrary choice of where the electrical potential is to equal zero. It can be shown that charge conservation is a consequence of the gauge symmetry of Maxwell's equations.[55]

Gauge symmetry is central to quantum mechanics and to the quantum field theories that underlie particle physics. In quantum mechanics its meaning is quite straightforward. The phase of a wave function, like the wave function itself, is unobservable. It is only the probability density that is observable. However, if a wave function undergoes a phase transformation such that $\psi \to \psi e^{i\delta}$, the probability density is unaffected:

$$|\psi|^2 = |\psi e^{i\delta}|^2. \tag{6.74}$$

Thus, the observables are conserved under gauge transformations.

2. The conservation of lepton number

In essence, the conservation of lepton number means that whenever a lepton of a certain generation is created or destroyed in a reaction, a corresponding antilepton from the same generation must be created or destroyed. The arithmetic is done by assigning elementary particles a *lepton quantum number*: +1 for the leptons, −1 for antileptons and 0 for all other particles.

Consider the electron capture mode of β decay. The process involves only first generation leptons – electrons and neutrinos:

$$\begin{array}{lccccccc} & p & + & e^- & \to & n & + & \nu \\ \text{Lepton Number} & 0 & & +1 & & 0 & & +1 \end{array}. \tag{6.75}$$

The particle emitted with the neutron must be a neutrino and not an antineutrino for lepton number to be conserved.

Lepton number is also conserved in decay processes such as $\mu^- \to e^- + \bar{\nu}_e + \nu_\mu$, in which muons of well-defined energies produce a distribution of electron energies balanced by a μ-neutrino and an electron antineutrino.

The conservation of lepton number has no clear theoretical basis nor has any simple symmetry yet been associated with it.

[55]The symmetry of an index-linked income under inflationary changes is an everyday example of gauge symmetry. Multiplying the base income by the index (gauge or phase) really does nothing. The value of the money is not changed; its observable purchasing power is unaffected. In contemporary terminology, the theory of general relativity is a *gauge theory*; its symmetry is a gauge symmetry. The *metric tensor* fulfils the role of the 'index' which ensures the invariance of theory's predictions.

3. The conservation of baryon number

Assigning elementary particles a *baryon quantum number*, $+1$ for the baryons, -1 for antibaryons and 0 for all other particles enables us to formulate the law that in any process baryon number is conserved. The stability of the proton can be attributed to this law. Since it is the lightest of all baryons, the hypothetical products of its decay would have to be non-baryons. Thus, the decay would violate the conservation of baryon number.

Recent attempts to unite the electromagnetic, weak and strong interactions into a *grand unified theory* or *GUT* have suggested that protons are in fact unstable and decay into leptons. However, the average lifetime predicted for the protons is $\sim 10^{30}$ years, which makes their decay hard to detect. Experimental evidence of proton decay has yet to be found.

The conservation of baryon number has no clear theoretical basis nor has any simple symmetry yet been associated with it.

A comprehensive elucidation of the remaining partially or incompletely conserved quantities – *isospin* and *parity* – is beyond the scope of this book. We will content ourselves with a brief outline of each.

4. The conservation of isospin

The observation that the strong interaction does not distinguish between nucleons – it treats protons and neutrons identically – underlies the concept of isospin. Instead of regarding protons and neutrons as totally different species, as far as strong interactions are concerned, they are regarded as being different isospin states of the same underlying *nucleon particle*. In a similar way, the three pions, π^+, π^- and π^0, can be regarded as three different states of the same particle when only strong interactions are involved.[56]

Isospin is mathematically similar to spin, though it has nothing whatsoever to do with angular momentum. Thus, isospin can be regarded as a vector, $\mathbf{I}$, in an abstract three-dimensional space, whose magnitude, I, and third component, I_3 or I_z, obey quantisation conditions similar to those of spin. The proton and neutron correspond to different orientations of the isospin vector: the proton corresponds to 'a nucleon with isospin up' ($I_3 = +½$); the neutron to 'a nucleon with isospin down' ($I_3 = -½$). The notion is analogous to the distinction we make between the two particles we designate as 'an electron with spin up' and 'an electron with spin down'.

Experiments indicate that both I and I_3 are conserved in processes that involve strong interactions. In processes involving weak interactions neither

[56]Since at close distances electromagnetic interactions are orders of magnitude weaker than strong interactions they can generally be ignored under such circumstances.

quantities are conserved. In those where strong and electromagnetic interactions are involved, only I_3 is conserved. Thus isospin conservation is not absolute.

The symmetry associated with the conservation of I is the rotational symmetry of the strong interaction in the three-dimensional abstract space. The conservation of I_3 is associated with the symmetry operation called *charge reflection* – changing the sign of all the electric charges, i.e. replacing $+q$ by $-q$, and *vice versa* – under which strong and electromagnetic interactions remain invariant.

If the only interaction we were concerned with was the strong interaction, the symmetry of neutrons and protons would be complete; neutrons could be substituted for protons, and *vice versa*, without any change occurring in the system. At the very close distances met within the nucleus this is true to a very good approximation. However, as the distance between the particles increases, and the effect of the electromagnetic interaction between the protons becomes dominant, the symmetry between the neutrons and protons breaks down. The isospin symmetry is said to be *broken* by the electromagnetic interaction.

The rationale that underlies the unification of the weak and electromagnetic interactions into the single *electroweak interaction* also hinges on the notion that apparently distinct particles, the regular photon and the $W^{\pm}$ and Z^0 bosons, can be regarded as differing forms of the same underlying quantum entity. However, their initial symmetric configuration is unstable and breaks down spontaneously, producing the asymmetry represented by the different properties they manifest. Restoring the symmetry requires the investment of energy. It is estimated that in excess of 1 TeV (10^{12} eV) is required for the symmetry of these particles to be made explicit. Such energies are unavailable in terrestrial laboratories but could have been available immediately after the big bang, before the initial symmetry was broken.[57]

5. The conservation of parity

In physical terms, parity refers to the operation of space reflection, symbolised by the letter P. Parity is conserved if the mirror image of a process is itself one which can occur in nature.[58] In quantum mechanics, the concept arises from considerations of the relation between the wave functions at the symmetric points, A and A′, in a system such as an harmonic oscillator (Fig. 6.46a).

[57] Spontaneous symmetry breaking is exhibited when a system passes from a state of unstable equilibrium to one of stable equilibrium. A pencil standing on its end is symmetrical, but it falls over, losing its symmetry, under the slightest perturbation.

[58] The parity of events and of their mirror images is one of the themes of Lewis Carroll's classic children's book *Through the Looking-Glass.*

Assuming the wave functions are real,

$$\psi_A = \pm\psi_{A'}. \tag{6.76}$$

When $\psi_A = +\psi_{A'}$, we say that the wave function has *even parity* or $P = +1$; when $\psi_A = -\psi_{A'}$, the wave function has *odd parity* or $P = -1$.

The strong and electromagnetic interactions both conserve parity. However, as we saw in the context of the β decay of radioactive nuclei,[59] it is not conserved by the weak interaction.

Another transformation that is not conserved by the weak interaction is *charge conjugation*, usually symbolised by the letter C, under which all the particles in a system are replaced by their corresponding antiparticles, without making any other changes. The link between the parity conservation and charge conjugation is illustrated by the decay of the π^+ pion into the muon, μ^+, and a neutrino, ν, and its charge conjugate decay process:

$$\begin{aligned} &\pi^+ \rightarrow \mu^+ + \nu \\ \text{Conjugate}\quad &\pi^- \rightarrow \mu^- + \bar{\nu} \end{aligned} \tag{6.77}$$

As Fig. 6.51 shows, the charge conjugate decay would require the antineutrino to have negative helicity, which is impossible; antineutrinos always have positive helicity. Thus, the conjugate decay cannot occur.

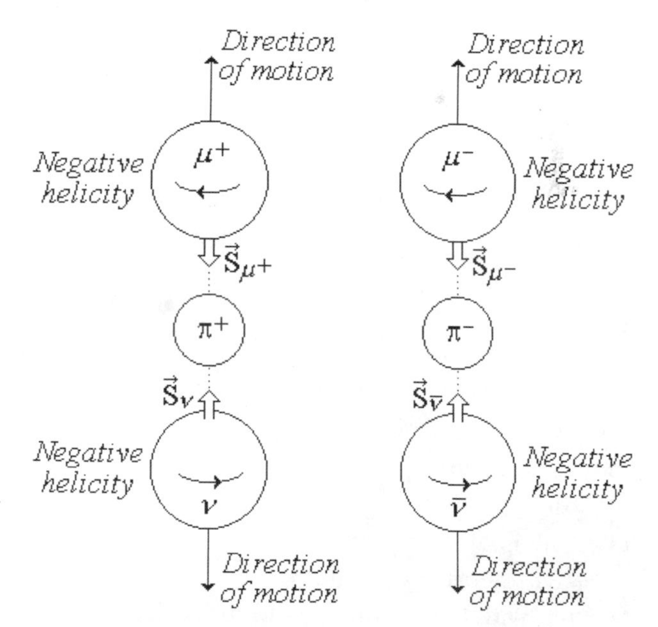

Fig. 6.51. The decay of the π^+ pion and its charge conjugate. The charge conjugate decay of the π^- pion cannot occur as this would require that the antineutrino possess a negative helicity.

[59] See Section 5.3.7.

However, suppose that the π^+ pion decay undergoes two symmetry operations, the P (parity) operation followed by the C (charge conjugate) operation, resulting in the combined operation CP. The π^- decay process that this gives can occur, for it assigns a positive helicity to the antineutrino (Fig. 6.52).

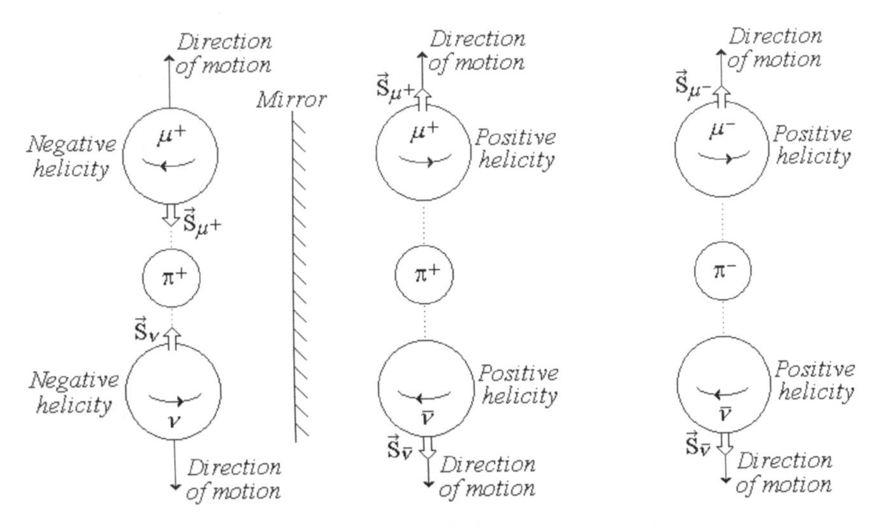

Fig. 6.52. The CP operation applied to π^+ decay. Two successive symmetry operations are applied to the π^+ decay. First, the P operation (reflection in a plane mirror) which changes the helicity of the particles, followed by the C (charge conjugation) which replaces each particle by its antiparticle. The π^- decay process this gives can occur, for it assigns a positive helicity to the antineutrino.

Although, in general, the weak interaction remains invariant under CP operations, an instance in which it does not conserve CP has been observed experimentally.[60] At this stage, the reader might well ask whether there is any symmetry operation under which all the known physical processes are invariably invariant.

There is good reason to believe that all physical laws are invariant under the combined symmetry operations of time reversal, T, parity, P, and charge conjugation, C, known as the CPT operation. The implication of this is that a violation of one of the parameters must be accompanied by a compensating violation of another parameter. If CP is violated, the time parameter must also be violated so that CPT may be conserved. Such a process would not be time symmetric, i.e. it could only occur in one time direction. Though we are familiar with this asymmetry in macroscopic systems, in which the second law

[60]The 1980 Nobel prize for physics was awarded to J.W. Cronin and V.L. Fitch, two American physicists, who observed the violation of CP symmetry in the decay of the unstable K^0 meson.

of thermodynamics holds sway, the import of its occurrence at the particle level is not fully understood.

If, indeed, there is an intrinsic asymmetry in nature, the observation that the universe is composed almost exclusively of matter, with a very small antimatter component could be explained. For if the symmetry between matter and antimatter were perfect, the material universe could not exist. Composed equally of matter and antimatter, which would interact and annihilate one another, it would vanish in a titanic burst of radiation. It is, perhaps, so that we could exist that a slight imperfection was allowed for at Creation.

Appendices to Part Six

6.6.1 The Probabilities of Stimulated and Spontaneous Emission

Consider a system comprising two energy levels, E_1 and E_2, that are occupied by N_1 and N_2 particles, respectively. Assuming the distribution of the particles between the two levels follows the Maxwell–Boltzmann rule, the ratio between their populations at thermal equilibrium will be

$$\frac{N_1}{N_2} = e^{(E_2 - E_1)/kT} = e^{hf/kT}, \tag{6.78}$$

where f is the frequency of the photons absorbed or emitted in transitions between the two levels.

If monochromatic radiation of frequency f and average photon density κ_f flows through the system, the number of particles, ΔN_2, in each unit volume that absorb a photon and 'jump' from the lower to the higher energy level in each second will be

$$\Delta N_2 = \kappa_f P_{Ab} N_1 = \kappa_f P_{Ab} N_2 e^{hf/kT}, \tag{6.79}$$

where P_{Ab} is the absorption transition probability per second per unit volume.

The excited particles in the higher energy state can subsequently return to the lower state either by the spontaneous or by the induced emission of a photon of frequency f. The total number, ΔN_1, of particles returning each second to the lower level in each unit volume will be

$$\Delta N_1 = \Delta N_{Sp} + \Delta N_{St}, \tag{6.80}$$

where ΔN_{Sp} and ΔN_{St} are the numbers returning by spontaneous and stimulated emission, respectively. The number, ΔN_{Sp}, of particles returning by spontaneous emission will depend only on the population, N_2, of the excited state and the spontaneous emission transition probability per second, P_{Sp}

$$\Delta N_{Sp} = P_{Sp} N_2. \tag{6.81}$$

The number, ΔN_{St}, returning by stimulated emission, i.e. as a result of an interaction with a photon of frequency f, will depend also on the average photon density, κ_f. Thus,

$$\Delta N_{St} = \kappa_f P_{St} N_2, \tag{6.82}$$

where P_{St} is the stimulated emission transition probability per second. Adding Eqs (6.68) and (6.69) gives

$$\Delta N_1 = P_{Sp} N_2 + \kappa_f P_{St} N_2. \tag{6.83}$$

At equilibrium the rate of absorption will equal the rate of emission, i.e.

$$\Delta N_1 = \Delta N_2$$

$$\kappa_f P_{Ab} N_2 e^{hf/kT} = P_{Sp} N_2 + \kappa_f P_{St} N_2, \tag{6.84}$$

from which we obtain for the average photon density

$$\kappa_f = \frac{P_{Sp}/P_{Ab}}{e^{hf/kT} - P_{St}/P_{Ab}}. \tag{6.85}$$

The dynamic radiative equibrilium in this simple system is comparable to that which prevails in a constant temperature enclosure, where radiation (cavity radiation) is continually absorbed and emitted from the electric oscillators on its walls. The cavity contains a gas of photons in equilibrium with its walls whose frequencies are set by those of the oscillators. The average photon density per unit, κ^f, is given by

$$\kappa_f = \frac{E_f}{hf}. \tag{6.86}$$

The energy density, E_f, of the radiation of frequency f is given by

$$E_f = \overline{E} \cdot \frac{\Delta n}{V}, \tag{6.87}$$

where $\overline{E}$ is the average energy of the oscillators and $\frac{\Delta n}{V}$ is the number of oscillators per unit volume of the cavity. From Eq. (2.14) the average energy of an oscillator is given by

$$\overline{E} = \frac{hf}{e^{hf/kT} - 1}, \tag{6.88}$$

and from Eq. (2.11) the number of vibrational modes in each unit volume of a cavity of volume V, is given by

$$\frac{\Delta n}{V} = \frac{8\pi}{\lambda^4} \cdot \Delta\lambda = \frac{8\pi f^2}{c^3} \cdot \Delta f. \tag{6.89}$$

Substituting these expressions in Eq. (6.86) gives

$$\kappa_f = \frac{8\pi f^2 / c^3}{e^{hf/kT} - 1}. \tag{6.90}$$

Comparing Eqs (6.85) and (6.90) we see that $\frac{P_{St}}{P_{Ab}} = 1$, i.e. the transition probabilities of stimulated emission and absorption are equal, and that the ratio, $\frac{P_{Sp}}{P_{St}}$, between the transition probability for spontaneous emission and stimulated emission is given by

$$\frac{P_{Sp}}{P_{St}} = \kappa_f \left(e^{hf/kT} - 1\right). \tag{6.91}$$

Part Six: Questions, Exercises and Problems

Questions

1. Given that the nuclear spin quantum number of ^{13}B is $1\frac{1}{2}$, how many spin states are there in this nucleus?
2. If the chemical shift, δ, of a certain nucleus is positive, is the diamagnetic shielding greater or less than in TMS?
3. What is the chemical shift in TMS?
4. The work function of a metal varies little with its temperature. Explain.
5. If the energy band in magnesium metal corresponding to the $3s$ energy level did not overlap that corresponding to the $3p$ level, the material would not be metallic. Explain.
6. Nuclear spin investigations show that the theory once held that nuclei are composed of electrons and protons is incorrect. Show that the most abundant isotope of nitrogen, ^{14}N, cannot be composed of electrons and protons.
7. Hall discovered that if a strip of conducting material is placed in a magnetic field whose direction is perpendicular to its length and an electric current flows along the strip, a potential difference is set up between the edges of the strip, i.e. perpendicular to the direction of the current. Show that the direction of this 'Hall potential' depends on the sign of the charge of the charge carriers in the strip.

Exercises

1. A sample containing hydrogen atoms exhibits resonance at a frequency of 35 MHz in a magnetic field of magnitude 0.8 T. Given that the magnetic moment of the proton is $1.41 \cdot 10^{-26}\,J \cdot T^{-1}$, calculate the strength of the local magnetic field of the protons in the sample.
2. When ^{7}Li nuclei are immersed in a magnetic field of strength 0.3384 T, resonance is obtained at a frequency 5.585 MHz. Calculate the value of the gyroscopic ratio of these nuclei.
3. In a magnetic field of strength 2.348 T, a nucleus exhibits resonance at a frequency of 100 MHz. What is the distribution between the two nuclear spin states in a sample of the nuclei at a temperature of 35°C?
4. At the nucleus of a hydrogen atom, the strength of the magnetic field generated by the $1s$ electron is 12.5 T. Calculate the energy gap between the nuclear spin states as a result of this field.
5. In a spectrometer operating at 60 MHz, the chemical shift of the proton in chloroform, $CHCl_3$, is 7.25 ppm.
   a. What is the difference between the resonance frequency of chloroform and TMS in this spectrometer?
   b. What would be the difference between the resonance frequency of chloroform and TMS in a spectrometer operating at 100 MHz?
6. In a certain crystal, the highest occupied energy band is full. The crystal is transparent to light of wavelength greater than 295 nm but opaque to light of shorter wavelengths. Calculate the energy gap between the two highest occupied bands in the crystal.

7. Germanium absorbs radiation of wavelength greater than 1.65 μm. Calculate the energy gap, E_g, in germanium.
8. Lead sulfide (PbS) can be used to detect infra-red radiation of wavelength greater than 3.5 μm. Calculate the energy gap, E_g, in lead sulfide.
9. The density of silver is $10.5 \cdot 10^3\,\text{kg/m}^3$ and its Fermi energy is 5.51 eV. On the assumption that each silver atom releases a single electron to its 'gas' of conduction electrons and that the average free path of a conduction electron is 200 times greater than the spacing between the atoms in the metal lattice, calculate the resistivity of silver.

Problems

1. An excited ^{57}Fe nucleus decays by the emission of a photon. The difference between the energy levels is 14.4 keV.
   a. Calculate the reduction in the photon's energy as a result of the recoil energy of the nucleus if the nucleus is free to move.
   b. Calculate the reduction in the photon's energy as a result of the recoil energy of the nucleus if the nucleus is set into a crystal of mass 1 gm.
2. The effect of Group V impurities on semi-conductors can be explained in terms of the Bohr model of hydrogen-like atoms as follows. When an atom of phosphorus is added to a silicon lattice, its fifth valence electron effectively 'sees' a nuclear charge of just $+e$; the rest of the nuclear charge is shielded by the atom's other electrons. Using the Bohr model and assuming the dielectric constant in the crystal equals 12, calculate:
   a. the first ionisation energy of a phosphorous atom in the lattice;
   b. the Bohr radius of the fifth valence atom of a phosphorus atom in a silicon lattice. Compare your answer to the average distance between the silicon atoms in the lattice, 0.234 nm.
3. Calculate the wavelength of the conduction electrons in copper given that their Fermi energy is 7.0 eV. How does your answer compare with the average distance between the copper atoms in the lattice, 0.26 nm? Under these circumstances is it reasonable to relate to the conduction electrons classically?

Supplementary Topics

A. The Mathematical Description of Wave Motion

A physical quantity that has a value at every point, such as the pressure in the atmosphere or the force exerted on a unit electric charge, constitutes a *field*. We say that there is a *pressure field* in the air or an *electric field* in the space around a charged body. The former is a *scalar field* since pressure is a scalar quantity; the latter a *vector field*.

When a bell is rung, sound is heard at distant points. The air in contact with the bell is disturbed and this disturbance is transmitted through the surrounding air as a variation in the air pressure. The propagation of these variations in the air pressure field constitutes a *wave*; sound is an air pressure wave. Similarly, when a lamp is lit, the light is soon seen far away; it is transmitted through space as an electromagnetic wave, a disturbance in the all-pervading electromagnetic field.

Waves are amenable to mathematical description and analysis. To avoid mathematical complexity, we shall restrict ourselves to waves that propagate without decay or distortion along the X-axis. The general equation that describes the magnitude of the physical quantity, ξ, associated with such a wave as a function of the coordinates x and t is:

$$\xi(x, t) = f(x \pm vt) \tag{Sup.1}$$

where v, called the *phase velocity*, is the velocity at which the waves propagate. The positive sign indicates a wave travelling in the negative x direction; the negative sign indicates a wave travelling in the positive x direction.

If the variations in the field are harmonic, the wave equation takes the form[1]

$$\begin{aligned} \xi(x, t) &= A\cos\frac{2\pi}{\lambda}(x \pm vt) \\ &= A\cos(kx \pm \omega t) \end{aligned} \tag{Sup.2}$$

[1]The choice of the cosine or sine function is arbitrary. A harmonic wave can just as well be described by the function $\xi(x, t) = A\sin\frac{2\pi}{\lambda}(x \pm vt)$.

where $k = 2\pi/\lambda$ and $\omega = 2\pi f$; λ and f are the wave's *wavelength* and *frequency*, respectively. A is the *amplitude* of the wave. The term $(kx \pm \omega t)$ is called the *phase.*

The intensity of a wave, I, is the amount of energy it transports each second across a unit area; it is directly proportional to the square, A^2, of its amplitude:

$$I \propto A^2. \tag{Sup.3}$$

A.1 The Principle of Superposition

When a number of waves meet at a point in space they combine to produce a compound wave, $\xi(x, t)$, that is the algebraic sum of the individual waves:

$$\xi(x,t) = \xi_1(x,t) + \xi_2(x,t) + \xi_3(x,t) + \cdots. \tag{Sup.4}$$

The waves are said to *interfere* at the point.

A corollary of the superposition principle is that the propagation of each individual wave is completely unaffected by its meeting with the others; after the encounter each wave continues on as though the meeting had never happened.

We shall illustrate the application of the superposition principle to the elucidation of three of the wave phenomena referred to in the text:

1. The phenomenon of beats;
2. Young's experiment;
3. Matter waves and probability densities.

In each case a different mathematical technique will be used:

1. Beats

Suppose that where they meet and interfere, the time variations of two harmonic waves of almost identical frequencies, f_1 and f_2, are given by $\xi_1(t) = A\cos 2\pi f_1 t$ and $\xi_2(t) = A\cos 2\pi f_2 t$. The compound wave, $\xi(t)$, produced at the point will be:

$$\begin{aligned}\xi(t) &= A\cos 2\pi f_1 t + A\cos 2\pi f_2 t \\ &= 2A\cos 2\pi\frac{(f_1 - f_2)}{2}t\cdot\cos 2\pi\frac{(f_1+f_2)}{2}t \\ &= 2A\cos 2\pi\left(\frac{\delta}{2}\right)t\cdot\cos 2\pi f t\end{aligned} \tag{Sup.5}$$

where $\delta = f_1 - f_2 \ll f$ and $f = \frac{f_1+f_2}{2} \approx f_1 \approx f_2$. To a very close approximation, this compound wave can be expressed as:

$$\xi(t) = A' \cos 2\pi ft, \tag{Sup.6}$$

i.e. as a harmonic wave of frequency f whose amplitude, $A' = 2A\cos 2\pi(\delta/2)t$, modulates harmonically with a frequency $\delta/2$.

The amplitude, A', of the compound wave equals zero whenever $\cos 2\pi(\delta/2)t = 0$, i.e. when $t = (2n+1)/2\delta$ ($n = 0, 1, 2, \ldots$); at these moments the component waves interfere destructively.

The amplitude of the compound wave has its maximum value, $A' = 2A$, i.e. when $t = n/\delta$ ($n = 0, 1, 2, \ldots$); at these moments the component waves interfere constructively. These periodic maxima in the compound wave's amplitude are called *beats*. They are illustrated in Fig. 1.

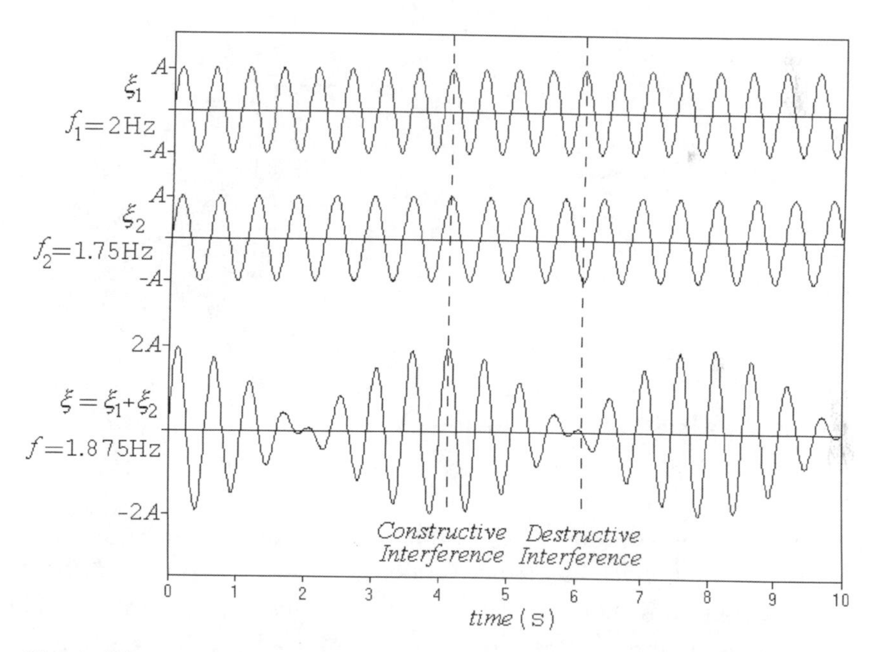

Fig. 1. The superposition of two harmonic waves of frequencies $f_1 = 2\,\text{Hz}$ and $f_2 = 1.75\,\text{Hz}$ produces a compound wave of frequency $f = \frac{2+1.75}{2} = 1.875\,\text{Hz}$, whose amplitude modulates with a frequency $f_{A'} = \frac{2-1.75}{2} = 0.125\,\text{Hz}$, i.e. every eight seconds. In each cycle of amplitude modulation two beats are produced. Thus the frequency of the beats is $f_{Beats} = 2 \cdot 0.125 = 0.25\,\text{Hz}$.

Since two beats occur during each cycle of amplitude modulation, their frequency, f_{Beats}, is twice that of the amplitude modulation:

$$f_{Beats} = 2\frac{\delta}{2} = \delta. \tag{Sup.7}$$

Thus, the frequency of the beats is equal to the difference between the neighbouring frequencies, f_1 and f_2, of the original component waves:

$$f_{Beats} = |f_1 - f_2|. \tag{Sup.8}$$

2. Young's experiment

Harmonic waves having the same frequency, f, such as those emitted from the two coherent point sources in Young's experiment, are most easily combined by the graphic technique of *rotating vectors* (*phasors*).

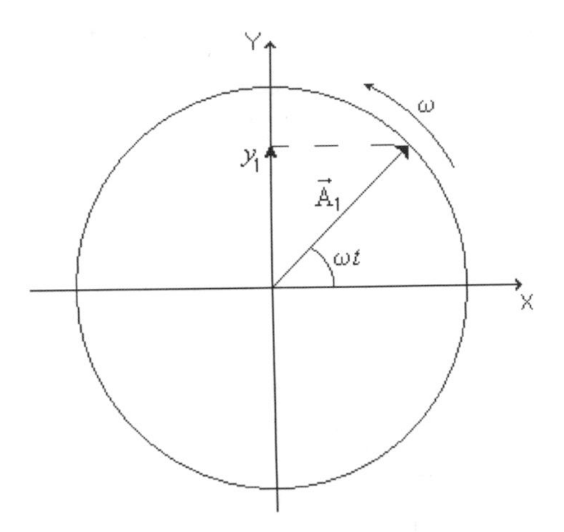

Fig. 2. The wave, y, is represented by the rotating vector $\vec{A}$ which at time t makes an angle ωt with the positive x-direction.

In general, the time dependency of any sinusoidal wave, $y = A \sin \omega t$, can be represented on a Cartesian coordinate system, such as that shown in Fig. 2, by a vector, $\vec{A}$, which rotates in an anti-clockwise direction around the origin with an angular velocity, $\omega = 2\pi f$. The quantity y is the component in the y-direction of the vector $\vec{A}$.

The result of the addition of two sinusoidal waves, $y_1 = A_1 \sin \omega t$ and $y_2 = A_2 \sin (\omega t + \phi)$, that share the same frequency but whose phases differ by an angle ϕ, can be found by the addition of the two vectors $\vec{A}_1$ and $\vec{A}_2$ (Fig. 3). Trigonometry gives the relationship:

$$A^2 = A_1^2 + A_2^2 + 2A_1A_2 \cos \phi. \tag{Sup.9}$$

In Young's experiment, the phase difference between the secondary waves arriving at the screen from each slit varies periodically from point to point. Thus, the amplitude and, likewise, the intensity of the compound wave also varies. At those points where $\cos\phi = +1$ the intensity is maximal; at those where $\cos\phi = -1$ it is minimal. This variation appears as the pattern of successive intensity maxima and minima across the screen that constitute the familiar interference pattern of bright and dark fringes.

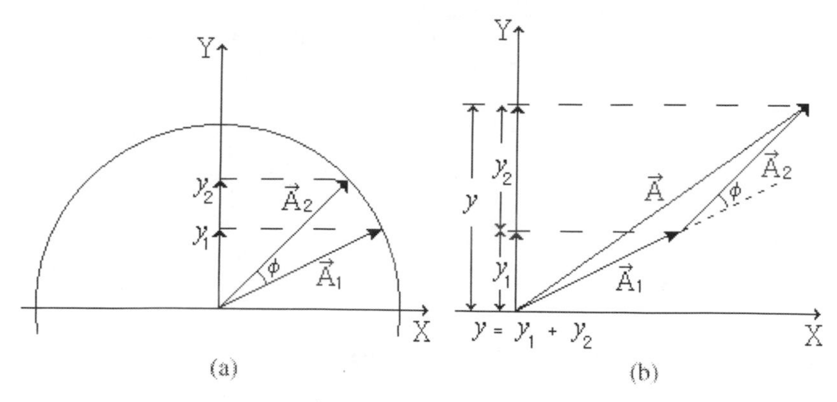

Fig. 3. (a) The two vectors, $\vec{A}_1$ and $\vec{A}_2$, which denote the waves y_1 and y_2. The angle ϕ between the vectors is the phase difference between the waves. (b) By the principle of superposition, $y = y_1 + y_2$. The compound wave, y, is therefore given by the vertical component of the vector $\vec{A} = \vec{A}_1 + \vec{A}_2$.

3. Matter waves and probability densities

Consider the plane defined by the X and Y axes (Fig. 4). We shall call the X-axis the *real* axis and the Y-axis the *imaginary* axis.[2] A *complex number*, z, is made up of a real and an imaginary part, i.e. it comprises both an x part and a y part. The real and imaginary parts are distinguished from one another by attaching the symbol i to the latter. This can be interpreted as meaning 'rotated through a right angle in the anti-clockwise direction' and thus it serves to distinguish Y-axis (imaginary) numbers from X-axis (real) numbers. Since two successive 90° rotations would convert a positive number on one of the axes to the corresponding negative number on the same axis, this is equivalent to asserting that $i^2 = -1$.

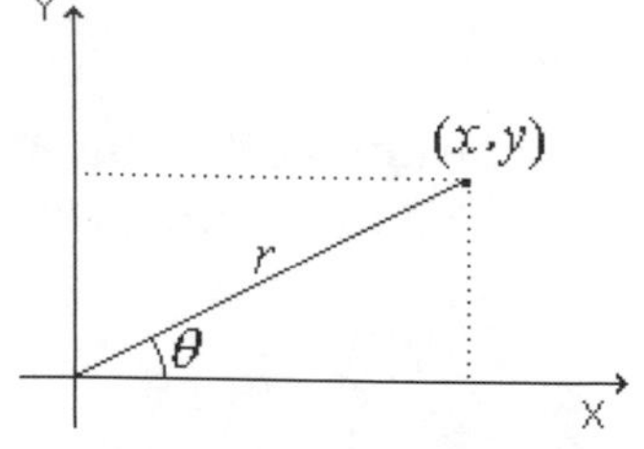

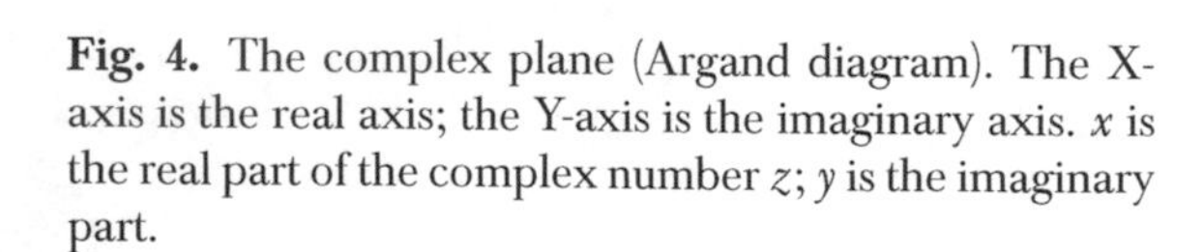
Fig. 4. The complex plane (Argand diagram). The X-axis is the real axis; the Y-axis is the imaginary axis. x is the real part of the complex number z; y is the imaginary part.

[2]The choice of which axis to call the *real* one and which the *imaginary* is arbitrary but is the conventional choice.

A complex number z can be written in a number of equivalent ways as follows:

$$\begin{aligned} z &= x + iy \\ &= r(\cos\theta + i\sin\theta) \\ &= re^{i\theta} \end{aligned} \tag{Sup.10}$$

where r and θ are the polar coordinates of the point (x, y) in the complex plane.

r is called the *modulus* or *absolute value* of the complex number z and is defined by:

$$|z| = r = \sqrt{x^2 + y^2}. \tag{Sup.11}$$

The complex number $Z^* = x - iy = re^{-i\theta}$ is called the *conjugate* of z. The square of the modulus, r^2, which by Pythagoras equals $x^2 + y^2$, is given by the product of the complex number, z, and its conjugate, z^*:

$$\begin{aligned} r^2 &= zz^* \\ &= (x + iy)(x - iy) \\ &= x^2 - ixy + ixy - i^2y^2 \\ &= x^2 + y^2. \end{aligned} \tag{Sup.12}$$

The real and imaginary parts of an imaginary number are analogous to the components of a vector, i.e. each can be considered independently of the other. For example, when adding complex numbers the real and imaginary parts are added separately:

$$\begin{aligned} z_1 + z_2 &= (x_1 + iy_1) + (x_2 + iy_2) \\ &= (x_1 + x_2) + i(y_1 + y_2). \end{aligned} \tag{Sup.13}$$

Thus, for example, the sinusoidal wave, $\xi(t) = A\cos(\omega t \pm \phi)$, formed at a point in space can be described by the real part of the complex number:

$$\begin{aligned} z &= A(\cos(\omega t \pm \phi) + i\sin(\omega t \pm \phi)) \\ &= Ae^{i(\omega t \pm \phi)} \end{aligned} \tag{Sup.14}$$

The imaginary part of this complex number is $Ai\sin(\omega t \pm \phi)$.[3]

The intensity, I, of a wave depends on the square of its amplitude, A^2. Thus, when the wave is described by either the real or imaginary part of a complex

[3]Alternatively, the sinusoidal wave $\xi(t) = A\sin(\omega t \pm \phi)$ could be described by the imaginary part of the complex number.

number, $z = Ae^{i(\omega t \pm \phi)}$, its intensity is specified by the product of the complex number, z, and its conjugate, z^*:

$$\begin{aligned} I &= zz^* \\ &= Ae^{i(\omega t \pm \phi)} Ae^{-i(\omega t \pm \phi)} \\ &= A^2. \end{aligned} \qquad \text{(Sup.15)}$$

Probability amplitudes (wave functions) are typically complex numbers, e.g. $\Psi = a + ib$ or $\Psi = Ae^{i\theta}$. Thus, the probability density, $|\Psi|^2 = \Psi\Psi^*$, of a matter wave is essentially its intensity.

In wave theory, the summations associated with the superposition principle often involve integrations which are most conveniently carried out using the complex representation. An advantage of this representation is that the complex quantity, z, that describes the wave can be split into its space dependent, $Ae^{+i\phi}$, and time dependent, $e^{i\omega t}$, parts:

$$z = Ae^{+i\phi} e^{i\omega t}. \qquad \text{(Sup.16)}$$

When two waves possessing the same angular frequency, ω, are superimposed, as in Young's experiment, e^{iwt} is a common factor in the complex number, z_{12}, that results from the addition of the two complex numbers, z_1 and z_2, whose imaginary parts describe the waves:

$$\begin{aligned} z_{12} &= z_1 + z_2 \\ &= (A_1 e^{+i\phi_1} + A_2 e^{+i\phi_2}) e^{i\omega t}. \end{aligned} \qquad \text{(Sup.17)}$$

The amplitude, A_{12}, of the resultant wave is given by the complex quantity:

$$A_{12} = A_1 e^{+i\phi_1} + A_2 e^{+i\phi_2}. \qquad \text{(Sup.18)}$$

The intensity, I, of the resultant wave is given by the product of the complex quantity, A_{12}, and its conjugate, A^*_{12}:

$$\begin{aligned} I &= A_{12} A^*_{12} \\ &= (A_1 e^{+i\phi_1} + A_2 e^{+i\phi_2})(A_1 e^{-i\phi_1} + A_2 e^{-i\phi_2}) \\ &= A_1^2 + A_2^2 + A_1 A_2 (e^{i(\phi_2 - \phi_1)} + e^{-i(\phi_2 - \phi_1)}) \\ &= A_1^2 + A_2^2 + 2A_1 A_2 \cos(\phi_2 - \phi_1). \end{aligned} \qquad \text{(Sup.19)}$$

This result is the same as that obtained above (Sup.9) using rotating vectors. Thus, in instances like Young's experiment, intensities and probability densities can be calculated by either technique, complex numbers or rotating vectors.

B. List of Physical Constants and Conversion Factors

Atomic mass unit	$\mathrm{u} = 1.66 \cdot 10^{-27}\,\mathrm{kg} = 931\,\mathrm{MeV}/c^2$
Avogadro's number – mole	$N_0 = 6.025 \cdot 10^{23}$
Bohr magneton	$\mu_\mathrm{B} = 9.2732 \cdot 10^{-24}\,\mathrm{J} \cdot \mathrm{T}^{-1}$
Bohr radius	$a_0 = 5.292 \cdot 10^{-11}\,\mathrm{m}$
Boltzmann's constant	$k = 8.62 \cdot 10^{-5}\,\mathrm{eV/K} = 1.38 \cdot 10^{-23}\,\mathrm{J/K}$
Electric charge unit	$e = 1.6 \cdot 10^{-19}\,\mathrm{C}$
Electron rest mass	$m_e = 9.109 \cdot 10^{-31}\,\mathrm{kg} = 0.00055\,\mathrm{u}$
Electronvolt	$1\,\mathrm{eV} = 1.6 \cdot 10^{-19}\,\mathrm{J}$
Joule	$1\mathrm{J} = 6.242 \cdot 10^{18}\,\mathrm{eV}$
Neutron rest mass	$m_n = 1.675 \cdot 10^{-27}\,\mathrm{kg} = 1.0087\,\mathrm{u}$
Permeability of free space	$\mu_0 = 4\pi \cdot 10^{-7}\,\mathrm{T} \cdot \mathrm{m} \cdot \mathrm{A}^{-1}$
Permittivity of free space	$\varepsilon_0 = 1/4\pi k = 8.854 \cdot 10^{-12}\,\mathrm{C}^2 \cdot \mathrm{N}^{-1} \cdot \mathrm{m}^{-2}$
Planck's constant	$h = 6.625 \cdot 10^{-34}\,\mathrm{J} \cdot \mathrm{s} = 4.14 \cdot 10^{-21}\,\mathrm{MeV} \cdot \mathrm{s}$
Proton rest mass	$m_p = 1.673 \cdot 10^{-27}\,\mathrm{kg} = 1.0073\,\mathrm{u}$
Rydberg's constant	$R_0 = 1.097 \cdot 10^{7}\,\mathrm{m}^{-1}$
Speed of light *in vacuo*	$c = 2.998 \cdot 10^{8}\,\mathrm{m/s}$
Stefan's constant	$\sigma = 5.670 \cdot 10^{-8}\,\mathrm{W} \cdot \mathrm{m}^{-2} \cdot \mathrm{K}^4$

C. The Greek Alphabet

Letters			Name	Letters			Name
Α	α	α	alpha	Ν	ν	ν	nu
Β	β	β	beta	Ξ	ξ	ξ	xi
Γ	γ	γ	gamma	Ο	ο	o	omicron
Δ	δ	δ	delta	Π	π	π	pi
Ε	ε	ε	epsilon	Ρ	ρ	ρ	rho
Ζ	ζ	ζ	zeta	Σ	σ	σ	sigma
Η	η	η	eta	Τ	τ	τ	tau
Θ	θ	θ	theta	Υ	υ	υ	upsilon
Ι	ι	ι	iota	Φ	φ	ϕ	phi
Κ	κ	κ	kappa	Χ	χ	χ	chi
Λ	λ	λ	lambda	Ψ	ψ	ψ	psi
Μ	μ	μ	mu	Ω	ω	ω	omega

Answers and Solutions to the Numerical Questions, Exercises and Problems

Part One

Questions

5. Hint: Use the relationship $(1+x)^n = 1 + nx$ that is valid when $x \ll 1$.

Exercises

1. 6.6 Hz
2. 6.4 s
3. 0.991 c
4. a. 357.1 ns b. 103 m c. 28.8 m
5. $x' = 138\,\text{km}$; $t' = -374\,\mu\text{s}$
6. 0.28 c
7. a. 1.58 s b. $2.28 \cdot 10^8$ m c. 0.948 s
8. $1.8 \cdot 10^{10}$ km
9. a. 2.57 MeV b. 2.06 MeV c. 2.52 MeV/c
10. a. 0.9988 c b. 0.145 c
11. $\sqrt{3}c/2$
12. 0.9771 c
13. 0.962 c

Problems

1. $\sqrt{3}c/2$
2. a. $1.7 \cdot 10^3$ eV b. 1.06 MeV
3. 6.5 MeV/c^2
4. 0.65 c
5. b. 15.2 mm
6. $2.55 \cdot 10^5$ V; 0.75 c
9. a. 0.8 c b. 0.988 c c. 0.9 c d. 0.988 c

Part Two

Exercises

1. 5270 K
2. 10.1 kW
3. 1:1.013
4. $2.685 \cdot 10^{-19}$ J; 1.68 eV
5. a. 5400 Å b. 3.9 eV
6. 4.45 eV
7. 0.157 V
8. $2.83 \cdot 10^{-19}$ J
9. $3.97 \cdot 10^{-7}$ m
10. $2.96 \cdot 10^{20}$
11. 124 kV
12. $5 \cdot 10^{18}$ Hz
13. $6 \cdot 10^{19}$
14. 2.8 Å

15. 2.9°
16. a. 9.6% b. 68%
17. $1.5 \cdot 10^{-12}$ m
18. 10^{-11} m
19. $6 \cdot 10^{-12}$ m, given $\Delta x \cdot \Delta p \approx h$
20. 5.1°
21. 6%, given $\Delta x \cdot \Delta p \approx h$
22. $1.46 \cdot 10^{-10}$ m
23. 2.1 MeV
24. 1.24 keV; 1.5 eV
25. 90.5 eV

Problems

1. 240 W
2. 5,820 K
3. a. $2.5 \cdot 10^{17}\,\mathrm{s}^{-1}$ b. 3.9 mA c. 0.39 V
6. a. $\Delta E_n = (2n+1)\dfrac{h^2}{8mL^2} \approx \dfrac{nh^2}{4mL^2}$
   $\Delta E_{n+1} \approx \Delta E_n$
   $\dfrac{(2n+1)}{n^2}; \dfrac{2}{n}$
8. 9.79 keV
10. 7.1 eV
14. ~4 K
15. 0.0306 nm; 0.0307 nm
16. $4.36 \cdot 10^6$ m/s; $4.46 \cdot 10^6$ m/s
   ee. 0.163 nm; 0.167 nm

Part Three

Exercises

1. 217.5 eV
2. 1.8751 μm
3. Hydrogen: 2.63 μm;
   He^+: 0.656 μm
4. 3.4 eV; 13.6 eV
5. 22.79 nm
6. 0.179 nm; 0.238 nm
7. 1.89 eV, 656.3 nm;
   10.21 eV, 121.6 nm;
   12.11 eV, 102.6 nm
8. a. $\sqrt{}12\hbar$ b. $-3\hbar \rightarrow +3\hbar$ c. 30°, 55°, 73°, 90, 107, 125°, 150°
9. $n > 3$, $m_l = -3 \rightarrow +3$, $m_s = \pm 1/2$
10. 0.0108
11. a. $n = 2$, $l = 0$, $m_l = 0$, $m_s = \pm 1/2$
    b. $n = 2$, $l = 1$, $m_l = -1, 0, +1$, $m_s = \pm 1/2$
12. b. 6.39 keV
13. Zn
14. 5.35 cm

Problems

1. $2.5 \cdot 10^{15}$ Hz
2. a. $2.58 \cdot 10^{-13}$ m b. 1,900 eV
3. a. 6.81 eV
9. $2.5 \cdot 10^7$
10. 105,000 K
11. 1.35 Å
14. 42.5 MHz

Part Four

Exercises

1. 0.05
2. 0.5
3. $0.0074J(J+1)$eV; $0.0037J(J+1)$eV
4. $1.86 \cdot 10^{-4}$ eV; $5.6 \cdot 10^{11}$ rad/s
5. 3.6 Å
6. $3.3 \cdot 10^{-47}$ kg·m^2, 1.42 Å
7. 317 N/m
8. (i) $1.5 \cdot 10^{-5}$ (ii) $7.8 \cdot 10^{-2}$
9. 516.2 N/m; 0.1861 eV

Problems

1. $6.18 \cdot 10^{12}$ Hz
3. $^{2}H^{35}Cl$: 2,159 cm^{-1};
   $^{1}H^{37}Cl$: 2,997 cm^{-1};
   $^{1}H^{35}Cl$: 3,000 cm^{-1}
4. $8.97 \cdot 10^{13}$ Hz

Part Five

Exercises

1. ~14 km
2. 7.1 MeV
3. 8 MeV
4. ^{16}N: 7.175 MeV; ^{16}O: 7.745 MeV;
   ^{16}F: 6.697 MeV
5. 12.9 MeV
7. 5.1 cm
8. e. $1.7 \cdot 10^{6}$; $3.2 \cdot 10^{-11}$ A
9. 8,000 years
10. 15 days
11. 10^{16}
12. 357
13. 880 gm
14. $3.2 \cdot 10^{10}$
15. 40 days
16. $8.6 \cdot 10^{9}$ years
17. $6 \cdot 10^{9}$ years
18. $t/34.6$
19. 4 grams

Problems

1. a. 0.847 MeV b. 434 MeV
4. a. 7.55 hours b. $5.3 \cdot 10^{6}$; $4.8 \cdot 10^{6}$
   c. 10.9 hours d. 168.7 hours
5. b. α: 183.1 s; β: 127.1 hours
   c. α: 0.0038; β: $2.2 \cdot 10^{-6}$
6. 87.5
8. 250 MeV

Part Six

Exercises

1. 22.4 mT
2. 2.165
3. $N_1/N_2 = 1 + 1.6 \cdot 10^{-5}$
4. $2.2 \cdot 10^{-6}$ eV

6. 4.2 eV
7. 0.75 eV
8. 0.355 eV
9. $3.5 \cdot 10^8 \Omega \cdot$ m; the actual value is about half of this.

Problems

1. a. $2 \cdot 10^{-3}$ eV b. $2 \cdot 10^{-25}$ eV
2. a. 0.094 eV b. 0.635 nm
3. 0.46 nm

Index